AA007204

5th Madrid Turbulence Workshop 2023

Journal of Physics: Conference Series Volume 2753

Madrid, Spain
29 May – 30 June 2023

ISBN: 978-1-7138-9738-5
ISSN: 1742-6588

Printed from e-media with permission by:

Curran Associates, Inc.
57 Morehouse Lane
Red Hook, NY 12571

Some format issues inherent in the e-media version may also appear in this print version.

Institute of Physics
Dirac House, Temple Back
Bristol BS1 6BE UK

Phone: 44 1 17 929 7481
Fax: 44 1 17 920 0979

techtracking@iop.org

Additional copies of this publication are available from:

Curran Associates, Inc.
57 Morehouse Lane
Red Hook, NY 12571 USA
Phone: 845-758-0400
Fax: 845-758-2633
Email: curran@proceedings.com
Web: www.proceedings.com

TABLE OF CONTENTS

Journal of Physics: Conference Series **2753** (2024) 011001 doi:10.1088/1742-6596/2753/1/011001

Fifth Madrid Summer School on Turbulence

Editor: Javier Jiménez

PREFACE

After a hiatus during the Covid-19 pandemic, the fifth biennial Summer School on Turbulence at the School of Aeronautics of the Universidad Politécnica de Madrid took place from May 29 to June 30, 2023. It was funded by the Caust grant of the European Research Council, whose goal is to identify causal relations in turbulence and in other fields of physics. As in the preceding four Schools [1–4], the present one aimed to provide a meeting place for theoreticians, experimentalists and simulators in which to develop and test new ideas on turbulence physics and structure, including new techniques such as machine learning and massive Monte Carlo simulations. Around sixty, mostly young, participants from thirty one international groups met for five weeks of collaborative work, primarily using the computational data archived in the receiving institution and, in many cases, also contributing their own. Although the format included some invited seminars and periodic plenary meetings, most of the work took place in small groups whose composition often changed during the workshop. These proceedings reflect the results of the work of these groups, which, in many cases, has later continued in the form of new collaborations.

erc

POLITÉCNICA
Ingeniamos el futuro

Journal of Physics: Conference Series **2753** (2024) 011001 doi:10.1088/1742-6596/2753/1/011001

Workshop organiser:

Javier Jiménez
School of Aeronautics, Universidad Politécnica de Madrid
28040 Madrid, Spain
javier.jimenezs@upm.es

Acknowledgements

The organizer is grateful to the European Research Council for their funding of this meeting, and to the School of Aeronautics of the Universidad Politécnica de Madrid for their help in hosting it. The meeting would not have been possible without the dedication of those participants who acted as hosts to the visiting groups, facilitating everything from everyday problems to computer access and data usage: Miguel P. Encinar, Adal Galván, Ricardo García-Mayoral, Jinyuan Liu, Adrián Lozano-Durán, Carlos Martínez-López, Kosuke Osawa and Alberto Vela-Martín. Very especially, we are indebted to Aurora García Sarrión, without whose administrative help nothing would have worked. The picture of the participants was graciously provided by Vanesa García Sáez, of the press department of the UPM.

References

[1] Jiménez J 2014 *J. Phys.: Conf. Series* **506**
[2] Jiménez J 2016 *J. Phys.: Conf. Series* **708**
[3] Jiménez J 2018 *J. Phys.: Conf. Series* **1001**
[4] Jiménez J 2020 *J. Phys.: Conf. Series* **1522**

Peer Review Statement

All papers published in this volume have been reviewed through processes administered by the Editors. Reviews were conducted by expert referees to the professional and scientific standards expected of a proceedings journal published by IOP Publishing.

- **Type of peer review:** Open Review
- **Conference submission management system:** Morressier
- **Number of submissions received:** 26
- **Number of submissions sent for review:** 26
- **Number of submissions accepted:** 24
- **Acceptance Rate (Submissions Accepted / Submissions Received × 100):** 92.3
- **Average number of reviews per paper:** 1.25
- **Total number of reviewers involved:** 30
- **Contact person for queries:**
  Name: Javier Jimenez
  Email: jjsendin@gmail.com
  Affiliation: U. Politecnica Madrid

Journal of Physics: Conference Series **2753** (2024) 012001 doi:10.1088/1742-6596/2753/1/012001

Forgetfulness of turbulent energy cascade associated with different mechanisms

Ryo Araki[1], Alberto Vela-Martín[2] and Adrián Lozano-Durán[3]

[1] Department of Mechanical and Aerospace Engineering, Faculty of Science and Technology, Tokyo University of Science, Yamazaki 2641, Noda-shi 278-8510, Japan

[2] Department of Aerospace Engineering, Universidad Carlos III de Madrid, Escuela Politécnica Superior, Avda. de la Universidad, 30. 28911 Leganés (Madrid) España

[3] Department of Aeronautics and Astronautics, Massachusetts Institute of Technology, Cambridge, Massachusetts 02139, USA

E-mail: araki.ryo@rs.tus.ac.jp

Abstract. The small scales of high Reynolds number turbulence exhibit the statistical universality by *forgetting* about the large scales during the energy cascade. In this study, we revisit this schematic description with an information-theoretic point of view. Our objective is to understand the importance of different cascade mechanisms in this process. To this end, we compute the "information flux", the information-theoretic measure of causality formulated by one of the authors, of the decomposed inter-scale energy transfer. For the kinetics, the ratio of the inertial-range inter-scale energy transfer between the strain self-amplification (SSA) and the vortex stretching (VS) mechanisms is 5 : 3. However, for the information-flux, the VS contribution is roughly twice as large as the SSA. We also find several deviations between the kinetics and the informatics, highlighting the nontrivial nature of information distributed among different scales and mechanics in turbulence. This result infers that the SSA is a less efficient mechanism than the VS in terms of information transfer, and contributes to losing the large-scale information.

1. Introduction

The hallmark of turbulence research is its statistical universality in the small scales, irrespective of the large-scale configurations [1, 2]. We often describe that the small-scale universality of turbulence emerges while the small scales *forget* about the large scales during the scale-local energy cascade process. Indeed, one finds the following schematic description in a standard turbulence textbook [3, § 5.2.1]:

> [The small scales of turbulence] do not retain any information which relates to their great-great-great-grand parents.

However, we must ask ourselves: To what extent is this typical schematic description true? Are the fine-scale vortices of developed turbulence entirely irrelevant to the large-scale vortices?

In a recent work by one of the authors [4], an information-theoretic framework was developed to investigate the causality among variables of a dynamical system. The authors found the predominance of a scale-local information flux from larger to smaller scales in developed turbulence. In another recent work by one of the other authors [5], information thermodynamics was applied to the fluctuating Navier–Stokes equations to derive the information-thermodynamic bound on information flow in turbulence. This work also revealed that the information is transferred towards small scales in developed turbulence along with the direct energy cascade. These results indeed contradict our schematic view introduced in the previous paragraph.

In order to reconcile the latter picture against the classical schematic of turbulence, we need to consider the information transfer mechanism of turbulence. In this study, we focus on the information of energy. Thus, we propose the following research question:

> Can we quantify the different physical mechanisms of the energy cascade from an information-theoretic point of view?

There have been multiple candidates for the predominant physical mechanism of the energy cascade in three-dimensional turbulence. In particular, several competing results have been reported in the last five years: vortex stretching (VS) [6, 7], strain self-amplification (SSA) [8–11], vortex reconnection [12], and elliptical instability [13, 14], to name a few. The problem is that different researcher(s) employ different quantifications of energy transfer based on their own assumptions to derive different conclusions. Thus, even though the concept of energy cascade is more than one hundred years old [15], the turbulence community has yet to reach a consensus on the main energy cascade mechanism.

In this manuscript, we investigate these questions from an information-theoretic point of view. By investigating the information-theoretic properties of different dynamics of the energy cascade, we can bring a new perspective to this long-standing question on the most important concept of turbulence.

2. Approach

In this study, we define the inter-scale energy transfer with the coarse-grained Navier–Stokes equations and employ the decomposition derived by Johnson [9, 10], as it has the least number of subjective assumptions among different definitions in the literature reviewed in §1. In this section, we first provide a derivation of the decomposition in §2.1 and then discuss the information-theoretic approach in §2.2.

2.1. Decomposed inter-scale energy transfer

We first consider the incompressible Navier–Stokes equations

$$\frac{\partial u_i}{\partial t} + u_j \frac{\partial u_i}{\partial x_j} = -\frac{\partial p}{\partial x_i} + \nu \nabla^2 u_i + f_i \quad \text{and} \quad \frac{\partial u_i}{\partial x_i} = 0, \tag{1}$$

where $\boldsymbol{u}$ is the velocity field, p is the pressure field, and $\boldsymbol{f}$ is the forcing field. The control parameter is the kinematic viscosity ν. We define the low-pass filtered velocity by

$$\bar{u}_i^\ell(\boldsymbol{x}) = \iiint_{-\infty}^{\infty} G_\ell(\boldsymbol{r}) u_i(\boldsymbol{x} + \boldsymbol{r})\, \mathrm{d}\boldsymbol{r}\,, \tag{2}$$

where $G_\ell(\boldsymbol{r})$ is a filter kernel function with the characteristic length scale ℓ. Similar to Johnson [9, 10], we employ the Gaussian filter kernel

$$G_\ell(\boldsymbol{r}) = \mathcal{N} \exp\left(-\frac{|\boldsymbol{r}|^2}{2\ell^2} \right), \quad (2\pi)^3 \mathcal{F}\{G_\ell\}(\boldsymbol{k}) = \exp\left(-\frac{1}{2} |\boldsymbol{k}|^2 \ell^2 \right). \tag{3}$$

Here, $\mathcal{F}\{\cdot\}$ denotes the Fourier transform and $\boldsymbol{k}$ is the wavenumber vector. By using this coarse-graining procedure, the subfilter-scale stress tensor can be expressed as

$$\tau_\ell(u_i, u_j) = \overline{u_i u_j}^\ell - \bar{u}_i^\ell \bar{u}_j^\ell. \tag{4}$$

Furthermore, the strain-rate tensor and the rotation-rate tensor are defined by

$$\bar{S}^\ell_{ij} = \frac{1}{2}\left[\bar{A}^\ell_{ij} + \bar{A}^\ell_{ji}\right], \quad \bar{\Omega}^\ell_{ij} = \frac{1}{2}\left[\bar{A}^\ell_{ij} - \bar{A}^\ell_{ji}\right], \tag{5}$$

respectively, where $A_{ij} = \partial u_i/\partial x_j$ denotes the velocity gradient tensor.

By using these quantities, the inter-scale energy transfer of the coarse-grained Navier–Stokes equations at scale ℓ is defined as

$$\Pi^\ell = -\left[\tau_\ell(u_i, u_j) - \frac{1}{3}\tau_\ell(u_k, u_k)\delta_{ij}\right]\bar{S}^\ell_{ij} \tag{6}$$

$$= \Pi^\ell_{s1} + \Pi^\ell_{\omega 1} + \Pi^\ell_{s2} + \Pi^\ell_{\omega 2} + \Pi^\ell_c, \tag{7}$$

where the square bracket in the first line extracts the anisotropic part of the subfilter stress tensor and the second line expresses the decomposition. The RHS terms of Eq. (7) are defined by:

$$\Pi^\ell_{s1} = -\ell^2 \bar{S}^\ell_{ij}\bar{S}^\ell_{jk}\bar{S}^\ell_{ki}, \tag{8}$$

$$\Pi^\ell_{\omega 1} = \frac{1}{4}\ell^2 \bar{S}^\ell_{ij}\bar{\omega}^\ell_i\bar{\omega}^\ell_j, \tag{9}$$

$$\Pi^\ell_{s2} = -\int_0^{\ell^2} \mathrm{d}\alpha\, \bar{S}^\ell_{ij}\tau_\beta\left(\bar{S}^{\sqrt{\alpha}}_{jk}, \bar{S}^{\sqrt{\alpha}}_{ki}\right), \tag{10}$$

$$\Pi^\ell_{\omega 2} = -\frac{1}{4}\int_0^{\ell^2} \mathrm{d}\alpha\, \bar{S}^\ell_{ij}\tau_\beta\left(\bar{\omega}^{\sqrt{\alpha}}_i, \bar{\omega}^{\sqrt{\alpha}}_j\right), \tag{11}$$

$$\Pi^\ell_c = \bar{S}^\ell_{ij}\int_0^{\ell^2} \mathrm{d}\alpha \left[\tau_\beta\left(\bar{S}^{\sqrt{\alpha}}_{jk}, \bar{\Omega}^{\sqrt{\alpha}}_{ki}\right) - \tau_\beta\left(\bar{\Omega}^{\sqrt{\alpha}}_{jk}, \bar{S}^{\sqrt{\alpha}}_{ki}\right)\right], \tag{12}$$

respectively. Their full derivation can be found in Appendix C of [10]. Note that Eq. (3.11) of [10] [(11) of the current manuscript] has a typo; the prefactor 1/4 is missing.

The terms with subscript "1" in the first and second lines (8–9) involve only quantities at the filter length scale ℓ. Thus, they represent the scale-local interactions. Here, Π^ℓ_{s1} and $\Pi^\ell_{\omega 1}$ are the SSA and the VS at scale ℓ, respectively.

The terms with subscript "2" in the third and fourth lines (10–11) represent the nonlinear interactions between the strain rate filtered at ℓ and the velocity gradient at scales smaller than ℓ. Note that while the integral range spans from 0 to ℓ^2, the integral variable α in the integrand appears within its square root. Thus, they represent the scale-nonlocal interactions. Similar to those with subscript "1", Π^ℓ_{s2} denotes the SSA, but this time the amplification of the small-scale strain by the larger-scale strain. The other term, $\Pi^\ell_{\omega 2}$, represents the stretching of the small-scale vorticity by the large-scale strain. The final term Π^ℓ_c denotes the interactions between the larger-scale strain and smaller-scale strain–vorticity covariance.

In summary, all the contributions in the decomposed inter-scale energy transfer (8–12) can be interpreted as the influence of the strain-rate tensor $\bar{S}^\ell_{ij}$ at scale ℓ on:

- Eq. (8): Squared strain-rate $\bar{S}^\ell_{jk}\bar{S}^\ell_{ki}$ at scale ℓ,
- Eq. (9): Enstrophy $\bar{\omega}^\ell_i\bar{\omega}^\ell_j$ at scale ℓ,
- Eq. (10): Subfilter stress tensor of the strain-rate $\tau_\beta\left(\bar{S}^{\sqrt{\alpha}}_{jk}, \bar{S}^{\sqrt{\alpha}}_{ki}\right)$ on scales $0 \le \alpha \le \ell^2$,
- Eq. (11): Subfilter stress tensor of the enstrophy $\tau_\beta\left(\bar{\omega}^{\sqrt{\alpha}}_i, \bar{\omega}^{\sqrt{\alpha}}_j\right)$ on scales $0 \le \alpha \le \ell^2$,
- Eq. (12): Covariance $\tau_\beta\left(\bar{S}^{\sqrt{\alpha}}_{jk}, \bar{\Omega}^{\sqrt{\alpha}}_{ki}\right)$ $\tau_\beta\left(\bar{\Omega}^{\sqrt{\alpha}}_{jk}, \bar{S}^{\sqrt{\alpha}}_{ki}\right)$ on scales $0 \le \alpha \le \ell^2$

2.2. *Information-theoretic causality*

Information theory stemmed from the science of message communication. Pioneered by Claude E. Shannon [16], it is now regarded as a growing area of science [17–19]. The original problem was how to recover the message sent through a noisy channel. To measure the average information content in a message, Shannon defined the Shannon entropy (SE),

$$H(Q) \equiv -\int \mathrm{d}q\, p(q) \log p(q), \tag{13}$$

for a random variable Q where $q \in Q$ obeys a probability distribution $p(q)$. For a joint probability distribution $p(q, q')$ with $q' \in Q'$, we can define the joint SE,

$$H(Q, Q') \equiv -\int \mathrm{d}q\, \mathrm{d}q'\, p(q, q') \log p(q, q'). \tag{14}$$

We can also define the conditional SE,

$$H(Q|Q') \equiv -\int \mathrm{d}q\, \mathrm{d}q'\, p(q, q') \log p(q|q'), \tag{15}$$

where $p(q|q') \equiv p(q, q')/p(q')$ is conditional probability distribution.

Here, we introduce the information flux first derived in one of the authors' recent study [4]. Suppose that we focus on a system consisting of N degrees of freedom and $N_Y \leq N$ denotes the number of observable variables. They are denoted by $\boldsymbol{Y} = [Y_1, Y_2, \ldots, Y_{N_Y}]$. Here, the information flux from variable Y_i^n (the superscript denotes nth time step) to Y_j^{n+1} can be quantified by

$$T^{Y}_{i \to j} = H\left(Y_j^{n+1} | \boldsymbol{Y}^n_{\not i}\right) - H\left(Y_j^{n+1} | \boldsymbol{Y}^n\right). \tag{16}$$

The RHS of this equation can be regarded as the improvement of the knowledge on the future state Y_j^{n+1} by knowing the current state Y_i^n. Note that $\boldsymbol{Y}_{\not i} = [Y_1, \ldots, Y_{i-1}, Y_{i+1}, \ldots, Y_{N_Y}]$ denotes the incomplete set of the observable variables. For the full derivation, readers may refer to [4, § IV.A] where a multivariable generalisation of (16) is discussed. We here note that the total information of Y_j^{n+1} cannot be explained only by the observable variables $\boldsymbol{Y}^n$ due to the unobserved variables. To quantify the information flux from them, we define the information leak

$$T^{Y}_{\mathrm{leak}, j} = H(Y_j^{n+1} | \boldsymbol{Y}^n). \tag{17}$$

2.3. *Lagrangian tracking of space-local inter-scale energy transfer*

In this work, we compute the information flux of the decomposed inter-scale energy transfer $T^{\Pi}_{i \to j}$ to quantify the information-theoretic importance of the different energy cascade mechanisms, particularly between the VS and the SSA. Furthermore, we employ the Lagrangian framework to closely track the physical structures of turbulence.

A two-dimensional schematic of our methodology is depicted in Fig. 1 with three length and time scales, denoted by ℓ_1, ℓ_2, ℓ_3 and $\Delta T, \Delta T_1, \Delta T_2$, corresponding from larger to smaller scales, respectively. Note that the definition of these quantities will be given in § 3.1. Our procedure consists of the following steps:

(i) Fix a point $\boldsymbol{x}_i$ in space at time $t = t_i$ and compute the average of the inter-scale energy transfer Π^{ℓ} for the length scale $\ell = \ell_1$ inside a sphere of radius with the same length scale $r = \ell_1$. This quantity is denoted as $\langle \Pi^{\ell_1} \rangle_{\ell_1}$.

(ii) Track the advection of $\boldsymbol{x}_i$ evaluated as $\Delta \boldsymbol{x}_1 = \langle \boldsymbol{u}(\boldsymbol{x}_i) \rangle_{\ell_1} \Delta T_1$. Here, the advection velocity is averaged inside a sphere of radius ℓ_1 and ΔT_1 denotes the corresponding time scale.

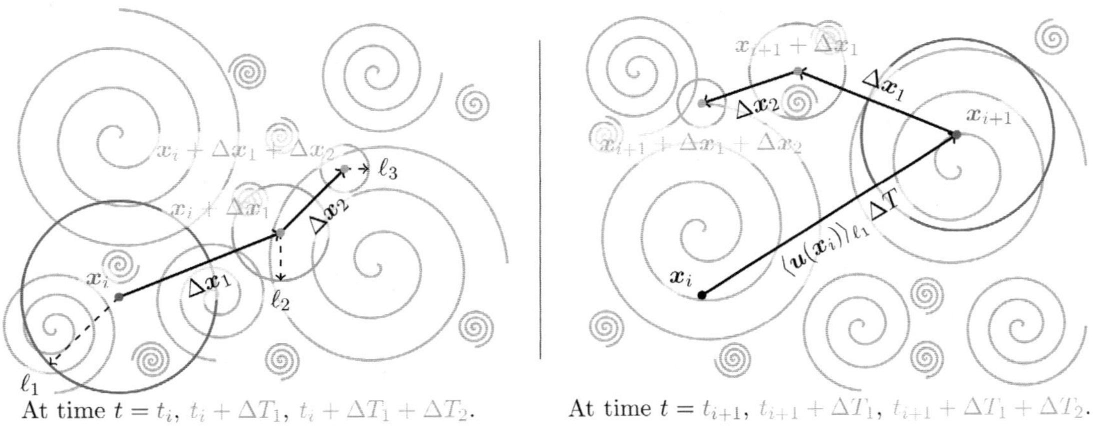

Figure 1: Schematic of Lagrangian temporal tracking of spatially local inter-scale energy transfer and its decomposition in a turbulent flow. (Left) At time $t = t_i$, compute the space-local average of the inter-scale energy transfer at the largest scale ℓ_1 centred at $\boldsymbol{x}_i$. Then, track the flow advection and compute the inter-scale energy transfer at smaller scales ℓ_2 and ℓ_3. The displacements $\Delta\boldsymbol{x}_1$ and $\Delta\boldsymbol{x}_2$ are evaluated by the advection due to the local velocity field during the characteristic time scales of the inter-scale energy transfer ΔT_1 and ΔT_2, respectively. (Right) Repeat the same process for $\boldsymbol{x}_{i+1}$, which is evaluated by the advection of $\boldsymbol{x}_i$ over the characteristic time scale of the energy cascade ΔT.

(iii) At $\boldsymbol{x}_i + \Delta\boldsymbol{x}_1$ and $t = t_i + \Delta T_1$, compute the space-local average of Π^{ℓ_2} inside the sphere of radius ℓ_2. This quantity is denoted as $\langle \Pi^{\ell_2} \rangle_{\ell_2}$.

(iv) Repeat the steps (ii) and (iii) for the smaller scales. Here, the advection distance reads $\Delta\boldsymbol{x}_2 = \langle \boldsymbol{u}(\boldsymbol{x}_i + \Delta\boldsymbol{x}_1) \rangle_{\ell_2} \Delta T_2$ and the integral domain is a sphere of radius ℓ_3. Note that the space-local average is performed at position $\boldsymbol{x}_i + \Delta\boldsymbol{x}_1 + \Delta\boldsymbol{x}_2$ and time $t = t_i + \Delta T_1 + \Delta T_2$.

(v) Compute the advection distance during the one energy cascade step by $\boldsymbol{x}_{i+1} = \boldsymbol{x}_i + \langle \boldsymbol{u}(\boldsymbol{x}_i) \rangle_{\ell_1} \Delta T$. Here, ΔT denotes the characteristic time scale of the energy cascade. Note that $\Delta T > \sum_i \Delta T_i$.

(vi) Repeat the steps (i–iv) from the new coordinate $\boldsymbol{x}_{i+1}$ at time $t = t_{i+1} = t_i + \Delta T$.

We iterate the above process with randomly distributed $\boldsymbol{x}_0$ in the flow domain to construct an ensemble of the Lagrangian dataset. We describe the details of the turbulence dataset and implementation of this procedure in §3.1.

3. Result

3.1. Numerical dataset

In this study, we investigate the homogeneous and isotropic turbulence (HIT) dataset obtained by Cardesa *et al.* [20]. Table 1 summarises the simulation parameters and statistical quantities. Further details of the flow configurations and statistics can be found in [20, 21].

Figure 2 shows the inter-scale energy transfer distribution over scale ℓ. Similar figures can be found in the recent works of Johnson [9, 10]. We plot the five contributions of Π^ℓ defined in (8–12). Here, we define the inertial range by $32 \leq 2\ell/\eta \leq 128$ to investigate this regime in the remainder of this manuscript. More specifically, we focus on the three length scales

$$2\ell_1/\eta = 128, \quad 2\ell_2/\eta = 64, \quad \text{and} \quad 2\ell_3/\eta = 32, \tag{18}$$

Table 1: Parameters of the dataset adapted from "HIT2" case in Table I of [20]. Re_λ is the Taylor-microscale-based Reynolds number. N^3 denotes the resolution of the simulation. L_0/η is the ratio of macroscopic and microscopic length scales defined by $L_0 \equiv \overline{K}^{3/2}/\overline{\epsilon}$ and $\eta \equiv (\nu^3/\overline{\epsilon})^{1/4}$, respectively. Note that the $\bar{\cdot}$ denotes the average of the time series. T_{sim}/T_0 is the simulation time T_{sim} normalised by macroscopic time scale $T_0 \equiv \overline{K}/\overline{\epsilon}$. $\Delta T/T_0$ is the cascade time ΔT, defined by (19), normalised by T_0. N_{samp} is the number of samples to compute the Lagrangian evolution of the space-local inter-scale energy transfer.

Re_λ	N^3	L_0/η	T_{sim}/T_0	$\Delta T/T_0$	N_{samp}
236	512^3	876	8	0.40	1000

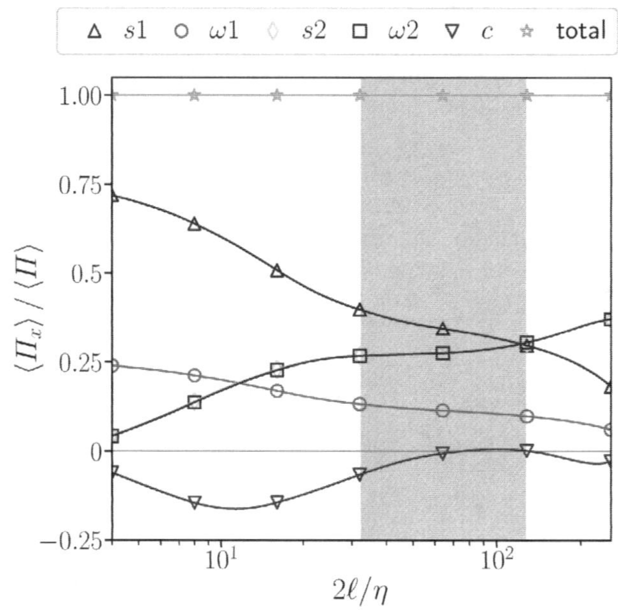

Figure 2: Distribution of decomposed inter-scale energy transfer components $\langle \Pi_x \rangle / \langle \Pi \rangle$ normalised by the total energy transfer over scale ℓ. The shaded region denotes the inertial range $32 \leq 2\ell/\eta \leq 128$.

from larger to smaller scales. Key findings of the inertial-range balance of the decomposed inter-scale energy transfer are the following [9, 10]:

(i) The scale-local contribution denoted by subscript 1 is approximately responsible for half of the total inter-scale energy transfer, while the scale-nonlocal part with subscript 2 accounts for the other half.

(ii) The strain-vorticity covariance Π_c^ℓ does not contribute to the inertial-range energy cascade (at least in space-average sense). There is a possible backscatter corresponding to the negative bump.

(iii) The SSA $\Pi_{s1}^\ell + \Pi_{s2}^\ell$ accounts for roughly 5/8 of the total inter-scale energy transfer while the VS $\Pi_{\omega1}^\ell + \Pi_{\omega2}^\ell$ occupies the remaining 3/8.

(iv) There is a balance between Π_{s2}^ℓ and $\Pi_{\omega2}^\ell$. However, they are not precisely equivalent.

Owing to the above results, Johnson [10] concluded that his work is:

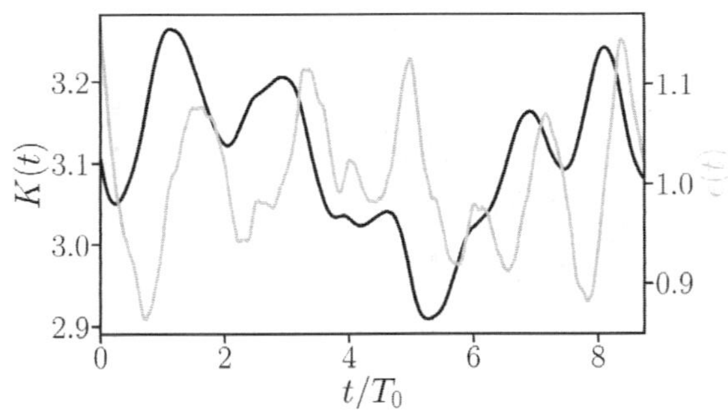

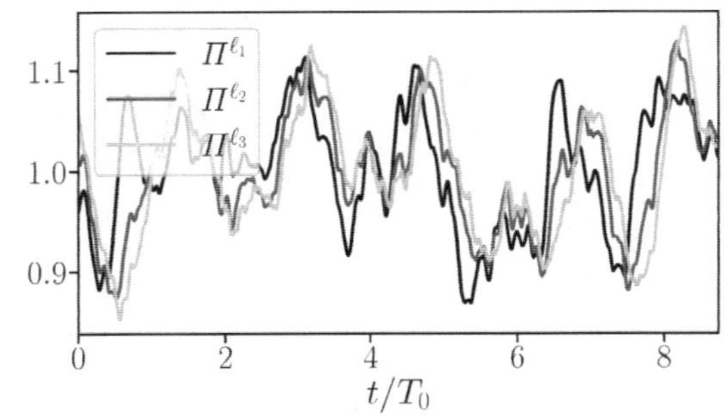

Figure 3: Time series of (a) the energy $K(t)$ and the energy dissipation rate $\epsilon(t)$ and (b) the inter-scale energy transfer $\Pi^{\ell}(t)$ at three different scales. Note that the time series in panel (b) are normalised by their mean values. For both panels, time is normalised by the characteristic time scale $T_0 = \overline{K}/\overline{\epsilon}$.

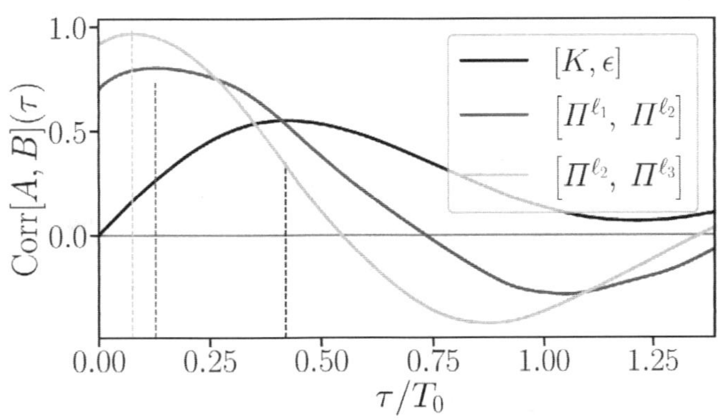

Figure 4: Two-time correlation function $\mathrm{Corr}[A, B](\tau)$ of three different combinations of the time series shown in Fig. 3. The vertical dashed lines denote their maximum correlation and the corresponding time delay determines the characteristic time scales: ΔT for K and ϵ (blue), ΔT_1 for Π^{ℓ_1} and Π^{ℓ_2} (red), and ΔT_2 for Π^{ℓ_2} and Π^{ℓ_3} (yellow). Note that $\Delta T > \Delta T_1 + \Delta T_2$.

highlighting the crucial role of strain-rate self-amplification in the energy cascade, calling into question the classical emphasis on vorticity stretching [22, 23].

The objective of the current study is to revisit these results from an information-theoretic point of view.

The flow is sustained by a deterministic force at the low wavenumbers (large scales). Consequently, the flow exhibits significant temporal fluctuations as shown in Fig. 3. Similar figures can be found in [20]. In Fig. 3(a), we plot the kinetic energy $K(t) = \langle \boldsymbol{u}^2 \rangle /2$ and the energy dissipation rate $\epsilon(t) = \nu \langle \boldsymbol{A}^2 \rangle$. They serve as the characteristic quantity of the largest and the smallest scales, respectively. We observe a correlation between the two quantities with a large time delay. In Fig. 3(b), we plot the inter-scale energy transfer rate $\Pi^{\ell}(t)$ at three different length scales in the inertial range, ℓ_1, ℓ_2, and ℓ_3 determined as (18). Indeed, we observe a clear correlation between the large- and small-scale time series with much smaller time delay in comparison to panel (a).

To quantify the characteristic time scale corresponding to these time-delayed fluctuations,

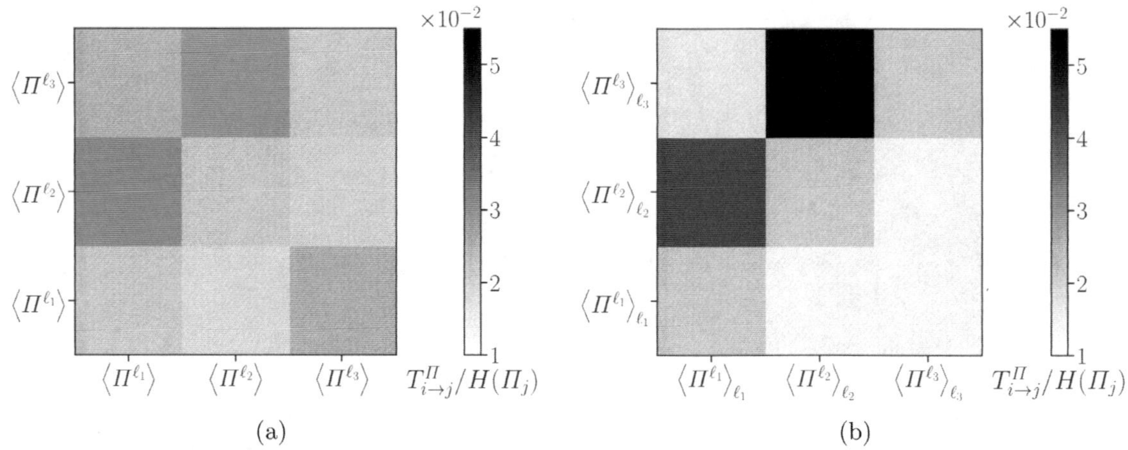

Figure 5: Information flux $T^{\Pi}_{i\to j}$ of the total inter-scale energy transfer Π^{ℓ} between two length scales. They are averaged in (a) the whole computational domain denoted by $\langle\cdot\rangle$ and (b) the space-local domain denoted by $\langle\cdot\rangle_{\ell}$. The space-local average is performed along with the Lagrangian frame depicted in Fig. 1. Note that the values are normalised by $H(\Pi_j)$, the entropy of the Π^{ℓ} at length scale ℓ_j. To compare the two panels, we employ the same colourmap range of $\left[10^{-2}, 5.5\times10^{-2}\right]$. Diagonal elements are shaded in pink to suppress the self-induced information flux.

we compute the two-time correlation function as shown in Fig. 4. We define the time scale of the energy cascade by the time instance where the correlation of K and ϵ attains its maximum value,

$$\Delta T \equiv \arg\max_{\tau} \mathrm{Corr}[K,\epsilon]\,(\tau). \tag{19}$$

To precisely track the energy cascade in the Lagrangian framework shown in Fig. 1, we define two more time scales from the inter-scale energy transfer time series,

$$\Delta T_1 \equiv \arg\max_{\tau} \mathrm{Corr}\left[\Pi^{\ell_1},\Pi^{\ell_2}\right](\tau) \quad \text{and} \quad \Delta T_2 \equiv \arg\max_{\tau} \mathrm{Corr}\left[\Pi^{\ell_2},\Pi^{\ell_3}\right](\tau). \tag{20}$$

3.2. Causality between decomposed inter-scale energy transfer

Figure 5 shows the information flux of the total inter-scale energy transfer Π^{ℓ}. Note that we evaluate the probability distributions with a direct binning approach to evaluate the information flux $T^{\Pi}_{i\to j}$. As we stated in the previous subsection, we investigate the inertial range $32 \le 2\ell/\eta \le 128$ of the dataset. Note that in Fig. 5 and the following figures, the bottom left corner corresponds to the large scale ℓ_1, and the top right corner describes the small scale ℓ_3. The entropy of the total inter-scale energy transfer reads $H(\Pi_j) = 2.46$ for $\ell_j = \ell_1$, 2.54 for ℓ_2, and 2.51 for ℓ_3. We observe more or less constant values for the inertial-range entropy of the inter-scale energy transfer. Note that we employ a shortened notation $\Pi_j = \left\langle \Pi^{\ell_j}\right\rangle_{\ell_j}$.

In Fig. 5(a), we plot the information flux of the global time series, which is similar to Fig. 6(a) of [4]. Indeed, we observe dominant scale-local information flux from larger to smaller scales as discussed in [4]. In Fig. 5(b), we compute a Lagrangian reinterpretation of panel (a) by tracking the space-local domain in the Lagrangian framework as schematically depicted in Fig. 1. Remarkably, we observe a more distinctive scale-local nature in this figure. It is thus an encouraging result for us to further extend the current Lagrangian approach.

Figure 6 shows the main result of this manuscript: information flux of the decomposed inter-scale energy transfer components defined in §2.1. First, we observe the robust scale-local

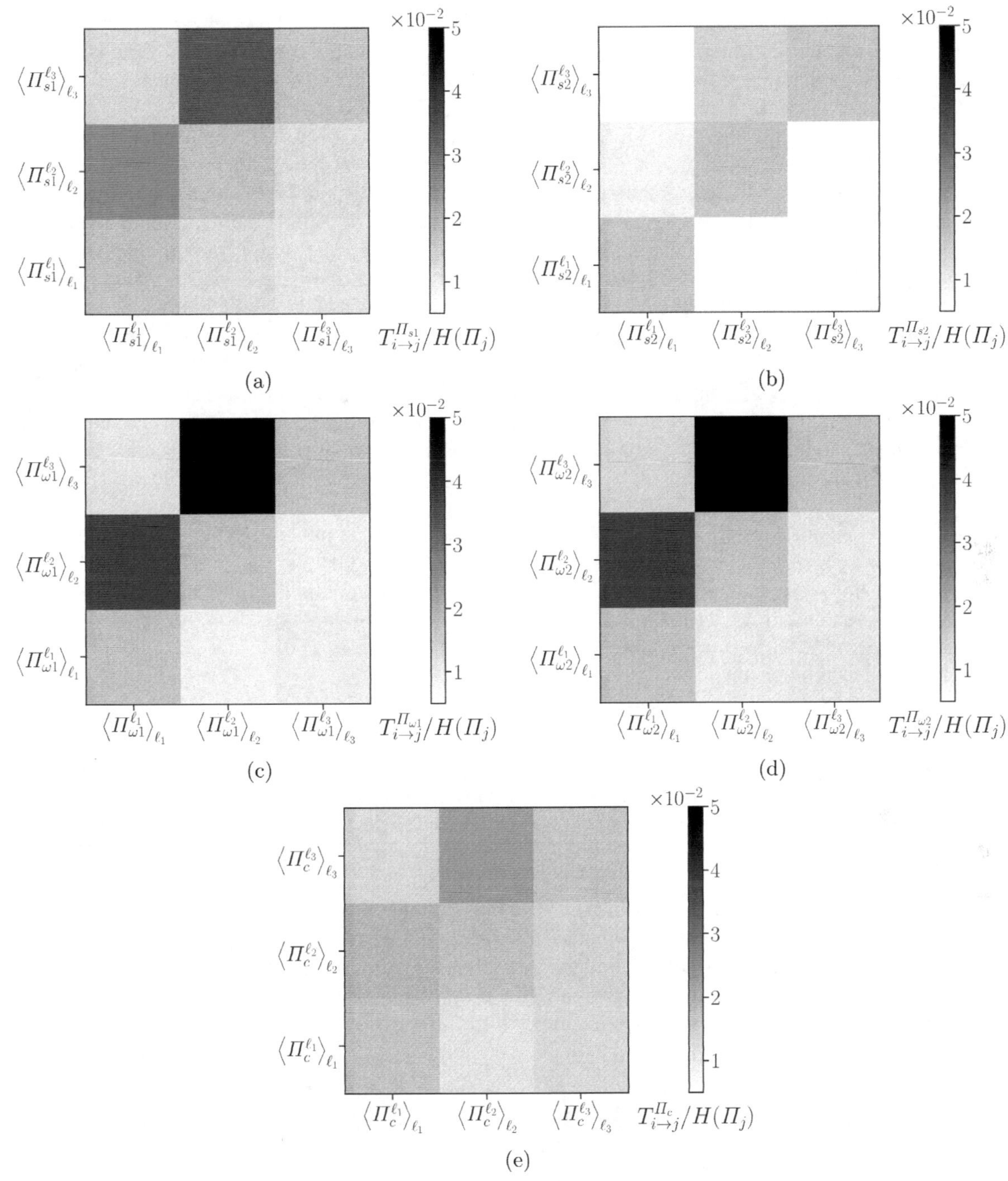

Figure 6: Information flux of the decomposed inter-scale energy transfer between two length scales. Each panel denotes: (a) $s1$, (b) $s2$, (c) $\omega1$, (d) $\omega2$, and (e) c contributions. For further details of the figure, see Fig. 5. To compare different panels, we normalise the values by the entropy of the total inter-scale energy transfer and employ the same colourmap range of $[5 \times 10^{-3}, 5 \times 10^{-2}]$.

information flux for both the SSA (top row) and VS (centre row). The contrast between the forward (large to small) and inverse (small to large) information flux is rather small for the strain-vorticity covariance shown in panel (e). It might be related to the backscatter of energy associated with this component (see negative bump of Π_c in Fig. 2).

Next, we compare the SSA and VS to find that the information flux associated with the VS is stronger than that of the SSA. Since the SSA is responsible for the majority of the inter-scale energy transfer (see Fig. 2), this result highlights the difference between the physical quantity and its information flux; The larger inter-scale energy transfer does not correspond to the larger information flux. In other words, we can state that the magnitude of the information flux of a physical quantity has nothing to do with the magnitude of the quantity itself.

Then, we focus on the scale-local (left column) and scale-nonlocal (right column) processes. For the SSA, the scale-local interaction is responsible for the stronger information flux than the nonlocal one. For the VS, we find that the magnitude of the information flux is comparable between the scale-local and nonlocal contributions. This observation emphasises the importance of the VS process, irrespective of scale dependency, in transferring large-scale information to smaller scales. Furthermore, by comparing the scale-nonlocal SSA and VS, we realise that the kinetically indistinguishable (see Fig. 2) mechanisms exhibit very different information-theoretic behaviour.

The above analysis may provide a clue to answer the question posed in §1: *How do the small scales of turbulence forget about the large scales and which physical mechanism is responsible for that?* Overall, Fig. 6 infers that the SSA tends to restrain the large-scale information along with the energy cascade compared to the VS. In other words, the SSA is a *less efficient* mechanism to transfer large-scale information than the VS.

4. Conclusions and perspectives

In this study, we investigate the information-theoretic nature of the energy cascade in the three-dimensional Navier–Stokes turbulence. The novelty of our approach is to measure the information flux (16) associated with the five different energy cascade mechanisms (8–12): strain self-amplification (SSA) and vortex stretching (VS) for scale-local and nonlocal interactions, and strain-vorticity covariance. It allows us to quantify the different mechanisms regarding information theory.

We investigate the existing HIT dataset to calculate the Lagrangian temporal evolution of the spatially local inter-scale energy transfer at three length scales in the inertial range. We find the scale-local nature of the information flux for the decomposed contributions. Importantly, there seems to be no clear relation between the magnitude of the inter-scale energy transfer and the information flux. Thus, while we have the following relation for the inter-scale energy transfer

$$\Pi_{s1} > \Pi_{s2} \approx \Pi_{\omega 2} > \Pi_{\omega 1} > \Pi_c \approx 0 \tag{21}$$

in the inertial range (see Fig. 2), the scale-local information flux exhibits the following relation

$$T^{\Pi_{\omega 1}} \approx T^{\Pi_{\omega 2}} > T^{\Pi_{s1}} > T^{\Pi_c} > T^{\Pi_{s2}} \tag{22}$$

as shown in Fig. 6. Our analysis infers that the SSA is a *less efficient* mechanism than the VS in terms of the information transfer from larger to smaller scales. In this sense, we may describe the SSA as a *forgetful* mechanism compared to the VS.

4.1. Some ideas to refine the analysis

We are considering improving the present project with the following approach. The first approach is to work on a larger dataset with a wider inertial range. However, it requires much longer post-process analysis time.

The second approach is to improve the evaluation of the information-theoretic quantities. Currently, we evaluate the probability distributions of each component of the decomposed inter-scale energy transfer, such as $p\left(\Pi_{s1}^{\ell_1}, \Pi_{s1}^{\ell_2}, \Pi_{s1}^{\ell_3}\right)$ for Π_{s1}, and their dimension corresponds to the number of scales in the inertial range. Once we obtain the probability distribution of all the decomposed inter-scale energy transfer components

$$p\left(\Pi_{s1}^{\ell_1}, \Pi_{s1}^{\ell_2}, \Pi_{s1}^{\ell_3}, \Pi_{\omega 1}^{\ell_1}, \dots, \Pi_{s2}^{\ell_1}, \dots, \Pi_{\omega 1}^{\ell_1}, \dots, \Pi_{c}^{\ell_1}, \dots\right), \tag{23}$$

it allows us to reduce the information leak due to the unobserved variable combinations and evaluate the information flux between different components (or different energy cascade mechanisms). To construct such a high-dimensional probability distribution, we need to implement the estimator-based probability distribution computation instead of the direct binning approach.

4.2. Future research questions

First, it would be interesting to investigate several different forcing configurations to quantify how the forced scales affect the information content of the inertial range. Indeed, in [11], it is argued that a specific forcing configuration, such as counter-rotating pairs of the large-scale vortex tubes, retains the flow in a transient state. It would be interesting if it is possible to quantify such *scars of the large scales* by the information-theoretic framework.

Next, we carefully avoid the term *cascade* for information flux in this manuscript because information is not an inviscid invariant. However, the information-theoretic analysis of developed turbulence indicates that a small amount of information seems to be conservatively transferred towards the smaller scales in the inertial range while the rest of the information may be trimmed off. We speculate that the latter reflects the large-scale configurations specific to each flow realisation, which is to be forgotten at the smaller scales. For the former case, it might be possible to define the conservative *information cascade*. It remains an open question how is the conservative information cascade related to the small-scale statistics of developed turbulence.

Similar to this point, it would be interesting to look at the information of conserved (i.e., energy) and not conserved (i.e., enstrophy) quantities in the inviscid limit. For example, how does the k^2 scaling of enstrophy flux [24, 25] contribute to its information content against the well-known k^0 scaling of inter-scale energy transfer in the inertial range? This question is directly related to quantifying the difference between 2D and 3D turbulence from the information-theoretic point of view.

Last, it has been shown that the small scales (e.g. $k\eta > 0.2$) of developed turbulence are enslaved to the larger scales [26–28]. It suggests that the information on scales below such a threshold is fully contained in the large scales. It is of interest to investigate the relation of this *chaos synchronisation* and the above-mentioned information-theoretic energy cascade picture.

Acknowledgments

This work was supported by the European Research Council under the Caust grant ERC-AdG-101018287 during the Fifth Madrid Turbulence Summer Workshop held at the Universidad Politécnica de Madrid on May-June, 2023, organised by Prof. Javier Jiménez. R. A. was supported by the Takenaka Scholarship Foundation. R. A. would like to thank Yuenong Ling at MIT for his generous help on the numerical implementation of information-theoretic tools and José I. Cardesa for his help on the numerical setup at the workshop. R. A. also would like to thank Tomohiro Tanogami for his insightful and crucial comments on the project.

References

[1] Kolmogorov A N 1941 *Doklady Akademii Nauk SSSR* **30**(4) 299–303 reprinted in Proc. R. Soc. Lond. A, **434**, pp. 9–13 (1991) and Selected Works of A. N. Kolmogorov, Vol. I: Mathematics and Mechanics, Kluwer Academic Publishers, pp. 312–318 (1991)
[2] Frisch U 1995 *Turbulence: The Legacy of A. N. Kolmogorov* (Cambridge University Press)
[3] Davidson P A 2015 *Turbulence: An Introduction for Scientists and Engineers* 2nd ed (Oxford University Press) URL https://books.google.fr/books?id=VblDCQAAQBAJ
[4] Lozano-Durán A and Arranz G 2022 *Physical Review Research* **4** 023195
[5] Tanogami T and Araki R 2022 *arXiv* (*Preprint* 2206.11163)
[6] Goto S, Saito Y and Kawahara G 2017 *Physical Review Fluids* **2** 064603
[7] Yoneda T, Goto S and Tsuruhashi T 2022 *Nonlinearity* **35** 1380
[8] Carbone M and Bragg A D 2020 *Journal of Fluid Mechanics* **883** R2
[9] Johnson P L 2020 *Physical Review Letters* **124** 104501
[10] Johnson P L 2021 *Journal of Fluid Mechanics* **922** A3
[11] McKeown R, Pumir A, Rubinstein S M, Brenner M P and Ostilla-Mónico R 2022 (*Preprint* ArXiv 2204.09023)
[12] Yao J and Hussain F 2020 *Journal of Fluid Mechanics* **883** A51
[13] McKeown R, Ostilla-Mónico R, Pumir A, Brenner M P and Rubinstein S M 2018 *Phys. Rev. Fluids* **3** 124702
[14] McKeown R, Ostilla-Mónico R, Pumir A, Brenner M P and Rubinstein S M 2020 *Sci. Advances* **6** eaaz2717
[15] Richardson L F 1922 *Weather Prediction by Numerical Process* (Cambridge University Press) available at https://archive.org/details/weatherpredictio00richrich
[16] Shannon C E 1948 *The Bell System Technical Journal* **27** 379–423
[17] MacKay D J C 2003 *Information Theory, Inference, and Learning Algorithms* (Cambridge University Press) ISBN 9780521642989 URL https://books.google.fr/books?id=AKuMj4PN_EMC
[18] Cover T M and Thomas J A 2012 *Elements of Information Theory* 2nd ed (Wiley) ISBN 9781118585771
[19] Falkovich G 2022 Physical Nature of Information URL https://www.weizmann.ac.il/complex/falkovich/sites/complex.falkovich/files/uploads/PNI22.pdf
[20] Cardesa J I, Vela-Martín A, Dong S and Jiménez J 2015 *Physics of Fluids* **27** 111702
[21] Cardesa J I, Vela-Martín A and Jiménez J 2017 *Science* **357** 782–784
[22] Taylor G I 1938 *Proc. Royal Soc. London. Series A: Math. Phys. Sci.* **164** 15–23
[23] Onsager L 1949 *Il Nuovo Cimento (1943-1954)* **6** 279–287
[24] Davidson P A, Morishita K and Kaneda Y 2008 *Journal of Turbulence* **9** N42
[25] Sadhukhan S, Samuel R, Plunian F, Stepanov R, Samtaney R and Verma M K 2019 *Phys. Rev. Fluids* **4** 084607
[26] Yoshida K, Yamaguchi J and Kaneda Y 2005 *Physical Review Letters* **94** 014501
[27] Lalescu C C, Meneveau C and Eyink G L 2013 *Physical Review Letters* **110** 084102
[28] Vela-Martín A 2021 *Journal of Fluid Mechanics* **913** R8

Transient growth of wavelet-based resolvent modes in the buffer layer of wall-bounded turbulence

Eric Ballouz[1], **Scott T. M. Dawson** [2], **H. Jane Bae**[3]

[1] Mechanical and Civil Engineering, California Institute of Technology, Pasadena CA 91125, USA

[2] Mechanical Materials, and Aerospace Engineering, Illinois Institute of Technology, Chicago IL 60616, USA

[3]Graduate Aerospace Laboratories, California Institute of Technology, Pasadena CA 91125, USA

E-mail: `eballouz@caltech.edu`

Abstract. In this work, we study the transient growth of the principal resolvent modes in the minimal flow unit using a reformulation of resolvent analysis in a time-localized wavelet basis. We target the most energetic spatial wavenumbers for the minimal flow unit and obtain modes that are constant in the streamwise direction and once-periodic in the spanwise direction. The forcing modes are in the shape of streamwise rolls, though pulse-like in time, and the response modes are in the form of transiently growing streaks. We inject the principal transient forcing mode at different intensities into a simulation of the minimal flow unit and compare the resulting nonlinear response to the linear one. The peak energy amplification scales quadratically with the intensity of the injected mode, and this peak occurs roughly at the same time for all forcing intensities. However, the larger energy amplification intensifies the magnitude of the nonlinear terms, which play an important role in damping the energy growth and accelerating energy decay of the principal resolvent mode. We also observe that the damping effect of the nonlinearities is less prominent close to the wall. Finally, we find that the principal resolvent forcing mode is more effective than other structures at amplifying the streak energy in the turbulent minimal-flow unit. In addition to lending support to the claim that linear mechanisms are important to near-wall turbulence, this work identifies time scales for the nonlinear breakdown of linearly-generated streaks.

1. Introduction

Near-wall turbulence is believed to be organized into streamwise rolls, and alternating low- and high-speed streamwise streaks [1, 2, 3, 4, 5]. These coherent structures are well-documented and their characterization is the subject of multiple works [6, 7, 8, 9, 10]. Additionally, many studies point to a quasi-periodic cycle, wherein the streamwise streaks are amplified by streamwise vortices, meander, then break down, which subsequently regenerates new quasi-streamwise vortices [11, 12, 13, 14, 15, 16, 17]. The cycle, also known as the self-sustaining process, can be more clearly observed in a minimal flow unit [18], where the domain is artificially restricted in the streamwise and spanwise directions in order to exclude the dynamics of the outer region of the channel. Despite the restriction, the study of the minimal flow unit reveals that the near-wall cycle can self-sustain even when motions at larger scales are inhibited [19].

Although nonlinear mechanisms play a role in the self-sustaining process, a lot of attention has been given to linear mechanisms and instabilities as the drivers of this process [13, 20]. One example is the Orr mechanism [21, 22], in which the mean shear near the wall tilts velocity perturbations forward in the streamwise direction and stretches vertical scales, intensifying the wall-normal velocity perturbations. Another example is lift-up [23], which occurs when wall-normal velocity perturbations transport slow-moving fluid near the wall away into the faster flow field farther away from the wall. Works like Del Alamo and Jiménez [10] and Pujals *et al.* [24] show that, even after removing the nonlinear term from the perturbation equations, linear transient growth via the mean shear generates the dominant (streaky) structures in wall-bounded turbulence. The linearized system additionally accounts for much of the energy spectra and reproduces the self-similar profile in the logarithmic region. Similarly, Lozano-Durán *et al.* [20] show through numerical experiments that the minimal flow unit can sustain turbulence without the nonlinear feedback between the velocity fluctuations and the mean velocity profile, except when the Orr-mechanism or push-over (momentum transfer from the spanwise mean shear into the streamwise velocity perturbation) are suppressed. The authors thus argue for linear transient growth as a prominent mechanism for transferring energy from the mean flow to turbulent fluctuations.

The linear amplification process linking streamwise vortices and streamwise streaks has also been fruitfully studied through the lens of resolvent analysis [25, 26, 27]. In resolvent analysis, the Navier-Stokes equations are reframed as a linear dynamical system for the velocity fluctuations, and the nonlinear term as external forcing acting on this system. The goal is then to solve for the spatial structure of the (nonlinear) forcing that generates the response (velocity) with the largest linear energy amplification. In the context of wall-bounded turbulent flows, the method is successful at identifying streamwise rolls as the most perturbing structures, and streamwise streaks as the most amplified structures. Furthermore, Bae *et al.* [28] demonstrate the role of these linearly-identified resolvent modes in transferring energy to coherent near-wall turbulent perturbations, even within a fully nonlinear turbulent flow. Their study shows that projecting out the contribution of the leading resolvent forcing mode from the nonlinear term at every time step interrupts the streak-regeneration process and greatly suppresses buffer layer turbulence.

In resolvent analysis, the resolvent operator is constructed from the linearized Navier-Stokes equations that are Fourier transformed in the homogeneous spatial directions and in time. For turbulent channel flow, this means the domain is usually assumed to be periodic in the streamwise and spanwise directions and statistically stationary in time [25, 27, 26, 29]. These resolvent modes, which are Fourier modes in time, lack transient linear growth information. In light of the importance of transient linear growth to wall-bounded turbulence, the aim of this work is to reformulate resolvent analysis in a time-localized basis, such as one constructed with wavelets rather than Fourier modes [30], and to further examine the interactions of these transient resolvent modes with the fully nonlinear turbulent flow in the minimal flow unit.

This paper is organized as follows. In §2.1, we compute resolvent modes in a time-localized wavelet basis as in Ballouz *et al.* [30], and constrain the forcing to a wavelet-shaped pulse. This formulation yields a forcing mode in the shape of streamwise rolls that is compactly supported in time, in addition to an optimal streak-like response that grows transiently before decaying. The justification for the choice of spatial wavenumbers and wavelets is given in §2.2. We then solve the fully nonlinear forced Navier-Stokes equations for the minimal flow unit at $Re_\tau = 186$, using the time-localized wavelet-based resolvent forcing mode as our forcing term. This step is detailed in §2.3. We track the evolution of this resolvent forcing mode as it generates and amplifies streamwise streaks, and compute relevant turbulent statistics, which we present in §3. Concluding remarks are given in §4.

2. Methods

In this work, we consider the flow in the minimal flow unit of size $L_x \times L_y \times L_z = 1.72\delta \times 2\delta \times 0.86\delta$, where δ is the channel half-height, and x, y and z are the streamwise, wall-normal and spanwise directions, respectively. We denote the velocity fluctuation field by $\boldsymbol{u} = [u, v, w]^T$, where u, v and w are the streamwise, wall-normal, and spanwise components respectively. The system is characterized by the friction Reynolds number $Re_\tau = \delta u_\tau / \nu \approx 186$, where u_τ is the friction velocity, and ν is the kinematic viscosity. The flow is periodic in the streamwise and spanwise directions, and the no-slip and no-penetration conditions hold at the walls of the channel.

For the direct numerical simulations (DNS) in this work, we discretize the streamwise and spanwise directions uniformly using $N_x = N_z = 32$ grid points, which results in streamwise and spanwise grid spacings of $\Delta x^+ \approx 10$ and $\Delta z^+ \approx 5$. In the wall-normal direction, we use a Chebyshev grid of size $N_y = 128$, which results in a wall-normal spacing of $\min(\Delta y^+) \approx 0.17$ near the wall and $\max(\Delta y^+) \approx 7.6$ at the centerline. Here, the superscript + denotes wall units normalized with u_τ and ν. We discretize the incompressible Navier-Stokes equations with a staggered, second-order-accurate, central finite difference method in space [31], and a fractional step method is used to compute pressure [32]. Time-advancement is performed with an explicit third-order-accurate Runge-Kutta method [33]. The DNS code has been validated in previous studies of turbulent channel flows [34, 35, 36]. Using this discretization, we obtain a mean streamwise velocity profile $U(y)$ by averaging DNS results of the unforced system in the homogeneous directions and time. This profile is used in the subsequent sections.

2.1. Spatio-temporal resolvent modes

We first compute the time-localized resolvent forcing modes and their corresponding transient responses for the minimal flow unit. The incompressible Navier-Stokes equations for the velocity fluctuations about the mean turbulent flow field $\boldsymbol{U}(y) = (U(y), 0, 0)$ are Fourier-transformed in the x- and z-directions, and wavelet-transformed in time, *i.e.*

$$\tilde{\partial}_t \tilde{\boldsymbol{u}} = -(\boldsymbol{U} \cdot \tilde{\nabla})\tilde{\boldsymbol{u}} - (\tilde{\boldsymbol{u}} \cdot \tilde{\nabla})\tilde{\boldsymbol{U}} - \frac{1}{\rho}\tilde{\nabla}\tilde{p} + \nu\tilde{\Delta}\tilde{\boldsymbol{u}} + \tilde{\boldsymbol{f}}, \;\; \tilde{\nabla} \cdot \tilde{\boldsymbol{u}} = 0, \tag{1}$$

where $(\tilde{\cdot})$ denotes a Fourier transform for a given streamwise and spanwise wavenumber pair (k_x, k_z) composed with a wavelet transform in time, ρ is density, and p is the pressure. These transformed quantities are functions of wall-normal position y, and the wavelet scale and shift parameters α and β , which respectively determine the time interval and frequency range of the wavelet mode onto which we project our physical quantities [30]. The nonlinear term, denoted by $\tilde{\boldsymbol{f}}$, is assumed to act as an exogenous forcing function on the velocity field. The modified spatial derivative operators are given by $\tilde{\nabla} = [\mathrm{i}k_x, \partial_y, \mathrm{i}k_z]^T$, where $\mathrm{i}^2 = -1$, and $\tilde{\Delta} = -k_x^2 + \partial_{yy} - k_z^2$, and the modified time derivative operator is defined as $\tilde{\partial}_t = \mathcal{W} \circ \partial_t \circ \mathcal{W}^{-1}$, where $\mathcal{W}$ is the discrete wavelet transform. The transform $\mathcal{W}$ projects an arbitrary function onto a set of shifted and rescaled wavelet and scaling functions [37, 38]. The choice of wavelet-scaling-function pair determines the properties of the transform, most notably its sparsity, which depends on the compactness of the chosen functions, and whether it is unitary, which depends on whether the wavelet and scaling function bases are orthonormal. For a discussion on the choice of wavelet transform, see §2.2.

We can rearrange (1) as

$$\begin{bmatrix} \tilde{\boldsymbol{u}}(y,\alpha,\beta) \\ \tilde{p}(y,\alpha,\beta) \end{bmatrix} = \widetilde{\mathcal{H}}^{(k_x,k_z)} \begin{bmatrix} \tilde{\boldsymbol{f}}(y,\alpha,\beta) \\ 0 \end{bmatrix}, \tag{2}$$

where the linear operator $\widetilde{\mathcal{H}}^{(k_x,k_z)}$ is the resolvent operator in this formulation. In the setting of a discretized spatial and temporal domain, $\widetilde{\mathcal{H}}^{(k_x,k_z)}$ is a $(4N_yN_t) \times (4N_yN_t)$ matrix, where N_t is the resolution in time. Indeed, this resolvent formulation targets all temporal basis functions, rather than one as in traditional resolvent analysis. The factor 4 is due to the fact that the left-hand side of equation (2) contains four variables, $\tilde{\boldsymbol{u}}$, and $\tilde{p}$. Note the dependence on the spatial parameters (k_x, k_z), which must be chosen to define the Fourier transform in the homogeneous directions. To discretize the wall-normal direction, we use the same Chebyshev grid as in the DNS. The spatial derivatives in y are similarly staggered second-order-accurate central finite differences, and the Fourier spatial derivatives in the streamwise and spanwise directions are computed with the modified wavenumbers that match the choice of differentiation scheme used in the DNS. We choose a second-order-accurate central finite difference matrix for the time derivative operator. A total time of $T = 22\ \delta/u_\tau$ is used to allow the resolvent response modes to decay to zero, allowing us to use periodic boundary conditions for the time derivative matrix. Though not shown, increasing the order of the finite difference operator does not strongly affect the resolvent modes.

We introduce the additional step of constraining the forcing along a wavelet-shaped pulse of any desired scale α and shift β using a temporal windowing matrix $\mathcal{B}$ [39, 40, 30]. We then take the singular value decomposition (SVD) of the combined operator

$$\widetilde{\mathcal{H}}^{(k_x,k_z)}\mathcal{B} = \sum_{j=1}^{\infty} \sigma_j \tilde{\boldsymbol{\psi}}_{\boldsymbol{j}}(y,\alpha,\beta) \otimes \tilde{\boldsymbol{\phi}}_{\boldsymbol{j}}(y,\alpha,\beta), \tag{3}$$

where we index the singular values $\{\sigma_i\}_{i=1}^{\infty}$ such that $\sigma_1 \geq \sigma_2 \geq ... \geq 0$. The right and left singular vectors $\{\tilde{\boldsymbol{\phi}}_{\boldsymbol{j}}\}_{j=1}^{\infty}$ and $\{\tilde{\boldsymbol{\psi}}_{\boldsymbol{j}}\}_{j=1}^{\infty}$ respectively define orthonormal bases for the spaces containing the nonlinear term (forcing) and the velocity and pressure fluctuations (response). For the SVD, we choose the inner product to be the kinetic energy seminorm, which we define for an arbitrary vector $\tilde{\boldsymbol{b}} = [\tilde{b}_u, \tilde{b}_v, \tilde{b}_w, \tilde{b}_p]^T$ to be

$$||\tilde{\boldsymbol{b}}||^2 = \frac{u_\tau}{\delta} \frac{1}{L_x(2\delta)L_z} \int_0^{L_x} \int_0^{2\delta} \int_0^{L_z} \int_{-\infty}^{+\infty} (|b_u|^2 + |b_v|^2 + |b_w|^2)\, \mathrm{d}t\, \mathrm{d}z\, \mathrm{d}y\, \mathrm{d}x, \tag{4}$$

where $\boldsymbol{b} = [b_u, b_v, b_w, b_p]^T$ is the inverse transform of $\tilde{\boldsymbol{b}}$. The kinetic energy amplification factor is given by the square of the singular values. The forcing modes are therefore ordered decreasingly according to the integrated kinetic energy amplification they undergo when acted on by $\widetilde{\mathcal{H}}^{(k_x,k_z)}\mathcal{B}$, and the response modes are the corresponding amplified coherent structures arising from this action. Thus, $\tilde{\boldsymbol{\phi}}_{\boldsymbol{1}} = [\tilde{\phi}_{1,u},\ \tilde{\phi}_{1,v},\ \tilde{\phi}_{1,w},\ 0]^T$ generates the largest linear energy amplification via the windowed resolvent operator, and $\sigma_1\tilde{\boldsymbol{\psi}}_{\boldsymbol{1}} = \sigma_1[\tilde{\psi}_{1,u},\ \tilde{\psi}_{1,v},\ \tilde{\psi}_{1,w},\ \tilde{\psi}_{1,p}]^T$ is the resulting optimally-amplified transient velocity and pressure fluctuation. For all j, $\boldsymbol{\phi}_{\boldsymbol{j}}$, the inverse transform of $\tilde{\boldsymbol{\phi}}_{\boldsymbol{j}}$, is shaped in time according to the wavelet or scaling function chosen by $\mathcal{B}$. As described in Bae *et al.* (2021), the resolvent modes occur in equivalent pairs of equal singular values due to the symmetry of the channel geometry. For the numerical experiments in §2.3, we linearly combine the two equivalent forcing modes (e.g. $\tilde{\boldsymbol{\phi}}_{\boldsymbol{1}}$ and $\tilde{\boldsymbol{\phi}}_{\boldsymbol{2}}$) to isolate the support of the forcing mode to the bottom half of the channel. Thus, upon injecting this mode into the DNS of the forced Navier-Stokes equations, only the bottom of half of the channel is subject to the forcing, allowing us to use the top half as a control system [28]. Henceforth, $\tilde{\boldsymbol{\phi}}_{\boldsymbol{1}}$ will refer to the linear combination of the first two equivalent forcing modes that limits the forcing to the bottom-half of the channel. $\tilde{\boldsymbol{\phi}}_{\boldsymbol{3}}$ is defined similarly with regards to the third and fourth forcing modes.

2.2. Choice of spatial and temporal scales

As in Bae *et al.* (2021), we choose to study the modes given by wavenumbers of $k_xL_x/(2\pi) = 0$ and $k_zL_z/(2\pi) = 1$, which account for a peak in the spectral energy content for the minimal flow unit at a wall-normal height of $y^+ = 15$. This is consistent with the idea that the minimal flow

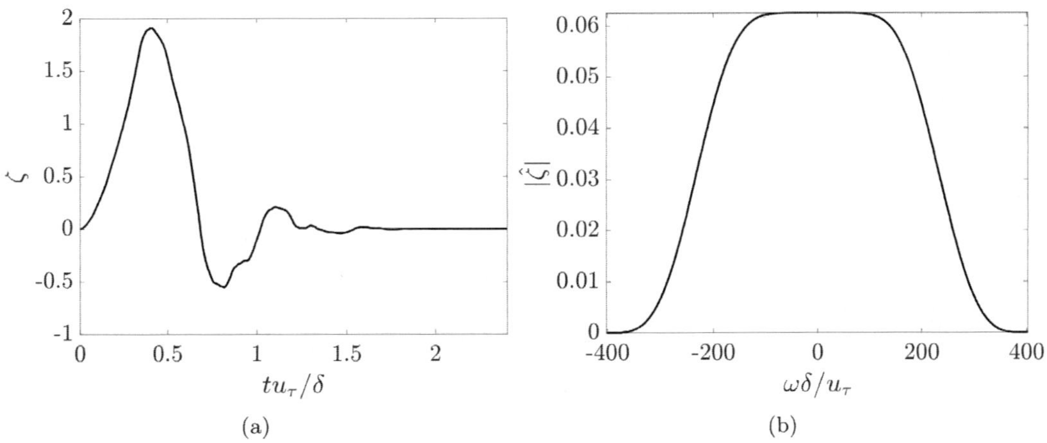

Figure 1: Daubechies-8 scaling function ζ in time (a) and frequency (b) domain.

unit is the smallest channel capable of sustaining turbulent streaks; the single streak thus appears to stretch the entire length of the domain hence the streamwise wavenumber of 0. Similarly, the domain is only large enough to contain one low- and high-speed streak pair in the spanwise direction, which yields the spanwise wavenumber of 1.

Traditional resolvent analysis in which the Navier-Stokes equations are Fourier-transformed in time reveals that a temporal frequency of $\omega = 0$ corresponds to the largest linear amplification [28]. However, since we constrain the forcing term to a pulse that is compactly-supported in time, resolvent analysis will not be able to target a single frequency, but will capture a wide range of frequencies. A trade-off exists between time and frequency localization, and the more precision we require in one domain, the less we preserve in the other [37, 38]. In this work, we use a single-level Daubechies-8 wavelet transform [41]. Using $\mathcal{B}$, we constrain the forcing term to the desired Daubechies-8 scaling function of arbitrary shift β (*i.e.* centered at an arbitrary time) denoted by $\zeta(t)$, since it is compactly-supported in time and its Fourier-spectrum is a quasi band-pass filter that encompasses the target frequency $\omega = 0$ resulting in the largest linear amplification via the traditional resolvent operator for the minimal flow unit. The scaling function satisfies $\left(\int_{-\infty}^{+\infty} |\zeta(t)|^2 \mathrm{d}t\right)\ u_\tau/\delta = 1$. The chosen scaling function is shown along with its Fourier spectrum in figure 1. We note that the simulations detailed in §2.3 resolve temporal wavenumbers up to $\omega\delta/u_\tau \approx 55,000$.

The obtained resolvent modes are shown in figure 2. Notably, the response modes exhibit transient energy growth and decay as seen in figure 2(a,b). The inverse transforms of the modes are shown in figure 2(c,d). The optimal transient nonlinear forcing mode appears in the shape of streamwise rolls, and the optimal velocity fluctuation response appears as predominantly streamwise streaks with alternating signs of the same magnitude. This supports the extensively-examined claim that streamwise streaks can be linearly generated by a linear lift-up mechanism, whereby slower moving fluid close to the wall is swept upwards into the faster moving mean flow farther away from the wall. The streak-shaped response mode grows in intensity before fading, showcasing the transient growth characteristic of the linearized Navier-Stokes equations for this system.

2.3. Forced direct numerical simulations

We first perform a DNS with a fixed mean flow given by $\boldsymbol{U} = [U(y), 0, 0]^T$, which was used to calculate the resolvent modes. From this simulation, we obtain snapshots with a frozen mean profile and use these to initialize the ensemble of the resolvent-forced simulations. Freezing the

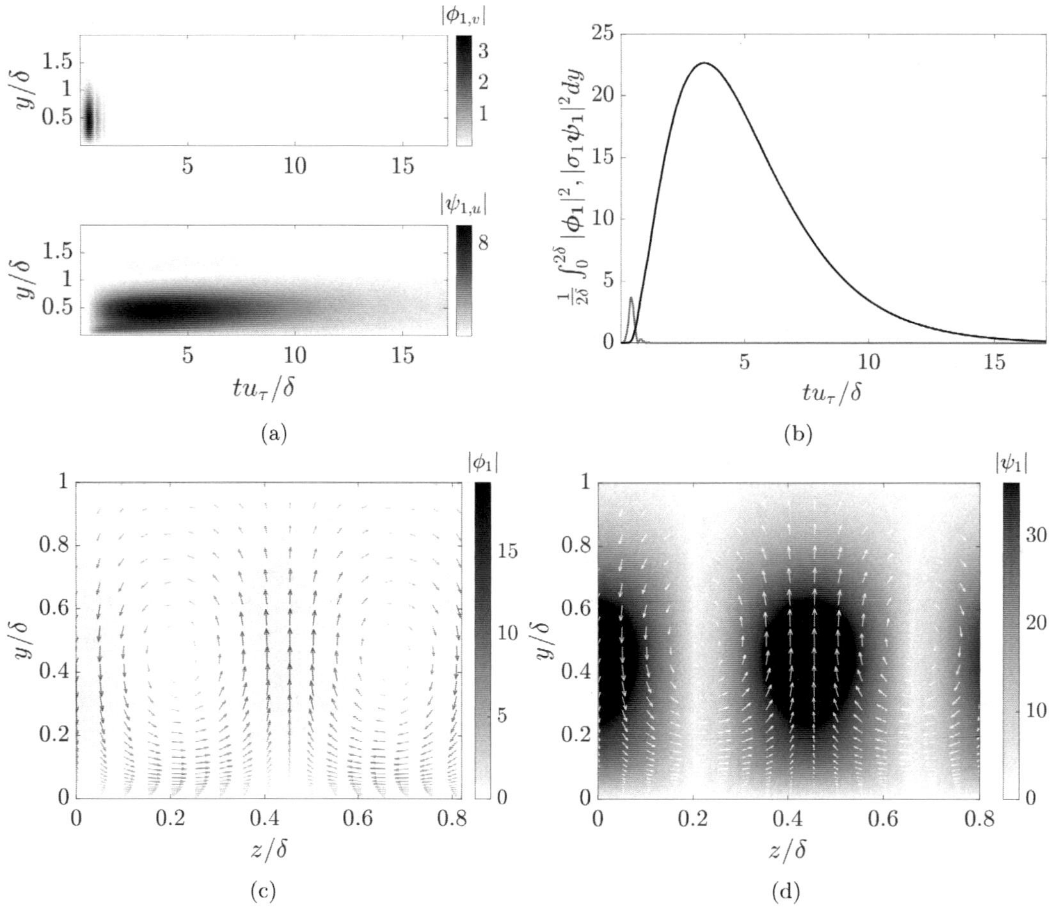

Figure 2: (a) Magnitudes of the wall-normal component of the principal forcing mode (top) and the streamwise component of the principal response mode (bottom). (b) Integrated energy of the principal forcing (red) and response (black) modes. (c) Principal forcing mode shown at peak amplitude ($tu_\tau/\delta \approx 0.40$). (d) Principal response modes at peak amplitude ($tu_\tau/\delta \approx 2.41$). In (c,d), the contours represent the streamwise magnitude; the arrows, which show the direction of the cross flow components, are colored according to their magnitudes $\sqrt{|\phi_{1,v}|^2 + |\phi_{1,w}|^2}$ in (c), or $\sqrt{|\psi_{1,v}|^2 + |\psi_{1,w}|^2}$ in (d).

mean profile ensures that the DNS mean profile matches the one used to compute the resolvent modes for all time. We do this by initializing the flow to have the desired mean streamwise profile of U, then by removing the steady-state contribution of the right-hand side of the Navier-Stokes equations. The initial snapshots from the fixed-mean simulation are separated by $\Delta t \approx 17.5\ \delta/u_\tau$, and amount to an ensemble size of approximately 990.

For each resolvent-forced simulation, we have a corresponding unforced fixed-mean simulation with the same initial condition. We denote the velocity fluctuations for the unforced simulations by $\boldsymbol{u}_0$, and the nonlinear terms in the momentum equations by $\boldsymbol{g}_0$. For the resolvent-forced DNS, we introduce a forcing term proportional to $\boldsymbol{\phi_1}$, the inverse transform of the principal resolvent forcing mode $\tilde{\boldsymbol{\phi_1}}$. The forcing, normalized so that $||\tilde{\boldsymbol{\phi_1}}||^2 = 1$, is scaled by a complex constant κ

with magnitude

$$|\kappa| := \gamma \left(\frac{\delta}{u_\tau^3} \frac{1}{L_x(2\delta)L_z} \int_{-\infty}^{+\infty} \int_0^{L_x} \int_0^{2\delta} \int_0^{L_z} \left| \frac{\partial \boldsymbol{u_0}}{\partial t} \right|_{t=0}^2 \zeta(t)^2 \mathrm{d}z\, \mathrm{d}y\, \mathrm{d}x\, \mathrm{d}t \right)^{1/2}, \tag{5}$$

where $\gamma \in \{1\%, 2\%, 5\%, 10\%\}$ such that the resolvent forcing mode is increasing the initial energy of the right-hand side by $\gamma\%$. Thus, $|\kappa|^2$ determines the integrated energy injected into the system by the forcing. We choose $\angle\kappa = \angle\langle \partial_t \hat{\boldsymbol{u}}_0^{(0,1)}, \phi_1 \rangle$, so the forcing mode is in phase with the right-hand side of the unforced flow-field. To test the optimality of $\boldsymbol{\phi_1}$ at forcing the turbulent channel, we repeat the case with $\gamma = 5\%$ using $\boldsymbol{\phi_3}$ and a randomly generated term $\boldsymbol{\phi_{\mathrm{rand}}} = \hat{\boldsymbol{\phi}}_{\mathbf{rand}}(y)\zeta(t)$, which we normalize and scale the same way. As in the unforced case, the mean profile of the forced simulation is fixed, and we run the forced DNS for a total time of $T = 5.69\,\delta/u_\tau$.

3. Results and discussion

In this section, we define the deviation operator Δ as the difference between the forced and unforced simulations, e.g., $\Delta\hat{u}^{(0,1)} = \hat{u}^{(0,1)} - \hat{u}_0^{(0,1)}$. The velocity deviation field is then given by $\hat{\boldsymbol{q}} = [\Delta\hat{u}^{(0,1)}, \Delta\hat{v}^{(0,1)}, \Delta\hat{w}^{(0,1)}]^T$, and the deviation in the nonlinear terms $\hat{\boldsymbol{f}} = \Delta\hat{\boldsymbol{g}}^{(0,1)}$, where $\hat{\boldsymbol{g}}$ is the nonlinear terms in the Navier-Stokes equations. We define the instantaneous streak energy as

$$\hat{E}_u^{(0,1)}(t) = \frac{1}{2\delta} \int_0^{2\delta} \frac{|\hat{u}^{(0,1)}|^2}{2} \mathrm{d}y, \tag{6}$$

and the deviation in total kinetic energy as $\Delta\hat{E}^{(0,1)}(t)$. Finally, we denote the ensemble average by $\overline{(\cdot)}$.

3.1. Transient energy growth and decay of streaks in the forced DNS

Figure 3(a) shows the streak energy contained in the $(0,1)$-mode as a function of time, for different resolvent-forcing amplitudes. For all cases, the energies grow and peak before decaying. The stronger the forcing, the faster the streak energy's growth rate, and the faster its decay. We note that the differences in decay rate are more dramatic compared to those in growth rate across the different forcing amplitude cases. The peak times, t_{peak}, defined as the times at which the energies reach their maxima, decrease slightly with forcing amplitude, but are relatively constant compared to the decay times, $\Delta t_{\text{decay}} = t_{10\%} - t_{\text{peak}}$, which we define as the time it takes for the energy to reduce from the peak to 10% of its peak (figures 3b, 3c). Indeed, the decay time varies widely and scales as $\Delta t_{\text{decay}} \sim |\gamma|^{-0.65}$. We note that all fully-coupled simulations decay significantly faster ($t \approx 1\delta/u_\tau$) than the linear response ($t \approx 15\delta/u_\tau$). In addition to energy growth and decay, we observe that the peak energy scales quadratically with the forcing amplitudes (figure 3d).

To measure the proportion of the linearly-amplified energy captured by the forced simulations, we compute an ensemble-averaged forcing efficiency, or effective amplification, which we define to be

$$\sigma_{\text{eff}} = \left(\frac{1}{u_\tau\delta} \int_0^T \frac{\overline{\Delta\hat{E}^{(0,1)}}}{|\kappa|^2/2} \mathrm{d}t \right)^{1/2}. \tag{7}$$

This is analogous to $\sigma_1 = \max_{\tilde{\boldsymbol{f}}} \|\tilde{\mathcal{H}}^{(0,1)}\mathcal{B}\tilde{\boldsymbol{f}}\|_2/\|\tilde{\boldsymbol{f}}\|_2$, where the numerator reflects the energy contained in the velocity field perturbations, and the denominator corresponds to the forcing amplitude. The computed σ_{eff} is shown in figure 4. The forcing efficiencies decrease with the intensity of the forcing, and all effective amplifications are lower than $\sigma_1 = 11.54$. This efficiency indicates that, for smaller resolvent forcing amplitudes, more of the forcing energy is linearly converted into streak energy.

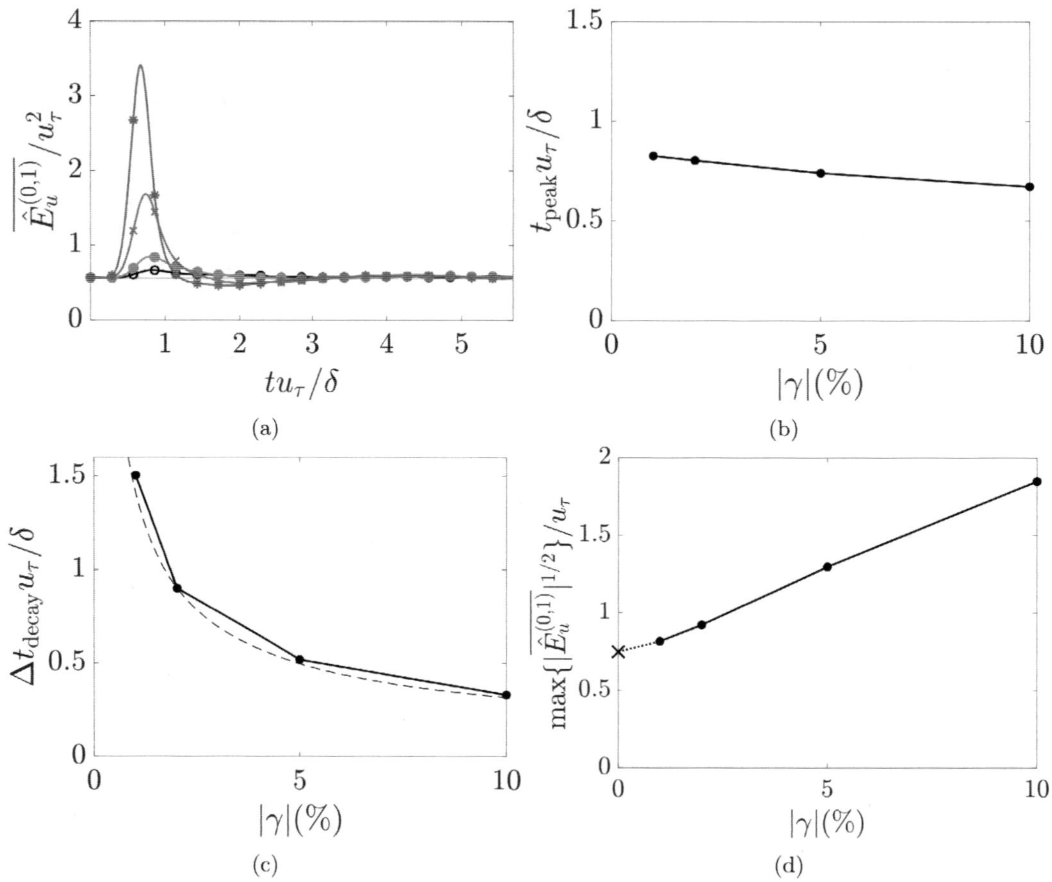

Figure 3: Average streak energy as a function of time (a), peak time (b), decay time (c), and streak energy peaks (d) as a function of γ. The cases plotted in (a) are $\gamma = 1\%$ (black $\circ$), $\gamma = 2\%$ (cyan $\bullet$), $\gamma = 5\%$ (purple $\times$), and $\gamma = 10\%$ (red $*$). The dashed line in (c) is $|\gamma|^{-0.65}$. In (d), $\times$ denotes the streak energy in the unforced case.

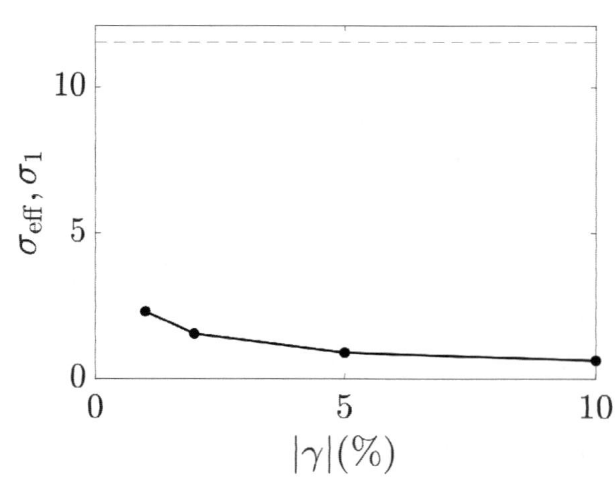

Figure 4: Effective amplification σ_{eff} (solid) and σ_1 (dashed).

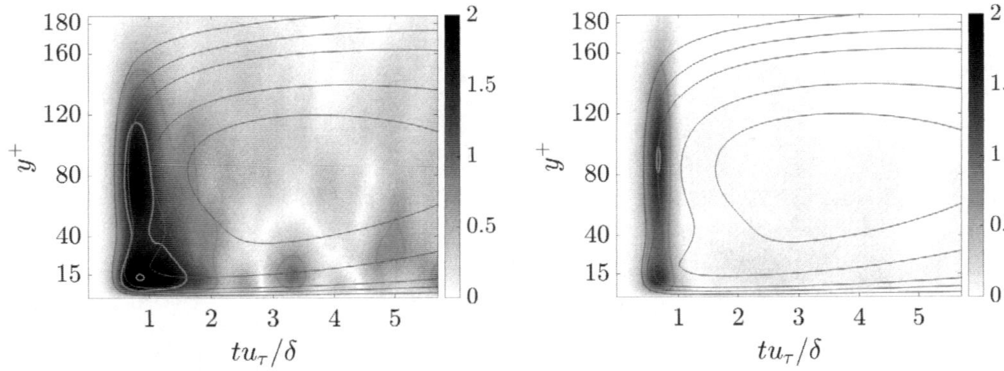

Figure 5: Average deviation in the streamwise velocity of the (0, 1)-Fourier mode, $\overline{|\Delta\hat{u}^{(0,1)}/\kappa|}/u_\tau$. The contours correspond to 7%, 15%, 25%, 75% and 90% of the maximum value of $\sigma_1\boldsymbol{\psi}_\mathbf{1}$. The lines represent the forced DNS case (blue), and the resolvent response (red).

3.2. Comparison of velocity deviations with linear response

To visualize the alignment of the velocity fluctuation fields with the linear response mode across all wall normal heights, we plot the contours of the streamwise velocity magnitude deviations for the (0, 1)-Fourier mode, along with the contours of the linear resolvent response modes (figure 5). At earlier times ($t < 0.7\delta/u_\tau$), the responses for both the $\gamma = 1\%$ and $\gamma = 10\%$ cases are very similar to the linear mode. The strongly-forced case, however, quickly reverts to the unforced channel flow statistics beyond an eddy turnover time unit. In contrast, the lightly-forced case of $\gamma = 1\%$ exhibits a longer-lasting velocity deviation, especially in the near wall region ($y^+ < 15$).

To more closely investigate how the agreement of the forced simulations and the optimal linear response varies with y^+, we show $\Delta\hat{u}^{(0,1)}$ divided by the scaling constant κ at two wall-normal heights, along with the linear response at those heights (figure 6). Though the initial growth rates are similar to the linear case and we obtain good collapse prior to $t \approx 0.7\delta/u_\tau$, the streak velocities peak earlier and decay more quickly for larger forcing amplitudes. For a given forcing amplitude, we notice that $\Delta\hat{u}^{(0,1)}$ diverges from the optimal linear response around the same time at both wall-normal locations plotted in figure 6. However, as y^+ moves closer to the wall, the growth rate of the linear response increases, and $\Delta\hat{u}^{(0,1)}$ manages to recover more of the linearly-amplified energy before decaying, which leads to a better agreement between the forced simulations and the optimal linear response in the near-wall region.

We quantify the linearity of the turbulent responses by projecting the deviation of the fluctuation field $\hat{\boldsymbol{q}}^{(0,1)}$ onto the response mode $\kappa\sigma_1\boldsymbol{\psi}_\mathbf{1}$. We similarly compute the projection of the deviation in nonlinear fluctuations $\hat{\boldsymbol{f}}^{(0,1)}$ on the principal resolvent forcing mode $\kappa\boldsymbol{\phi}_\mathbf{1}$. For a system governed by the linearized Navier-Stokes equations, we expect the projection of both $\hat{\boldsymbol{q}}^{(0,1)}$ onto $\kappa\sigma_1\boldsymbol{\psi}_\mathbf{1}$ and $\hat{\boldsymbol{f}}^{(0,1)}$ onto $\kappa\boldsymbol{\phi}_\mathbf{1}$ to be equal to 1. The results are shown in figure 7. The magnitude of the projection of $\hat{\boldsymbol{q}}^{(0,1)}$ onto $\gamma\sigma_1\boldsymbol{\psi}_\mathbf{1}$ decreases with the intensity of the forcing (figure 7a). The magnitude trend reflects the results in figure 6, which shows that the linear transient growth behavior lasts for shorter times as the forcing amplitude increases. The angle of the projection also varies widely for different values of γ.

Compared to the velocity fluctuation projection, the magnitude of the projection of the nonlinear term is much lower and relatively constant with γ. This shows that, across all forcing cases, the injected resolvent mode accounts for very little of the Reynolds stresses in the system. Considering this result, it is remarkable that the forcing successfully produces significant energy growth that tracks the optimal linear response to the observed degree.

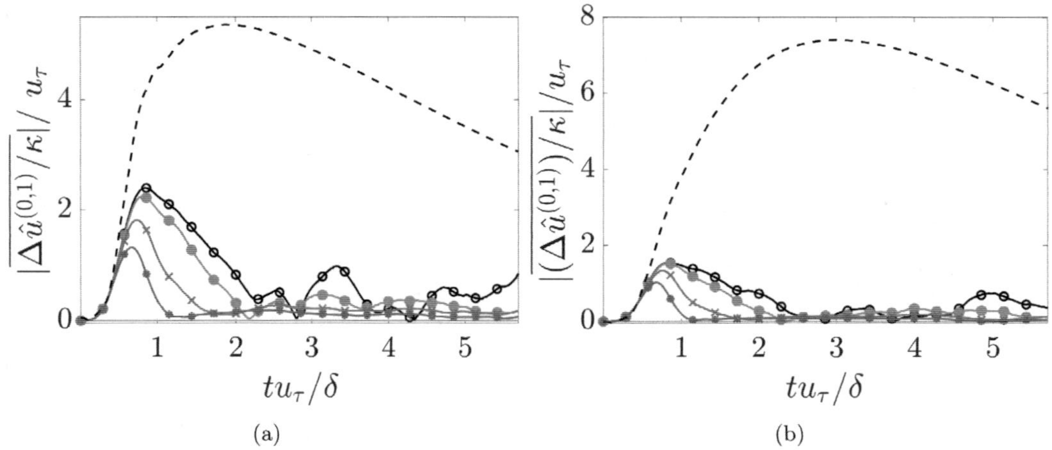

Figure 6: Average deviation in the streamwise velocity of the $(0, 1)$-Fourier mode at $y^+ \approx 16$ (a) and 39 (b). The cases plotted are $\gamma = 1\%$ (black ∘), $\gamma = 2\%$ (cyan •), $\gamma = 5\%$ (purple ×), $\gamma = 10\%$ (red $*$), and the linear response mode $\sigma_1 \boldsymbol{\psi_1}$ (black, dashed).

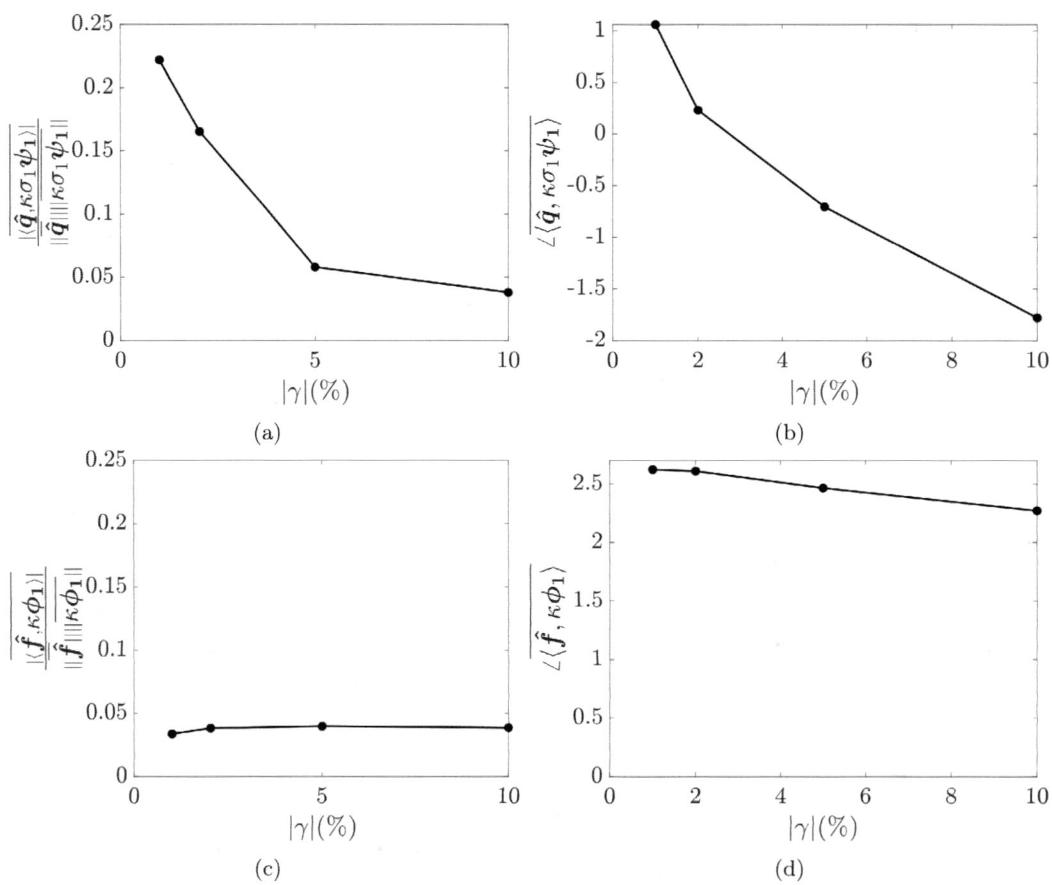

Figure 7: Magnitude (a) and phase (b) of the average projection of the velocity field deviation onto the principle resolvent response mode $\boldsymbol{\psi_1}$. Magnitude (c) and phase (d) of the average projection of the deviation in the nonlinear terms onto the principle resolvent forcing mode $\boldsymbol{\phi_1}$.

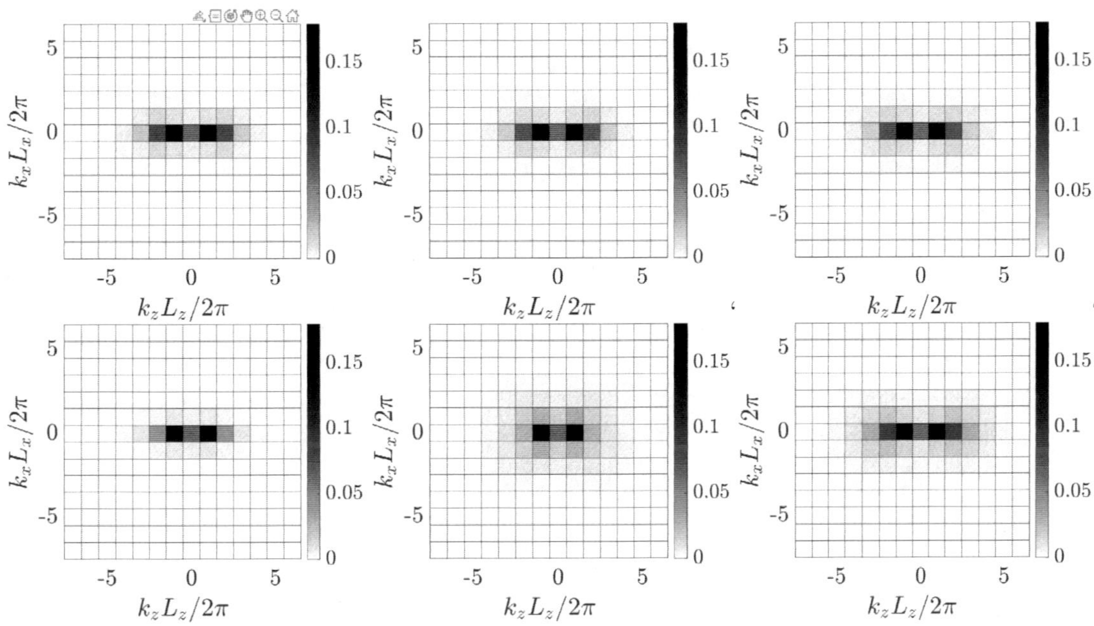

Figure 8: Streamwise spectral energy content $\overline{|\hat{u}|^2}/(2u_\tau^2)$ at $y^+ \approx 16$ for $\gamma = 1\%$ (top) and $\gamma = 10\%$ (bottom). The spectra shown are at times $t = 0.5\delta/u_\tau$ (left), $1\delta/u_\tau$ (middle) and $2\delta/u_\tau$ (right). The energy contained in the $(0,0)$–mode (red) is excluded for clarity.

3.3. Spectra

We compute the streamwise and spanwise spectra at a wall-normal height of $y^+ \approx 16$ for $\gamma = 1\%$ and $\gamma = 10\%$, before the energy peak ($t = 0.5\,\delta/u_\tau$), during energy decay ($t = \delta/u_\tau$), and after decay ($t = 2\,\delta/u_\tau$). The results are shown figure 8. Across all times, the $(0,\pm1)$–mode is the most dominant one, accounting for $30\% - 70\%$ of the total turbulent energy, followed by the $(0,\pm2)$– mode accounting for $6\% - 18\%$. During the energy decay of the $(0,1)$–mode, the strongly-forced case exhibits energy transfer to secondary modes, especially the $(\pm1,\pm1)$–modes. These multiscale effects are not as prominent for the lightly-forced case.

The energy cascade to smaller scales is clearly visible in figures 9a and 9b, which plot the energy content in the $(0,\pm1)$–mode and the smaller scales as function of time for the different forcing amplitude cases at $y^+ \approx 16$ and $y^+ \approx 39$. Across all forcing amplitudes, the rise of the energy content of smaller scales coincides with the start of the energy decay of the $(0,1)$–mode, *i.e.*, the interrupted growth and faster decrease of $\overline{|\hat{u}^{(0,1)}|^2}/(2u_\tau^2)$ compared to the optimal linear response. At both wall normal heights shown in figures 9a and 9b, we observe that a higher amplification triggers an earlier and larger energy leak into the smaller spatial scales. Furthermore, for a given forcing amplitude, the energy content of the small scales rises faster and peaks earlier farther away from the wall.

To better visualize the cross-scale energy transport timescales, we plot the streamwise energy peak times for the $(0,\pm1)$–mode and the smaller scales as a function of y^+ and for different forcing amplitudes (figures 9c). For all wall-normal heights, increasing the forcing amplitude indeed causes the energy of both the $(0,\pm1)$–mode and the smaller scales to peak earlier. For a given forcing amplitude, the peak time for the $(0,\pm1)$–mode is constant with y^+ in the near wall region, increases slightly with y^+ in the buffer layer, and plateaus again in the outer region of the flow. The variation in the energy peak times for the smaller scales is more dramatic: the peak time is also constant in the near-wall region, but decreases significantly with y^+ within the

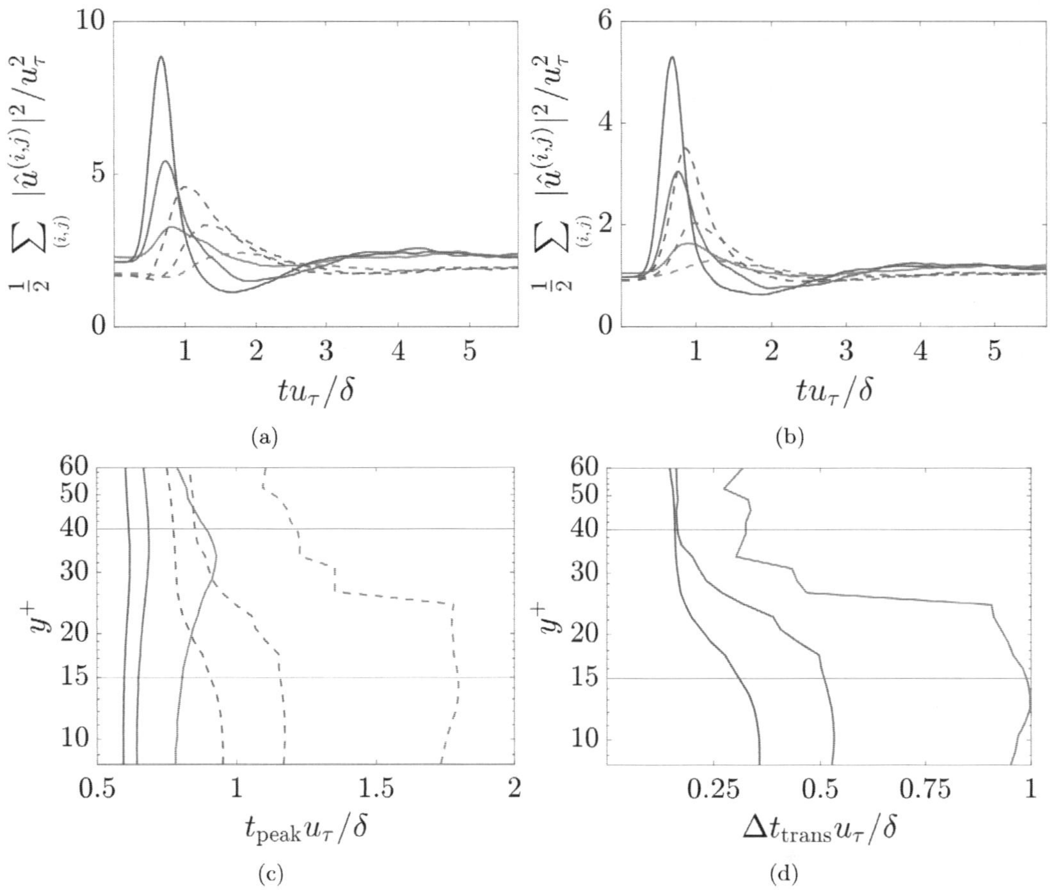

Figure 9: Streamwise spectral energy content at $y^+ \approx 16$ (a) $y^+ \approx 39$ (b) and for the $(0, \pm 1)$–Fourier modes (solid lines) and the other fluctuation wavenumbers (dashed lines). (c) Energy peak times for the $(0, \pm 1)$–mode (solid) and the smaller scales (dashed) cases. (d) Timescale for the cross-scale energy transport. Colors indicate $\gamma = 2\%$ (blue), $\gamma \approx 5\%$ (purple), and $\gamma = 10\%$ (red). The horizontal black lines in (c) and (d) represent $y^+ = 15$ and $y^+ = 40$, which delineate the buffer layer.

buffer layer, and levels off in the outer region.

We define the timescale for cross-scale energy transport, Δt_{trans}, as the time delay between the energy peaks of the $(0, \pm 1)$–modes and the smaller scales; its dependence on y^+ is mostly determined by the energy peak time for the smaller scales (figure 9d). Interestingly, though a larger forcing amplitudes accelerates the energy transport from the $(0, \pm 1)$–mode to smaller scales for all wall normal heights, the sensitivity of Δt_{trans} to forcing amplitude decreases as y^+ moves farther away from the wall. Indeed, in the outer regions of the flow, Δt_{trans} converges to a value of approximately $0.18\,\delta/u_\tau$ for high forcing amplitude.

3.4. Optimality of resolvent forcing

We finally compare the results of forcing using the principal resolvent forcing mode $\boldsymbol{\phi_1}$, the first suboptimal mode $\boldsymbol{\phi_3}$, and the random mode $\boldsymbol{\phi_{\text{rand}}}$. We see that the streak energy grows to higher peak when the minimal flow unit is forced by $\boldsymbol{\phi_1}$ than when forced by $\boldsymbol{\phi_3}$ (figure 10). In both cases, the minimal flow unit is much more responsive compared to the case with random

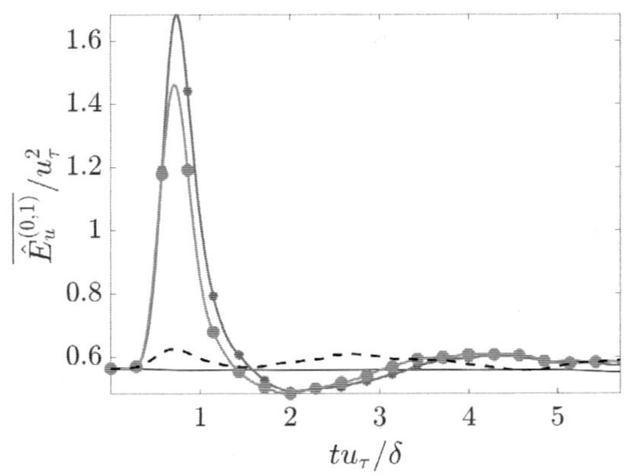

Figure 10: Average streak energy for forcing terms aligned with $\boldsymbol{\phi_1}$ (red $*$), $\boldsymbol{\phi_3}$ (cyan $\bullet$), and $\boldsymbol{\phi_{\text{rand}}}$ (dashed black). The unforced case is shown in black.

forcing. This suggests that resolvent analysis identifies a forcing structure to which the minimal flow unit is indeed sensitive, even when governed by the fully nonlinear Navier-Stokes equations. We note, however, that the advantage the optimal forcing mode is significantly reduced as the effective amplifications, σ_{eff}, in DNS forced by $\boldsymbol{\phi_1}$ and $\boldsymbol{\phi_3}$ differ only by a factor of 1.03, whereas $\sigma_1/\sigma_3 = 2.16$.

4. Conclusions

In this work, we studied the growth of time-localized resolvent modes in the minimal flow unit at $Re_\tau \approx 186$. We formulated resolvent analysis in a wavelet-basis in time that endows the resolvent modes with transient information, and obtained a linearly optimal time-localized forcing mode and its corresponding transient response mode.

Resolvent analysis ignores the feedback between the velocity fluctuations and the nonlinear terms; we thus tested the optimality of the resolvent forcing within a DNS of a minimal flow unit by numerically injecting the principal resolvent forcing mode into the flow at varying amplitudes. This allowed us to investigate the interactions between the transiently growing linear response mode and the nonlinear effects of the turbulent flow. We compared the resulting flow to one forced by the first suboptimal forcing mode and another forced by a spatially-random forcing. The principal resolvent forcing produces a larger transient energy growth than the suboptimal mode, but the energy amplification was notably lower in both cases compared to the linearized case. Both systems were significantly more amplified than the random forcing case. In all cases, the injected forcing term is negligible compared to the newly-induced Reynolds stresses, and it is remarkable that it still manages to produce a significant energy amplification and a perturbation to the velocity field that tracks the optimal linear response for a short time.

The nonlinearities of turbulence interrupt the initial algebraic energy growth driven by the linear dynamics of the flow. This is seen in all cases forced by the principal resolvent mode, though the amplitude of the resolvent forcing affects how closely the turbulent trajectory behaves like the optimal resolvent response mode. Across all forcing amplitudes, the initial growth phase is similar, and the systems peak at roughly the same time. However, the higher the forcing amplitude, the faster the decay of the system back to the unforced turbulent system. The more intense Reynolds stresses in the high-amplitude-forcing cases are more effective at damping the effects of the initial forcing. Additionally, the forced DNS flow fields are closer to the resolvent response mode in the near-wall region. The spectral energy content becomes increasingly nonlinear and multi-scale during the decay phase due to a transfer of energy to the non-forced spatial scales. This cross-scale energy transfer is more prominent and occurs faster for the high-amplitude-forcing cases. The nonlinear effects interrupt the initial linear energy growth mechanism and lead to streak breakdown.

5. Acknowledgments

This work was supported in part by the European Research Council under the Caust grant ERC-AdG-101018287 and AFOSR grant FA9550-22-1-0109.

References

[1] Klebanoff P S, Tidstrom K D and Sargent L M 1962 *J. Fluid Mech.* **12** 1–34
[2] Kline S J, Reynolds W C, Schraub F A and Runstadler P W 1967 *J. Fluid Mech.* **30** 741–773
[3] Smith C and Metzler S 1983 *J. Fluid Mech.* **129** 27–54
[4] Blackwelder R F and Eckelmann H 1979 *J. Fluid Mech.* **94** 577–594
[5] Johansson A V, Her J Y and Haritonidis J H 1987 *J. Fluid Mech.* **175** 119–142
[6] Bakewell Jr H P and Lumley J L 1967 *Phys. Fluids* **10** 1880–1889
[7] Landahl M 1980 *J. Fluid Mech.* **98** 243–251
[8] Butler K M and Farrell B F 1993 *Phys. Fluids A* **5** 774–777
[9] Chernyshenko S and Baig M 2005 *J. Fluid Mech.* **544** 99–131
[10] Del Alamo J C and Jimenez J 2006 *J. Fluid Mech.* **559** 205–213
[11] Kim H T, Kline S and Reynolds W C 1971 *J. Fluid Mech.* **50** 133–160
[12] Hamilton J M, Kim J and Waleffe F 1995 *J. Fluid Mech.* **287** 317–348
[13] Panton R L 2001 *Prog. Aerosp. Sci.* **37** 341–383
[14] Jiménez J 2018 *J. Fluid Mech.* **842** P1
[15] Smits A J, McKeon B J and Marusic I 2011 *Annu. Rev. Fluid Mech.* **43** 353–375
[16] Adrian R J 2007 *Phys. Fluids* **19**
[17] Robinson S K 1991 *Annu. Rev. Fluid Mech.* **23** 601–639
[18] Jiménez J and Moin P 1991 *J. Fluid Mech.* **225** 213–240
[19] Jiménez J and Pinelli A 1999 *J. Fluid Mech.* **389** 335–359
[20] Lozano-Durán A, Constantinou N C, Nikolaidis M A and Karp M 2021 *J. Fluid Mech.* **914** A8
[21] Jiménez J 2013 *Phys. Fluids* **25**
[22] Orr W M F 1907 *Proc. R. Ir. Acad.* vol 27 (JSTOR) pp 69–138
[23] Hwang Y and Cossu C 2010 *Phys. Rev. Lett.* **105** 044505
[24] Pujals G, García-Villalba M, Cossu C and Depardon S 2009 *Phys. Fluids* **21** 015109
[25] McKeon B J and Sharma A S 2010 *J. Fluid Mech.* **658** 336–382
[26] Moarref R, Sharma A S, Tropp J A and McKeon B J 2013 *J. Fluid Mech.* **734** 275–316
[27] McKeon B 2017 *J. Fluid Mech.* **817** P1
[28] Bae H J, Lozano-Duran A and McKeon B J 2021 *J. Fluid Mech.* **914** A3
[29] Bae H J, Lozano-Durán A, Bose S T and Moin P 2018 *Phys. Rev. Fluids* **3**(1) 014610
[30] Ballouz E, Lopez-Doriga B, Dawson S T and Bae H J 2023 *AIAA SCITECH 2023 Forum* p 0676
[31] Orlandi P 2000 *Fluid flow phenomena: a numerical toolkit* vol 55 (Springer Science & Business Media)
[32] Kim J and Moin P 1985 *J. Comput. Phys.* **59** 308–323
[33] Wray A A 1990 *NASA Ames Research Center, California, Report No. MS* **202**
[34] Lozano-Durán A and Bae H 2016 *Annual Research Briefs. Center for Turbulence Research* **2016** 97–103
[35] Bae H J, Lozano-Duran A, Bose S and Moin P 2018 *Phys. Rev. Fluids* **3** 014610
[36] Bae H J, Lozano-Durán A, Bose S T and Moin P 2019 *J. Fluid Mech.* **859** 400–432
[37] Mallat S 1999 *A wavelet tour of signal processing* (Elsevier)
[38] Najmi A H 2012 *Wavelets: A concise guide* (JHU Press)
[39] Jeun J, Nichols J W and Jovanović M R 2016 *Phys. Fluids* **28** 047101
[40] Kojima Y, Yeh C, Taira K and Kameda M 2020 *J. Fluid Mech.* **885** R1
[41] Daubechies I 1992 *Ten lectures on wavelets* (SIAM)

Numerical experiments on scalar transport and mixing in turbulent boundary layers

Gabriele Boga, Alfonso Giancola and Andrea Cimarelli

Dipartimento di Ingegneria Enzo Ferrari, Università di Modena e Reggio Emilia, 41125 Modena, Italy

E-mail: gabriele.boga@unimore.it

Abstract. In this work, we present numerical experiments aimed at dynamically establishing the separate role of the inner and outer cycles on the scalar transport in the configuration of a temporally evolving boundary layer. The experiments are based on the study of the evolution of passive scalars driven by velocity fields where inner and outer cycles are alternately suppressed. Two different approaches are implemented. In the first, the discrimination between inner and outer cycle activities is based on the scale dimension of the involved motions. The second instead, discriminates on the basis of the distance from the wall of the turbulent motions. The two approaches depict the same scenario. Both the inner and outer cycles appear to be autonomous and, in a sense, independent, since their dynamics remain qualitatively unaltered despite facing two different conditions. The outer cycle faces a free boundary at the top and simply rescales according to what is supplied by the inner cycle. The inner cycle, on the other hand, resides between the wall and the outer region. As a result, the reduction of the scalar fluxes in the outer region due to the suppression of the outer cycle causes a damping in the near-wall region activities.

1. Introduction

Wall flows are among the most technologically relevant flows and despite numerous studies (see [1] for a partial review) their dynamic is not yet fully understood. In this family of flows, it is common to identify different regions based on their features. One of the categorisations distinguishes a near-wall region, a bulk region and an overlap region in between. In the near-wall region, the production of turbulent fluctuations exceeds their dissipation rate. The excess is transported towards the wall and towards the bulk region, where is finally dissipated. The overlap region lying between these two layers constitutes a region of local equilibrium between production and dissipation. The near-wall layer is certainly crucial to the dynamics of attached shear flows: it is dominated by intense interacting structures ([2] and [3]) and it is the seat of the highest rate of turbulent energy production and of the maximum turbulent intensities. In this region, turbulence is sustained by a self-sustaining and autonomous cycle [4] which we will refer to as the near-wall cycle. However, for high Reynolds numbers, also the outer overlap layer is thought to become relevant being the site of self-sustaining mechanisms of large flow structures following mixed inner/outer scaling [5]. It is then of overwhelming importance to assess the combined role played by inner and outer turbulence self-sustaining mechanisms in determining the overall momentum and heat transfer in wall turbulence at relatively high Reynolds numbers. In order to identify the separate role of these two layers on mixing and transport that is otherwise concealed by the non-linearity of the flow evolution, we perform numerical experiments where scalar fields dynamically evolve under the separate action of the sole near-wall cycle and of the sole outer cycle. The choice of addressing the scalar field evolution is crucial to decouple the effect of the two layers. More specifically, we analyse the evolution of passive scalars in the configuration of a temporally evolving boundary layer ([6], [7] and [8]). The flow configuration of the boundary layer is chosen as it represents a canonical flow for the study of wall turbulence and additionally exhibits the presence of a Turbulent/Non-Turbulent Interface (TNTI). For this reason, the boundary layer is also a case study for the turbulent entrainment process ([9] and [10]) that has already been studied with a similar approach in the flow configuration of a temporally evolving planar jet in [11]. The Schmidt number for the scalars is set to $Sc = 1$ in order to rely on the Reynolds analogy linking the scalar and momentum transports. Hence, these experiments aim at dynamically establishing the separate role played by the inner and outer layers in determining the overall transport mechanisms in wall turbulence. The paper is organized as follows. The methodology is covered in section 2, the results are presented and discussed in section 3 and the paper concludes with final remarks in section 4.

2. Simulations and experimental set-up

In the following section we present the methodology used to carry out the experiments, starting with the presentation of the chosen flow configuration, the description of the experimental set-up and a final brief section regarding the filtering operation used in the experiments.

2.1. Temporal boundary layer

The flow selected for the numerical experiments is a temporal boundary layer. This flow has already been investigated through direct numerical simulations and presents all the main phenomena observed in the classical turbulent boundary layer [7]. This type of flow consists in a moving wall at constant speed U_w and scalar concentration Θ_w at the bottom of an initially quiescent fluid. The initial velocity and scalar profiles are designed to mimic the wake of a wall-mounted trip wire with diameter D,

$$u^*(x^*, y^*, z^*, t^* = 0) = (1 + c')\,\frac{U_w}{2}\left[1 + \tanh\left(\frac{D - z^*}{2h}\right)\right] \tag{1}$$

$$\theta^*(x^*, y^*, z^*, t^* = 0) = \frac{\Theta_w}{2}\left[1 + \tanh\left(\frac{D - z^*}{2h_\theta}\right)\right] \tag{2}$$

where $h = 54\nu/U_w$ and $h_\theta = 54\alpha/\Theta_w$ can be understood as the shear-layer thickness of the wall-mounted trip wire and $c'(x^*, y^*, z^*) \in [-0.05, 0.05]$ is a superimposed white noise used to speed up transition. Here, ν is the kinematic viscosity and α is the scalar diffusivity. The Reynolds number based on the tripping thickness is set to $Re_D = U_w D/\nu = 500$ and the Schmidt number is set to $Sc = \nu/\alpha = 1$. In the above initial conditions, the asterisk is used to denote dimensional quantities. In the rest of the paper, equations and quantities, unless otherwise specified, are expressed in dimensionless form by using D, U_w and Θ_w.

The boundary layer develops in time rather than in space (figure 1), preserving statistical homogeneity in the streamwise (x) as well as in the spanwise (y) direction. The advantages of this type of flow over a spatially developing boundary layer are the additional direction of statistical homogeneity and the possibility of using domains of reduced streamwise extension.

Periodic boundary conditions are applied in the streamwise and spanwise directions, while a Dirichlet no-slip boundary condition and a free-slip impermeable boundary condition are applied on the bottom and the top of the domain respectively. The domain extension in the streamwise, spanwise and wall-normal directions is respectively $(Lx, Ly, Lz) = (600, 300, 150)$ discretised in $(Nx, Ny, Nz) = (512, 512, 384)$ points. It is important to note that the domain dimensions are

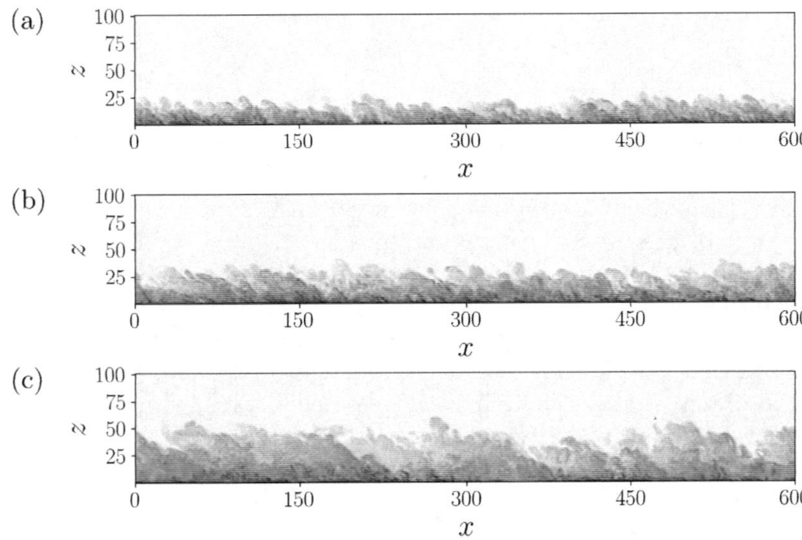

Figure 1. Sections of the scalar field at different time instants corresponding respectively to (a) $Re_\tau = 500$, (b) $Re_\tau = 750$ and (c) $Re_\tau = 1000$. The scalar concentration increases from light to dark colors.

chosen to guarantee that at the final friction Reynolds number the boundary layer thickness is $\delta < L_z/3$ in order to have reasonably small confinement effects from the top boundary. Here, the boundary layer thickness δ is defined such that $\langle u \rangle (\delta) = 0.01U_w$. The domain dimensions in terms of boundary layer thickness and the grid resolution in terms of friction units are reported in table 1. The final friction Reynolds number reached by the simulations is $Re_\tau = u_\tau \delta/\nu = 1000$, where $u_\tau = \sqrt{\tau_w/\rho}$ is the friction velocity and τ_w and ρ are respectively the wall shear stress and the density.

The set of equations is solved using the massively-parallel open-source code CaNS [12], which uses a pressure-projection method with a second-order finite-difference scheme for the spatial discretization on a staggered grid. A mixed time integration is used to avoid the very small time-step required by the highly stretched grid. For the diffusion terms in the vertical direction z an implicit time integration is used, employing the Crank-Nicholson scheme. Conversely, all the other terms are integrated explicitly in time with a three-step Runge–Kutta method with a $CFL = 0.7$. In this preliminary work, we opted for low resolution simulations to test the

Table 1. Domain extension in terms of boundary layer thickness and grid resolution in friction units for three different time instants corresponding to three different friction Reynolds numbers. The subscript $\cdot_{|w}$ and $\cdot_{|\delta}$ indicates a quantity measured respectively at the wall ($z = 0$) and at $z = \delta$.

Re_τ	$(Lx, Ly, Lz)/\delta$	Δx^+	Δy^+	$\Delta z^+_{\|w}$	$\Delta z^+_{\|\delta}$
500	(26.73, 13.36, 6.68)	26.06	13.03	0.14	7.86
750	(16.95, 8.47, 4.24)	24.78	12.39	0.13	10.46
1000	(12.37, 6.18, 3.09)	23.83	11.91	0.13	12.47

Journal of Physics: Conference Series **2753** (2024) 012003 doi:10.1088/1742-6596/2753/1/012003

potential of the method used. We verified that the classical statistics obtained with these settings are reliable when compared to those obtained from higher resolution simulations presented in [8].

2.2. Numerical experiments

As mentioned above, the inner and outer cycles interact with each other and their combined role governs the exchange of momentum and scalar concentration. Understanding the role played by these two cycles is therefore of great interest, both from a flow control and modelling perspective. In this regard, the present work aims to dynamically establish the role played by these two mechanisms by analysing the evolution of two additional passive scalars transported by two modified velocity fields where the near-wall and outer cycles are alternatively suppressed. The two modified velocity fields are obtained through a filtering operation performed at runtime. It is important to note that the filtering operation is not directly applied to the scalar fields, but it is only used to obtain the two modified velocity fields transporting the two alternative scalars. These velocity fields are not integrated over time but are obtained at each time-step from the total velocity. The resulting set of equations is:

$$
\begin{aligned}
&\frac{\partial u_i}{\partial x_i} = 0 \\
&\frac{\partial u_i}{\partial t} + \frac{\partial u_i u_j}{\partial x_j} = -\frac{\partial p}{\partial x_i} + \frac{1}{Re}\frac{\partial^2 u_i}{\partial x_j \partial x_j} \\
&\frac{\partial \theta}{\partial t} + \frac{\partial \theta u_j}{\partial x_j} = \frac{1}{ReSc}\frac{\partial^2 \theta}{\partial x_j \partial x_j} \\
&\frac{\partial \theta}{\partial t} + \frac{\partial \theta u_j}{\partial x_j} = \frac{1}{ReSc}\frac{\partial^2 \theta}{\partial x_j \partial x_j} \\
&\frac{\partial \theta_{out}}{\partial t} + \frac{\partial \theta_{out} u_j^{out}}{\partial x_j} = \frac{1}{ReSc}\frac{\partial^2 \theta_{out}}{\partial x_j \partial x_j} \\
&\frac{\partial \theta_{inn}}{\partial t} + \frac{\partial \theta_{inn} u_j^{inn}}{\partial x_j} = \frac{1}{ReSc}\frac{\partial^2 \theta_{inn}}{\partial x_j \partial x_j}
\end{aligned}
\qquad (3)
$$

where θ is the passive scalar transported by the complete velocity field u_i, θ_{out} is the scalar transported by velocity field retaining the sole outer cycle u_i^{out} and θ_{inn} is the scalar field transported by the velocity field retaining the sole inner cycle u_i^{inn}. The initial condition for the two scalar fields θ_{out} and θ_{inn} is taken from the reference case θ at a time instant corresponding to $Re_\tau = 500$.

2.3. Filtering

Two different filtering approaches have been tested. For the sake of clarity, in the present section we will describe only the first approach, the results of which are presented in section 3, while we leave the description and the results of the second approach in Appendix A. In both approaches, the filter used is a 2D Gaussian filter applied in the homogeneous directions x and y,

$$\overline{u}_i\left(x,y,z,t\right) = \int_{-\infty}^{+\infty}\int_{-\infty}^{+\infty} u_i\left(x+r_x, y+r_y, z, t\right) G\left(r_x, r_y\right) dr_x dr_y, \qquad (4)$$

$$G\left(r_x, r_y\right) = \frac{1}{2\pi\Delta_x\Delta_y} e^{-\frac{1}{2}\left(\frac{r_x^2}{\Delta_x^2}+\frac{r_y^2}{\Delta_y^2}\right)} \qquad (5)$$

where $\overline{u}_i$ is the resulting filtered velocity and Δ_x and Δ_y are the characteristic lengths of the filter in the streamwise and spanwise directions. In the first approach, the above filtering operation is applied to the whole domain. The velocity field retaining the sole outer cycle is defined as $u_i^{out} = \overline{u}_i$ and, hence, it contains only the large-scale motions, which are those mainly contributing to the outer cycle dynamics. On the other hand, the velocity field retaining the sole inner cycle u_i^{inn} is defined as $u_i^{inn} = u_i - \overline{u}_i + \langle u_i \rangle$, and hence, it contains only the small-scale motions that are dominant in the near-wall cycle, being large-scale motions merely an imprinting from the outer flow. The action of the average velocity $\langle u_i \rangle(z,t)$ (being $\langle \cdot \rangle$ the average operator in the homogeneous directions x and y) is added explicitly, since $\langle u_i - \overline{u}_i \rangle = 0$. The filter lengths are chosen on the basis of the premultiplied spectra shown in figure 2 and are respectively set to $\Delta_x^+ = 600$ and $\Delta_y^+ = 100$. The portions of the spectra filtered out in the spanwise direction to obtain u_i^{inn} and u_i^{out} are shown in shaded grey in figure 2(a) and (b), respectively measured in the buffer layer and in the outer region. Note that the shaded grey areas shown in figure 2 are merely indicative since the Gaussian filter does not prescribe a sharp cut-off scale in spectral space.

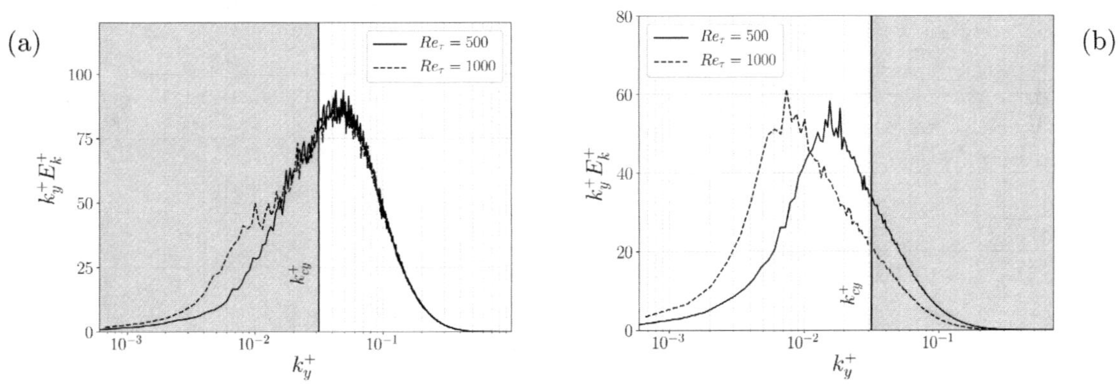

Figure 2. Premultiplied spectra of the turbulent kinetic energy E_k in the spanwise wavenumber k_y respectively (a) in the buffer layer at $z^+ = 20$ and (b) in the outer region at $z = 0.3\delta$. The shaded areas are indicative of the filtered regions of the spectra (a) in u_{inn} and (b) in u_{out}, being $k_{cy}^+ = 2\pi/\Delta_y^+$ the wavenumber corresponding to the spanwise filter length in inner units.

3. Results

Let us start analysing the scalar field topology at the final friction Reynolds number $Re_\tau = 1000$ reached during the experiments. In figure 3 the instantaneous scalar contours are shown on a wall-normal (x, z) section. The trace of the large engulfing scales is clearly visible in θ_{out} and, at first glance, the thickness of the boundary layer seems comparable to that of the reference case θ. Concurrently, the effect of the lack of the small scales can be seen in the presence of large unmixed regions. On the contrary, the field θ_{inn} appears to be more mixed and exhibits a slower growth rate, which can be attributed to the lack of large-scale engulfment events.

We now proceed to analyse the data from a statistical point of view, starting with the temporal evolution of two integral quantities reported in figure 4, the Stanton number

$$St = -\frac{1}{ReSc} \left. \frac{\partial \langle \theta \rangle}{\partial z} \right|_w \tag{6}$$

and the boundary layer thickness based on the scalar concentration δ_θ, defined such that $\langle \theta \rangle(\delta_\theta) = 0.01\Theta_w$. In order to better understand the reason behind the observations, we report

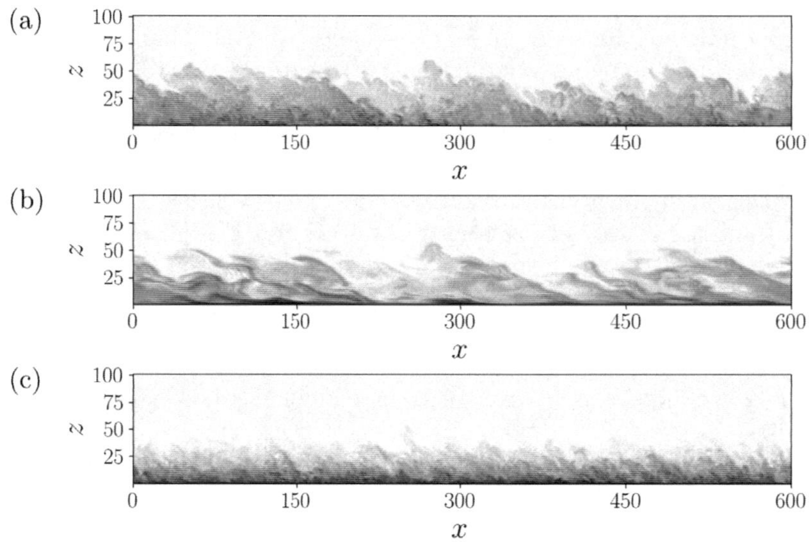

Figure 3. Sections of the scalar fields of the three experiments at the same time instant, corresponding to $Re_\tau = 1000$. Respectively (a) θ, (b) θ_{out} and (c) θ_{inn}. The reference case (a) is duplicated here in order to simplify comparisons.

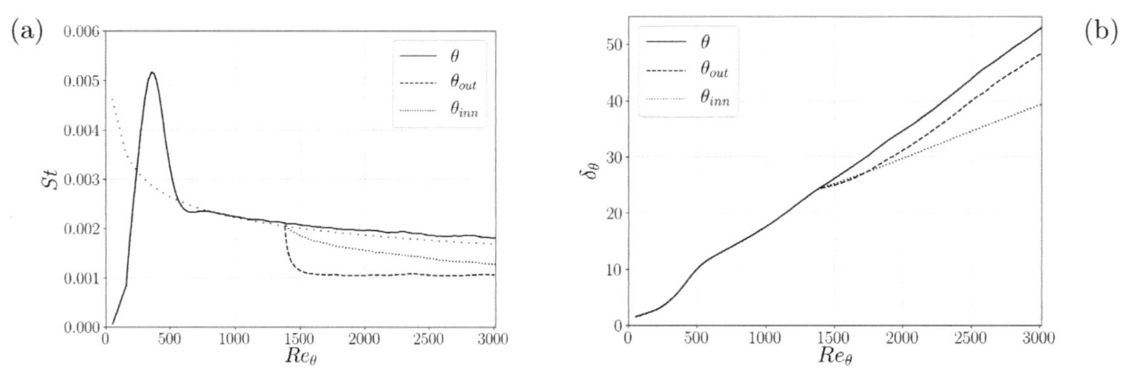

Figure 4. (a) Temporal evolution of the Stanton number and its theoretical law $St = 0.0125 Re_\theta^{-1/4}$ in loosely dotted line $(\cdots\cdots)$ and (b) temporal evolution of the boundary layer thickness based on the scalar concentration δ_θ.

the probability density function of the scalar fields at some locations of interest for the observed quantities (i.e. in the viscous sub-layer for St and at the mean scalar interface for δ_θ) in figure 5. A reduction in St can be noticed in both the experiments. The reduction observed in θ_{out} (dashed line in figure 4(a)) can be interpreted as the effect of the direct suppression of the main turbulent motions involved in the near-wall cycle, leading to an almost immediate decrease. This hypothesis is supported by the lower anisotropy observed in the near-wall region fluctuations that can be visualized as the limited extension of the left tail of the θ_{out} pdf reported in figure 5(a). On the other hand, the field θ_{inn} (dotted line) preserves the scales directly contributing to the wall scalar transfer. In this case, the reduction in St can be attributed to the increase in the mean scalar concentration resulting from the suppression of the outer cycle, thus justifying the more gradual decrease with respect to θ_{out}. Support for this can be found in the fact that θ_{inn}

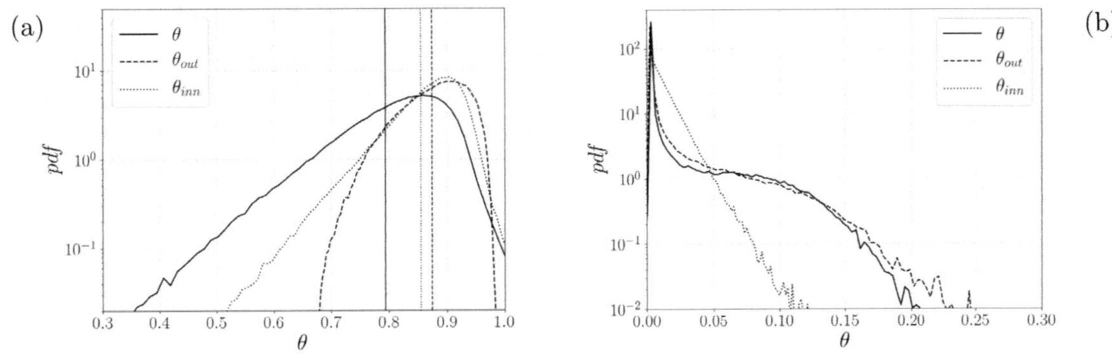

Figure 5. Probability density function of θ (a) at $z^+ = 5$ and (b) at $z = \delta_\theta$ at the final $Re_\tau = 1000$. The vertical lines in (a) indicate the respective mean values of the scalars.

shows a similar pdf to that of θ, the difference being its shift towards higher values (figure 5(a)). We now consider the temporal evolution of the boundary layer thickness δ_θ reported in figure 4(b). After an initial readjustment, the trend of $\delta_{\theta_{out}}$ resumes a growth rate similar to that of the reference δ_θ. This almost unchanged growth rate can be explained through the following two considerations. The first is that large-scale anisotropic motions that mainly contribute to the entrainment process ([11]) are preserved in u_i^{out}, whose trace can be seen in the right tail of the θ_{out} pdf reported in figure 5(b). The second consideration regards the measured observable δ_θ. In fact, the position of the mean interface is not sensitive to the threshold used. For this reason, the reduction in the scalar concentration flux supplied by the inner region in θ_{out} has little influence on the evolution of this observable. On the contrary, the suppression of the large anisotropic motions in u_i^{inn} directly acts on the entrainment process at the interface of the boundary layer, resulting in the slower growth rate of $\delta_{\theta_{inn}}$.

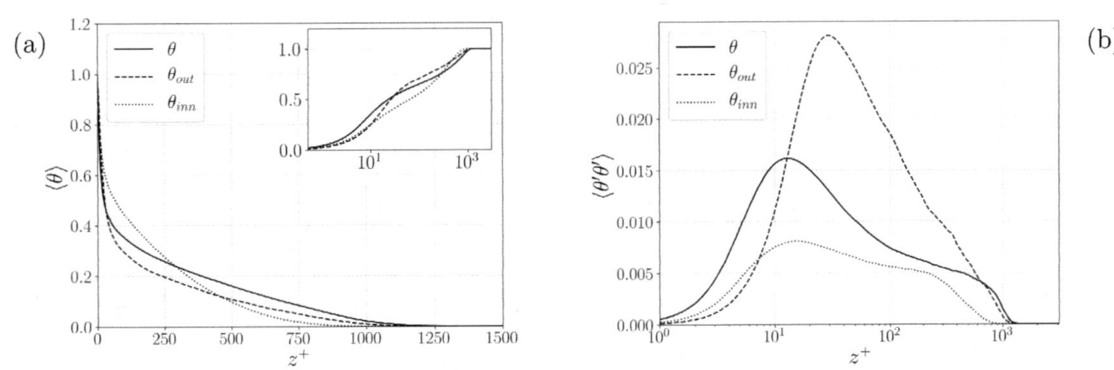

Figure 6. (a) Mean scalar profiles, respectively $\langle\theta\rangle$ in the main panel and $1 - \langle\theta\rangle$ in semi-logarithmic axes in the inset panel and (b) variances at $Re_\tau = 1000$.

For a deeper understanding of the dynamics observed so far, we now move to examine the mean profiles and the variances of the scalar fields reported in figure 6. Both profiles $\langle\theta_{out}\rangle$ and $\langle\theta_{inn}\rangle$ show a higher scalar concentration in the near-wall region, for $z^+ < 20$. This is in accordance with the reduction observed in the St for the two reasons explained above, namely: the reduced scalar flux towards the outer region caused by the suppression of the inner cycle in θ_{out} and the blockage effect caused by the suppression of the outer cycle in θ_{inn} (that will be shown in figure 7). Despite a different distribution of the scalar concentration $\langle\theta_{out}\rangle$, a region resembling a logarithmic behaviour can still be observed as shown in the inset of figure 6(a). The lower scalar flux supplied by the near-wall region does not seem to influence the outer dynamics of θ_{out}. On the other hand, θ_{inn} presents a qualitatively different profile. The suppression of the outer cycle, with the associated reduction of the mean scalar flux, results in a high scalar concentration region extending from the near-wall up to a good portion of the outer region.

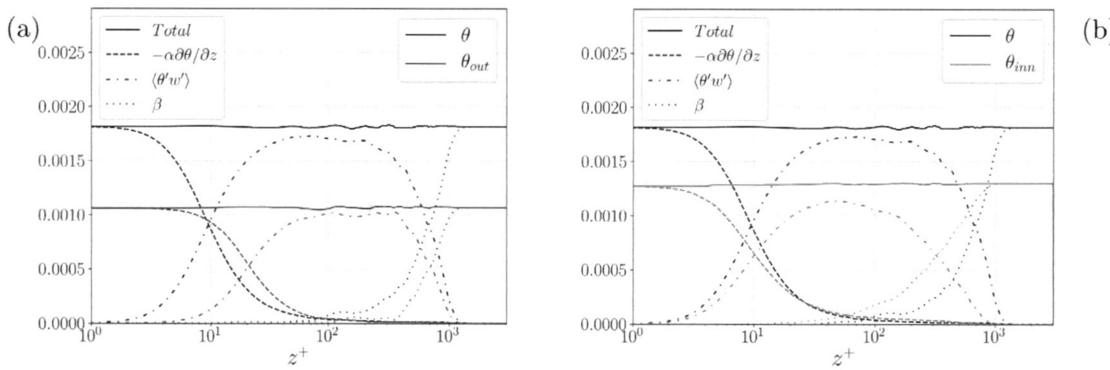

Figure 7. Mean fluxes of (a) θ_{out} and (b) θ_{inn} compared with θ at $Re_\tau = 1000$.

In order to quantify the reduction of the fluxes, we introduce the following equation:

$$St = \langle\theta' w'\rangle - \frac{1}{ReSc}\frac{\partial\langle\theta\rangle}{\partial z} + \beta \quad \text{with} \quad \beta = \frac{d}{dt}\int_0^z \langle\theta\rangle dz \tag{7}$$

obtained by integrating in the wall-normal direction the equation for the mean scalar concentration. To note that, for $z \gg \delta_\theta$ the above equation simply reads $St = \beta\,(z \to \infty)$, where $\beta\,(z \to \infty)$ is the rate of entrainment of scalar concentration. The terms of equation (7) of the two experiments are shown and compared with the reference case in figure 7. At first glance, one can see a macroscopic similarity in the shape of the fluxes with a global reduction in accordance with the reduction observed in the St. Taking a closer look to the turbulent flux, it can be seen that $\langle\theta'_{out}w'_{out}\rangle$ appears to be more active at a higher distance from the wall, while $\langle\theta'_{inn}w'_{inn}\rangle$ appears to be more active closer to the wall, as expected. Since $\langle\theta' w'\rangle$ and β must sum up to St in the outer region (where $-k\partial\theta/\partial z$ is negligible), these different profiles of $\langle\theta' w'\rangle$ determine also the shape of β. The high activity of $\langle\theta'_{out}w'_{out}\rangle$ in the outer region results in a steep profile of β_{out}, meaning that the scalar flux is mainly sustained by the entrainment process occurring away from the wall and near the interface. On the contrary, the lower activity of $\langle\theta'_{inn}w'_{inn}\rangle$ in the outer region results in a less steep profile of β_{inn}, suggesting that the main scalar flux is sustained by the homogenization in the core of the boundary layer, thus by the mixing effect of the small quasi-isotropic scales. Despite the similarity between the mean profiles of θ_{out} and θ, their variances, reported in figure 6(b), present different characteristics. In particular, $\langle\theta'_{out}\theta'_{out}\rangle$ exhibits a more pronounced peak located at a greater distance from the wall with respect to $\langle\theta'\theta'\rangle$. This difference can be attributed to the presence of large unmixed regions initiated by the large-scale anisotropic motions and left unmixed by the absence of the small mixing scales. Conversely, the profile of $\langle\theta'_{inn}\theta'_{inn}\rangle$ resembles the reference profile, although it appears to be damped, indicating a more uniform field. Both $\langle\theta'_{out}\theta'_{out}\rangle$ and $\langle\theta'_{inn}\theta'_{inn}\rangle$ present lower values in the near-wall region with respect to the reference case.

In order to take a closer look into the dynamics of the scalar fluctuations we introduce the scalar variance budget equation:

$$\frac{\partial\langle\theta'\theta'\rangle}{\partial t} + \frac{\partial}{\partial z}(\Phi_u + \Phi_\alpha) = \Pi - \epsilon_\theta \tag{8}$$

where the turbulent production Π and dissipation ϵ_θ terms are defined as:

$$\Pi = -2\langle\theta' w'\rangle\frac{\partial\langle\theta\rangle}{\partial z} \quad \text{and} \quad \epsilon_\theta = \frac{2}{ReSc}\langle\frac{\partial\theta'}{\partial x_j}\frac{\partial\theta'}{\partial x_j}\rangle \tag{9}$$

and the turbulent Φ_u and diffusive Φ_α fluxes as:

$$\Phi_u = \langle\theta'\theta' w'\rangle \quad \text{and} \quad \Phi_\alpha = \frac{-1}{ReSc}\frac{\partial\langle\theta'\theta'\rangle}{\partial z} \tag{10}$$

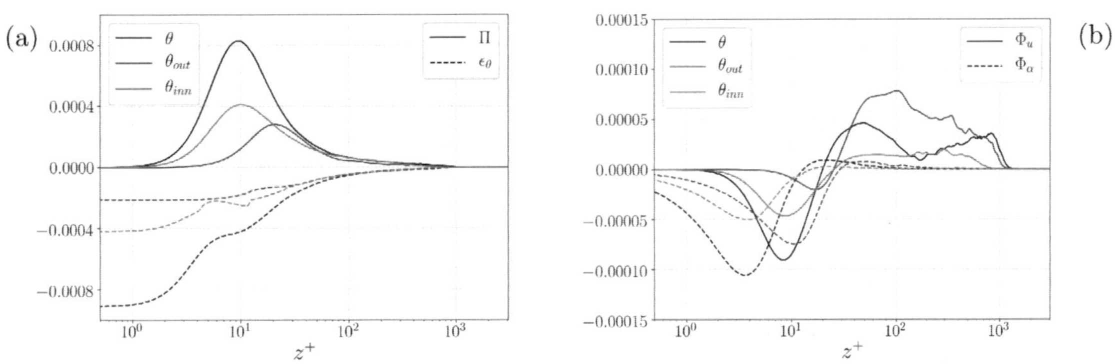

Figure 8. Scalar variance (a) production Π, dissipation ϵ_θ and (b) turbulent and diffusive fluxes Φ_u and Φ_α at $Re_\tau = 1000$.

The picture shown by the scalar variance budget in figure 8 can be interpreted as follows. Both the scalar variance production and the turbulent flux towards the near-wall region are strongly reduced in θ_{out}, thus justifying the lower scalar variance observed in the near-wall region. In addition to this reduction, a shift to higher wall distances can also be observed, reinforcing the idea that large-scale motions are not directly active in the near-wall region, beside the imprinting they give on the flow and the modulation with which they act on the small scales. The diffusive flux and the dissipation adapt themselves to this scenario. On the other hand, the inner dynamics of θ_{inn} appear qualitatively similar to the reference case, even if lower in intensity. This is in accordance with the idea that the inner cycle finds itself in a condition bounded by the outer dynamics but remains qualitatively unaltered. In particular, the suppression of the outer cycle limits the turbulent flux of fluctuations directed toward the wall, and the increase of the mean scalar value limits the possibility of a large variance because of the intrinsic limitation $\theta \leq 1$.

4. Conclusions

An attempt to decouple the dynamics of the inner and outer cycles in the flow configuration of a temporally evolving boundary layer is carried out. The attempt is based on studying the evolution of two additional passive scalars driven by two modified velocity fields. The use of the two scalar fields is the key point that allows us to separate the two effects, as it permits us to study a field on which no direct manipulation has been applied. The velocity fields used to transport these two additional scalars are obtained at each timestep by manipulating the total velocity u_i in order to retain only features of the inner (u_i^{inn}) and outer (u_i^{out}) cycles respectively. More specifically, the two velocity fields are obtained by suppressing scales that are mostly contributing to the inner or outer cycle activities based on their dimension.

The outer cycle dynamics observed through θ_{out} show many similarities with the reference case θ. In both the approaches tested, it seems possible to identify a logarithmic-like region in the mean profile $\langle\theta_{out}\rangle$. Furthermore, the growth rate of the scalar boundary layer thickness $\delta_{\theta_{out}}$ is unaltered, indicating that the entrainment process at the boundary layer interface is not very sensitive to the near-wall region dynamics. At the same time, in the region directly affected in this experiment (i.e. the inner region), the effect of inner cycle suppression can be observed. In this region, both a reduction in the scalar transfer at the wall (caused by the damping of the scalar fluctuations) and a reduction of turbulent production and transport mechanisms occurs. In particular, the reduction of the mean scalar flux in the near-wall region also implies a reduction in the whole domain due to the conservation law. Summarising, by suppressing the small scales, the outer cycle appears to maintain its dynamics unaltered and simply rescales the scalar flux according to what is supplied by the flux coming from the inner cycle. From a modelling perspective, this conclusion may be relevant for reduced-order approaches such as wall-modelled LES and RANS.

The inner cycle, on the other hand, is confined on both sides, facing the wall at its lower boundary and the outer cycle on the other side. The suppression of the outer cycle appears to have the effect of a bottleneck for the inner cycle. This interpretation is supported by the fact that the pdf of θ_{inn} in the near-wall region shows a similar distribution but with a shift towards higher values (reason for the gradual reduction of the Stanton number). Other symptoms of an almost unchanged dynamic can be found in the production and turbulent transport mechanisms in the near-wall region. Despite appearing rescaled, these profiles show a very similar shape to the reference case θ. In the outer region of θ_{inn}, the suppression of the large anisotropic scales has the effect of reducing the rate of entrainment and the generation of a more homogeneous scalar field.

In conclusion, both the inner and outer cycles appear to have quite independent dynamics, with the difference being that, the outer cycle simply rescales to lower scalar values depending on what is supplied from the inner cycle, facing a free region on its top bound. On the other hand, the inner cycle dynamics is confined to operate in between the wall and the outer cycle, the suppression of which can act as a bound, limiting the inner cycle to operate in a narrower and narrower range of scalar values due to the intrinsic limitation $\theta \leq 1$.

Acknowledgments

This work was performed in part during the Fifth Madrid Turbulence Workshop, funded by the European Research Council under the Caust grant ERCAdG-101018287. The authors would like to thank Prof. Jiménez for organizing the event.

Appendix A. Effect of the filtering procedure

In order to show the robustness of the present results on the type of filtering procedure adopted, we present here the results obtained through a different filtering approach. This second approach tested uses the filtering operation only in sub-portions of the domain to obtain the two modified velocity fields. The field u_i^{out} is obtained by lowpass filtering the near-wall region ($z^+ < 60 = z_c^+$) in order to suppress the inner cycle and leave the outer cycle motions unaltered. On the other hand u_i^{inn} is obtained by lowpass filtering the outer region ($z^+ > 60 = z_c^+$) and leaving the velocity in the near-wall region unchanged. The filter lengths vary gradually among the two regions through the use of a smoothing function $\widetilde{\Delta_i} = \Delta_i 0.5(1 \pm \tanh(z_c/(2h_\Delta))(1 - z/z_c))$ with Δ_i being the reference Gaussian filter lengths used in the streamwise and spanwise directions. In both experiments the width of the hyperbolic tangent is set to $h_\Delta = 0.2$. In the experiment θ_{out} (where the inner cycle is suppressed) Δ_i is expressed in inner units, while in the experiment

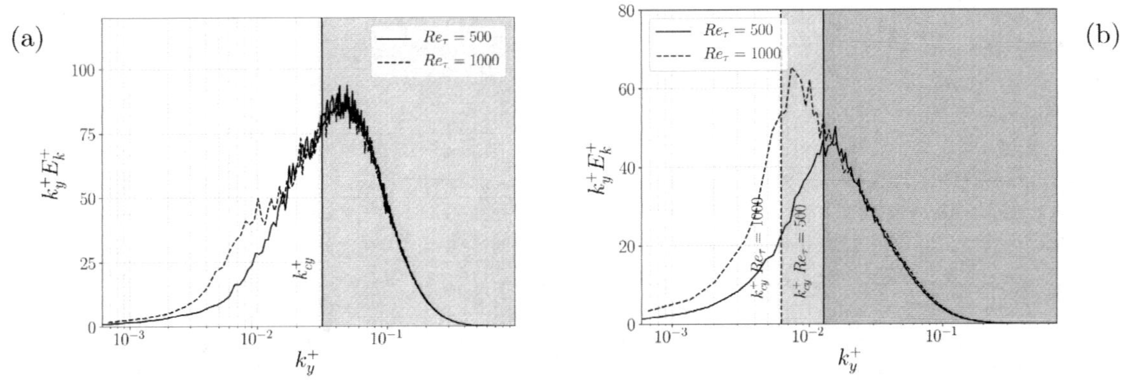

Figure A1. Premultiplied spectra of the turbulent kinetic energy E_k in the spanwise wavenumber k_y respectively (a) in the buffer layer at $z^+ = 20$ and (b) in the outer region at $z^+ = 200$. The shaded areas indicate the filtered regions of the spectra (a) in u_{out} and (b) in u_{inn}, being $k_{cy}^+ = 2\pi/\Delta_y^+$ the wavenumber corresponding to the spanwise filter length in inner units. In (b), k_{cy}^+ is reported at the initial $Re_\tau = 500$ and final $Re_\tau = 1000$.

θ_{inn} (where the outer cycle is suppressed), Δ_i is expressed in outer units. The filter lengths are chosen to remove most of the energy content in the region where they are applied. By looking at the premultiplied spectra, they have been respectively set to $\Delta_x^+ = 600$ and $\Delta_y^+ = 100$ for u_i^{out}, and $\Delta_x = 1.5\delta$ and $\Delta_y = 0.5\delta$ for u_i^{inn}. In order to ensure u_i^{inn} and u_i^{out} to be divergence free, the pressure projection method is applied to the two velocity fields after the filtering operation. The portions of the spectra filtered out in the spanwise direction to obtain u_i^{inn} and u_i^{out} are shown in shaded grey in figure A1(a) and (b), respectively measured in the buffer layer and in the outer region.

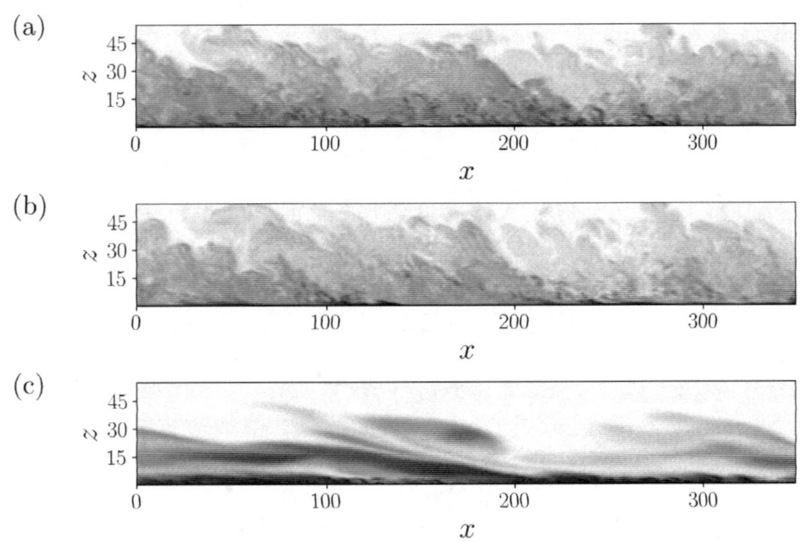

Figure A2. Sections of the scalar fields of the three experiments at the same time instant, corresponding to $Re_\tau = 1000$. Respectively (a) θ, (b) θ_{out} and (c) θ_{inn}. The reference case (a) is duplicated here in order to simplify comparisons.

We now proceed to present the results of these experiments. All the figures reported are maintained as in section 3 in order to have a clear comparison between the data. In figure 3 the instantaneous scalar contours are shown on a wall-normal section. In figure A2, the instantaneous scalar contours are shown on a wall-normal section. The scalar fields of the two experiments θ_{out} and θ_{inn} show a very different topology if compared to those obtained with the approach presented in section 3. In particular, θ_{out} looks macroscopically almost unaltered, with a rescaling toward lower scalar values. Looking closer, in the near-wall region (which is difficult to distinguish) it is possible to note substantial differences resulting from the absence of small-scale motions. On the contrary, the field θ_{inn} appears completely different with respect to the reference case due to the application of the filtering operation with large filter lengths ($\Delta_x = 1.5\delta$ and $\Delta_y = 0.5\delta$). Despite this topological difference, we will see below that the statistics of interest for this work and the main conclusions obtained in section 3 remain valid.

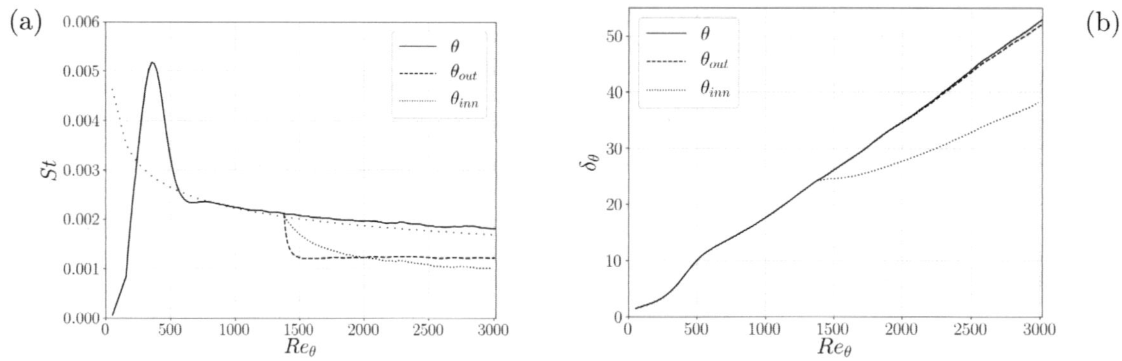

Figure A3. (a) Temporal evolution of the Stanton number and its theoretical law $St = 0.0125Re_\theta^{-1/4}$ in loosely dotted line ($\cdots\cdots$) and (b) temporal evolution of the boundary layer thickness based on the scalar concentration δ_θ.

As shown in figures A3(a) and 4(a), the Stanton number evolutions resulting from the two approaches are qualitatively similar. Accordingly, also the pdf profiles measured in the viscous sub-layer reported in figures A4(a) and 5(a) describe a similar scenario.

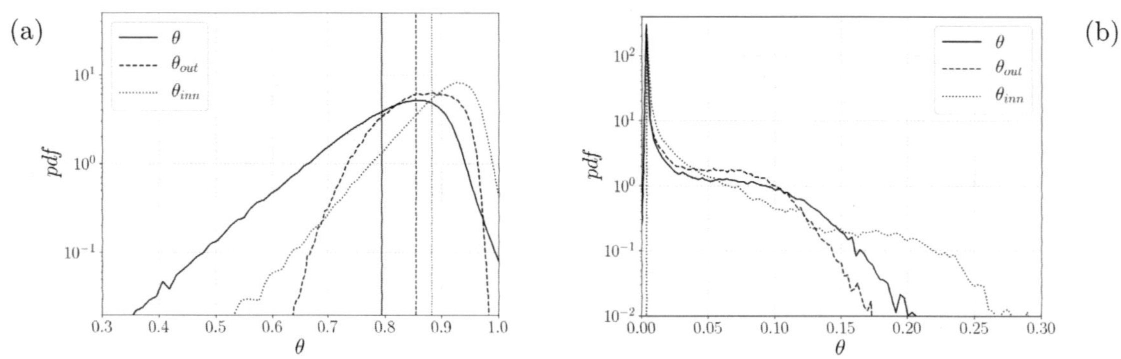

Figure A4. Probability density function of θ (a) at $z^+ = 5$ and (b) at $z = \delta_\theta$ at the final $Re_\tau = 1000$. The vertical lines in (a) indicate the respective mean values of the scalars.

A similarity can also be found in the temporal evolution of the boundary layer thickness as shown in figures A3(b) and 4(b) and in the relative pdf profiles measured at the mean scalar interface (figures A4(b) and 5(b)). One notable difference is in the pdf distribution of θ_{inn} (dotted line). The presence of an extended right tail in the pdf of θ_{inn} in figure A4(b) is caused by the intersection between the plane located at the mean scalar interface position and the protruding bulges generated by the large anisotropic fluctuations. These bulges contain a wide range of scalar values and remain unmixed due to the absence of the small scales. The extension of this tail is more extended with respect to the first approach because of the larger filter lengths used. A minor change can also be observed in the evolution of $\delta_{\theta_{inn}}$ in figure A3(b) that is nearly identical to that of the reference δ_θ in figure A3(b). In the present approach, in fact, the velocity field in the outer region is not modified ($u_i^{inn} = u_i$) and the small difference between $\delta_{\theta_{inn}}$ and δ_θ is only the result of the lower scalar concentration flux coming from the inner cycle. This effect is comparable to a change of threshold in the definition of the mean interface position, which is therefore a very small effect.

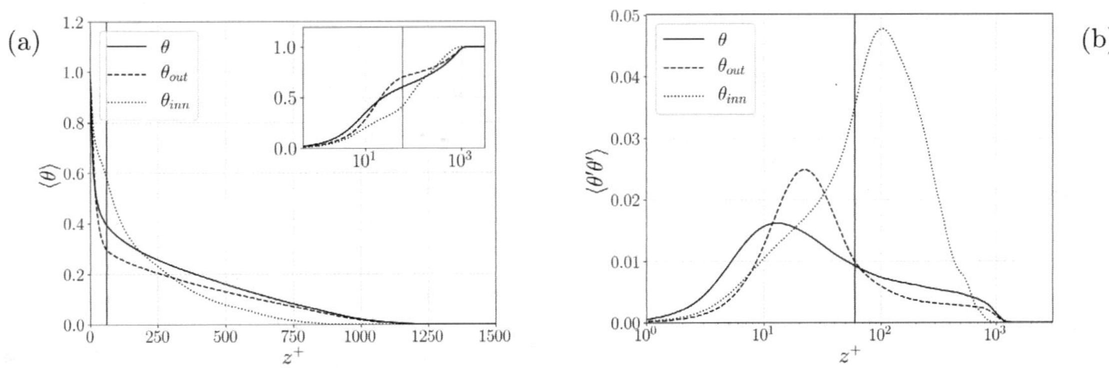

Figure A5. (a) Mean scalar profiles and (b) variances at $Re_\tau = 1000$. The solid vertical line is located at $z_c^+ = 60$.

The mean profiles reported in figure A5 show a behaviour in accordance with the one observed in figure 6, while the scalar variances describe a scenario opposite to the one observed in section 3 and in accordance with what is shown in figure A2. The highly unmixed outer region of θ_{inn} causes a peak in $\langle\theta'_{inn}\theta'_{inn}\rangle$ that have a great impact on the scalar variance fluxes, as we will see.

The mean fluxes reported in figure A6 are in good accordance with what is observed in figure 7.

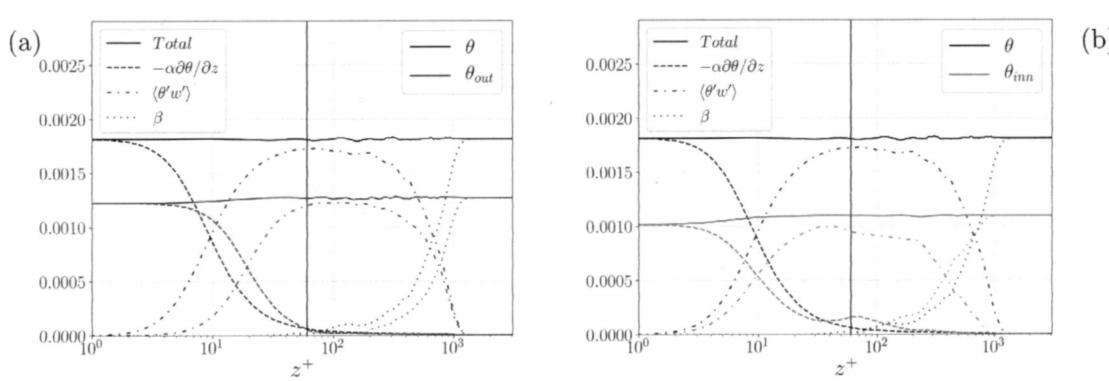

Figure A6. Mean fluxes of (a) θ_{out} and (b) θ_{inn} compared with θ at $Re_\tau = 1000$. The solid vertical line is located at $z_c^+ = 60$.

Finally, the scalar variance budgets show some macroscopic differences due to the more intrusive nature of the approach here presented. However, if we limit our analysis far enough from the region affected by the filtering operation (i.e. $z^+ \ll 60$ for θ_{inn} and $z^+ \gg 60$ for θ_{out}) we can note that the main conclusions drawn in section 3 find support in these graphs.

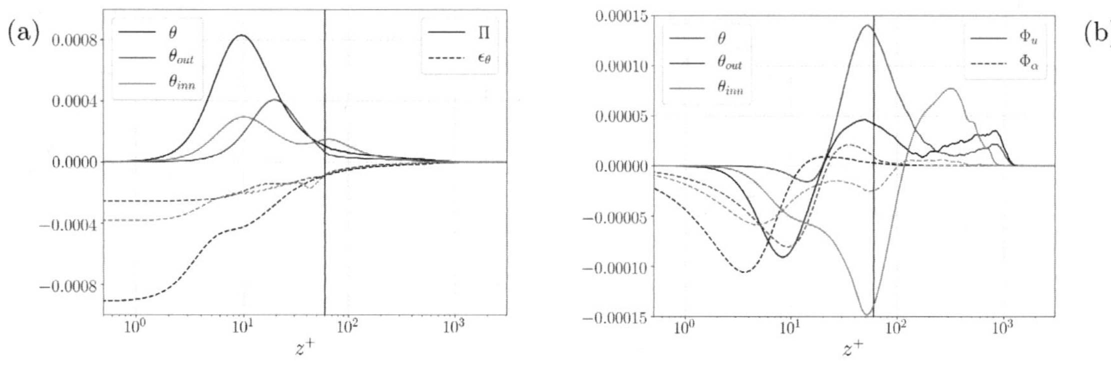

Figure A7. Scalar variance (a) production Π, dissipation ϵ_θ and (b) turbulent and diffusive fluxes Φ_u and Φ_α at $Re_\tau = 1000$. The solid vertical line is located at $z_c^+ = 60$.

To conclude, despite the topological differences shown by the corresponding scalar fields of the two presented approaches, the results obtained lead to the same conclusions in both experiments. Confirmation of this can be found in the similarities between the evolution of the integral quantities, in the mean profiles, in the turbulent fluxes of both the mean scalar concentration and its variance and the production and dissipation of the fluctuations (if confined in the regions of interest, namely the inner region for θ_{inn} and the outer region for θ_{out}).

References

[1] Smits A J, McKeon B J and Marusic I 2011 High–Reynolds number wall turbulence *Annu. Rev. Fluid Mech.* **43** 353-75
[2] Robinson S. K. 1991 Coherent motions in the turbulent boundary layer *Annu. Rev. Fluid Mech.* **23** 601–39
[3] Townsend A A 1976 *The Structure of Turbulent Shear Flows* (Cambridge: Cambridge University Press)
[4] Jiménez J and Pinelli A 1999 The autonomous cycle of near-wall turbulence *J. Fluid Mech.* **389** 335-59
[5] Mizuno Y and Jiménez J 2013 Wall turbulence without walls *J. Fluid Mech.* **723** 429-55
[6] Li Q, Schlatter P, Brandt L and Henningson D S 2009 DNS of a spatially developing turbulent boundary layer with passive scalar transport *Int. J. Heat Fluid Flow* **30** 916-29
[7] Kozul M, Chung D and Monty J 2016 Direct numerical simulation of the incompressible temporally developing turbulent boundary layer *J. Fluid Mech.* **796** 437-72
[8] Cimarelli A, Boga G, Pavan A, Costa P and Stalio E 2023 Energy cascade phenomena in temporal boundary layers *Flow Turbul. Combust.* 1-17
[9] da Silva C B, Hunt J C R, Eames I and Westerweel J 2014 Interfacial layers between regions of different turbulence intensity *Annu. Rev. Fluid Mech.* **46** 567–90
[10] Zhang X, Watanabe T and Nagata K 2023 Reynolds number dependence of the turbulent/non-turbulent interface in temporally developing turbulent boundary layers *J. Fluid Mech.* **964** A8
[11] Cimarelli A and Boga G 2021 Numerical experiments on turbulent entrainment and mixing of scalars *J. Fluid Mech.* **927** A34
[12] Costa P 2018 A FFT-based finite-difference solver for massively-parallel direct numerical simulations of turbulent flows *Comput. Math. Appl.* **76**(8) 1853-62

Are the dynamics of wall turbulence in minimal channels and larger domain channels equivalent? A graph-theoretic approach

Ahmed Elnahhas[1], Emma Lenz[2], Parviz Moin[1], Adrián Lozano-Durán[3] and H. Jane Bae[2]

[1] Center for Turbulence Research, Stanford University, Stanford CA, 94305 USA
[2] Graduate Aerospace Laboratories, California Institute of Technology, Pasadena CA, 91125 USA
[3] Department of Aeronautics & Astronautics, Massachusetts Institute of Technology, Cambridge MA, 02139 USA

E-mail: ahmed97@stanford.edu

Abstract.
This work proposes two algorithmic approaches to extract critical dynamical mechanisms in wall-bounded turbulence with minimum human bias. In both approaches, multiple types of coherent structures are spatiotemporally tracked, resulting in a complex multilayer network. Network motif analysis, i.e., extracting dominant non-random elemental patterns within these networks, is used to identify the most dominant dynamical mechanisms. Both approaches, combined with network motif analysis, are used to answer whether the main dynamical mechanisms of a minimal flow unit (MFU) and a larger unconstrained channel flow, labeled a full channel (FC), at $Re_\tau \approx 180$, are equivalent. The first approach tracks traditional coherent structures defined as low- and high-speed streaks, ejections, and sweeps. It is found that the roll-streak pairing, consistent with the current understanding of self-sustaining processes, is the most significant and simplest dynamical mechanism in both flows. However, the MFU has a timescale for this mechanism that is approximately 2.83 times slower than that of the FC. In the second approach, we use semi-Lagrangian wavepackets and define coherent structures from their energetic streak, roll, and small-scale phase space. This method also shows similar motifs for both the MFU and FC. It indicates that, on average, the most dominant phase-space motifs are similar between the two flows, with the significant events taking place approximately 2.21 times slower in the MFU than in the FC. This value is more consistent with the implied timescale ratio of only the slow speed streaks taking part in the roll-streak pairing extracted using the first multi-type spatiotemporal approach, which is approximately 2.17 slower in the MFU than the FC.

1. Introduction

It has long been hypothesized that wall-bounded turbulence can be decomposed into a set of coherent structures that coexist and interact [1]. This concept can be interpreted kinematically by synthesizing the turbulent flow field as a superposition of building block eddies; the attached eddy model (AEM) adopts this approach for high-Re_τ flows using a hierarchy of eddies scaling with distance from the wall [2, 3]. A more comprehensive model would account for the dynamics of the building block eddies, both in isolation as well as for their interactions. The most famous dynamic coherent structure model of wall-bounded turbulence is the roll-streak self-sustaining process (SSP) [4, 5, 6, 7, 8, 9, 10, 11, 12]. This SSP is primarily extracted from simulations of minimal flow units (MFU) at low Re_τ, which isolate a single scale of wall-bounded turbulence, as well as a single instance of the SSP at a time by enforcing severe periodicity [4, 5, 6, 12]. These MFU simulations have also been extended to larger scales by isolating structures in the logarithmic region of the flow where similar dynamical cycles were observed [7, 8, 9, 10, 11]. Taking the AEM, one can theorize that a "dynamic"-AEM can be constructed based on the findings from MFUs of the buffer and logarithmic regions of the flow by superposing SSPs at each scale of motion and neglecting strong interscale interactions. However, recent experimental observations of inner/outer modulation and large-scale streak aggregation suggest more to the story [13, 14, 15, 16]. Hence, a few questions immediately arise concerning the current understanding of wall-bounded turbulence: Are the dynamics extracted from MFUs relevant to wall-bounded turbulent flows in large unbounded domains at the same Re_τ? Or do interactions between different instances of SSPs heavily influence each other, perhaps altering their timescales? Are other, possibly dominant, dynamic cycles suppressed in the artificially restricted domains of MFUs? Finally, what is the importance of interscale interactions of different-sized SSPs at larger Re_τ? To understand the dynamical interactions amongst the same scales of motion and inter-scale, we need to find a systematic and algorithmic way to extract and study these dynamics. In this study, we propose such a framework and apply it first to study whether the dynamics of the MFU exist unaltered in large domain channel simulations, henceforth referred to as full channels (FC), at $Re_\tau = 180$, or not.

Recently, coherent structure identification and temporal tracking algorithms have been developed to understand the basic dynamical properties of select eddy definitions, such as momentum-carrying structures, velocity streaks, and energy-containing eddies [17, 18, 19, 20]. These algorithms are based on spatial and temporal clustering of grid points that satisfy certain thresholding and proximity criteria. The result is a directed graph where the nodes represent the coherent structures, and the edges represent temporal connections between the structures. Therefore, advanced graph processing tools can extract information about these coherent structures' spatiotemporal properties. Extending the clustering and tracking algorithms to track several definitions of eddies simultaneously, along with the interconnections between eddy types, results in a higher dimensional graph, where the eddy type acts as the new dimension, called a multilayered network [21, 22]. We use network motifs and graphlet search algorithms [23, 24] to analyze and compare dynamically significant patterns extracted from the buffer layer MFU and large domain channel flows.

A network motif of a complex network is a pattern occurring at a significantly higher frequency than in a random network of the same size [23]. Network motifs have been identified in various biological networks, such as those responsible for gene expression and neuronal cell interactions [25]. They have also been identified in larger-scale ecological and engineering networks [23]. Since our directed multilayer network has time embedded in its construction, detecting motifs accounts for the most significant dynamical patterns between eddies of various types directly. Figure 1(a) illustrates the proposed approach. Each node in the network represents the full spatiotemporal history of a coherent structure, and the edges represent the interactions between two coherent structures at some point during their evolution. In the context of the current question, finding the most dominant interaction pattern compared to a set of random networks helps us isolate the key dynamical mechanism in both the FC and the MFU and compare their similarity. Applications of network analysis to fluid flow are only recent phenomena [18, 26, 27, 28, 29, 30, 31, 32, 33, 34], and applications to the study of the dynamics of coherent structures have not been attempted in fully developed wall-bounded turbulent flows. The results of this spatiotemporal approach are supplemented with motifs extracted from an energetic phase-space constructed using semi-Lagrangian wavepackets illustrated in figure 1(b).

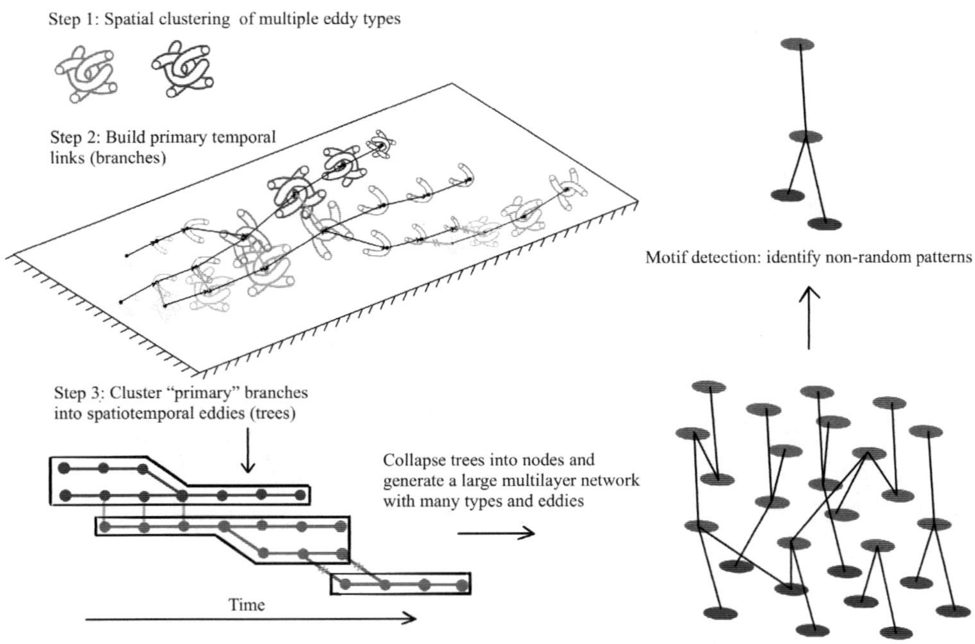

(a)

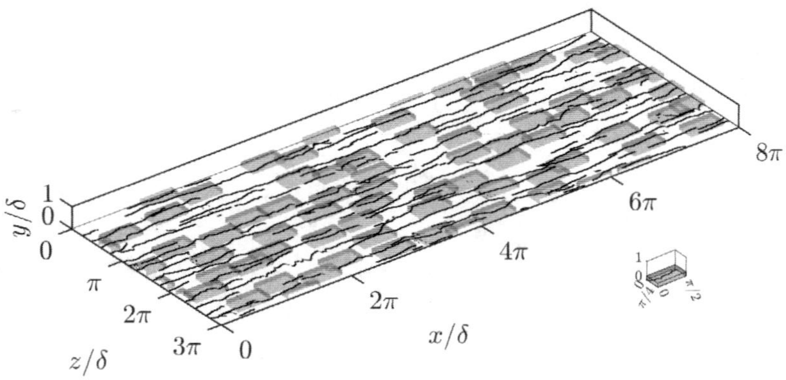

(b)

Figure 1: (a) The proposed multi-type spatiotemporal tracking and motif detection methodology. Different types of eddies are clustered simultaneously, and the resulting multilayer network is searched for patterns called network motifs. The lighter structures represent earlier times in the evolution of a spatiotemporal eddy. Green lines represent the intertype connections transferred from single-time connections to branches to trees. Orange cross-lines indicate intra-type branch links that are not considered simultaneously "primary" for tree clustering. (b) The proposed semi-Lagrangian wavepacket approach where the greyed areas are MFU-sized boxes, which are allowed to advect and collect quantities for the phase-space-temporal dynamical analysis. Black lines are the spanwise centroids of low-speed streaks projected onto the wall. The MFU is shown for comparison.

The rest of the paper is organized as follows. Section 2 describes the datasets utilized to perform the analysis. In section 3, the multi-type coherent structure tracking and motif detection algorithms are described, and the results of their application to the MFU and the FC at $Re_\tau = 180$ are discussed. In section 4, the semi-Lagrangian wavepackets and phase-space temporal motif analysis approach is explained, and the results of applying it to the same datasets are discussed. Finally, conclusions and future work are presented in section 5.

2. Simulation details

Let u_τ and ν denote the flow's friction velocity and kinematic viscosity, respectively. The two flows analyzed are direct numerical simulations (DNS) of an MFU and an FC inside a channel of half-height δ. The friction Reynolds number for both flows is $Re_\tau \equiv \delta u_\tau/\nu \approx 180$. In this flow, x, y, and z are the streamwise, wall-normal, and spanwise directions, respectively, corresponding to their respective mean and fluctuating velocities U and u, V and v, and W and w. The domain of the MFU has a size of $\{0.564\pi \times 2 \times 0.282\pi\}\delta$, and that of the FC has a size of $\{8\pi \times 2 \times 3\pi\}\delta$. The resolution in inner units of both simulations are $\{\Delta x^+, \Delta y^+_{min}, \Delta y^+_{max}, \Delta z^+\} = \{6.05, 0.17, 7.52, 6.05\}$, which is consistent with prior DNS simulations. Flow fields are stored every $\Delta t^+ \approx 1$ for $T^+ = 2000$ and $T^+ = 25000$ inner time units for the FC and MFU, respectively. These correspond to approximately $11\delta/u_\tau$ and $135\delta/u_\tau$, respectively.

3. Multi-type spatiotemporal coherent-structure tracking and motif detection

3.1. Multilayer network generation

The multi-type coherent-structure tracking algorithm is a three-step process, schematically illustrated in figure 1(a):

Step 1: In this work, the eddies considered are the low- and high-speed streaks and wall-normal streamwise momentum flux eddies, namely ejections and sweeps, interchangeably referred to as Q2 and Q4, respectively [17, 20]. At each time instant, each of these eddies is defined as the set of connected points in space, Ω, satisfying the following inequalities

$$\{(x,y,z) : |uv| > \alpha_{uv} u'(y) v'(y) \quad \& \quad \begin{cases} \left\{\left(\frac{\int_\Omega u\, dV_\Omega}{\int_\Omega dV_\Omega} < 0\right) \& \left(\frac{\int_\Omega v\, dV_\Omega}{\int_\Omega dV_\Omega} > 0\right)\right\} \implies \text{Ejections (Q2)} \\ \left\{\left(\frac{\int_\Omega u\, dV_\Omega}{\int_\Omega dV_\Omega} > 0\right) \& \left(\frac{\int_\Omega v\, dV_\Omega}{\int_\Omega dV_\Omega} < 0\right)\right\} \implies \text{Sweeps (Q4)} \end{cases}, \tag{1}$$

$$\{(x,y,z) : \sqrt{u^2 + w^2} > \alpha_{str} u'(y) \quad \& \quad \begin{cases} \{u < 0\} \implies \text{Slow Streak} \\ \{u > 0\} \implies \text{Fast Streak} \end{cases}, \tag{2}$$

where $u'(y)$ and $v'(y)$ are the root-mean-squared streamwise and wall-normal velocity fluctuations extracted from the corresponding simulation, i.e., the MFU and the FC, $\alpha_{\{uv,str\}}$ are scaling coefficients for the thresholds, and V_Ω is the physical volume associated with the set of points, Ω. The clustering process is known to have a percolation crisis, and we use the values $\{\alpha_\tau, \alpha_{str}\} = \{1.30, 1.75\}$, which maximize the number of clusters identified per timestep on average. Each region, Ω, is labeled as a coherent structure at some instant in its evolution. Once subdivided into these four sets, contiguous time snapshots construct their temporal evolution.

Step 2: The details of the structure tracking in time are described in [18]. In summary, object volumetric overlap between timesteps is used to construct a graph with edges representing connections in time and objects at different time steps representing nodes. While previously this was done for only a single type of object, this procedure is repeated N^2 times where N is the

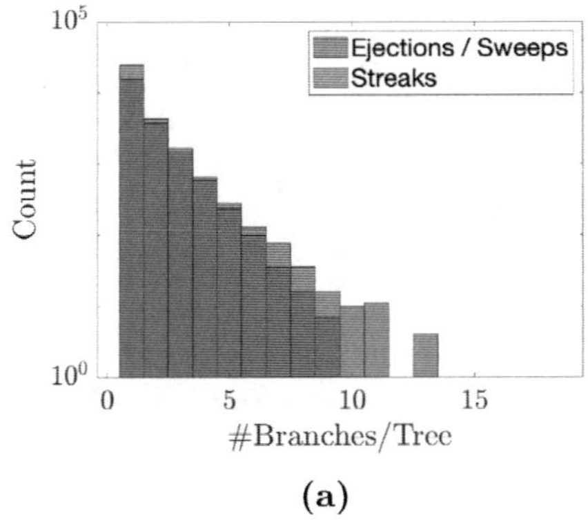

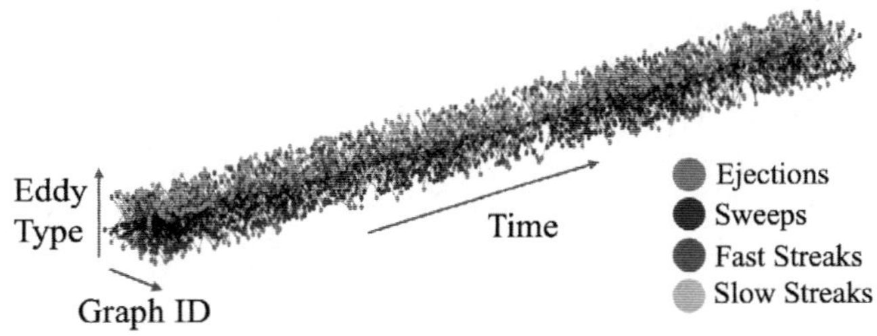

Figure 2: (a) Histogram indicating the distribution of how many individual attached eddy branches are grouped into trees using the simultaneous primary branch interaction metric in Equation (3). (b) A multilayer network with each node representing the full spatiotemporal evolution of an attached eddy tree placed at its temporal centroid. The links between the nodes indicate an interaction or overlap at some point during the lifetime of the two linked spatiotemporal eddies. Only a subset of the complete network in the FC is shown.

number of eddy types. The graph is first clustered into branches, which are the links between objects that mutually experience the least relative change from one timestep to the next.

Step 3: The branches are then further partitioned to generate subgraphs or trees, which are collections of branches that interact in a "primary" way. We define a primary branch connection using the relative overlap metric from branch i to j of the same eddy type and vice versa, which is defined as follows

$$O_{i\to j} = \frac{\int V_i \cap V_j dt_i}{\int V_i dt_i} \quad \text{and} \quad O_{j\to i} = \frac{\int V_j \cap V_i dt_j}{\int V_j dt_j}, \tag{3}$$

where the integration is over the lifetime of the source branch and V_i is the volume of a branch at a given time across its lifetime. The intersecting volume is computed at the same absolute time for both branches. The branch link j to branch i with the highest overlap metric is considered the primary branch in the positive or negative time directions. If two branches are each other's primary links, they are clustered into a subgraph or tree. This process is applied recursively, and each resulting tree is subsequently referred to as a spatiotemporal eddy. An illustration of the partitioning when two branches do not point to each other as primary branch connections is represented with the crossed orange lines in figure 1(a).

During step 2 and step 3 of the process, the physical connections between eddies of different types are transferred from single-time connections at the end of step 1 to connections between branches and, finally, trees at the end of steps 2 and 3, respectively. This results in a large, complex, multilayered network that encodes spatiotemporal interactions between eddies of the same type and those between eddies of different types. These intertype connections are represented with green lines in the schematic illustration of figure 1(a).

After the three-step process, many of the spatiotemporal eddies are small structures existing in the center of the channel, which, if tracked, result in an overly large/complex multilayer network. To ease the subsequent analysis, we limit ourselves to interactions between attached eddies that live for a minimal amount of time, i.e., those who, at some point during their lifetime, had a minimal wall-normal height (on either wall) of $y^+_{min} < 20$ [18, 20], and have lived for longer than $T^+_{life} - 20$. Furthermore, they must have interacted with another attached eddy during their

Table 1: The average percentage of the domain's volume occupied by certain types of eddies that satisfy particular criteria in both the full channel and minimal flow unit. $\mathcal{A}$: $y^+_{min} < 20$. $\mathcal{T}$: $T^+_{life} > 20$. $\mathcal{I}$: Interaction with another $\mathcal{A}$–valid eddy. $\mathcal{S}_1^{\{F,M\}}$: Involved in most significant roll-streak motif. $\mathcal{F}_2^M$: Involved in similar motif only presented in MFU. The two values for $\mathcal{S}_1^M$ and $\mathcal{F}_2^M$ eddies are averages performed over the entire temporal extent of the simulation and only over the times when constituent eddies in the motifs are alive, respectively.

Minimal Flow Unit			Full Channel		
Filter	Q2 & Q4	Streaks	Filter	Q2 & Q4	Streaks
$\mathcal{A}$	5.12%	2.55%	$\mathcal{A}$	5.19%	2.43%
$\mathcal{A}$ & $\mathcal{T}$	4.90%	2.52%	$\mathcal{A}$ & $\mathcal{T}$	4.52%	2.15%
$\mathcal{A}$ & $\mathcal{T}$ & $\mathcal{I}$	4.43%	2.31%	$\mathcal{A}$ & $\mathcal{T}$ & $\mathcal{I}$	4.16%	2.10%
$\mathcal{A}$ & $\mathcal{T}$ & $\mathcal{I}$ & $\mathcal{S}_1^M$	1.09-3.87%	0.72-2.44%	$\mathcal{A}$ & $\mathcal{T}$ & $\mathcal{I}$ & $\mathcal{S}_1^F$	2.04%	1.01%
$\mathcal{A}$ & $\mathcal{T}$ & $\mathcal{I}$ & $\mathcal{F}_2^M$	2.56-4.14%	1.10-2.15%	—	—	—

lifetime. The volumes occupied by the ejections, sweeps, slow, and fast streaks that satisfy these criteria and subsets of them in both the MFU and the FC are presented in table 1. The applied criteria only restrict the minority of the attached eddies by volume, approximately $11.5-17.5\%$, from participating in the subsequent motif analysis. Figure 2(a) showcases the result of the step 3 partitioning of branches into trees using the overlap metric for attached eddies, where the size of the trees in terms of branches is exponentially decaying, and no single tree contains more than $\mathcal{O}(10)$ branches. Given this partitioning, figure 2(b) shows a subset of the multilayer network that emerges due to all the interactions of attached spatiotemporal ejections, sweeps, and slow and fast streaks in the FC where each dot contains within it the entirety of the detailed spatiotemporal evolution of each of the spatiotemporal eddies.

3.2. Motif detection

The resulting multilayer network, at the tree level, is fed into an open-source motif detection code, FAst Network MOtif Detection (FANMOD) [35], which uses the exhaustive enumerate subgraph (ESU) algorithm to enumerate all isomorphic subgraphs, s, of a certain size in our multilayer network [36]. The frequency of each isomorphic subgraph s, given by $m(s)$, is compared with statistics from a set of randomly generated networks to determine over-representation in the original network, i.e., significant motifs.

In this case, the randomly generated networks are constructed from our original network by randomly switching edges between nodes in the network while maintaining its size, $|G|$. The sizes of the original network for each flow are $|G|_{FC} = 31855$, and $|G|_{MFU} = 1251$. Since their criteria are mutually exclusive, ejections and sweeps cannot form connections, as is the case between fast and slow streaks. This constraint is reflected in the random networks generated by maintaining the local colored degree vector of each node during the randomization [37]. This means that, for example, if a sweep were originally connected to two fast streaks and one slow streak, it would remain connected to two fast streaks and one slow streak after the randomization. We can, therefore, interpret the randomization as a reordering in space and time of the spatiotemporal eddies while maintaining their allowable interactions.

The frequency of each subgraph in the random graph is given by $m_r(s)$. The mean and variance of $m_r(s)$ are computed from an ensemble of random graphs and denoted $\overline{m_r(s)}$ and $Var(m_r(s))$, respectively. Enough random graphs were generated so that the mean and the variance converged. This was done with 100 and 1000 random graphs for the FC and MFU, respectively. A significance score was determined for each isomorphic subgraph based on the random mean and standard deviation. Furthermore, to account for shifts in the scores associated with the differences in the sizes of the networks, a normalization based on the law of large numbers utilizing $|G|$ is added. The significance score is, therefore,

$$\widetilde{W}(s) = \frac{m(s) - \overline{m_r(s)}}{(Var(m_r(s))|G|)^{1/2}}. \tag{4}$$

3.3. Results

The results of the motif analysis indicate that there are no 3-node motifs, i.e., none where significantly different than random, and we focus our discussions on 4-node motifs. The schematics in table 2 illustrate the results for the 4-node motifs. The motifs are labeled as $\{\mathcal{S},\mathcal{F}\}_{\#}^{\{F,M\}}$ with $\mathcal{S}$ or $\mathcal{F}$ denoting motifs ordered by significance score and frequency, respectively, and the superscripts F and M denoting the flow, FC and MFU, respectively. The left portion of the schematics in the table shows the top six to seven motifs in descending order of significance score. The two flows share the most significant motifs, where the top six in the MFU are found in the top seven of the FC. However, besides the most dominant motif, the order of the other five is shifted around in the MFU with respect to the FC, as denoted by the two-headed black arrows. The most dominant 4-node motif for both flows is a pairing of a slow and a fast streak interacting with an ejection and a sweep, consistent with the roll-streak SSP, labeled from now on as $\mathcal{S}_1^{\{F,M\}}$. This likely indicates that the roll-streak pairing is, in fact, the simplest form of spatiotemporal organization exhibited in these single-scale wall-bounded flows. The right portion of the schematics in the table shows the top six motifs in descending order of frequency. The two flows share five of the top six between them. However, their orders are shifted around in a similar manner to the top six most dominant motifs. Furthermore, motif $\mathcal{F}_2^M$ strongly resembles the structure of $\mathcal{S}_1^{\{F,M\}}$ except for a missing link between slow streaks and sweeps. This is likely due to the confined domain restricting the emergence of the structures simultaneously. Whether this confinement leads to some ordering in the emergence and death of the constituent eddies of this motif is examined further below.

Furthermore, table 2 reports the significance scores and frequency of occurrence of each motif. Note that while the frequency of the most significant motif in both flows seems to be low compared to all possible 4-node induced subgraphs in either G_{FC} or G_{MFU}, that does not mean that the constituent eddies are few or not substantial. In fact, table 1 shows that approximately 50% of the attached eddies by volume in the FC participate in this motif. For the MFU, the volume occupied by these eddies averaged over the entire temporal extent of the simulation indicates that only 25 – 30% of them participate in the motif. However, there are times when no eddies participating in this motif are present in the simulation domain. Thus, if we condition the average on the times when the eddies participating in the motif are present, i.e., alive, the average volume of ejections and sweeps is around 87% of their unconditioned average volume, and that of streaks is even larger than the unconditioned average volume of attached streaks throughout the simulation. A plausible explanation is as follows: In the MFU, on average, only a single instance of an SSP can exist at a time, and once the large-scale streaky structures break down in what is known as bursting [38], the current tracking procedure sees no attached structures. While this also happens in the case of the FC, since there is always at least one instance of this motif active in the domain, conditionally averaging over the times when a constituent eddy is alive is redundant. However, for the MFU, there are time periods where we are averaging over zero volume due to the presence of only small scales. Overall, this

Table 2: The dynamical 4-node motifs detected using the ESU algorithm implemented in FANMOD in both the FC and the MFU using the aforementioned random network model. The left portion shows motifs ordered in descending order of significance score, indicating the strongest non-random events in both domains. The right portion shows motifs ordered in descending order of frequency. The reported motifs exclude any with a significance score below $\widetilde{W}(s) < 0.05$. The top and bottom rows of motif schematics represent those extracted from the FC and the MFU, respectively. Above and below the motif schematics, the significance score, $\widetilde{W}(s)$, and frequency percentage, $\mathcal{F}(s)$, i.e., the percentage of occurrences in the original networks with respect to the total number of all 4-node subgraphs present in the MFU and FC networks, which were 1811927 and 7057, respectively, are presented along with their labels. The bounding red box indicates that the most non-random dynamical pattern, $\mathcal{S}_1^{\{F,M\}}$, is the roll-streak pairing and is consistent amongst both flow domains. The bounding dashed purple box shows a similar motif to $\mathcal{S}_1^{\{F,M\}}$, $\mathcal{F}_2^M$, that only emerges in the MFU. ● → Ejections. ● → Sweeps. ● → Fast streaks. ● → Slow streaks.

	Significance Ordering							Frequency Ordering					
FC	$\mathcal{S}_1^F$	$\mathcal{S}_2^F$	$\mathcal{S}_3^F$	$\mathcal{S}_4^F$	$\mathcal{S}_5^F$	$\mathcal{S}_6^F$	$\mathcal{S}_7^F$	$\mathcal{F}_1^F$	$\mathcal{F}_2^F$	$\mathcal{F}_3^F$	$\mathcal{F}_4^F$	$\mathcal{F}_5^F$	$\mathcal{F}_6^F$
$\widetilde{W}(s)$	4.71	3.76	3.50	3.47	3.13	3.04	2.56	0.13	0.13	0.13	0.13	0.13	0.13
$\mathcal{F}(s)$	0.26	0.19	0.16	0.14	0.15	0.09	0.09	7.28	7.10	6.10	4.95	3.53	3.36

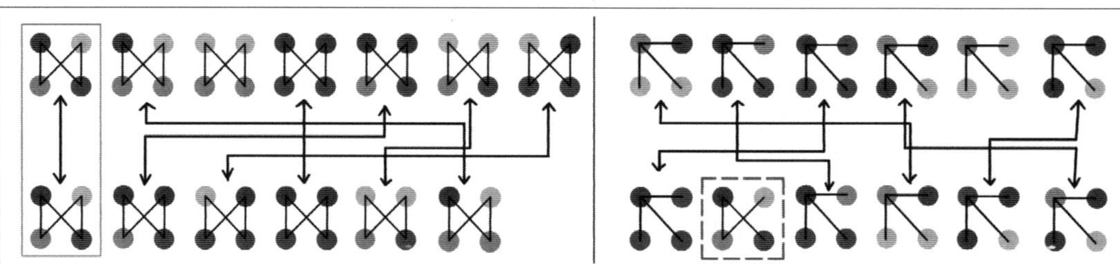

MFU	$\mathcal{S}_1^M$	$\mathcal{S}_2^M$	$\mathcal{S}_3^M$	$\mathcal{S}_4^M$	$\mathcal{S}_5^M$	$\mathcal{S}_6^M$	—	$\mathcal{F}_1^M$	$\mathcal{F}_2^M$	$\mathcal{F}_3^M$	$\mathcal{F}_4^M$	$\mathcal{F}_5^M$	$\mathcal{F}_6^M$
$\widetilde{W}(s)$	1.54	1.30	0.83	0.79	0.73	0.38	—	0.37	0.05	0.37	0.37	0.37	0.37
$\mathcal{F}(s)$	0.72	0.47	0.34	0.35	0.14	0.10	—	15.20	8.86	8.67	3.83	3.49	3.25

implies that in subsequent analyses, we should include the simultaneous tracking of small scales in the motifs we search for to find the full cycle of the SSP. Finally, it is interesting to note that all the most frequent yet marginally significant motifs, except for $\mathcal{F}_2^M$, in both the FC and MFU have an equal significance score in each flow, respectively. This indicates that their frequencies are linked during the construction of the random networks and that they are most likely kinematically constrained to one another. We, therefore, focus our dynamical analyses on the eddies participating in each of the motif instances of $\mathcal{S}_1^{\{F,M\}}$, albeit including data from motif $\mathcal{F}_2^M$ in the case of the MFU due to its similarity to $\mathcal{S}_1^M$.

Table 3 shows the average lifetimes of each motif, $\overline{T^+}_{life}$, defined as the time from when the first participating eddy emerges to when the final participating eddy dies, in both the FC and MFU for motifs $\mathcal{S}_1^{\{F,M\}}$ and $\mathcal{F}_2^M$. Simultaneously, it shows the average advection velocity, $\overline{u^+}_{adv}$, of the participating eddies in the streamwise direction. It is observed that $\overline{T^+}_{life}$ is approximately 2.83 times longer in the MFU compared to its FC counterpart. The advection

Table 3: The average lifetime in inner units, $\overline{T^+}_{life}$, of the instances of the most significant motifs in each of the two flows, $\mathcal{S}_1^{\{F,M\}}$, along with the average advection velocity in inner units, $\overline{u^+}_{adv}$, of the constituent eddies taking part in these significant motifs. In the case of the MFU, the eddies taking part in $\mathcal{F}_2^M$ are also included in these averages due to the similarity of the motif and the lack of samples.

	$\overline{T^+}_{life}$	
Motif Identifier	Minimal Flow Unit	Full Channel
$\mathcal{S}_1^{F,M}$	601	212
	$\overline{u^+}_{adv}$	
Eddy Type	Minimal Flow Unit	Full Channel
Ejections	4.50	14.07
Sweeps	11.01	14.30
Fast Streaks	9.28	11.86
Slow Streaks	6.00	13.00

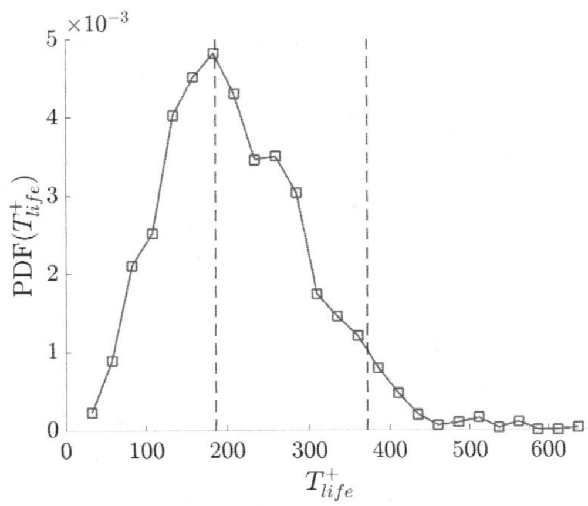

Figure 3: PDF of the lifetime of the roll-streak motif instances, $\mathcal{S}_1^F$, extracted from the FC. The two vertical dashed lines are placed at δ/u_τ and $2\delta/u_\tau$, respectively.

velocities of the constituent eddies of these motifs, $\overline{u^+}_{adv}$, range from 1.3 to 3.1 times slower in the MFU compared to its FC counterpart, with the primary reduction present in the ejections and the slow streaks associated with them. This indicates that while the main dynamical mechanism in both flows is the same, they differ in timescales substantially. In fact, it will be shown that the reduction in the speed of slow streaks in the MFU of approximately 2.17 times compared to those in the FC, is consistent with phase-space motifs extracted using the semi-Lagrangian wavepacket basis analysis to come, which employs the presence of slow streaks as a conditioning variable to determine the advection velocity of the wavepackets. Figure 3 shows the probability density function (PDF) of T^+_{life} for the $\mathcal{S}_1^F$ motif, highlighting that the most probable instances of this motif happen over a single eddy turnover timescale, δ/u_τ, with a majority ($> 95\%$) not lasting longer than two eddy turnover times.

Table 4: The percentage of time that ejections, sweeps, fast streaks, and slow streaks emerge and die in a particular order for motifs $\mathcal{S}_1^F$ for the FC, and $\mathcal{S}_1^M$ and $\mathcal{F}_2^M$ for the MFU.

FC	Emergence(%)				Death(%)			
Eddy Type	1st	2nd	3rd	4th	1st	2nd	3rd	4th
Ejections	14.2	20.7	24.5	40.6	35.1	27.2	21.8	15.9
Sweeps	28.8	25.3	21.6	24.3	32.8	23.2	21.9	22.1
Fast Streaks	37.6	26.1	22.1	14.2	11.4	20.1	28.8	39.7
Slow Streaks	19.4	31.8	27.9	20.9	20.7	29.5	27.5	22.3
MFU	**Emergence(%)**				**Death(%)**			
Eddy Type	**1st**	**2nd**	**3rd**	**4th**	**1st**	**2nd**	**3rd**	**4th**
Ejections	49.3	24.7	17.9	8.1	9.6	15.8	17.9	56.7
Sweeps	13.0	18.9	29.2	38.9	42.9	27.5	19.4	10.2
Fast Streaks	25.3	29.6	35.2	9.9	15.2	38.3	28.3	18.2
Slow Streaks	12.4	26.8	17.8	43.0	32.3	18.3	34.5	14.9

Furthermore, we note that the motifs presented in table 2, and in particular $\mathcal{S}_1^{\{F,M\}}$, are not partitioned into their $2 \times 4!$ possible emergence/death orderings. That is because no sufficiently prominent ordering was found in the FC. The emergence ordering that occurred the most, 9.91% of the time, was fast streak → sweep → slow streak → ejection. The death order that occurred the most, 8.53% of the time, was ejection → sweep → slow streak → fast streak. Therefore, we comment on the ordering trends for each individual eddy type, regardless of which permutation this ordering belongs to. Table 4 reports these statistics. Overall, in the FC, ejections are more likely to emerge last and die first, sweeps are more likely to emerge at any order and die at any order, albeit with a slightly higher tendency to die first similar to the ejections. On the other hand, fast streaks are more likely to emerge first and die last, whereas slow streaks are slightly more likely to emerge and die in the middle ordering of the motif. It is apparent that the average emergence and death orderings for the fast streaks and the ejections are inverted with respect to each other.

In the MFU, the emergence and death orderings of the ejections are inverted with respect to those of the FC, where they now appear first and die last. This indicates their longer timescales, consistent with their slower advection velocities. Sweeps, however, tend to appear last and die first. Fast streaks are approximately equally likely to appear first, second, or third and die second or third. Finally, slow streaks appear last, with the sweeps, and die either first or third. These differences between the two flows are probably due to the confinement effects of periodic boundary conditions, which the MFU is subject to.

Finally, streaks, being larger coherent structures than ejections and sweeps, are associated with multiple ejections and sweeps events during their lifetime. This was verified spatiotemporally by examining their type-dependent degree distribution, which is not shown here for brevity. As such, it is possible that a clearer picture of order would appear if, rather than individual motif instances, clusters of individual motif instances are examined [39].

4. Semi-Lagrangian wavepackets and phase-space temporal motifs

4.1. Defining the size of the semi-Lagrangian wavepackets

In addition to coherent structures, integral quantities of the flow also provide insight into the flow dynamics. While we can collect statistics on integral quantities directly in the MFU, we require an equally sized domain of integration in the FC to make meaningful comparisons between the two flows. To do so, we utilize semi-Lagrangian wavepackets in the FC to collect statistics on these integral quantities. These wavepackets will be advected downstream following an advection speed to track the relevant structures. To minimize the effect of shearing, we choose the domain to have $y^+ \le 40$, as the structures in this region are known to advect at a similar advection velocity [18]. In the wall-parallel directions, the wavepackets, and hence the domain of integration, are sized to match the extents of the MFU. Therefore, we consider wavepackets of size $(L_x^+, L_y^+, L_z^+) = (320.5, 40, 160.2)$. As the FC domain is considerably larger than the MFU, we can track many wavepackets simultaneously in each flow field, as seen in figure 1(b).

4.2. Production–dissipation trajectories and advection speed

We first consider trajectories of production and dissipation,

$$\langle\!\langle P \rangle\!\rangle = -\frac{1}{L_y}\int_0^{L_y} \langle uv \rangle \frac{\partial U}{\partial y}\,\mathrm{d}y, \quad \langle\!\langle \varepsilon \rangle\!\rangle = \frac{1}{L_y}\int_0^{L_y} 2\nu \langle S_{ij} S_{ij} \rangle\,\mathrm{d}y, \tag{5}$$

in the minimal and full channels, where S_{ij} is the rate-of-strain tensor, and $\langle\cdot\rangle$ indicates averages in the streamwise and spanwise directions in the domain of integration.

The production and dissipation are calculated at each timestep for the MFU and the different instances of wavepackets in the FC. The joint PDF of those two quantities is shown in figure 4(a,b) for the two cases. The PDF is compiled as a histogram over 50×50 bins, and the arrows represent the mean evolution velocity of all the states within a particular bin. These results are consistent with those observed in [38]. We also observe that the mean evolution velocities of the MFU and FC are similar; however, the PDF of the FC has a larger spread of production and dissipation values than the MFU, indicating a much more intense production and dissipation events in the FC.

The production-dissipation space is then divided into quadrants following [38], with the origin at the mean production and dissipation, and the axes determined by the principal axes of the PDF. Although the FC has different mean production-dissipation values and principal axes compared to the MFU, the results were not sensitive to the choice of the mean values and principal axes, so the quadrants based on the MFU were used for consistency.

The probability of moving from one quadrant to another, referred to as the transition probability, was determined following the method of [27] and is shown in figure 4(c,d). The transition probabilities only consider the relative probability of moving to different quadrants and do not consider the probability of staying in the same quadrant. Interestingly, figure 4(c,d) show a high probability of moving from *I* to *III* and from *IV* to *II*, whereas (a,b) do not show arrows pointing in those directions. This is not necessarily a contradiction because figure 4(c,d) consider only the probability of moving from one quadrant to the other, while figure 4(a,b) also consider the speed of evolution of the state. To quantify the error between transition probabilities in the MFU and FC, the transition probabilities are assembled into a vector, and the error is defined as the norm of the difference in the probability vectors between the MFU and FC.

It was noted that the choice of advection speed has a significant effect on the calculated quantities for the full channel. To determine the best advection method, we compare the production–dissipation plot for the MFU and FC using different advection methods and compare the transition probabilities. The advection velocity that minimized error was the average velocity

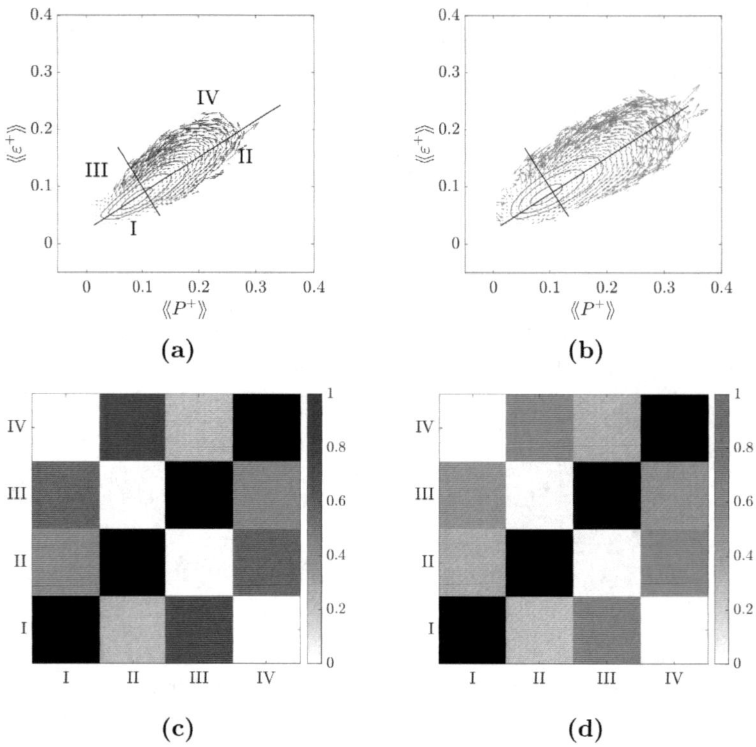

Figure 4: Production-dissipation of the (a) MFU and (b) FC. Contours are PDF at 10%, 40%, and 70% of occurrences, and arrows show the mean trajectory computed by taking averages over 50 × 50 bins. Perpendicular lines show quadrant divisions. (c) Relative transition probability between quadrants for the (c) MFU and (d) FC. Horizontal entries are the initial quadrant, and vertical entries are the next quadrant.

of only the regions of the flow contained in low-speed streaks, defined as

$$\{(x,y,z) : \sqrt{u^2 + w^2} > \alpha u_\tau \quad \& \quad u < 0\}, \tag{6}$$

at $y^+ = 40$, with $\alpha = 3.4$ [20]. Other definitions of advection velocities were considered, including average velocity over the whole box, average velocity at several y-locations, and average velocity conditioned to negative u-fluctuations, but all had higher errors in the production-dissipation cycle. Interestingly, [18] shows that these structures advect with the mean velocity at $y^+ = 40$. However, in the case of production-dissipation cycles, this velocity resulted in higher error than the velocity conditioned to streaks at $y^+ = 40$. For the remainder of the section, we use the definition given by equation (6). This leads to an average advection velocity in the streamwise direction of $10.7u_\tau$ which is similar, but not exactly equal to, the advection velocity of slow streaks extracted in the FC using coherent structure tracking as presented in table 3. Note that the advection velocity reported in table 3 is of the volumetric centroid of the streak, i.e., its group velocity, rather than the average of the velocity field occupied by the region of the slow streak only at $y^+ = 40$.

4.3. Temporal motif of energy contained in streak, roll, and small scales

Additionally, we track energy in eddies related to the self-sustaining process. The magnitude of the streaks is defined by the magnitude of the streamwise velocity captured in the $(k_x, k_z) \equiv$

$(k_x^* L_x/(2\pi), k_z^* L_z/(2\pi)) = (0,1)$ mode of the MFU domain,

$$\langle\langle \hat{u}_{0,1} \rangle\rangle = \frac{1}{L_y} \int_0^{L_y} (\hat{u}(k_x = 0, k_z = 1)^2)^{1/2} \, \mathrm{d}y, \tag{7}$$

where k_x^* and k_z^* are the dimensional wavenumbers. Similarly, the magnitude of the rolls is defined by the wall-normal and spanwise velocity magnitude of the $(k_x, k_z) = (0,1)$ mode of the MFU domain,

$$\langle\langle \hat{r} \rangle\rangle = \frac{1}{L_y} \int_0^{L_y} \left(\hat{v}(k_x = 0, k_z = 1)^2 + \hat{w}(k_x = 0, k_z = 1)^2\right)^{1/2} \mathrm{d}y. \tag{8}$$

Finally, the magnitude of the small scales is defined as the sum of all velocity magnitude contained in smaller scales,

$$\langle\langle \hat{s} \rangle\rangle = \frac{1}{L_y} \int_0^{L_y} \left(\sum_{(k_x,k_z)\neq(0,0),(0,\pm1)} \hat{u}(k_x, k_z)^2 + \hat{v}(k_x, k_z)^2 + \hat{w}(k_x, k_z)^2 \right)^{1/2} \mathrm{d}y. \tag{9}$$

The definitions are consistent with the definition of a streak and roll in an MFU, where a single pair of infinitely long low- and high-speed streaks are present.

The magnitude of streaks, rolls, and small scales are analyzed similarly to production and dissipation. Distribution of the quantities and average trajectories are shown in figure 5(a,b). Since three quantities were considered, the phase space was divided into eight octants, with the origin at the mean value of all three for the MFU. Principal axes of the data were identified, but they were very closely aligned with the Cartesian axes, so Cartesian axes were used to divide the space into octants, as shown in figure 5(a,b). The octants were named with a 3-digit binary key, with 0 for below-average magnitude and 1 for above-average magnitude. The order of entries is (streaks, rolls, small scales), so for example, octant $(0,1,1)$ represents low magnitude contained in streaks, high magnitude contained in rolls, and high magnitude contained in small scales. Transition probabilities were calculated between the octants, shown in 5(c,d). Generally, transition probabilities in the MFU and FC have many similarities. The MFU has a larger spread of probability values, and it has more asymmetry in the direction of movement between octants. For example, $(0,1,0) \to (0,0,0)$ has a significantly higher probability than $(0,0,0) \to (0,1,0)$, but in the FC, both directions have a similar probability.

We then form a graph using the trajectories. Each time the trajectory enters a new octant, it is defined as an event. The PDF of the duration of single events is shown in 6(a). Events in the MFU had longer durations than the FC, with the average duration of an MFU event being 2.21 times as long as the average FC event. This value is close to the ratio of lifetimes of the spatiotemporal motifs reported in table 3 of 2.83 and is even closer to the ratio of timescales implied by the advection velocities of the slow streaks between the two flows of approximately 2.17. The trajectory can be recast as a series of events, which then form the graph, where each event is a node and the transition from one event to the next event is an edge between the two nodes. Some events occurred for short durations, especially in the FC, so events lasting $\Delta t^+ \leq 5$ were removed, and the trajectory was reassembled. This corresponds to the region of figure 6(a) where the duration of events in the FC started to plateau. It was found that the motif results were not sensitive to a cutoff value of $\Delta t^+ = 10$. The resulting graph sizes were 17,952 events for the MFU and 90,579 events for the FC.

The graphs were analyzed for motifs following the method of [23]. Because each node in the graph had one incoming and one outgoing connection, the method could be simplified and was much less computationally expensive than the method described in section 3. Due to the restricted nature of the trajectories, there were 8^n possible isomorphic subgraphs of n nodes.

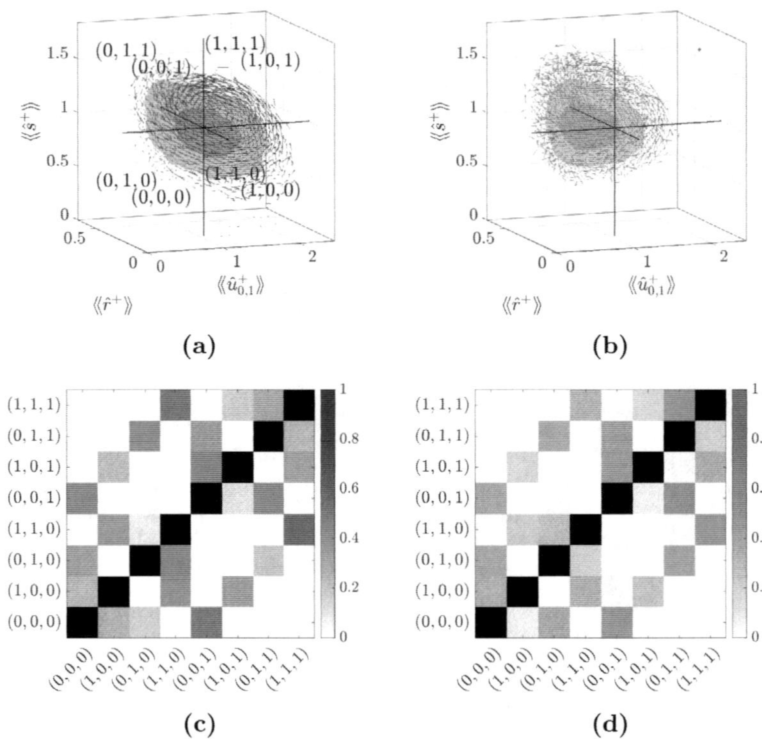

Figure 5: Magnitude of the streak, roll, and small scales for the (a) MFU and (b) FC. Contours are at 10% and 50% of occurrences, and arrows show the mean trajectory. Black lines show octant divisions. Relative transition probability between octants for the (c) MFU and (d) FC. Horizontal entries are the initial octant, and vertical entries are the next octant.

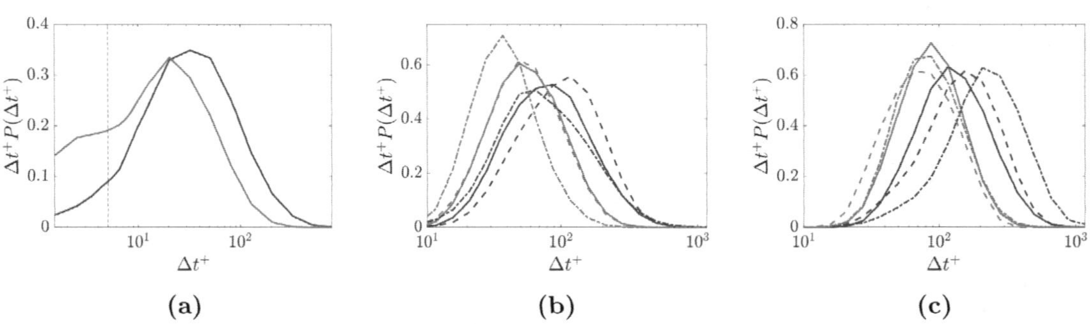

Figure 6: (a) PDF of duration of single events for the MFU (blue) and FC (red). The vertical line at $\Delta t^+ = 5$ indicates the filter cutoff. (b) PDF of duration of two-event subgraphs for the MFU (blue) and FC (red). Solid lines represent all two-event subgraphs, dashed lines represent the $(1,1,0) \rightarrow (1,1,1)$ motif, and dot-dashed lines represent the $(1,0,0) \rightarrow (1,0,1)$ motif. (c) PDF of duration of three-node subgraphs for the MFU (blue) and FC (red). Solid lines represent all three-event subgraphs, dashed lines represent the $(0,0,1) \rightarrow (1,0,1) \rightarrow (1,0,0)$ motif, and dot-dashed lines represent the $(0,0,0) \rightarrow (1,0,0) \rightarrow (1,1,0)$ motif.

To analyze the trajectory for n-node motifs, the trajectory was broken down into sequences of n event-induced subgraphs, each classified as one of the 8^n isomorphic subgraphs. The number of times each isomorphic subgraph occurred was then tabulated. The significance of each of the induced subgraphs were computed following equation 4.

The first five 2-node and 3-node subgraphs with the highest significance are reported in tables 5 and 6. Generally, motifs in the MFU had a higher significance $\widetilde{W}(s)$ compared to the FC. This was expected since the MFU is more dynamically restricted, so the dynamics were frequently repeated.

When considering 2-node motifs, all identified motifs in the MFU, and three of the top five motifs in the FC, involve changing energy in the small scales. subgraphs $(1,1,0) \to (1,1,1)$ and $(1,0,0) \to (1,0,1)$ were among the five most significant motifs in both the MFU and FC. These motifs represent physical phenomena, with the flow initially containing high energy in large-scale structures and moving to a state with energy in both large and small-scale structures, reflective of the energy cascade.

The 2-node motifs provide insight into the directionality of the trajectories. For example, in the MFU, $(0,0,0) \to (0,0,1)$ and $(0,0,1) \to (0,0,0)$ have very similar significance values, which suggests that both directions have similar importance to the dynamics. However, $(0,1,1) \to (0,1,0)$ has much higher significance than $(0,1,0) \to (0,1,1)$, with significance scores $\widetilde{W}_{(0,1,1)\to(0,1,0)} = 0.3276$ and $\widetilde{W}_{(0,1,0)\to(0,1,1)} = 0.0236$. This matches the physical understanding of the SSP because $(0,1,1) \to (0,1,0)$ reflects the step in the SSP where streamwise rolls are formed from small-scale structures by a nonlinear process. $(0,1,0) \to (0,1,1)$ is not understood as a part of the SSP, so it has a lower significance. Interestingly, there are variations in directionality between the MFU and FC. For example, in the FC, $(1,0,0) \to (1,0,1)$ and $(1,0,1) \to (1,0,0)$ are similar, with

$$\frac{\widetilde{W}_{(1,0,0)\to(1,0,1)}}{\widetilde{W}_{(1,0,1)\to(1,0,0)}} = 1.95.$$

In the MFU, $(1,0,0) \to (1,0,1)$ is much more significant than $(1,0,1) \to (1,0,0)$, with

$$\frac{\widetilde{W}_{(1,1,0)\to(1,1,1)}}{\widetilde{W}_{(1,1,1)\to(1,1,0)}} = 4.54.$$

When considering the ratios of the number of occurrences, rather than the significance, similar values are found. This suggests that the directionality of the movement between these octants is more significant in the MFU than the FC.

When considering 3-node motifs, several very rare subgraphs had high significance because they were also very rare in the randomly generated networks. Therefore, we did not report subgraphs that occurred fewer than 40 times. These rare subgraphs tended to contain diagonal movements or back-and-forth movements between octants. The five subgraphs with the highest significance meeting the occurrence threshold are reported in table 6. Again, there is strong similarity between the MFU and FC, and the MFU tended to have higher significance values.

Motifs $(0,0,1) \to (1,0,1) \to (1,0,0)$ and $(0,0,0) \to (1,0,0) \to (1,1,0)$ were among the five most significant motifs in both the MFU and FC. While most subgraphs had similar significance rankings between the MFU and FC, there were a few that were very different. For example, $(0,0,0) \to (0,0,1) \to (1,0,1)$ had the third highest significance in the MFU, with $\widetilde{W}_{(0,0,0)\to(0,0,1)\to(1,0,1)} = 0.0410$. However in the FC, $(0,0,0) \to (0,0,1) \to (1,0,1)$ was the 349th most significant subgraph, with a negative significance value of $\widetilde{W}_{(0,0,0)\to(0,0,1)\to(1,0,1)} = -.0013$. This suggests that there are dynamical patterns that are very significant to the MFU but not the FC. This is similar to the re-ordering of the most significant, as well as most common, yet dominant, spatiotemporal motifs in table 2. However, in general, the similarities between the motifs indicate that there is significant dynamical similarity between the MFU and FC.

Table 5: Five most significant 2-node motifs. Bold-face motifs are common between the MFU and FC. $\widetilde{W}(s)$ is the significance score, and Occ., or occurrence, is the number of times the subgraph occurs in the real trajectory.

Minimal Flow Unit			Full Channel		
s	$\widetilde{W}(s)$	Occ.	s	$\widetilde{W}(s)$	Occ.
$\mathbf{(1,1,0) \to (1,1,1)}$	0.370	1669	$\mathbf{(1,1,0) \to (1,1,1)}$	0.326	2664
$(0,1,1) \to (0,1,0)$	0.328	816	$(1,1,0) \to (1,0,0)$	0.244	1181
$(0,0,0) \to (0,0,1)$	0.297	1109	$(1,1,1) \to (1,1,0)$	0.229	2024
$(0,0,1) \to (0,1,1)$	0.290	1141	$\mathbf{(1,0,0) \to (1,0,1)}$	0.191	526
$\mathbf{(1,0,0) \to (1,0,1)}$	0.279	610	$(1,1,1) \to (1,0,1)$	0.178	888

The duration of significant motifs was compared to the duration of all subgraphs taken together, and the results are shown in figure 6(b,c). Duration was determined by summing the duration of the events making up the subgraph. In particular, motifs that were among the top five most significant in both the MFU and FC were considered. For two node motifs, this was $(1,1,0) \to (1,1,1)$ and $(1,0,0) \to (1,0,1)$, and for three node motifs, this was $(0,0,1) \to (1,0,1) \to (1,0,0)$ and $(0,0,0) \to (1,0,0) \to (1,1,0)$. The FC tended to have faster times than the MFU, and this was seen in both the duration of all events, and the duration of individual subgraphs. Even though the same subgraphs were occurring with similar significance values in the MFU and FC, they tended to have shorter durations in the FC. Further, the durations of the significant subgraphs did not scale proportionally between the MFU and FC. For example, $(0,0,1) \to (1,0,1) \to (1,0,0)$ was generally slower than the background turbulence (represented by the solid lines in figure 6) in the FC, but faster than the background turbulence in the MFU.

Table 6: Five most significant 3-node motifs, excluding those with fewer than 12 occurrences. Bold-face motifs are common between the MFU and FC. $\widetilde{W}(s)$ is the significance score, and Occ., or occurrence, is the number of times the subgraph occurs in the real trajectory.

Minimal Flow Unit			Full Channel		
s	$\widetilde{W}(s)$	Occ.	s	$\widetilde{W}(s)$	Occ.
$\mathbf{(0,0,1) \to (1,0,1) \to (1,0,0)}$	0.0466	79	$(0,1,1) \to (1,1,1) \to (1,1,0)$	0.0178	910
$(0,1,0) \to (1,1,0) \to (1,0,0)$	0.0432	283	$\mathbf{(0,0,0) \to (1,0,0) \to (1,1,0)}$	0.0167	282
$(0,0,0) \to (0,0,1) \to (1,0,1)$	0.0410	138	$(1,0,1) \to (1,1,1) \to (1,0,1)$	0.0156	127
$(1,1,1) \to (1,1,0) \to (0,1,0)$	0.0305	89	$\mathbf{(0,0,1) \to (1,0,1) \to (1,0,0)}$	0.0153	106
$\mathbf{(0,0,0) \to (1,0,0) \to (1,1,0)}$	0.0296	261	$(0,0,1) \to (1,0,1) \to (1,1,1)$	0.0132	206

5. Conclusions

Interpreting wall-bounded turbulence as a collection of interacting coherent structures is a long-standing paradigm. However, it needs a consistent framework or an algorithmic approach to extract and analyze important dynamical events. In this work, we attempt to do so by employing coherent-structure tracking, semi-Lagrangian wavepackets, and analysis techniques from network science and graph theory. As a proof of concept, we had two aims. First, to extract the signature of the self-sustaining process algorithmically. Second, to analyze the dynamics of the extracted self-sustaining process in both a minimal flow unit and a large unrestricted domain to understand how they differ.

To do so, we first extended the coherent structure tracking algorithms of [18] to allow for simultaneously tracking multiple sets of eddies. In this study, we limited ourselves to tracking attached ejections, sweeps, and slow and fast streaks simultaneously. The spatiotemporal interactions between these eddies are encoded into a complex multilayer network, which is then searched for significantly non-random patterns given an appropriate model for a random network, referred to as network motifs [23]. This analysis extracted the roll-streak pairing as the most significant and perhaps simplest non-random dynamical pattern in both flows, satisfying the first goal. However, it was detected that other significant, as well as frequent, yet dominant patterns were ordered differently between the two flows. This indicates that while the most prominent dynamical mechanism is the same between the MFU and the FC, there are others that are substantially different in both frequency and significance between the two flows. Furthermore, even for the most significant dynamical pattern, $\mathcal{S}_1^{\{F,M\}}$, the average timescale was 2.83 longer in the MFU than the FC.

We simultaneously employed semi-Lagrangian wavepackets to analyze the phase-space dynamics of entire MFU-sized regions advecting within the FC, and compared them to those of the MFU. Within the octant space defined by streak, roll, and small-scale magnitude, graphs were constructed from the sequence of transition events from one octant to the next. Overall, it was found that both the 2-node and 3-node motifs were mostly similar between the MFU and the FC, with 3-node motifs representing transitions from one scale to another. However, a few key differences were found. First, the significance scores, $\widetilde{W}(s)$, were higher for the MFU compared to FC, presumably due to the restricted dynamics of the MFU. Second, the average timescales of each motif were once again faster in the FC compared to MFU, with the average duration of an MFU event being 2.21 slower than that of an FC event, consistent with the motif lifetimes extracted using spatiotemporal tracking, and in particular with the implied ratio of timescales for the slow speed streaks of 2.17.

Overall, an algorithmic multi-eddy-type spatiotemporal tracking approach to detecting key dynamical mechanisms in wall-bounded turbulence was proposed, developed, and tested on a control experiment. It was found that the roll-streak SSP is the lowest form of spatiotemporal organization, and that its timescale in the MFU is substantially slower compared to its unrestricted counterpart. This key finding was corroborated with a semi-Lagrangian wavepacket approach combined with temporal motif extraction in phase space. In the future, we aim to extend this approach to detect key interscale dynamics at higher Re_τ, and extract higher-order mechanisms by comparing to more complex random network models that constrain more than just the colored-degree distribution.

Acknowledgments
A. Elnahhas and P. Moin acknowledge support from NASA's Transformational Tools and Technologies project under grant no. 80NSSC20M0201. E. Lenz acknowledges support from the NSF Graduate Research Fellowship under grant no. DGE-1745301. All authors acknowledge insightful comments from José Cardesa on an earlier version of the manuscript. All authors acknowledge the European Research Council (ERC) under the Caust grant ERC-AdG-101018287, and Prof. J. Jiménez for the participation in the 2023 Madrid Summer Turbulence Workshop.

References

[1] Robinson S K 1991 Coherent motions in the turbulent boundary layer *Annu. Rev. Fluid Mech.* **23** 601–639
[2] Townsend A A 1976 *The Structure of Turbulent Shear Flow* (Cambridge, UK: Cambridge University Press)
[3] Marusic I and Monty J P 2019 Attached eddy model of wall turbulence *Annu. Rev. Fluid Mech.* **51** 49–74
[4] Jiménez J and Moin P 1991 The minimal flow unit in near-wall turbulence *J. Fluid Mech.* **225** 213–240
[5] Hamilton J M, Kim J and Waleffe F 1995 Regeneration mechanisms of near-wall turbulence structures *J. Fluid Mech.* **287** 317–348
[6] Jiménez J and Pinelli A 1999 The autonomous cycle of near-wall turbulence *J. Fluid Mech.* **389** 335–359
[7] Flores O and Jiménez J 2010 Hierarchy of minimal flow units in the logarithmic layer *Phys. Fluids* **22** 71704
[8] Hwang Y and Cossu C 2010 Self-sustained process at large scales in turbulent channel flow *Phys. Rev. Letters* **105** 044505
[9] Hwang Y and Cossu C 2011 Self-sustained processes in the logarithmic layer of turbulent channel flows *Phys. Fluids* **23** 061702
[10] Bae H J and Lozano-Durán A 2019 A minimal flow unit of the logarithmic layer in the absence of near-wall eddies and large scales *Annu. Res. Briefs*, Center for Turbulence Research, Stanford University, 247–256
[11] Bae H J, Lozano-Durán A and McKeon B 2021 Nonlinear mechanism of the self-sustaining process in the buffer and logarithmic layer of wall-bounded flows *J. Fluid Mech.* **914** A3
[12] Lozano-Durán A, Bae H J and Encinar M P 2020 Causality of energy-containing eddies in wall turbulence *J. Fluid Mech.* **882** A2
[13] Hutchins N and Marusic I 2007 Evidence of very long meandering features in the logarithmic region of turbulent boundary layers *J. Fluid Mech.* **579** 1–28
[14] Hutchins N and Marusic I 2007 Large-scale influences in near-wall turbulence *Phil. Trans. Royal Soc. A: Math. Phys. Eng. Sci.* **365** 647–664
[15] Marusic I, Mathis R and Hutchins N 2010 Predictive model for wall-bounded turbulent flow *Science* **329** 193–196
[16] Zhou Z, Xu C X and Jiménez J 2022 Interaction between near-wall streaks and large-scale motions in turbulent channel flows *J. Fluid Mech.* **940** A23
[17] Lozano-Durán A, Flores O and Jiménez J 2012 The three-dimensional structure of momentum transfer in turbulent channels *J. Fluid Mech.* **694** 100–130
[18] Lozano-Durán A and Jiménez J 2014 Time-resolved evolution of coherent structures in turbulent channels: characterization of eddies and cascades *J. Fluid Mech.* **759** 432–471
[19] Dong S, Huang Y, Yuan X and Lozano-Durán A 2020 The coherent structure of the kinetic energy transfer in shear turbulence *J. Fluid Mech.* **892** A22
[20] Bae H J and Lee M 2021 Life cycle of streaks in the buffer layer of wall-bounded turbulence *Phys. Rev. Fluids* **6** 064603
[21] Boccaletti S, Bianconi G, Criado R, Del Genio C I, Gómez-Gardeñes J, Romance M, Sendiña-Nadal I, Wang Z and Zanin M 2014 The structure and dynamics of multilayer networks *Phys. Reports* **544** 1–122
[22] Kivelä M, Arenas A, Barthelemy M, Gleeson J P, Moreno Y, Porter A M and Estrada E 2014 Multilayer networks *J. Comp. Networks* **2** 203–271
[23] Milo R, Shen-Orr S, Itzkovitz S, Kashtan N, Chklovskii D and Alon U 2002 Network motifs: Simple building blocks of complex networks *Science* **298** 824–827
[24] Ribeiro P, Paredes P, Silve M E P, Aparicio D and Silva F 2021 A survey on subgraph counting *ACM Comp. Surveys* **54** A28
[25] Alon U 2007 Network motifs: theory and experimental approaches *Nature Rev. Gen.* **8** 450–461
[26] Taira K, Nair A G and Brunton S L 2016 Network structure of two-dimensional decaying isotropic turbulence *J. Fluid Mech.* **795** R2
[27] Schmid P J, García-Gutierrez A and Jiménez J 2018 Description and detection of burst events in turbulent flows *J. Phys.: Conf. Ser.* **1001**, Third Madrid Summer School for Turbulence, 012015
[28] Iacobello G, Scarsoglio S, Kuerten J G M and Ridolfi L 2019 Lagrangian network analysis of turbulent mixing *J. Fluid Mech.* **865** 546–562
[29] Krueger P S, Hahsler M, Olinick E V, Williams S H and Zharfa M 2019 Quantitative classification of vortical flows based on topological features using graph matching *Proc. Royal Soc. A: Math. Phys. Eng. Sci.* **475** A8
[30] Fernex D, Noack B R and Semaan R 2021 Cluster-based network modeling: From snapshots to complex dynamical systems *Science Adv.* **7** 5006–5022
[31] Iacobello G, Ridolfi L and Scarsoglio S 2021 Large-to-small scale frequency modulation analysis in wall-bounded turbulence via visibility networks *J. Fluid Mech.* **918** A13
[32] Li H, Fernex D, Semaan R, Tan J, Morzyński M and Noack B R 2021 Cluster-based network model *J. Fluid Mech.* **906** A21

[33] Perrone D, Kuerten J G M, Ridolfi L and Scarsoglio S 2021 Network analysis of Reynolds number scaling in wall-bounded lagrangian mixing *Phys. Rev. Fluids* **6** 124501

[34] Taira K and Nair A G 2022 Network-based analysis of fluid flows: Progress and outlook *Prog. Aero. Sci.* **131** 100823

[35] Wernicke S and Rasche F FANMOD: a tool for fast network motif detection *Bioinformatics.* **22** 1152–1153

[36] Wernicke S Efficient detection of network motifs *IEEE/ACM trans. comp. biology & bioinformatics* **3** 347–359

[37] Ribeiro P and Silva F 2014 Discovering colored network motifs *Complex Networks V: Proceedings of the 5th Workshop on Complex Networks* 107–118

[38] Jiménez J, Kawahara G, Simens M P, Nagata M and Shiba M 2005 Characterization of near-wall turbulence in terms of equilibrium and "bursting" solutions *Phys. Fluids* **17** 015105

[39] Benson A R, Gleich D F and Leskovec J 2016 Higher-order organization of complex networks *Science*, **353** 163–166

Turbulent activity in the near-wall region of adverse pressure gradient turbulent boundary layers

Taygun R Gungor[1], Yvan Maciel[1], Ayse G Gungor[2]

[1] Department of Mechanical Engineering, Université Laval, Quebec City, QC, G1V 0A6 Canada

[2] Faculty of Aeronautics and Astronautics, Istanbul Technical University, 34469 Maslak, Istanbul, Turkey

E-mail: taygun-recep.gungor.1@ulaval.ca

Abstract.

Two direct numerical simulation (DNS) databases are investigated to understand the effect of the outer-layer turbulence on the inner layer's structures and energy transfer mechanisms. The first DNS database is the non-equilibrium adverse-pressure-gradient (APG) turbulence boundary layer (TBL) of Gungor et al. [1]. Its Reynolds number and the inner-layer pressure gradient parameter reach above 8000 and 10, respectively. The shape factor spans between 1.4 and 3.3, which indicates the flow has various velocity defect situations. The second database is the same flow as the first one but the outer layer turbulence is artificially eliminated in this flow. Turbulence is removed above 0.15 local boundary layer thickness. For the analysis, we chose four streamwise positions with small, moderate, large, and very-large velocity defect. We compare the wall-normal distribution of Reynolds stresses, two-point correlations and spectral distributions of energy, production and pressure strain. The results show that the inner layer turbulence can sustain itself when the outer-layer turbulence does not exist regardless of the velocity defect or the pressure gradient. The two-point correlations of both cases show that outer large-scale structures affect the inner layer structures significantly. The streamwise extent of the correlation contours scales with pressure-viscous units. This shows the importance of the pressure gradient's effect on the inner-layer structures. The spectral distributions demonstrate that the energy transfer mechanisms are probably the same in the inner layer regardless of the velocity defect, which suggests the near-wall cycle may exist even in very-large defect APG TBLs where the mean shear in the inner layer is considerably lower than small-defect APG TBLs.

1. Introduction

When a turbulent boundary layer (TBL) is exposed to an adverse pressure gradient (APG), it decelerates. A sustained and/or strong deceleration leads the mean velocity profile and consequently the mean shear profile to change dramatically. The mean shear increases significantly in the outer layer. As a result of this, the outer layer turbulence becomes dominant and an outer maximum for the streamwise Reynolds normal stress ($\langle u^2 \rangle$) emerges when the defect is large enough [2, 3]. In contrast, the mean shear is attenuated in the inner layer and therefore the inner-layer turbulent activity becomes weaker and loses its importance [2, 4]. An important outcome is that the well-known inner-peak of the streamwise Reynolds stresses vanishes in large defect APG TBLs [5, 3].

In the inner layer, the turbulent structures are not affected heavily by the effect of APG when the defect is small [6, 7]. The spectral distributions of energy are very similar in canonical flows and small-defect APG TBLs [8, 1]. Contrary to this, the spatial features and spectral characteristics of structures become different in large-defect APG TBLs. The near-wall streaks are irregular and less streaky and furthermore, the sweeps/ejections are weaker in large-defect APG TBLs [9, 10, 11]. Moreover, the inner peak in the $\langle u^2 \rangle$-spectra does not exist [6, 1]. The near-wall cycle may therefore be modified or hidden among the footprints of the large-scale outer-layer structures that are energetic in large-defect TBLs.

Indeed, inner-layer structures in large-defect APG TBLs are affected heavily by the outer-layer energetic structures. The one-dimensional energy spectra show that most of the energy of the u-structures in the inner layer resides at the same wavelengths as outer-layer large-scale structures [8, 1]. Gungor et al. [1] demonstrated that even the two-dimensional energy spectra do not show any significant energy at wavelengths that are associated with the inner layer streaks (streamwise wavelength $\lambda_x^+ \approx 1000$ and spanwise wavelength $\lambda_z^+ \approx 120$, where superscript $^+$ refers to friction-viscous units). The most energetic structures in the inner-layer have very wide wavelengths ($\lambda_z^+ \approx 500$), which is very similar to outer-layer energetic structures. These findings suggest that the inner layer is dominated by imprints of large-scale outer-layer structures.

Besides the structures' characteristics, the existence of the near-wall cycle in large-defect APG TBLs is another crucial topic to address. To this date, however, there has been no conclusive evidence of whether the near-wall cycle in large-defect APG TBLs exists or not. Unfortunately, only a few studies focused on this issue. Jiao et al. [12] suggested that the classical near-wall cycle is weakened heavily or dissolved completely under the effect of APG. They found a new mechanism of near-wall turbulence production initiated by the wall-normal nonlinear transport of the outer wall-normal velocity fluctuations to the near-wall region. It is important to highlight that their study is based on a Couette flow and not a TBL. The inner-layer pressure-gradient parameter ($\beta_i = (\nu/(\rho u_\tau^3))(dP_w/dx)$, where P_w is the pressure at the wall, ν is viscosity and ρ is density) reaches up to 0.137 in their case but the Reynolds number is $Re_\tau \approx 57$ when $\beta_i = 0.137$. In contrast to this, Gungor et al. [1] suggested that the near-wall cycle still exists in large-defect APG TBLs even though the spectral signature of the streaks disappears. They hypothesized that the near-wall cycle or another self-sustaining process with similar features may exist without the streaks. Even in canonical flows, Jiménez [13] showed later that the near-wall cycle was possible without streaks and that the latter might be byproducts of the bursting process. Moreover, Gungor et al. [1] claimed that the inner-layer streaks may be still there but we simply cannot see or detect them because the large-scale outer-layer structures that dominate the flow in the inner layer hide the streaks and streaks' signature.

The present study aims to investigate the effect of the outer-layer turbulence on the inner-layer turbulent activity in APG TBLs. For this, we employ two direct numerical simulation (DNS) databases. The first one is a non-equilibrium APG TBL taken from the literature [1]. Its Reynolds number based on momentum thickness (Re_θ) and β_i reaches above 8000 and 10, respectively. The shape factor spans between 1.4 and 3.3. The second one is a novel non-equilibrium APG TBL designed based on the first one. This database's most important feature is that the outer-layer turbulence is artificially removed to investigate the inner layer without the effect of the outer layer. We compared the wall-normal distribution of Reynolds stresses, the two-point correlations and the spectral distributions of energy, production, and pressure-strain of these two cases.

2. Databases

The first database, generated using DNS, is described in detail in Ref.[1]. As mentioned above, it is a non-equilibrium APG TBL that is exposed to a strong APG. At the beginning of the domain, the flow is a ZPG TBL but the effect of the pressure gradient leads to an increasing

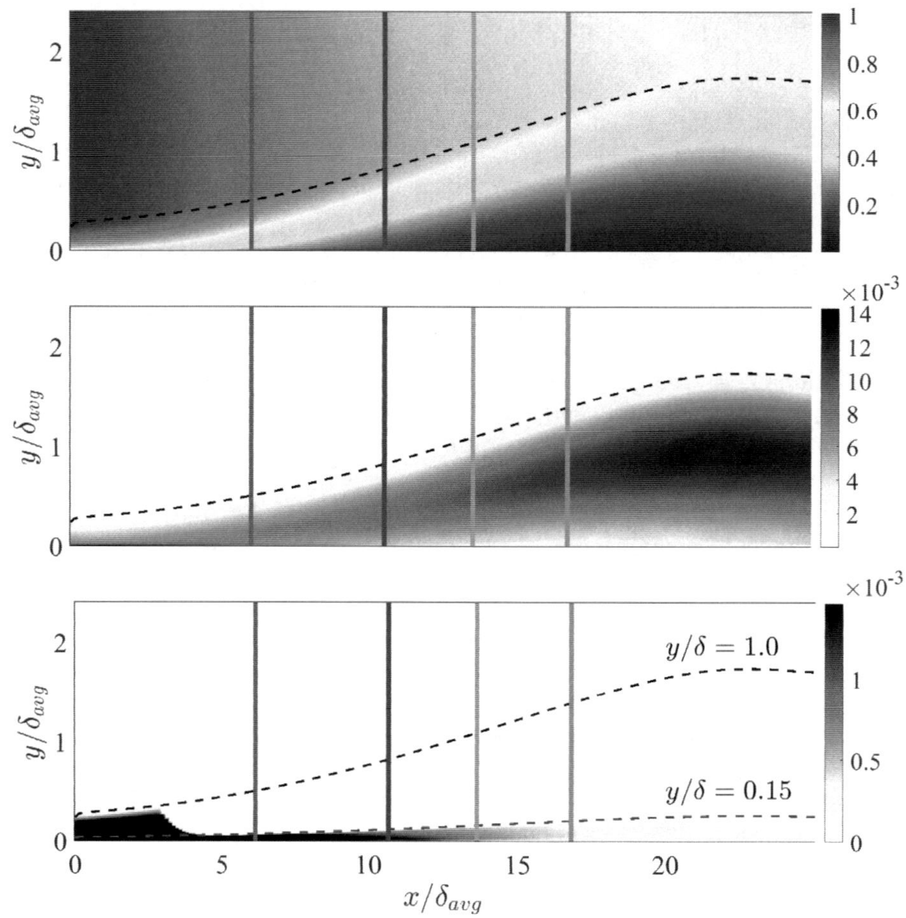

Figure 1. The streamwise distribution of the mean flow (top) and $\langle u^2 \rangle$ of the original case (middle) and the manipulated case (bottom) as a function of x and y. The axes are normalized with δ_{avg} and the levels are normalized with the edge velocity at the inlet. The straight lines represent the four streamwise positions (SP1 to SP4). The dashed lines indicate δ and 0.15δ. Only a part of the domain in the wall-normal direction is shown.

defect in the mean velocity profile. Towards the end of the domain, the flow evolves into a TBL at the verge of separation due to the APG. The Reynolds number based on momentum thickness (Re_θ) and the shape factor (H) span between 1400-8200 and 1.4-3.3, respectively and β_i reaches above 10. Figure 1(a, b) shows the mean velocity and $\langle u^2 \rangle$ as a function of x and y, where the axes are normalized with the boundary layer thickness averaged in the region of interest (δ_{avg}). The mean velocity distribution shows that the boundary layer decelerates and the boundary layer thickness (δ) grows in height as the flow develops. As for the turbulence, $\langle u^2 \rangle$ is high in the inner layer at the beginning of the domain. But this changes and the maximum turbulence is found in the outer layer as the defect increases.

The second APG TBL database is a manipulated APG TBL generated using the same DNS case mentioned above. The flow is basically the same as the previous one with a major exception. The outer layer turbulence is artificially removed above the inner layer in the zone that corresponds to the wake layer in canonical wall flows, that is above the overlap layer. There

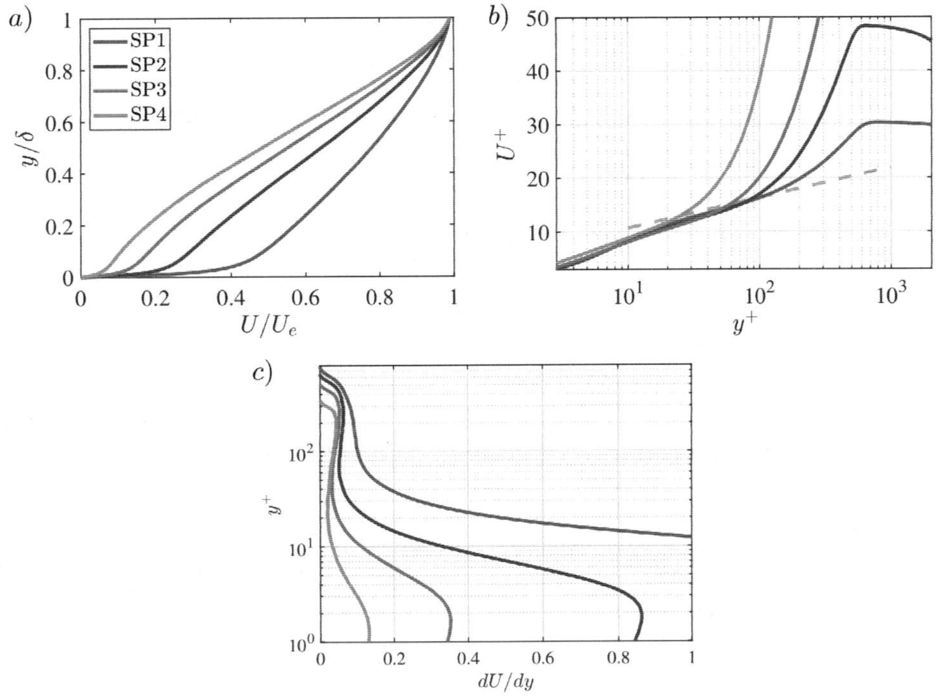

Figure 2. The mean velocity profiles as a function of y. The levels and axes are normalized with outer scales (a) and friction-viscous scales (b). The gray line is the logarithmic law with $\kappa = 0.41$ and $B = 5.1$. The mean shear profile as a function of y^+ (c). The mean shear is not normalized.

is no commonly accepted layer definition for APG TBLs [14]. Therefore, we have assumed the inner layer, the zone where viscous effects are important, to be between the wall and 0.15δ, where δ is the local boundary thickness. The fluctuations and therefore the Reynolds stresses are zero in the region above 0.15δ as will be discussed later in the paper.

The DNS code utilizes spectral discretization in the homogeneous spanwise direction (More information on the code and or the DNS setup can be found in references [15, 16, 3]). Therefore, the instantaneous flow field is stored as a function of x, y and k_z where k_z is the wavenumber in the spanwise direction. To eliminate turbulence in the outer layer, all spanwise modes with the exception of the zeroth mode, which is the mean flow, are set to zero at each time step as follows

$$\widehat{U}_i(x, y, k_z; y > 0.15\delta, k_z = 1 : n_z) = 0, \tag{1}$$

where n_z is the number of spanwise modes and $\widehat{U}_i$ is the Fourier transform of the velocity signal. The issue with eliminating turbulence is that once the turbulence is removed, the mean velocity changes significantly. However, we want to prevent this from happening because we want to keep both flows identical to isolate the effect of outer-layer turbulence as much as possible. To keep both flows identical, the mean velocity (averaged in time and spanwise direction) from the original case is imposed to the zeroth mode of the velocity of the manipulated case at each time step as

$$\widehat{U}_i(x, y, k_z; k_z = 0) = U_{i,original}(x, y) \tag{2}$$

Table 1. The main properties of the streamwise positions

Name	Re_θ	Re_τ	H	C_f	β_i
SP1o/SP1m	3004	645	1.65	0.00217	0.0278
SP2o/SP2m	4740	574	2.14	0.00087	0.0831
SP3o/SP3m	5796	458	2.63	0.00035	0.1928
SP4o/SP4m	6734	340	3.11	0.00013	0.6184
SPH293	6342	383	2.91	0.00019	0.3832
SPH300	6466	356	2.97	0.00016	0.5049

Figure 1(c) shows the spatial evolution of $\langle u^2 \rangle$ in the manipulated case as a function of x/δ_{avg} and y/δ_{avg}. The distribution shows the region where the outer layer turbulence activity is removed. The elimination of turbulence starts at approximately $3\delta_{avg}$ at the inlet to avoid some numerical issues.

We choose four streamwise positions for comparison of these two cases, as shown with straight lines in figure 1. The main parameters of these four stations are given in Table 1. These positions correspond to small, moderate, large and very-large velocity defect situations. Figure $2(a, b)$ presents the outer-scaled and inner-scaled mean velocity profiles as a function of wall-normal distance (y). The outer-scaled mean profiles demonstrate the dramatic changes in the mean velocity. Whereas the profile at SP1 resembles canonical wall flows, the profiles progressively change with increasing velocity defect. The inner-scaled profiles show that none of the flows follow the traditional logarithmic law. Despite this, there is a clear trend that the deviation from the logarithmic law increases with increasing velocity defect [1]. Figure $2(c)$ presents the mean shear profiles without any normalization because there are not appropriate scales for mean shear levels that we can use for both small- and large-defect APG TBLs. As the profiles indicate, the mean shear in the inner layer is inversely proportional to the velocity defect.

3. Results

3.1. The wall-normal distribution of Reynolds stresses

Here we consider the streamwise and tangential component of the Reynolds stresses. Figure $3(a, b, c, d)$ presents the inner-scaled Reynolds stress profiles as a function of y^+ for the original and manipulated cases. In addition, outer-scaled Reynolds-stress profiles are also given in figure $3(e, f)$ so that the differences between the original and manipulated cases can be seen more clearly. The profiles show that the elimination of turbulence activity in the outer layer is performed successfully. The boundary where the turbulence is removed is different in all stations because its location is chosen based on local δ and not the friction-viscous length scale. These results show that the inner-layer turbulence can sustain itself without any effect from outer turbulence. This does not necessarily mean that the canonical-flow near-wall cycle exists in the near-wall region. But it nonetheless shows that there is a mechanism sustaining the near-wall turbulence.

One interesting result is that the Reynolds stress levels have the same order of magnitude in the manipulated case when normalized with friction-viscous scales, as seen in figure $3(a, b)$. In contrast, in the original case, figure $3(c, d)$ shows that friction-viscous scales fail to scale Reynolds stresses even for moderate velocity defect cases. This result suggests that although the pressure force may affect the production of inner turbulence for large-defect TBLs, it is still the mean shear that drives it. But the situation would necessarily change for even larger defect cases, those very close to or at separation, where mean shear is almost null near the wall. It is rather difficult to compare maxima in the original flow because there is no inner peak

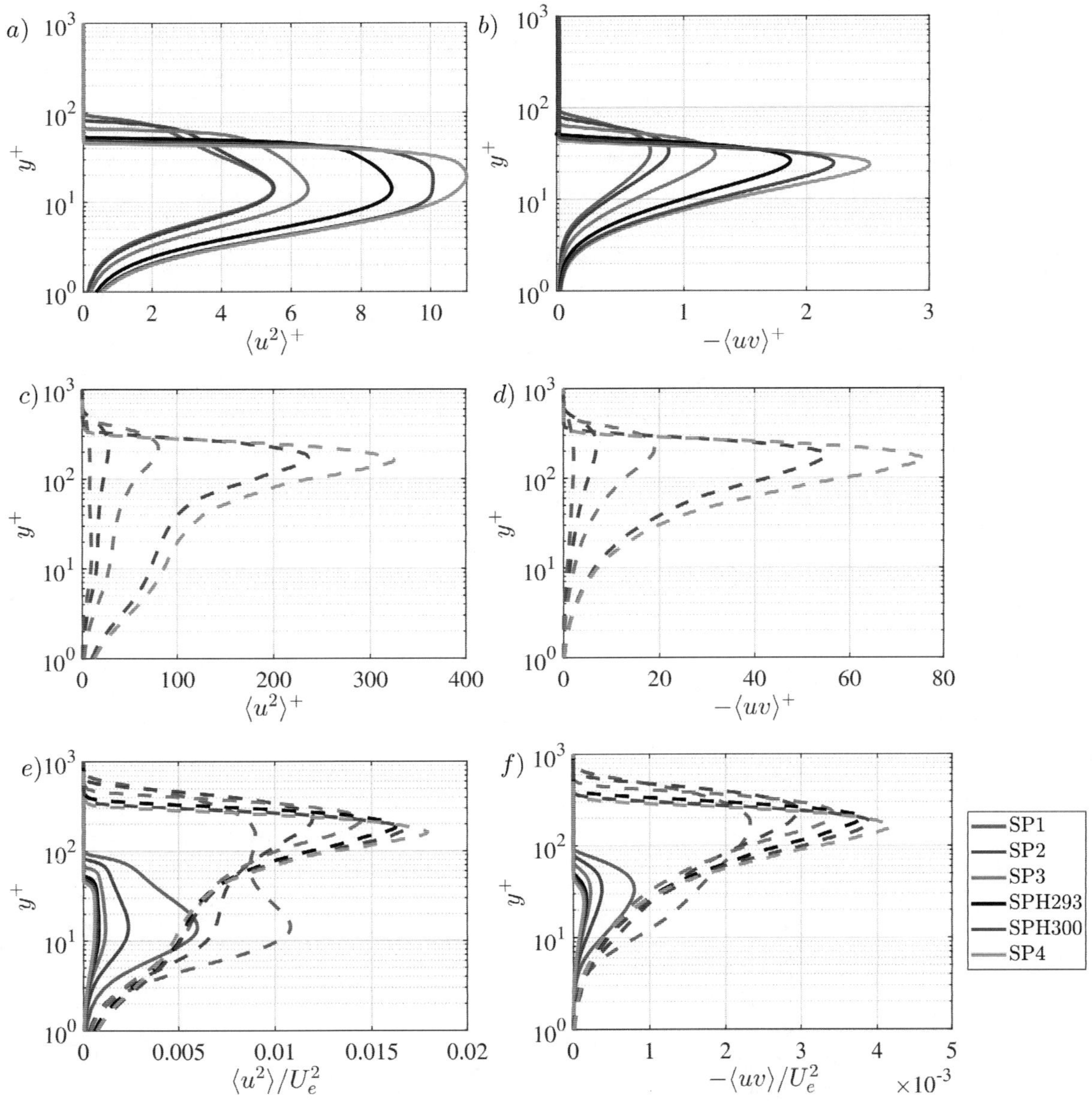

Figure 3. The $\langle u^2\rangle^+$ and $\langle uv\rangle^+$ profiles (a, b, c, d) and the $\langle u^2\rangle/U_e^2$ and $\langle uv\rangle/U_e^2$ profiles (e, f) as a function of y^+ for the four streamwise positions. The dashed and straight lines are for the original and manipulated cases, respectively.

except for the first position. However, a rough comparison between SP1o and SP2o, as shown in figure 3(c, d), shows that the levels of $\langle u^2\rangle^+$ increase from approximately 10 at SP1 to 16 at the plateau of SP2 whereas the levels are almost the same in SP1m and SP2m. This change of order of magnitude reflects the fact that the wall-normal transport of outer turbulence to the near-wall region is important for large-defect APG TBLs.

Besides the Reynolds stress levels, the location of the inner peak of the $\langle u^2\rangle^+$ profiles is the same in the first three positions of the manipulated case, $y^+ \approx 13$. Furthermore, the $\langle u^2\rangle^+$ profiles have similar shapes up to SP3, but start changing further downstream as the defect increases. To further examine this, we consider two additional streamwise positions: SPH293 and SPH300 where H is 2.93 and 2.97, and β_i is 0.38 and 0.50, respectively. It is clear that the profile keeps its shape at SPH293 but then a plateau replaces the peak at SPH300. A maximum appears again at $y^+ \approx 25$ at SP4. A similar trend of change exists for the $\langle uv \rangle^+$ and production of $\langle u^2 \rangle$ (not given here) profiles too. The peak changes from $y^+ \approx 35$ to 24 with increasing defect for $\langle uv \rangle$. As for the production, the production peak is located at $y^+ \approx 10$ for the first three positions of the manipulated case, but it is at $y^+ \approx 25$ in SP4 as consistent with the Reynolds-shear stresses. It is not clear if these differences in profiles stem from the fact that the turbulent/non-turbulent interface gets closer to the maximum of $\langle u^2 \rangle$ as the defect increases, thereby maybe influencing the turbulent structures.

3.2. Spatio-temporal velocity fields

Figure 4 shows the spatio-temporal evolution of streamwise velocity fluctuations at $y^+ \approx 13$ for all streamwise positions of the original and manipulated cases. We employ the temporal data that we collected at these positions and transform the data into a spatial representation using Taylor's frozen turbulence hypothesis with the local mean velocity. This way is preferred over the temporal representation because it is convenient to compare the physical length of structures. It is also important to state that the figures do not show the whole spanwise extent for the first three streamwise positions.

The spatio-temporal evolution in the original case shows that the flow changes considerably from SP1 to SP4. The structures' spanwise width increases significantly with increasing defect. Moreover, they become less streaky and more meandering, especially at SP3 and SP4. In contrast with the original case, the flow field does not change much in the manipulated case. The structures are more compact in large-defect cases, a consequence of using friction-viscous scales, but they are streaky regardless of the defect. As for comparing both cases, the flow fields are similar in the first position. This is expected considering that outer layer turbulence is still weak at SP1o. However, the flow fields become different from each other progressively as the defect increases from SP1 to SP4. In the last two positions, SP3 and SP4, the flow fields are completely different from each other. Neither the structures' features nor their size are similar. The massive difference at SP3/4 is expected because it was noted that the outer layer structures dominate the flow in large-defect APG TBLs [1]. What we are seeing in the original flow are the footprints of outer layer turbulence.

3.3. Two-point correlations

The two-point correlations are examined to investigate the spatial features of $\langle u^2 \rangle$- and $\langle uv \rangle$-carrying structures in the inner layer of APG TBLs. We compare the original and manipulated cases for the four streamwise positions, as described earlier. The two-point cross-correlation coefficient for two generic variables is defined as follows,

$$C_{ab}(\boldsymbol{r}, \boldsymbol{r}') = \frac{\langle a(\boldsymbol{r}) b(\boldsymbol{r}') \rangle}{\sigma_a(\boldsymbol{r}) \sigma_b(\boldsymbol{r}')} \tag{3}$$

where $\boldsymbol{r}$ is the reference point, $\boldsymbol{r}'$ is the moving point, a and b are generic variables, and σ is the standard deviation. For $\langle u^2 \rangle$- and $\langle uv \rangle$-carrying structures, we consider the auto-correlation of the streamwise component (C_{uu}) and the cross-correlation of u and v (C_{uv}). Figure 5 presents C_{uu} and C_{uv} for two cases at $y^+ \approx 13$, where the contour levels are 0.5, 0.1, and -0.1. It should be noted that the two-point correlation does not necessarily indicate the size or shape of the structures [17], but low-level contours are associated with large-scale structures.

The correlation contours are considerably different between the original and manipulated cases for C_{uu} and the difference is much more pronounced in large-defect cases, consistent with the spatio-temporal distributions of figure 4. The contours are both streamwise elongated at SP1

Journal of Physics: Conference Series **2753** (2024) 012005 doi:10.1088/1742-6596/2753/1/012005

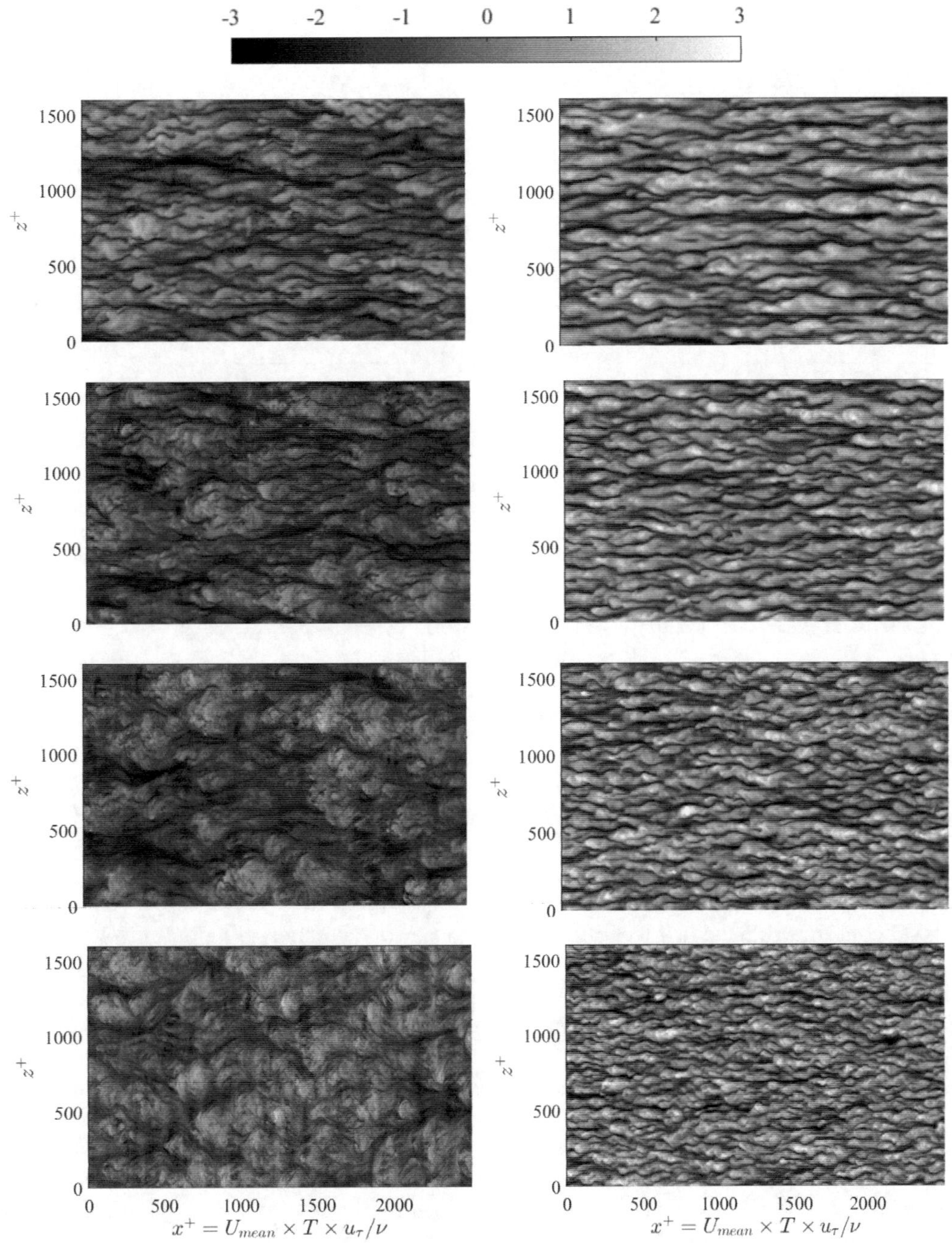

Figure 4. The spatio-temporal evolution of the streamwise velocity fluctuations at $y^+ \approx 13$ for the four streamwise positions of both cases as a function of time and z. The rows are for, from top to bottom, SP1, SP2, SP3, and SP4. The columns are for the original (left) and manipulated (right) cases. The time axis is transformed into x through Taylor's frozen turbulence hypothesis using the local mean velocity. Axes are normalized with friction-viscous scales. The fluctuations are normalized with the standard deviation of the fluctuations.

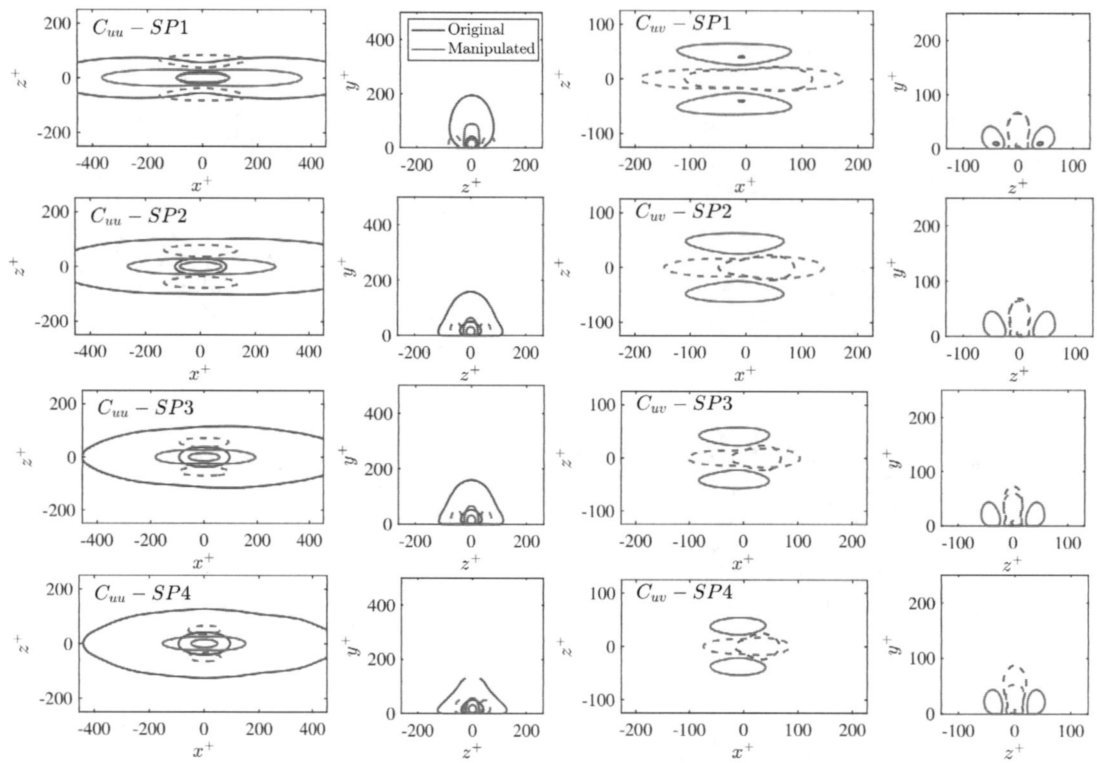

Figure 5. The two-point correlations C_{uu} (two columns at the left) and C_{uv} (two columns at the right) at $y^+ \approx 13$. The contour levels are for 0.5 and 0.1 for straight lines, and -0.1 for the dashed lines. The rows are for SP1, SP2, SP3, and SP4.

as seen in the xz-plane. The high-level contour (0.5) is almost identical in both cases. However, the low-level contours differ considerably from one case to the other. The extent of the contours is much larger in the original case than in the manipulated case. Both high and low-level contours become wider in the original case with increasing velocity defect. Furthermore, they are no longer streamwise elongated and the low-level contours lose their distinctive wasp shape that is narrow at the center. The correlation contours' shape becomes an oval. In contrast, the contours remain almost identical, with more minor variations in their size in the manipulated case.

The dramatic differences between the two cases exist in the zy-plane too. The shape of the contours is completely different in both cases, even at SP1 which is not the case for the contours on the xz-plane. As the defect increases, the contours of the original case change significantly and adopt a triangle-like shape where the contours are broader below the reference point and narrower above. This was already reported before for the same database for SP3 [18]. But here we also consider SP4 which has a considerably higher velocity defect. This triangle-like shape does not exist in small-defect APG TBLs or ZPG TBLs as reported before. Despite these major changes in the original case, the contours remain almost identical and conserve their distinctive shape in the manipulated case. They are completely different in both cases regardless of the streamwise positions. Furthermore, as is the xz-plane, the contours are much more compact in the manipulated case than in the original case.

In addition to the shape and size of the contours, the other major difference, maybe the most important one, is the presence of negative contours at both sides of the positive contours in the spanwise direction in the manipulated case. This spatial organization of the contours indicates the low- and high-speed streak couples. The two-point correlations show that the streak couples exist from SP1 to SP4. The streak couples have the expected distance of $\Delta z^+ \approx 50$ at SP1 but this distance diminishes at large defect cases. That streak couple exist regardless of the defect is an important finding since it shows that the streaks and therefore the near-wall cycle may exist even in flows with very-large velocity defect, where the mean shear is considerably lower than the one in canonical flows.

It should be noted that the lack of positive-negative contour couples in the original flow case does not indicate that the streak pairs are absent, because outer turbulence leaves its signature in the two-point correlations in this case. The positive-negative contour couples do not exist for ZPG TBL cases either [18]. However, it is well-known that the streaks and the near-wall cycle exist in the inner layer of canonical flows.

The reason for the differences in the size and shape of the contours between the two cases, as described above, is probably the interference of outer-layer structures in the inner layer. As discussed earlier, the outer layer turbulence becomes dominant with increasing velocity defect as figure 3 shows. Due to this, the imprints of large-scale outer-layer structures probably become more relevant in the inner layer in large-defect cases and therefore modify more importantly the coherency in the inner layer.

The differences for the cross-correlation of u and v between the two cases are milder than for C_{uu}. We start by discussing the meaning of cross-correlations because it is not as straightforward as auto-correlations. Sillero et al. [19] stated that C_{uv} is more like a shorter version of C_{uu} and C_{vu} is like a longer version of C_{vv}. This means the cross-correlation does not represent the $\langle uv\rangle$-carrying structures. The best interpretation of C_{uv}, as Sillero et al. [19] pointed out, is the extent of a streak associated with a sweep or ejection. This is even more useful for our analysis because it directly features the streaks, ejections and sweeps that are playing a role in the turbulence production while C_{uu} indicates the energy-carrying structures. The cross-correlation contours have the negative-positive couples for all defect cases in the manipulated case like the auto-correlation contours and retain some of its features.

As discussed above, the correlation contours conserve their shape in the manipulated case to some extent. Because of this, we can check the scaling of the correlation contours. Figure 6 presents C_{uu} and C_{uv} of the manipulated case with normalized using the friction-viscous length scale ($l^+ = \nu/u_\tau$) and pressure-viscous length scale ($l^{pi} = \nu/u^{pi}$, see equation 4 for u^{pi} where p is the pressure.).

$$u^{pi} = \left(\frac{\nu}{\rho}\frac{dp}{dx}\right)^{1/3} \tag{4}$$

The correlation contours of both C_{uu} and C_{uv} become smaller with increasing velocity defect when the axes are normalized with friction-viscous units. This is expected since u_τ is no longer a valid velocity scale for large-velocity defect cases (it is zero at separation). This is true even if we saw earlier on that friction-viscous scales can scale the position of the maximum of $\langle u^2\rangle$ at least up to the large-defect case of SP4, because these are two different scaling considerations. On the other hand, pressure-viscous units scale the streamwise extent of the contours for C_{uu} and C_{uv} very well. It shows that the structure's streamwise length is affected by the pressure gradient. But such a scaling cannot be used for ZPG TBLs since $u^{pi} = 0$.

3.4. Spectral distributions of energy, production and inter-component energy transfer

To further investigate the presence of the near-wall cycle, we examine energy transfer mechanisms through the spectral distributions of energy, production and pressure strain. Production is the extraction of energy from the mean flow to $\langle u^2\rangle$ and the pressure strain governs the inter-component energy transfer between the three normal components. Here, to simplify

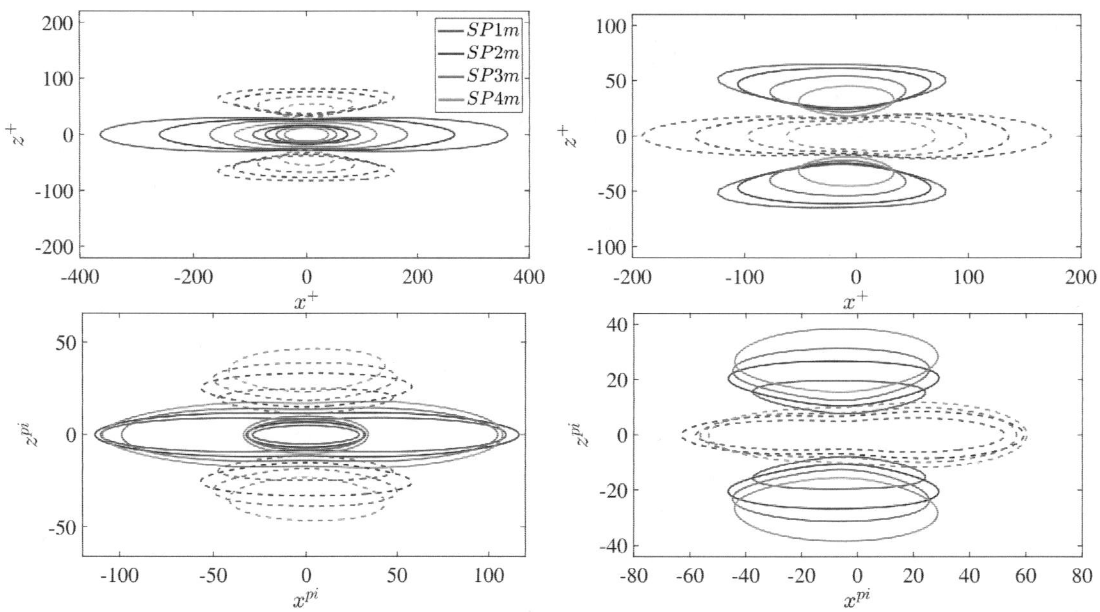

Figure 6. The two-point correlations C_{uu} (left) and C_{uv} (right) at $y^+ \approx 13$. Axes are normalized with friction-viscous (top) and pressure-viscous (bottom) units. The contour levels are for 0.5 and 0.1 for straight lines, and -0.1 for the dashed lines.

the discussion, we only consider the spectra related to $\langle u^2 \rangle$. For analyzing the energy transfer mechanisms, we employ the transport equations for the two-point correlations tensor as explained in Gungor et al. [1]. Figure 7 presents the spectral distributions of the three streamwise positions (SP2 is not given) at $y^+ \approx 13$ along with the ones for the channel flow of Lee & Moser [20], which is included for comparing both APG TBL cases with canonical flows.

The spectral distributions of energy, production and pressure-strain in the channel flow and SP1o are very similar to each other as reported before [1]. The minor difference is that the energy spectra at SP1o have a tail which is the signature of the imprints of the outer-layer structures. Regarding the production and pressure strain, they are also very similar, despite the minor differences in their size. Production structures are slightly shorter and narrower than the most energetic structures in both cases. As the defect increases from SP1 to SP4, the energy spectra become completely different and the energy resides at very large structures, the size of which is similar to those in the outer layer. The difference in production and pressure strain is not as dramatic as the difference in the energy spectra. The structures' aspect ratio remains relatively close. Furthermore, the pressure strain structures are still shorter than the production structures while having a very similar spanwise width. Despite these minor similarities, the shape of the spectra of production and pressure strain at SP3 and SP4 has changed.

In the small-defect case of the manipulated case, the spectra of production, pressure strain and energy are very similar to the ones in channel flow and SP1o. Interestingly, the channel flow case and SP1m are more similar to each other than SP1o and SP1m. For instance, the energy spectra at SP1m do not have the tail at large wavelengths of SP1o. This shows the effect of elevated outer-layer activity on the inner-layer even if SP1 is a small-defect APG case.

In contrast with the original case, the spectral distributions remain very similar in the manipulated case as the defect increases from SP1 to SP4. The shape of the distributions, the relative size and the aspect ratio of structures are all almost identical. There is a minor shift towards the smaller wavelengths with increasing defect but this is, at least in part, because

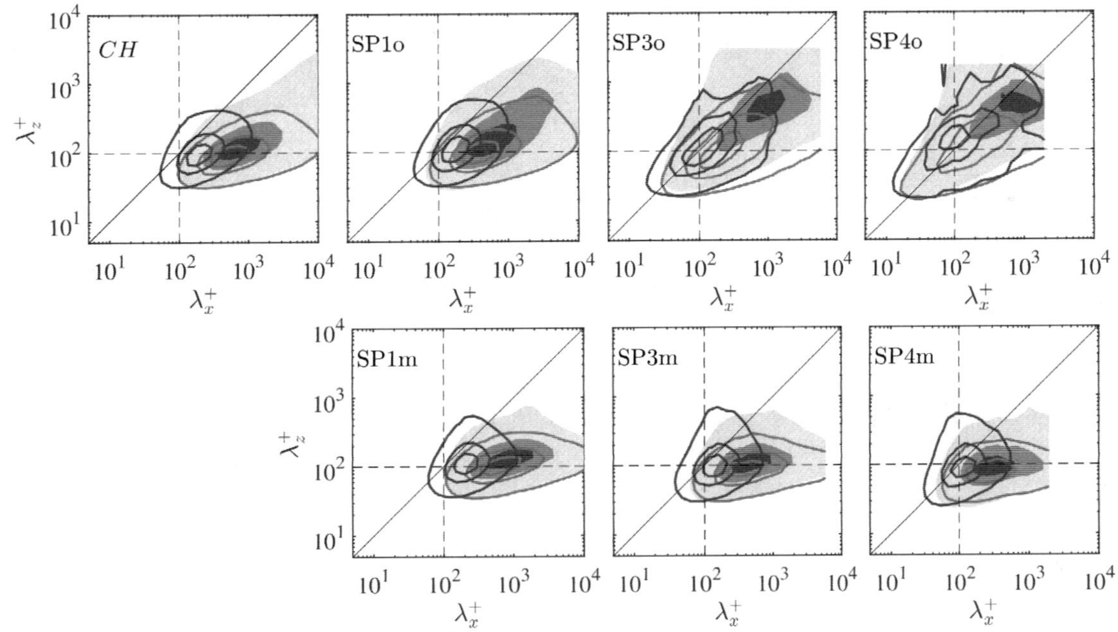

Figure 7. The spectral distibutions of energy (shaded), production (red), and pressure strain (purple) at $y^+ \approx 13$. Top row from left to right: The channel flow case of Lee and Moser [20], SP1o, SP3o, and SP4o. The bottom row from left to right: SP1m, SP3m, and SP4m. The levels are [0.1 0.5 0.8] of maxima of spectra.

friction-viscous scales are used. The difference between the original and manipulated cases at large-defect is massive. It is much more pronounced than the small-defect case. Even though the inner contours of production and pressure-strain are at similar wavelengths (for instance, λ_z^+ of the pressure-strain peak is 100 at SP1o and 150 for SP1m.), there is an intense energy and energy transfer at wavelengths that are associated with very-large structures in SP3/4. This is expected considering the outer flow dominates the inner layer in SP3/4 and those large-scale structures are probably the imprints of large-scale outer-layer structures.

The most important outcome of the similarity of production and pressure-strain structures in the manipulated case as discussed above is that energy transfer mechanisms are probably the same or very similar at all streamwise positions. This means the near-wall cycle probably exists even if the defect is as large as at SP4, where β_i and H are approximately 0.62 and 3.11. It also indicates that the near-wall cycle may survive even when the mean shear is considerably lower than the one in canonical flows.

One interesting point is that energy spectra have a clear peak and the energy resides at wavelengths that are associated with the inner-layer streaks regardless of the velocity defect. It is still not clear if these streaks are convected from the upstream or generated locally. However, the fact that the spectra of production and pressure-strain are also very similar suggests the latter.

4. Conclusion

This study examines the turbulence in the near-wall region of APG TBLs to further understand the effect of outer-layer turbulence on inner-layer turbulence activity and dynamics in low shear environment. For this, we employ two non-equilibrium APG TBLs. The first one is the non-equilibrium APG TBL of Gungor et al. [1]. The other one is a novel APG TBL database that is identical to the first one with a major change: the outer layer turbulence is artificially removed. The major findings of this study are listed below.

(i) The inner-layer turbulence sustains itself regardless of the mean shear in the inner-layer when the outer layer does not exist.
(ii) The streak couples continue to exist in the near-wall region even when the flow is very close to separation in the manipulated flow whereas they cease to exist in large-defect cases of the original flow.
(iii) It is important to state that it is not clear if these streaks in the near-wall region of the manipulated flow are generated locally or simply convected from upstream where the shear is still high.
(iv) The pressure-viscous length scales the streamwise extent of C_{uu} and C_{uv} contours while friction-viscous length does not. Despite this, neither friction-viscous nor pressure-viscous length scale the spanwise extent of the contours.
(v) The spectral characteristics of energy, production and pressure-strain are very similar to those of canonical wall flows in the inner layer regardless of the velocity defect when the outer layer is removed.
(vi) This suggests that energy transfer mechanisms such as production or inter-component energy are similar in the inner layer and therefore the near-wall cycle may still exist even when the flow is very close to separation.

Our analysis shows that the streaks and the near-wall cycle exist in the large-defect APG TBLs when the outer layer is removed. However, it is not clear if the structures playing a role in the near-wall cycle co-exist with the imprints of large-scale outer-layer structures or simply vanish. If they do co-exist, then how do they interact with the imprints of large-scale structure? More work is required to further investigate this.

5. Acknowledgments

This work was supported in part by the European Research Council under the Caust grant ERC-AdG-101018287. The authors would like to thank Prof. Javier Jiménez for organizing the Fifth Madrid Turbulence Workshop. The computational resources were provided by Calcul Québec (www.calculquebec.ca) and the Digital Research Alliance of Canada (www.alliancecan.ca). This work was also funded in part by NSERC of Canada.

References

[1] Gungor T R, Maciel Y and Gungor A G 2022 *Journal of Fluid Mechanics* **948** A5
[2] Skåre P and Krogstad P 1994 *Journal of Fluid Mechanics* **272** 319–348
[3] Gungor A G, Maciel Y, Simens M P and Soria J 2016 *Int. J. Heat and Fluid Flow* **59** 109–124
[4] Elsberry K, Loeffler J, Zhou M and Wygnanski I 2000 *Journal of Fluid Mechanics* **423** 227–261
[5] Maciel Y, Rossignol K and Lemay J 2006 *Experiments in Fluids* **41** 573–590
[6] Lee J 2017 *Journal of Fluid Mechanics* **810** 323–361
[7] Harun Z, Monty J, Mathis R and Marusic I 2013 *Journal of Fluid Mechanics* **715** 477–498
[8] Kitsios V, Sekimoto A, Atkinson C, Sillero J, Borrell G, Gungor A G, Jiménez J and Soria J 2017 *Journal of Fluid Mechanics* **829** 392–419
[9] Skote M and Henningson D S 2002 *Journal of Fluid Mechanics* **471** 107–136
[10] Maciel Y, Gungor A G and Simens M 2017 *International Journal of Heat and Fluid Flow* **67** 95–110
[11] Maciel Y, Simens M and Gungor A G 2017 *Flow, Turbulence and Combustion* **98** 1–20
[12] Jiao Y, Chernyshenko S I and Hwang Y 2022 *Journal of Fluid Mechanics* **941** A37
[13] Jiménez J 2022 *Journal of Fluid Mechanics* **945** R3
[14] Maciel Y, Wei T, Gungor A G and Simens M 2018 *Journal of Fluid Mechanics* **844** 5–35
[15] Simens M, Jiménez J, Hoyas S and Mizuno Y 2009 *Journal of Computational Physics* **228** 4218–4231
[16] Borrell G, Sillero J and Jiménez J 2013 *Computers & Fluids* **80** 37–43

[17] Hutchins N and Marusic I 2007 *Journal of Fluid Mechanics* **579** 1–28
[18] Gungor T R 2023 *Coherent structures and energy transfer in decelerated turbulent boundary layers* Ph.D. thesis Istanbul Technical University, Laval University
[19] Sillero J A, Jiménez J and Moser R D 2014 *Physics of Fluids* **26** 105109
[20] Lee M and Moser R 2019 *Journal of Fluid Mechanics* **860** 886–938

Journal of Physics: Conference Series **2753** (2024) 012006 doi:10.1088/1742-6596/2753/1/012006

Spectral analysis of the correlation between temperature and wall-normal velocity in Rayleigh-Bénard convection at moderate Rayleigh numbers

Myoungkyu Lee

Department of Mechanical Engineering, University of Houston
Engineering Building 1, 4226 MLK Blvd Houston, TX 77204, USA

E-mail: leemk@uh.edu

Abstract. Turbulent Rayleigh-Bénard convection holds significant scientific and engineering implications, with one of the most debated topics in convection research being the ultimate scaling of the Nusselt number at extremely high Rayleigh numbers. A widely adopted method for investigating this issue is direct numerical simulation (DNS); however, achieving the ultimate Rayleigh number with DNS can be prohibitively expensive. To overcome this limitation, one approach is to employ a small simulation domain size for high Rayleigh number cases. This study explores the impact of simulation domain size by conducting DNS with a large geometrical aspect ratio. Specifically, we perform spectral analysis on the covariance of wall-normal velocity and temperature, along with its transport equations. The results reveal that the characteristic length scale in the intermediate length-scale region linearly increases with wall-normal distance. Moreover, the spectral analysis indicates that large-scale structures play a crucial role in the dynamics of the covariance of wall-normal velocity and temperature. However, the study suggests that these large-scale structures have a minimal impact on the scaling of the Nusselt number. This is attributed to the fact that large-scale structures remain in regions away from the wall, and no processes transfer them from large to small-scale structures. As a consequence, the dynamics of the covariance of wall-normal velocity and temperature in the close vicinity of the wall experience little influence from large-scale dynamics.

1. Introduction

The investigation of Rayleigh-Bénard (RB) convection holds significant importance in addressing scientific and engineering challenges. This phenomenon involves the movement of energy through fluid flow driven by a thermal gradient. Insights gained from studying Rayleigh-Bénard convection extend to understanding geophysical and astrophysical processes, such as mantle convection and atmospheric dynamics [1]. Despite numerous research findings on this topic, several unexplored aspects of Rayleigh-Bénard convection persist.

One crucial question in Rayleigh-Bénard convection pertains to how heat flux changes with temperature difference and fluid properties [1, 2, 3]. The Nusselt number, Nu, serves as a nondimensional measure of heat flux from the hot plate to the cold plate. Nu is generally scaled with the Rayleigh number, Ra, and Prandtl number, Pr. Here, Ra represents the ratio of buoyancy to thermal and viscous diffusion, while Pr represents the ratio of viscous diffusion to thermal diffusion. Due to the difficulty in directly measuring heat flux, numerical simulations have become a frequent approach to studying the ultimate scaling of Nu in Rayleigh-Bénard convection. Specifically, direct numerical simulation (DNS), a first-principle-based simulation technique, has been employed extensively and facilitated by advancements in computing power.

However, conducting DNS at very high Ra demands exceptionally expensive computing power. As Ra increases, the length-scale becomes smaller, necessitating more grid points to capture all the small scales. Consequently, many DNS studies of Rayleigh-Bénard convection utilize small domain sizes, which is barely enough to capture the self-sustaining dynamics of turbulence [4, 5, 6] . A notable recent development is DNS of RB convection at moderate Ra values, employing large simulation domains [7, 8, 9]. The results reveal pronounced structures of large-scale motions. It is noteworthy that similar strong influences of large-scale motions have been observed in other types of turbulence, particularly when fluid motion is driven by boundary conditions, as seen in turbulent Couette flows [10].

In this particular investigation, we focus on the covariance of temperature and wall-normal velocity, denoted as $\langle w\theta\rangle$. Here, w represents normalized wall-normal velocity, and θ signifies normalized temperature. The angular brackets, $\langle\cdot\rangle$, indicate averaging over homogeneous direction and time. In canonical RB convection, the Nu is a linear combination of $\langle w\theta\rangle$ and the mean temperature gradient. Consequently, the impact of domain size on $\langle w\theta\rangle$ holds significance in the scaling of Nu.

Various research methods are available for exploring the dynamics of Rayleigh-Bénard convection [11, 12, 13]. In this study, we employ spectral analysis on the budget equation of $\langle w\theta\rangle$ transport. The transport equation serves as the basis for our investigation into the spectral behavior of the terms within it. A similar approach has been undertaken in turbulent channel flows using DNS data at high Reynolds numbers [14, 15]. This prior work demonstrates that the analysis technique can offer detailed insights into the roles of different motions at various length scales and wall-normal locations. In other words, this analysis provides a comprehensive map of the life cycle as a function of wavelength and wall-normal location.

The subsequent sections of this work unfold as follows: In § 2, we furnish a detailed account of our simulation data and analysis techniques. Moving on to § 3, we present comprehensive profiles and spectral densities of $\langle w\theta\rangle$, along with detailed explanations of the terms in its transport equation. The concluding remarks and open questions are addressed in the final section, § 4.

2. Method

2.1. Formulations

In the following discussion, x (or x_1) and y (or x_2) directions are homogeneous directions. z (or x_3) direction denotes the wall-normal direction. The corresponding velocities are U (or U_1) and V (or U_2) for homogenous directions and W (or W_3) for wall-normal direction. Also, $\langle\cdot\rangle$ denotes the expectation or average of a quantity in homogeneous directions and time, and the lower case denotes the fluctuations of a quantity from its expectations. i.e. $U = \langle U\rangle + u$ and $\langle u\rangle = 0$.

The normalized governing equations for the incompressible Rayleigh-Bénard flows are:

$$\begin{aligned}
\frac{\partial U_k}{\partial x_k} &= 0 \\
\frac{\partial U_i}{\partial t} &= -\frac{\partial U_i U_k}{\partial x_k} + \sqrt{\frac{Pr}{Ra}}\frac{\partial^2 U_i}{\partial x_k \partial x_k} - \frac{\partial p}{\partial x_i} + \Theta\delta_{i3} \\
\frac{\partial \Theta}{\partial t} &= -\frac{\partial \Theta U_k}{\partial x_k} + \frac{1}{\sqrt{Pr\,Ra}}\frac{\partial^2 \Theta}{\partial x_k \partial x_k}.
\end{aligned} \tag{1}$$

Here, Prandtl number is the ratio of kinematic viscosity, ν to thermal diffusivity, α, i.e. $Pr = \nu/\alpha$. Also, Rayleigh number, Ra, is defined as $Ra = \beta g \Delta_T H^3/(\nu\alpha)$ where β, g, Δ_T and H are the thermal expansion coefficient, the gravitational acceleration constant, the temperature difference between top and bottom planes and the distance between planes, respectively. x_i are normalized by H, and Θ is the temperature normalized by Δ_T. Finally, t, U_i and p are time, velocity and pressure normalized by $\sqrt{H/(g\beta\Delta_T)}$, $\sqrt{g\beta\Delta_T H}$ and $\rho_0 g\beta\Delta_T H$, respectively, where ρ_0 is the reference density.

We will now derive and define terms in the evolution equation for the spectral density of $\langle\theta^2\rangle$. Initially, the Reynolds decomposition is applied to u, p and θ, yielding zero expectations for the velocity components in the plane Rayleigh-Bénard convection.

$$U_i = \langle U_i\rangle + u_i \tag{2a}$$
$$P = \langle P\rangle + p \tag{2b}$$
$$\Theta = \langle \Theta\rangle + \theta \tag{2c}$$

$$\langle U_i\rangle = 0 \tag{3}$$

From the continuity,

$$\frac{\partial u_k}{\partial x_k} = 0. \tag{4}$$

Using (3) and (4), we obtain the evolution equations for u_i and θ:

$$\frac{\partial u_i}{\partial t} = -\frac{\partial u_i u_k}{\partial x_k} + \frac{\partial\langle u_i u_3\rangle}{\partial x_3} + \sqrt{\frac{Pr}{Ra}}\frac{\partial^2 u_i}{\partial x_k \partial x_k} - \frac{\partial p}{\partial x_i} + \theta\delta_{i3} \tag{5}$$
$$\frac{\partial \theta}{\partial t} = -\frac{\partial \theta u_k}{\partial x_k} + \frac{\partial\langle \theta u_3\rangle}{\partial x_3} + \frac{1}{\sqrt{PrRa}}\frac{\partial^2 \theta}{\partial x_k \partial x_k} - u_3\frac{\partial\langle\Theta\rangle}{\partial x_3}. \tag{6}$$

Now, we will introduce the θ at the separation distance as $\tilde{\theta} = \theta(x + r_x, y, z + r_z)$ where r_x and r_z are separation distances in x and z directions, respectively. The evolution equation for $\tilde{\theta}$ is obtained from (6):

$$\frac{\partial \tilde{\theta}}{\partial t} = -\frac{\partial \widetilde{\theta u_k}}{\partial \tilde{x}_k} + \frac{\partial\langle \theta u_3\rangle}{\partial x_3} + \frac{1}{\sqrt{PrRa}}\frac{\partial^2 \tilde{\theta}}{\partial \tilde{x}_k \partial \tilde{x}_k} - u_3\frac{\partial\langle\Theta\rangle}{\partial x_3}. \tag{7}$$

The evolution equation for the two-point correlation function of w and $\tilde{\theta}$, denoted as $R^{w\theta}(r_x, r_y, z) = \langle w\tilde{\theta}\rangle$, is obtained by multiplying w and $\tilde{\theta}$ with (5) and (7), followed by taking expectations of their summation.

$$\frac{\partial R^{w\theta}}{\partial t} = \left\langle \tilde{\theta}\frac{\partial w}{\partial t}\right\rangle + \left\langle w\frac{\partial \tilde{\theta}}{\partial t}\right\rangle = R_P^{w\theta} + R_T^{w\theta} + R_{\alpha,u}^{w\theta} + R_{\alpha,\theta}^{w\theta} + R_{\Pi}^{w\theta} \tag{8}$$

where

$$R_P^{w\theta} = \left\langle \theta\tilde{\theta} \right\rangle - \langle w\tilde{w}\rangle \frac{\partial \langle\Theta\rangle}{\partial z} \qquad \text{(Production)}$$

$$R_T^{w\theta} = -\left\langle \tilde{\theta}\frac{\partial w u_k}{\partial x_k} \right\rangle - \left\langle w\widetilde{\frac{\partial \theta u_k}{\partial x_k}} \right\rangle \qquad \text{(Turbulent transport)}$$

$$R_{\alpha,u}^{w\theta} = \sqrt{\frac{Pr}{Ra}} \left\langle \tilde{\theta}\frac{\partial^2 w}{\partial x_k \partial x_k} \right\rangle \qquad \text{(Viscous transport)}$$

$$R_{\alpha,\theta}^{w\theta} = \frac{1}{\sqrt{PrRa}} \left\langle w\widetilde{\frac{\partial^2 \theta}{\partial x_k \partial x_k}} \right\rangle \qquad \text{(Thermal transport)}$$

$$R_\Pi^{w\theta} = -\left\langle \tilde{\theta}\frac{\partial p}{\partial z} \right\rangle. \qquad \text{(Pressure transport)}$$

These terms can be further decomposed. First, the production term can be separated into production by thermal fluctuation and velocity fluctuations.

$$R_P^{w\theta} = R_{P,u}^{w\theta} + R_{P,\theta}^{w\theta}$$

$$R_{P,u}^{w\theta} = -\langle w\tilde{w}\rangle \frac{\partial \langle\Theta\rangle}{\partial z} \qquad \text{(Production by velocity fluctuation)}$$

$$R_{P,\theta}^{w\theta} = \langle \theta\tilde{\theta}\rangle \qquad \text{(Production by thermal fluctuation)}$$

We then decompose the turbulent transport term into components related to wall-normal transport and interscale transfer.

$$R_T^{w\theta} = R_{T\perp}^{w\theta} + R_{T\parallel}^{w\theta}$$

$$R_{T\perp}^{w\theta} = -\frac{1}{2}\frac{\partial \langle \tilde{\theta} w w\rangle}{\partial z} - \frac{1}{2}\frac{\partial \langle w\widetilde{\theta w}\rangle}{\partial z} \qquad \text{(Turbulent transport in wall-normal direction)}$$

$$R_{T\parallel}^{w\theta} = -\left\langle \tilde{\theta}\frac{\partial w u_k}{\partial x_k} \right\rangle - \left\langle w\widetilde{\frac{\partial \theta u_k}{\partial x_k}} \right\rangle + \frac{1}{2}\frac{\partial \langle \tilde{\theta} w w\rangle}{\partial z} + \frac{1}{2}\frac{\partial \langle w\widetilde{\theta w}\rangle}{\partial z} \qquad \text{(Interscale transfer)}$$

Note that

$$\int_0^H R_{T\perp}^{w\theta} \mathrm{d}z = 0, \quad \forall (r_x, r_y)$$

and

$$\lim_{\substack{r_x \to 0 \\ r_y \to 0}} R_{T\parallel}^{w\theta} = 0, \quad \forall z.$$

Finally, the viscous and thermal diffusion terms are decomposed into wall-normal transport terms and the remaining terms. For the evolution equations of turbulent kinetic energy and temperature variances, the remaining terms pertain to dissipation. Therefore, we will denote this remaining term as "ϵ," a common notation for dissipation.

$$R^{w\theta}_{\alpha,u} = R^{w\theta}_{D,u} - R^{w\theta}_{\epsilon,u}$$

$$R^{w\theta}_{D,u} = \sqrt{\frac{Pr}{Ra}}\frac{\partial^2\langle w\tilde{\theta}\rangle}{\partial z^2} \qquad \text{(Wall-normal transposrt by viscous diffusion)}$$

$$R^{w\theta}_{\epsilon,u} = -\sqrt{\frac{Pr}{Ra}}\left\langle \tilde{\theta}\frac{\partial^2 w}{\partial x_k \partial x_k}\right\rangle + \sqrt{\frac{Pr}{Ra}}\frac{\partial^2\langle w\tilde{\theta}\rangle}{\partial z^2}$$

$$R^{w\theta}_{\alpha,\theta} = R^{w\theta}_{D,\theta} - R^{w\theta}_{\epsilon,\theta}$$

$$R^{w\theta}_{D,\theta} = \frac{1}{\sqrt{PrRa}}\frac{\partial^2\langle w\tilde{\theta}\rangle}{\partial z^2} \qquad \text{(Wall-normal transposrt by thermal diffusion)}$$

$$R^{w\theta}_{\epsilon,\theta} = -\frac{1}{\sqrt{PrRa}}\left\langle w\widetilde{\frac{\partial^2\theta}{\partial x_k \partial x_k}}\right\rangle + \frac{1}{\sqrt{PrRa}}\frac{\partial^2\langle w\tilde{\theta}\rangle}{\partial z^2}$$

Note that

$$\int_0^H R^{w\theta}_{D,u}\mathrm{d}z = \int_0^H R^{w\theta}_{D,\theta}\mathrm{d}z = 0, \quad \forall(r_x, r_y).$$

In the final step, we compute the Fourier transform of all terms in terms of r_x and r_y.

$$\begin{aligned}\frac{\partial E^{w\theta}}{\partial t} &= \int_{-\infty}^{\infty}\int_{-\infty}^{\infty}\frac{\partial R^{w\theta}}{\partial t}\mathrm{e}^{-\sqrt{-1}(r_x k_x + r_y k_y)}\mathrm{d}r_x\mathrm{d}r_y \\ &= E^{w\theta}_{P,u} + E^{w\theta}_{P,\theta} + E^{w\theta}_{T\perp} + E^{w\theta}_{T\parallel} + E^{w\theta}_{D,u} - E^{w\theta}_{\epsilon,u} + E^{w\theta}_{D,\theta} - E^{w\theta}_{\epsilon,\theta} + E^{w\theta}_{\Pi}\end{aligned} \tag{9}$$

Note that integration of each term in (9) will results in the terms in the budget equation of $\langle w\theta\rangle$.

$$\frac{\partial\langle w\theta\rangle}{\partial t} = \int_0^{\infty}\int_0^{\infty} E^{w\theta}\mathrm{d}k_x\mathrm{d}k_z = P^{w\theta}_u + P^{w\theta}_\theta + T^{w\theta} + D^{w\theta}_u - \epsilon^{w\theta}_u + D^{w\theta}_\theta - \epsilon^{w\theta}_\theta + \Pi^{w\theta}.$$

where

$$T^{w\theta} = \int_0^{\infty}\int_0^{\infty} E^{w\theta}_{T\perp}\mathrm{d}k_x\mathrm{d}k_z$$

and

$$\int_0^{\infty}\int_0^{\infty} E^{w\theta}_{T\parallel}\mathrm{d}k_x\mathrm{d}k_z = 0.$$

2.2. Simulations

We conduct simulations for three cases with different Rayleigh numbers (Ra). The Prandtl number (Pr) is held constant at unity, resulting in the relationship $\sqrt{Pr/Ra} = 1/\sqrt{PrRa}$. While Pr is a crucial nondimensional parameter in Rayleigh-Bénard convection, its detailed investigation falls beyond the scope of this study. The simulation parameters are outlined in table 1.

We utilize a spectral-Galerkin method with a 3/2 dealiasing technique in the homogeneous directions. In the wall-normal direction, we employ seventh-order basis splines with the locations of knots determined by the following function.

$$y^{\text{knot}}_n = \frac{1}{2}\left(\frac{\sin\left[\left(\frac{2(n-1)}{N_\perp - 7} - 1\right)\frac{\eta\pi}{2}\right]}{\sin\left(\frac{\eta\pi}{2}\right)} + 1\right) \tag{10}$$

Table 1. Simulation parameters. L: simulation domain size in the homogeneous directions, H: distance between walls, $N_{\parallel}$: number of grid points in spectral domain, $N_{\perp}$: number of collocation points in wall-normal direction, η: grid-stretch parameter in the wall-normal direction, $Re_{\parallel}$: horizontal Reynolds number, $Re_{\perp}$: wall-normal Reynolds number, Re_T: total Reynolds number, δ_θ: thermal boundary layer thickness

Case	Ra	L/H	$N_{\parallel}$	$N_{\perp}$	η	Nu	$Re_{\parallel}$	$Re_{\perp}$	Re_T	$\delta_\theta \times 10^2$
1E5	1×10^5	10π	512	96	0.97	4.34	55.6	40.1	68.6	11.52
6E6	6×10^6	5π	768	256	1.00	13.62	425.3	299.4	520.1	3.672
4E8	4×10^8	5π	2048	512	1.00	46.37	3141.7	1962.4	3704.3	1.078

where $n = 1, 2, \cdots, N_{\perp} - 6$. We use Greville abscissae to obtain the collocation points in wall-normal direction [16]. The information regarding simulation software specifics can be located in [17, 18]. Also, table 1 includes Nusselt number, the homogeneous ($Re_{\parallel}$), wall-normal ($Re_{\perp}$) and total Reynolds numbers (Re_T), and thermal boundary layer thickness (δ_θ) to compare the simulation result with values from [19].

$$Nu = \langle \theta w \rangle - \frac{1}{\sqrt{PrRa}} \frac{\partial \langle \Theta \rangle}{\partial z} \tag{11}$$

$$Re_{\parallel} = \left[\frac{Ra}{Pr} \int_0^1 (u^2 + v^2) \mathrm{d}z \right]^{1/2} \tag{12a}$$

$$Re_{\perp} = \left[\frac{Ra}{Pr} \int_0^1 w^2 \mathrm{d}z \right]^{1/2} \tag{12b}$$

$$Re_T = \left[\frac{Ra}{Pr} \int_0^1 (u^2 + v^2 + w^2) \mathrm{d}z \right]^{1/2} \tag{12c}$$

$$\delta_\theta = \frac{1}{2Nu} \tag{13}$$

The variations as functions of Ra follow the same trend (not shown) although some Ras of current work are different from Ras in [19].

The statistical convergence is investigated with Nu, which is the sum of mean temperature gradient and covariance of temperature and wall-normal velocity. The value of Nu should be constant in the statistically stationary flow. The relative residual of (11) is defined as

$$\mathcal{R}_\theta(z) = \left| \frac{Nu(z)}{Nu_{z=0}} - 1 \right|. \tag{14}$$

We use the method from [20] to estimate the statistical uncertainty. $\mathcal{R}_\theta(z)$ is smaller than 0.001 and estimated uncertainties for all cases (not shown). Therefore, we conclude that the dataset for the current study is statistically converged.

3. Result

In this section, we present the results of a detailed spectral analysis applied to the transport equation governing the covariance of velocity and temperature. Our exploration initiates with

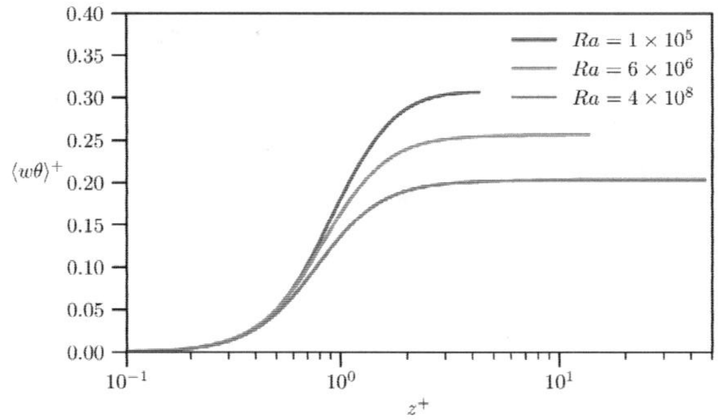

Figure 1. the profile of $\langle w\theta \rangle$

an examination of the covariance of velocity and temperature, followed by an in-depth analysis of the terms within the transport equation as delineated in Section 2.1. The normalization of velocity and temperature involves scaling by the viscous and thermal dissipation at the wall, denoted by the superscript "+." Additionally, the wall-normal distance is normalized by the thermal boundary layer thickness. To be specific, the length-scale is normalized by δ_θ and temperature and velocities are normalized by Θ_η and W_η, respectively, where

$$\delta_\theta = \frac{\Theta_{z=1} - \Theta_{z=0}}{2 \left. \dfrac{\partial \langle \Theta \rangle}{\partial z} \right|_{\text{wall}}}, \quad \Theta_\eta = \delta_\theta \left[\sqrt{PrRa} \left\langle \frac{\partial \theta}{\partial x_k} \frac{\partial \theta}{\partial x_k} \right\rangle \right]^{1/2}_{\text{wall}}, \quad \text{and} \quad W_\eta = \left[\frac{1}{2} \sqrt{\frac{Pr}{Ra}} \left\langle \frac{\partial u_j}{\partial x_k} \frac{\partial u_j}{\partial x_k} \right\rangle \right]^{1/4}_{\text{wall}}.$$

3.1. Covariance of velocity and temperature

figure 1 depicts the covariance profile between wall-normal velocity and temperature. The data suggests that normalizing with dissipations at the wall is effective in the near-wall region, specifically up to $z^+ = 0.5$. Beyond this point, the covariance $\langle w\theta \rangle$ undergoes logarithmic increases, eventually reaching a plateau after $z^+ = 2$.

The κ-premultiplied spectral density of $\langle w\theta \rangle$ is presented in figure 2. Generally, the small-scale portion in the near-wall region exhibits similarity across different Ra values. In all cases, the peaks at small scales occur at $\lambda^+ = 10$ and $z^+ = 2$. Strong peaks in the spectral density are consistently observed at the center of the domain, with a corresponding length-scale of $\lambda/H = 5 \sim 10$, aligning with observations by [9].

Notably, at $Ra = 4 \times 10^8$, the length-scale of large-scale structures demonstrates linear growth with z. Additionally, the values at the peaks of the large-scale structures generally exhibit a gradual decrease with Ra, which is intricately linked to the normalization values by viscous and thermal dissipations.

At higher Ra, an expected behavior is the growth of the peak at the large scale in λ^+ due to the decrease in δ_θ with Ra because $\lambda^+ = \lambda/\delta_\theta$. This decrease in δ_θ is proportional to $1/(2Nu)$. Therefore, the linear growth behavior at very high Ra depends on the ultimate scaling of Nu at infinite Ra. Nonetheless, there is speculation that the dominant length-scale between large-scale structure and small-scale structure grows linearly with z since δ_θ serves as the normalizing parameter for both z and λ. To confirm this speculation, DNS at higher Ra with a large aspect ratio is required.

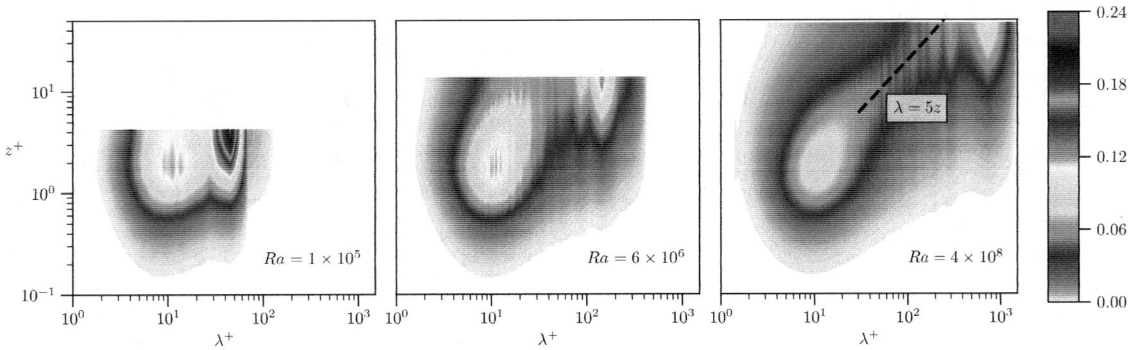

Figure 2. κ-premultiplied spectral density of $\langle w\theta \rangle$, $E^{w\theta}$

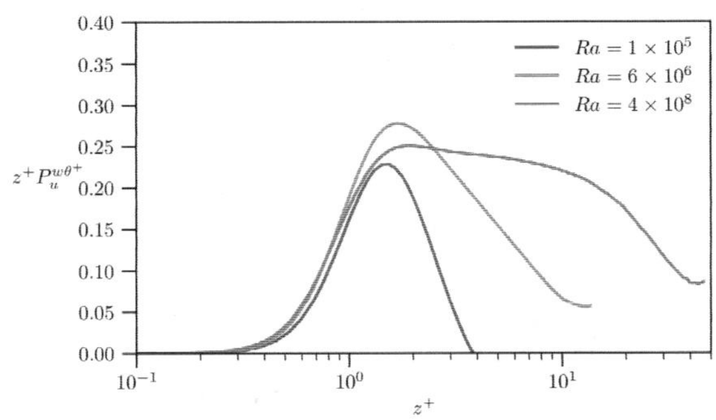

Figure 3. z-premultiplied profile of $w\theta$ production by velocity flucutation, $P_u^{w\theta}$

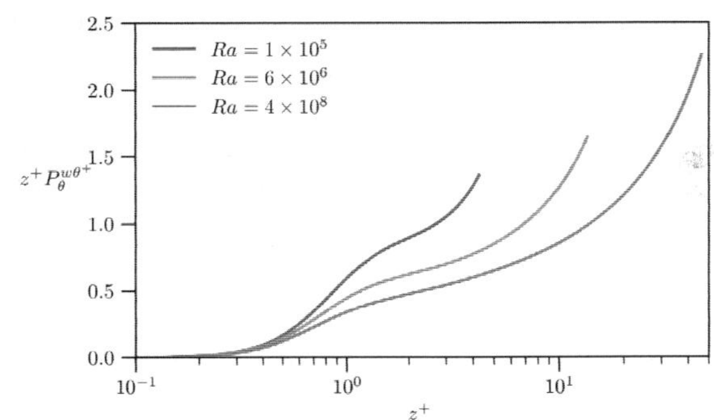

Figure 4. z-premultiplied profile of $w\theta$ production by thermal flucutation, $P_\theta^{w\theta}$

3.2. Production of $\langle w\theta \rangle$

The production term of $\langle w\theta \rangle$ comprises two components: production by velocity fluctuation and thermal fluctuation. The profiles of these two production components are depicted in figures 3 and 4. Generally, production by velocity fluctuation is significantly smaller than production by thermal fluctuation. This is attributed to the temperature gradient term in production by velocity fluctuation, which approaches zero at the center of the domain. In the z^+-premultiplied profile, the case with $Ra = 4 \times 10^8$ exhibits two distinct slopes in different regions. Specifically, $P_u^{\theta w}$ experiences a gradual decrease between $z^+ = 2$ and $z^+ = 10$, followed by a rapid decrease beyond $z^+ = 10$. Conversely, the production by thermal fluctuation demonstrates much stronger contributions. It exhibits growth solely with z, and its magnitude increases with Ra.

The spectral densities of production by velocity and thermal fluctuation are presented in figure 5. In all cases, the spectral density of $P_u^{\theta w}$ exhibits peaks at $\lambda^+ = 10$ and $z^+ = 1.5$. Notably, $z^+ = 1.5$ corresponds to the location where the profile of $P_u^{\theta w}$ attains peak values. With increasing Ra, the structure of $P_u^{\theta w}$ spectra experiences a gradual growth. At $Ra = 4 \times 10^8$, the structure of $P_u^{\theta w}$ spectra displays a distinct growth rate around $z^+ = 10$. It is speculated that $z^+ = 10$ marks the end of the "inner" region, where the flow is influenced by the presence of walls. Consequently, at even higher Ra, it is anticipated that $P_u^{\theta w}$ spectra will exhibit behavior where $\lambda \sim z$, based on observations in figure 5. Once again, confirmation of this speculation requires DNS data at higher Ra. The spectral density of production by thermal fluctuation, $P_\theta^{w\theta}$, exhibits significantly higher quantities than production by velocity fluctuations. $P_\theta^{w\theta}$ spectral

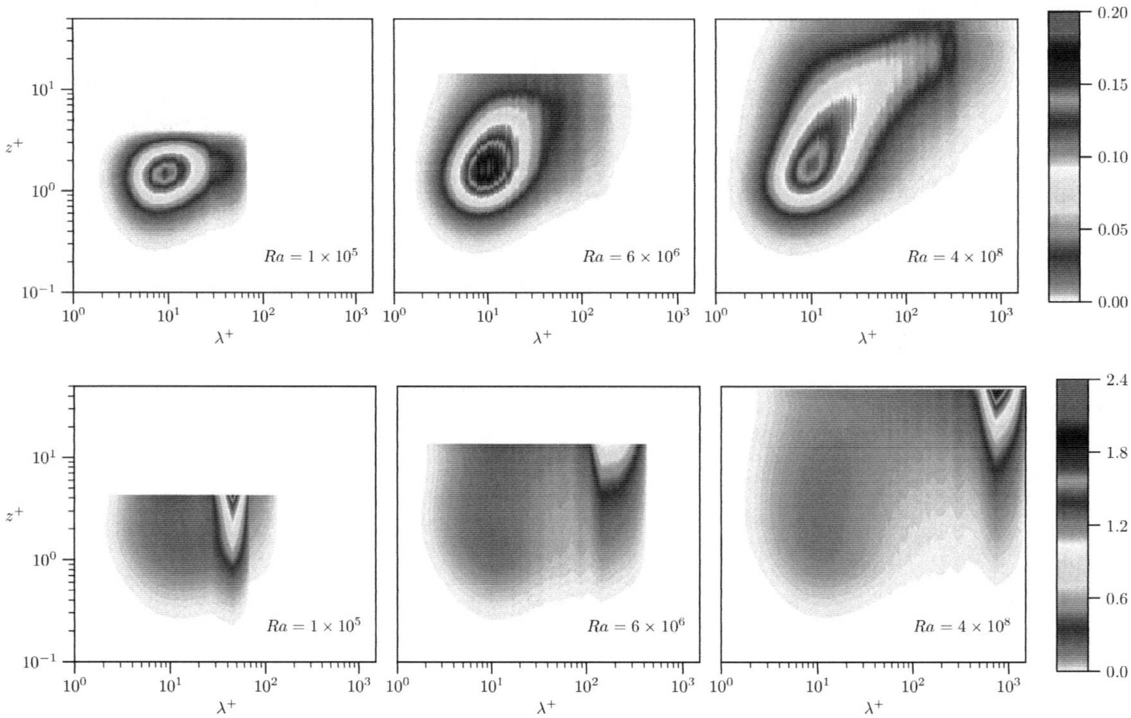

Figure 5. κ-z-premultiplied spectral density of $w\theta$ production. (Top) $w\theta$ production by velocity flucutation, $E^{w\theta}_{P,u}$ (Bottom) $w\theta$ production by thermal flucutation, $E^{w\theta}_{P,\theta}$

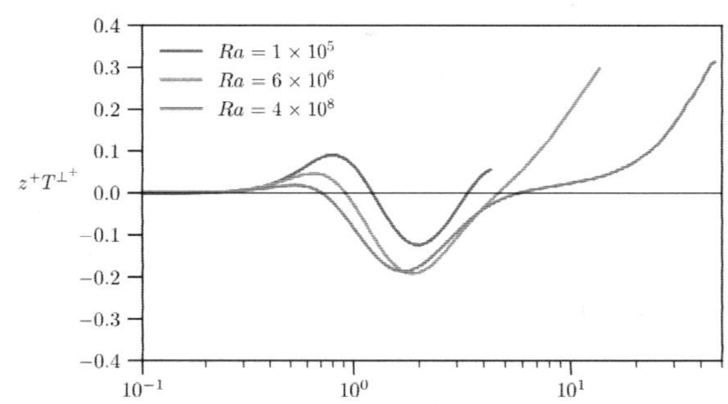

Figure 6. z-premultiplied profile of $w\theta$ transport by turbulence, $T^{w\theta}$

density displays two peaks, with the large-scale peak being strongest at the center of the domain. It is noteworthy that these peak locations align with the peaks of spectral densities of $\langle w\theta \rangle$. The small-scale structure peaks in the near-wall region are challenging to discern because the spectral densities of $P^{w\theta}_{\theta}$ are dominated by the large-scale structure. However, the peak values of $P^{w\theta}_{\theta}$ in the near-wall region are comparable to the peak values of $P^{w\theta}_{u}$, and their locations coincide.

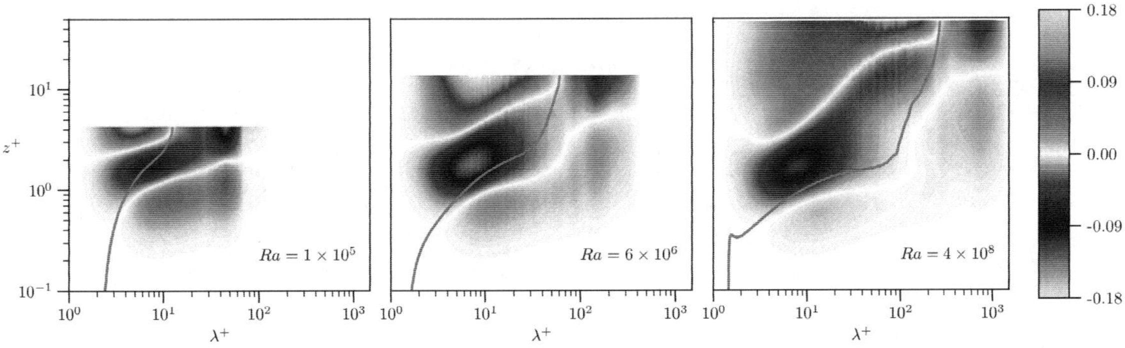

Figure 7. κ-z-premultiplied spectral density of $T^{w\theta}_{\perp}$, $E^{w\theta}_{T\perp}$

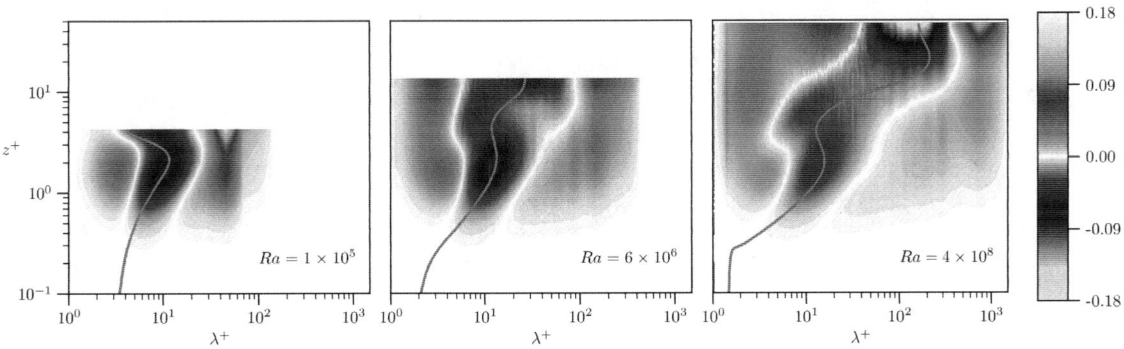

Figure 8. κ-z-premultiplied spectral density of $T^{w\theta}_{\parallel}$, $E^{w\theta}_{T\parallel}$

3.3. Turbulent transport of $\langle w\theta \rangle$

The transport profile of $\langle w\theta \rangle$ by turbulence is illustrated in figure 6. In the low Ra case, the majority of $\langle w\theta \rangle$ is transported in the direction towards the wall. However, as Ra increases, the amount of $\langle w\theta \rangle$ transport towards the wall rapidly diminishes, while $\langle w\theta \rangle$ is increasingly transported to outer flows.

The spectral density of $T^{w\theta}_{\perp}$, $E^{w\theta}_{T\perp}$, is depicted in figure 7. The red lines in the figure indicate where the flux of $E^{w\theta}_{T\perp}$ is zero:

$$\int_0^z E^{w\theta}_{T\perp} \mathrm{d}z = 0$$

In this context, the region above the red line at any λ suggests that turbulence transports $\langle w\theta \rangle$ away from the wall, while below the red line signifies turbulence transport towards the wall. Generally, small-scale structures are transported away from the wall, whereas large-scale structures are transported towards the wall. For large-scale structures ($\lambda/H = 3$ and beyond), turbulence primarily transports from the domain center to the near-wall region. The donor region, highlighted in blue in figure 7, indicates a separation of scales. At $Ra = 4 \times 10^8$, a significant portion of $\langle w\theta \rangle$ between $\lambda^+ = 40$ and $\lambda^+ = 200$ is transported to the center of the domain.

The inter-scale transfer of $\langle w\theta \rangle$ by turbulence is presented in figure 8. The red line in the figure denotes where the flux in interscale is zero:

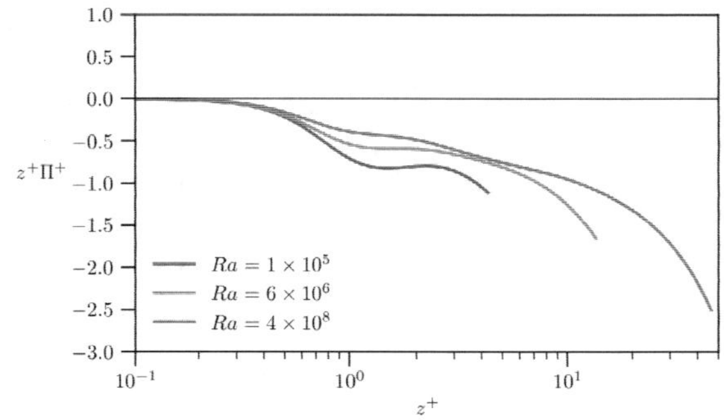

Figure 9. z-premultiplied profile of $w\theta$ transport by pressure, $\Pi^{w\theta}$

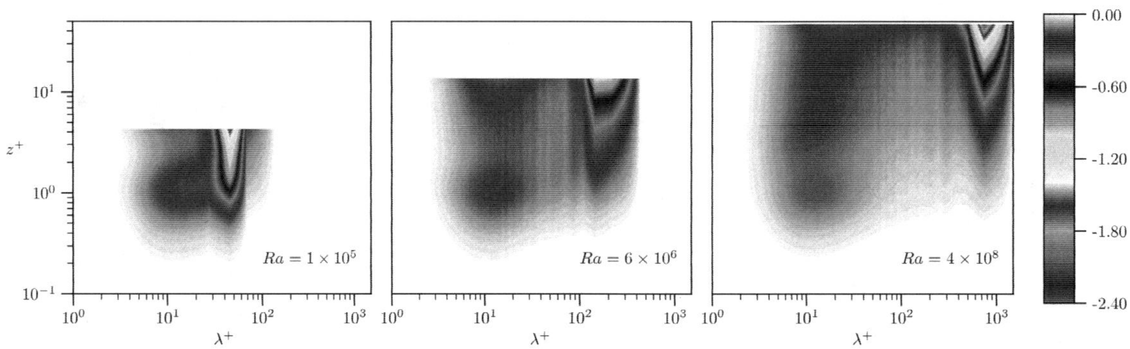

Figure 10. κ-z-premultiplied spectral density of $\Pi^{w\theta}$, $E_{\Pi}^{w\theta}$

$$\int_0^{\kappa} E_{T\parallel}^{w\theta} \mathrm{d}\kappa = 0$$

Similar to $E_{T\perp}$, the left of the red line signifies that $\langle w\theta \rangle$ is transferred from large- to small-scales, and the opposite direction is true for the right of the red line. It is intriguing that all cases exhibit transfer from small- to large-scale transfer. The recipient part of the large-scale structure shows peaks at the center of the domain, coinciding with the peaks of spectral densities of $\langle w\theta \rangle$ and $P_{\theta}^{w\theta}$. Therefore, the large-scale structure of $\langle w\theta \rangle$ arises not only from production but also from inter-scale transfer.

Interestingly, at $Ra = 4 \times 10^8$, the recipient part of figure 7 and the donor part of figure 8 exhibit peaks at the center of the domain and $\lambda^+ = 100$. This indicates that there are two different paths of $\langle w\theta \rangle$ transport at the large-scale structure in the outer flow region. For example, the structure at $\lambda^+ = 100$ and $z^+ = 10$ can be directly transferred to $\lambda^+ = 500$ and $z^+ = 10$. It can also be transported by going to the center of the domain and transferred from $\lambda^+ = 100$ to $\lambda^+ = 500$. Finally, it is transported from the center of the domain to $z^+ = 10$.

3.4. Pressure transport of $\langle w\theta \rangle$

This section delves into the role of pressure, with a focus on the profile of pressure transport of $\langle w\theta \rangle$, $\Pi^{w\theta}$, as depicted in Figure 9. Across all Ra values, $\Pi^{w\theta}$ consistently exhibits a negative trend, and its negative contribution to $\langle w\theta \rangle$ intensifies with z. Remarkably, at the center of

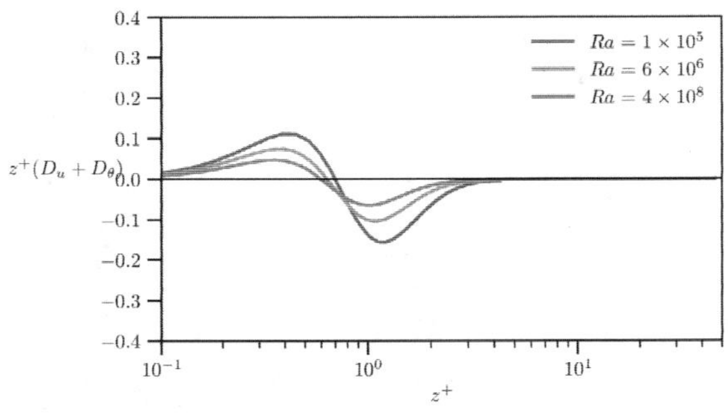

Figure 11. z-premultiplied profile of $w\theta$ transport by viscous and thermal diffusion

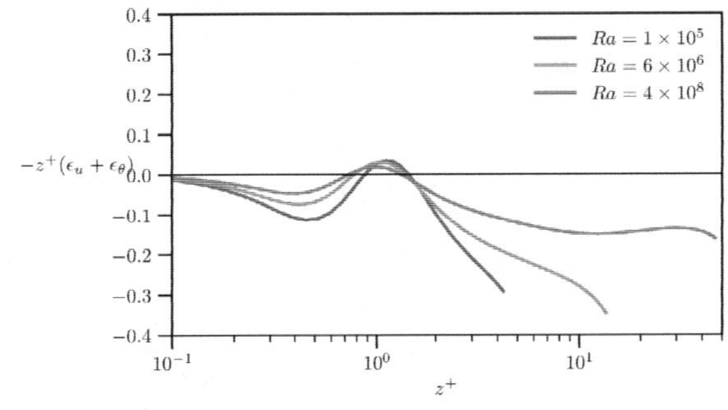

Figure 12. z-premultiplied profile of viscous and thermal dissipations of $w\theta$

the domain, the value of $\Pi^{w\theta}$ surpasses the sum of production terms. This phenomenon arises because turbulence transports $\langle w\theta \rangle$ towards the center of the domain, as detailed in the previous section.

The spectral density of $\Pi^{w\theta}$ in Figure 10 exhibits similar structures to the spectral density of $P_\theta^{w\theta}$ with opposite signs. It demonstrates a strong contribution from large-scale structures. The magnitude of the large-scale structure in $\Pi^{w\theta}$ at $\lambda/H \approx 6$ in $Ra = 4 \times 10^8$ flows is slightly stronger than the magnitude of large-scale structure in $P^{w\theta}$. Again, this is because turbulence transfers $\langle w\theta \rangle$ from small- to large-scale. The negative sign in the pressure transport term is not yet understood, and further research is required to elucidate this aspect.

3.5. Effect of viscous and thermal diffusion in $\langle w\theta \rangle$ transport

Finally, this section explores the impact of the viscous and thermal diffusion term. The diffusion term can be decomposed into transport in the wall-normal direction and the remaining terms. The profile of wall-normal transport in figure 11 reveals that this term exclusively transports $\langle w\theta \rangle$ towards the wall. Corresponding to the remaining terms for turbulent kinetic energy or temperature variance are referred to as viscous and thermal dissipations. However, naming the remaining terms as dissipation is unclear since the value of the remaining term can be both positive and negative. Generally, the values of these remaining terms are negative, indicating that viscous and thermal diffusion primarily degrade $\langle w\theta \rangle$. It is important to note that the magnitude of destruction of $\langle w\theta \rangle$ is much smaller than $\Pi^{w\theta}$.

The spectral density of wall-normal transport and the remaining diffusion term are displayed in Figure 13. The red lines in the figure indicate where the flux of $E_{D,\theta}^{w\theta} + E_{D,u}^{w\theta}$ is zero:

$$\int_0^z (E_{D,\theta}^{w\theta} + E_{D,u}^{w\theta}) \mathrm{d}z = 0$$

. As expected, the spectral density of wall-normal transport by the diffusion term is dominant in the near-wall region, and the corresponding length-scale of the structure remains in small-scale. Similarly, the spectral density of the remaining diffusion term predominantly exhibits negative values, and the length-scale of its structure is in small-scales. Notably, there are small regions where the values are positive, corresponding to the peaks in the spectral density of $\langle w\theta \rangle$ in the near-wall region. The relationship between these peaks is currently unknown.

4. Conclusion and discussion

In this study, we investigate the spectral densities of $\langle w\theta \rangle$ and terms in the transport equation. Generally, all terms, except for the diffusion term, prominently exhibit the influence of large-scale

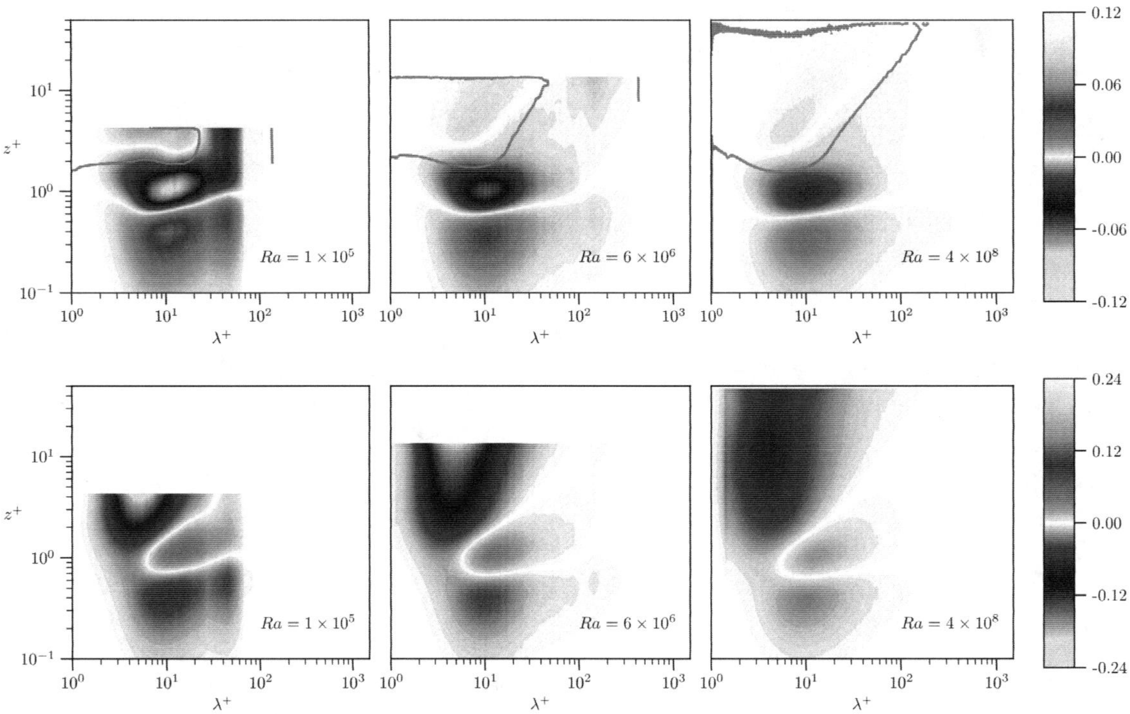

Figure 13. κ-z-premultiplied spectral density of wall-normal transport and dissipation of $w\theta$. (Top) $E^{w\theta}_{D,u} + E^{w\theta}_{D,\theta}$ (Bottom) $E^{w\theta}_{\epsilon,u} + E^{w\theta}_{\epsilon,\theta}$

motions in the outer flow region. Specifically, the large-scale structure arises from the production by thermal fluctuation and is further amplified by scale transfer from small to large scales. The scale separation between small-scale structures and large-scale structures is heightened with increasing Ra, underscoring the pivotal role of a large domain size in comprehending the dynamics of Rayleigh-Bénard convection.

This leads us to consider the validity of studies on Nu scaling at infinite Ra using DNSs with small-domain sizes, which may not adequately represent the large-scale structures. The current data indicates that the large-scale structure is generated in outer flows and is counteracted by pressure terms. Additionally, scale transfer occurs from small to large scales, with these mechanisms primarily operating in large scales and having minimal contribution in the near-wall flows. Moreover, Nu is directly influenced by $\langle w\theta \rangle$. Therefore, we cautiously propose that utilizing a small domain size for DNS may have a very limited impact on the change of Nu at the same Ra, even though the dynamic behavior of $\langle w\theta \rangle$ differs significantly between simulation results with large and small domains. It is essential to note that if there is a dominant scale transfer from very large-scale to small-scale, then this speculation is invalid because using a small domain essentially misses important information from unrepresented large-scale motion. However, in this case, such a dominant scale transfer from large- to small-scale is not observed. Certainly, further study is required to validate this speculation.

Acknowledgements

This work was performed in part during the Fifth Madrid Summer Workshop, funded by the European Research Council under the Caust grant ERCAdG-101018287. Also, this research used resources of the Argonne Leadership Computing Facility, a U.S. Department of Energy (DOE) Office of Science user facility at Argonne National Laboratory and is based on research supported by the U.S. DOE Office of Science-Advanced Scientific Computing Research Program, under Contract No. DE-AC02-06CH11357.

References

[1] Lohse D and Shishkina O 2023 *Physics Today* **76** 26–32
[2] Grossmann S and Lohse D 2000 *Journal of Fluid Mechanics* **407** 27–56
[3] Lohse D and Xia K Q 2010 *Annual Review of Fluid Mechanics* **42** 335–364
[4] Scheel J D, Kim E and White K R 2012 *Journal of Fluid Mechanics* **711** 281–305
[5] Iyer K P, Scheel J D, Schumacher J and Sreenivasan K R 2020 *Proceedings of the National Academy of Sciences* **117** 7594–7598
[6] Kawano K, Motoki S, Shimizu M and Kawahara G 2021 *Journal of Fluid Mechanics* **914** A13
[7] Hartlep T, Tilgner A and Busse F H 2003 *Physical Review Letters* **91** 064501
[8] Hardenberg J v, Parodi A, Passoni G, Provenzale A and Spiegel E A 2008 *Physics Letters, Section A: General, Atomic and Solid State Physics* **372** 2223–2229
[9] Stevens R J A M, Blass A, Zhu X, Verzicco R and Lohse D 2018 *Physical Review Fluids* **3** 041501(R)
[10] Lee M and Moser R D 2018 *Journal of Fluid Mechanics* **842** 128–145 ISSN 0022-1120
[11] Xie Y C, Cheng B Y C, Hu Y B and Xia K Q 2019 *Journal of Fluid Mechanics* **878** R1
[12] Togni R, Cimarelli A and Angelis E D 2020 *Journal of Fluid Mechanics* **782** 380–404
[13] Motoki S, Kawahara G and Shimizu M 2022 *Philosophical transactions. Series A, Mathematical, physical, and engineering sciences* **380** 20210037
[14] Lee M and Moser R D 2015 *Journal of Fluid Mechanics* **774** 395–415
[15] Lee M and Moser R D 2019 *Journal of Fluid Mechanics* **860** 886–938
[16] Johnson R W 2005 *Applied Numerical Mathematics* **52** 63–75
[17] Lee M, Malaya N and Moser R D 2013 Petascale direct numerical simulation of turbulent channel flow on up to 786K cores Proceedings of the International Conference on High Performance Computing, Networking, Storage and Analysis (New York, New York, USA: ACM Press) pp 1–11
[18] Lee M, Ulerich R, Malaya N and Moser R D 2014 *Computing in Science & Engineering* **16** 24–31
[19] Krug D, Lohse D and Stevens R J A M 2020 *Journal of Fluid Mechanics* **887** A2
[20] Oliver T A, Malaya N, Ulerich R and Moser R D 2014 *Physics of Fluids* **26** 035101

Journal of Physics: Conference Series **2753** (2024) 012007 doi:10.1088/1742-6596/2753/1/012007

Coherence analysis of geophysical wakes in flow past a three-dimensional hill

Jinyuan Liu[1] **and Sutanu Sarkar**[1,2]

[1]Department of Mechanical and Aerospace Engineering, UC San Diego, La Jolla, CA 92093, USA

[2]Scripps Institution of Oceanography, La Jolla, CA 92037, USA

E-mails: `jinyuanliu@ucsd.edu`, `ssarkar@ucsd.edu`

Abstract. Wakes generated by topographies are ubiquitous in nature. The formation and evolution mechanism of wake-generated eddies is essential to understanding the effect of topography on oceanic bottom flow. In this study, we analyze the temporally resolved numerical database [1] of stratified wakes past an isolated three-dimensional hill, via proper orthogonal decomposition (POD) of the vertical vorticity field. We study the dynamics of the near wake ($x/D < 3$) and the far wake ($x/D \geq 3$) separately, where x is the downstream distance from the center of the hill. The aim is to understand how rotation influences the change of coherence from the near to the far wake and also the coherence across heights.

1. Introduction and motivation

Eddies are found everywhere in the ocean and the atmosphere and their origins are diverse. In this work we focus on eddies generated by a known source - an isolated seamount in deep water. Figure 1 illustrates the physical setting. A conical hill with base diameter D and height $h = 0.3D$ on the bottom of the seafloor interacts with a steady current with velocity U_∞ in the x-direction. The background density is linearly stratified and the reference frame is rotating at a frequency of $\Omega_c = f_c/2$, where f_c is the Coriolis frequency.

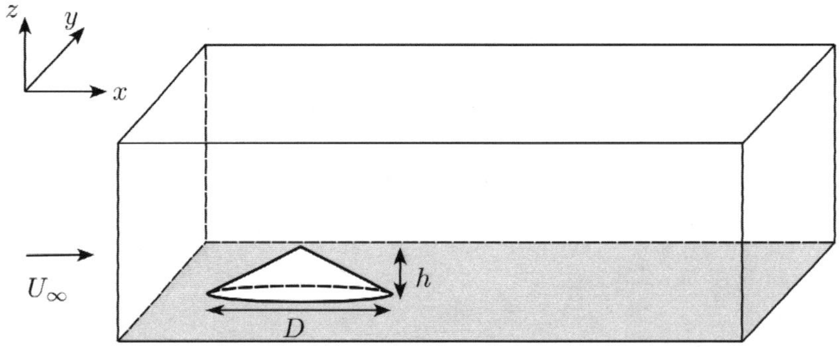

Figure 1: Flow configuration.

We use a temporally resolved LES database [1]. The flow is governed by the incompressible

Navier-Stokes equations under the Boussinesq approximation:

$$\nabla \cdot \boldsymbol{u} = 0 \tag{1}$$

$$\frac{\partial \boldsymbol{u}}{\partial t} + \nabla \cdot (\boldsymbol{u}\boldsymbol{u}) + f_c \hat{\boldsymbol{e}}_z \times (\boldsymbol{u} - U_\infty \hat{\boldsymbol{e}}_x) = -\frac{1}{\rho_0}\nabla p^* + \nabla \cdot \boldsymbol{\tau} - \frac{\rho^* g}{\rho_0}\hat{\boldsymbol{e}}_z \tag{2}$$

$$\frac{\partial \rho}{\partial t} + \nabla \cdot (\rho \boldsymbol{u}) = \nabla \cdot \boldsymbol{J}_\rho \tag{3}$$

where the stress and the scalar flux are

$$\boldsymbol{\tau} = (\nu + \nu_{\text{sgs}})(\nabla \boldsymbol{u} + \boldsymbol{u}\nabla), \ \boldsymbol{J}_\rho = (\kappa + \kappa_{\text{sgs}})\nabla \rho, \tag{4}$$

with ν being the kinematic viscosity and κ being the scalar (density) diffusivity. The sub-grid scales (denoted by subscript 'sgs') are estimated with the WALE model [2]. Here $\boldsymbol{u} = (u, v, w)$ is the velocity vector in (x, y, z) directions, p is pressure, and ρ is density. Unit vectors $\hat{\boldsymbol{e}}_x$ and $\hat{\boldsymbol{e}}_z$ are in x and z directions, respectively. The pressure and density involved in the momentum equations (2) are their deviations (p^* and ρ^*) from the hydrostatic balance, where a vertical pressure gradient balances the unperturbed background density field, and the geostrophic balance, where a spanwise pressure gradient balances the Coriolis force due to the uniform freestream. The equations are solved with a central finite-difference solver on a Cartesian grid and the object is handled with an immersed boundary method [3, 4]. For more numerical details the readers are referred to [5, 1].

Table 1 gives a summary of the simulation parameters. The flow is controlled by the following non-dimensional numbers: the rotation Rossby number $Ro = U_\infty/(f_c D)$, where f_c is the Coriolis frequency; the stratification Froude number $Fr = U_\infty/(Nh)$, where N is the buoyancy frequency; the Reynolds number $Re = U_\infty D/\nu$. The buoyancy frequency is defined by $N^2 = -(g/\rho_0)\mathrm{d}\rho_b/\mathrm{d}z$ where ρ_b is the background density and ρ_0 is a constant reference density. We use negative Coriolis frequency ($f_c < 0$, i.e. in the Southern Hemisphere) that results in negative Rossby numbers. The Froude number is kept at $Fr = 0.15$ representing moderately strong stratification, so that most of the flow is around the obstacle. The Reynolds number is $Re = U_\infty D/\nu = 10\,000$ in the present simulation. Rotation strength is varied such that $Ro = \infty, -0.75, -0.15$, representing dynamics of non-rotational, submesoscale, and mesoscale motions. An additional induced parameter, the Burger number defined as $Bu = (Ro/Fr)^2 = (Nh/f_c D)^2$, is an important measure of the relative importance of rotation to stratification and will also be used in this work.

Figure 2 presents the visualization of the vertical vorticity (ω_z). It can be seen that stratified hill wakes are very different from their unstratified counterpart, which was studied in [6]. Being restricted by the potential energy barrier, the fluid below $O(Frh)$ from the top of the hill will

Table 1: Parameters of the simulated cases. Each case is named according to their Burger number. The number of snapshots is $N_t = 2000$, which covers more than 30 vortex shedding periods in each case.

Case	$Bu = (Ro/Fr)^2$	Ro	Fr	Re	N_t	Color
BuInf	∞	∞				red
Bu25	25	-0.75	0.15	10 000	2000	green
Bu1	1	-0.15				blue

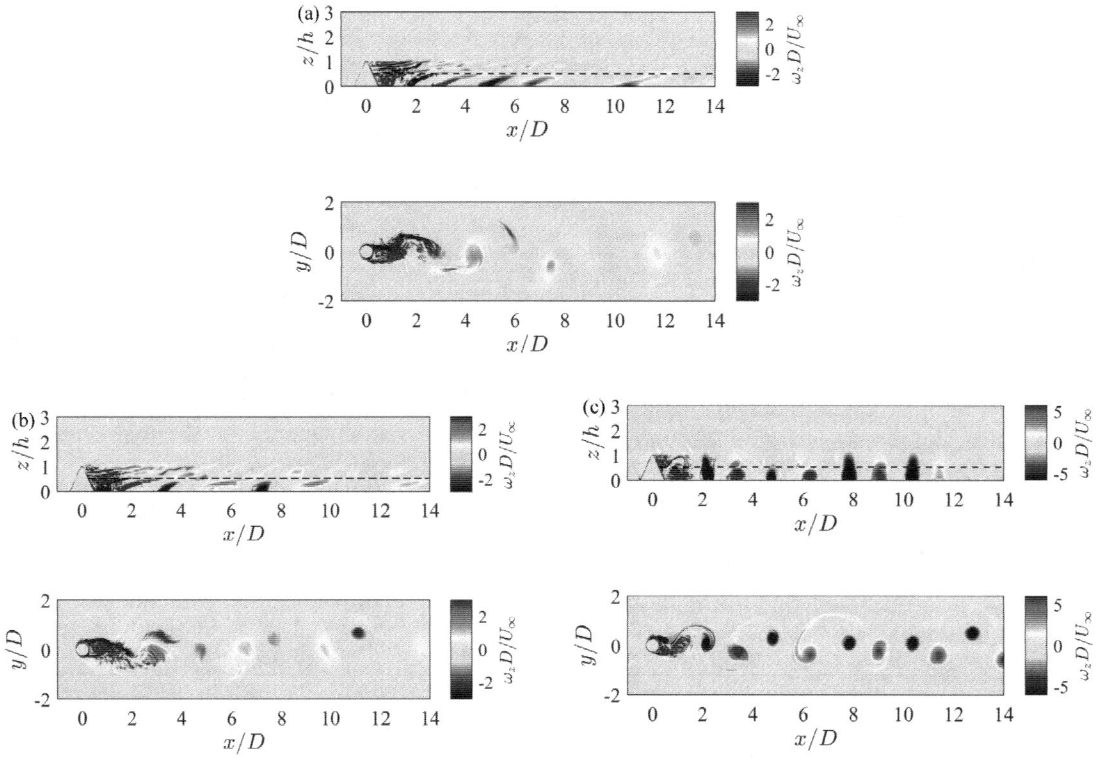

Figure 2: Flow visualizations of the vertical vorticity ω_z: (a) BuInf, (b) Bu25, (c) Bu1. Each subfigure shows ω_z on a vertical plane ($y = 0$) and on a horizontal plane ($z/h = 0.50$). Note that in (c), the magnitude of ω_z (in convective units) is much greater than in the other two cases, hence its colorbar range is different. The black dashed line in each vertical plane corresponds to $z/h = 0.50$.

not have sufficient energy to flow over the hill but can only bypass it on the sides. As a result, well-defined patterns of Kármán vortex shedding are found in horizontal planes. As the strength of rotation increases (Burger number decreases), the vertical structures of vorticity change from slanted 'tongues' to upright 'cones'.

It can also be seen that the near wake (herein $x/D < 3$) and the far wake ($x/D > 3$) exhibit different behavior. Stronger turbulence is generated near the body whereas the far wake contains predominantly coherent vortex shedding (VS) motion that persist to the end of the computational domain. It was also found that the non-hydrostatic pressure (deviation from the hydrostatic balance) is greater in the near wake and negligible in the far wake (not shown). This motivates us to separate their physics and modelling. It was found in all three cases in table 1 that there are global VS modes and a planar VS frequency that doesn't depend on vertical location [1]. This suggests a vertical correlation scale much greater than the vertical scale $U_\infty/N = 0.15h$ restricted by stratification.

In this work, we perform POD on the near and far wake of [1] separately, analyze the different limit cycles corresponding to the near and the far wake, and unveil the influence of rotation on the system. The phase portraits of POD coefficient and the correlations between planes at different heights are also studied to shed light on possible reduced-order modelling strategies.

2. Proper orthogonal decomposition

The POD problem [7, 8, 9] is to solve the following eigenvalue problem

$$\mathcal{R}\phi_i(\boldsymbol{x}) = \int_{\mathcal{A}} \boldsymbol{R}(\boldsymbol{x}, \boldsymbol{x}')\boldsymbol{W}(\boldsymbol{x}')\phi_i(\boldsymbol{x}')\,\mathrm{d}\boldsymbol{x}' = \lambda_i\phi_i(\boldsymbol{x}), \tag{5}$$

with $\{\phi_i\}_{i=1}^{\infty}$ as the basis functions which are mutually orthogonal over the domain $\mathcal{A}$. Here $\boldsymbol{R}(\boldsymbol{x}, \boldsymbol{x}') = \langle \boldsymbol{q}(\boldsymbol{x}, t)\boldsymbol{q}(\boldsymbol{x}', t)\rangle$ is the spatial correlation tensor of a zero-mean generic signal $\boldsymbol{q}(\boldsymbol{x}, t)$ and $\boldsymbol{W}(\boldsymbol{x})$ is a symmetric weight matrix that makes the following inner product,

$$(\boldsymbol{q}_1, \boldsymbol{q}_2)_{\boldsymbol{W}} = \int_{\mathcal{A}} \boldsymbol{q}_2^{\mathrm{T}}(\boldsymbol{x})\boldsymbol{W}(\boldsymbol{x})\boldsymbol{q}_1(\boldsymbol{x})\,\mathrm{d}\boldsymbol{x}, \tag{6}$$

suitable for defining an energy norm ($\|\boldsymbol{q}\|_{\boldsymbol{W}} = (\boldsymbol{q}, \boldsymbol{q})^{1/2}$). Typically, $\boldsymbol{W}$ contains the constants of numerical quadrature in a discrete system.

In this work, the wake is divided into two separate regions: the near wake (A1, $x/D < 3$), and the far wake (A2, $x/D \geq 3$). The domain decomposition can be seen in figure 4. The POD problem is solved on $\mathcal{A} = \mathcal{A}_1$ and $\mathcal{A} = \mathcal{A}_2$ separately. The simulation data is arranged into

$$\boldsymbol{Q} = [\boldsymbol{q}_1, \boldsymbol{q}_2, \cdots, \boldsymbol{q}_{N_t}] \in \mathbb{R}^{N_d \times N_t} \tag{7}$$

where each $\boldsymbol{q}_i \in \mathbb{R}^{N_d}$ is one snapshot, N_d is the degree of freedom of the data, and $N_t = 2000$ is the number of snapshots. The dimension of the discrete eigenvalue problem is then $\min(N_d, N_t) = N_t$ in the present database and hence the POD is performed in the temporal domain via the method of snapshot [10].

The discrete estimate of the correlation tensor is

$$\boldsymbol{R} = \langle \boldsymbol{q}\boldsymbol{q}^{\mathrm{T}}\rangle = \frac{1}{N_t - 1}\boldsymbol{Q}\boldsymbol{Q}^{\mathrm{T}} = \frac{1}{N_t - 1}\sum_{i=1}^{N_t} \boldsymbol{q}_i\boldsymbol{q}_i^{\mathrm{T}} \tag{8}$$

where the factor $N_t - 1$ is the Bessel's correction such that the expectation of the sampled (co)variance converges to the (co)variance of the random signals. The discrete eigenvalue problem becomes

$$\boldsymbol{R}\boldsymbol{W}\boldsymbol{\Phi} = \boldsymbol{\Phi}\boldsymbol{\Lambda}, \tag{9}$$

or, equivalently, in the formulation of Sirovich's method of snapshots [10],

$$\frac{1}{N_t - 1}\boldsymbol{Q}^{\mathrm{T}}\boldsymbol{W}\boldsymbol{Q}\boldsymbol{\Psi} = \boldsymbol{\Psi}\boldsymbol{\Lambda}, \tag{10}$$

where the POD eigenmodes are then recovered as $\tilde{\boldsymbol{\Phi}} = \boldsymbol{Q}\boldsymbol{\Psi}\boldsymbol{\Lambda}^{-1/2}$ such that the eigenfunction expansion is $\boldsymbol{R} = \tilde{\boldsymbol{\Phi}}\boldsymbol{\Lambda}\tilde{\boldsymbol{\Phi}}^{\mathrm{T}}$ with the orthogonality condition for the eigenmodes being $\tilde{\boldsymbol{\Phi}}^{\mathrm{T}}\boldsymbol{W}\tilde{\boldsymbol{\Phi}} = \boldsymbol{I}$. For convenience, the tilde will be dropped from now on. The POD expansion is then

$$\boldsymbol{q}(\boldsymbol{x}, t) = \sum_{i=1}^{N_t} a_i(t)\phi_i(\boldsymbol{x}), \tag{11}$$

where the time-dependent amplitude $a_i(t) = \sqrt{\lambda_i}\psi_i(t)$ is the rescaled eigenfunctions in the temporal domain and (11) can be regarded as either an eigen-expansion in space or in time.

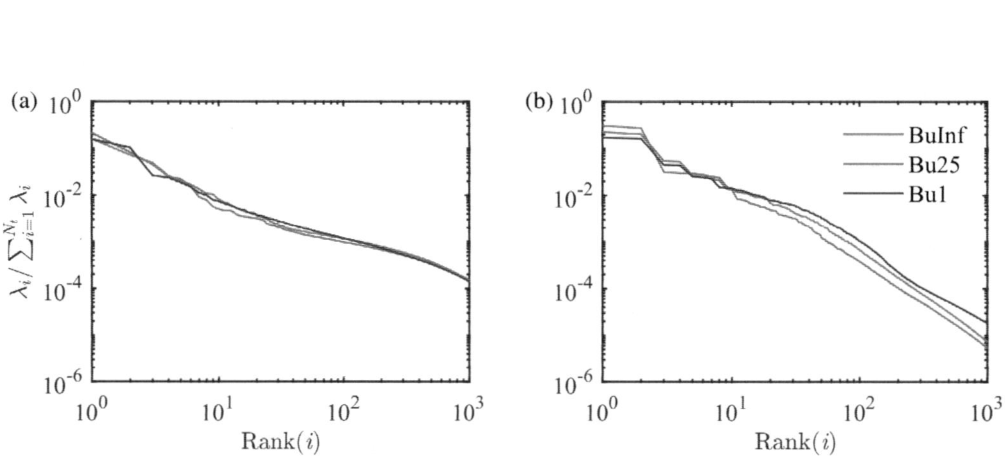

Figure 3: POD eigenspectra at $z/h = 0.25$, for BuInf (red), Bu25 (green), and Bu1 (blue): (a) the near wake; (b) the far wake. In each case and plane, the eigenvalues are normalized by the sum of all eigenvalues. The first thousand eigenvalues are shown. The first two modes account for greater than 30% of the total vertical enstrophy in all cases.

3. Results

3.1. POD eigenspectra and eigenmodes

In this work, POD is applied to the vertical component of vorticity (ω_z) in selected horizontal planes. Since the mean is removed before POD, the variance of the signal that is represented by POD is interpreted as the area integral of the fluctuation part of the vertical enstrophy ($\int_A \omega_z'^2(\boldsymbol{x})\,\mathrm{d}\boldsymbol{x}$). Figure 3 shows the POD eigenspectra normalized by total vertical fluctuation enstrophy in near and far wakes, respectively. The spectra at different Burger numbers are quite similar, despite the increasing enstrophy in smaller scales (higher POD rank) in the far wake, as rotation strengthens (Bu decreases). POD eigenvalues are found to occur as pairs in the far wake but not in the near wake, presumably because the far wake is advection dominated and less dissipative than the near wake, and hence can be better described by travelling-wave like modes (e.g., in figure 4) that imply the degeneracy of eigenvalues. In a laminar cylinder wake, POD eigenvalues occur in pairs and the corresponding eigenmodes differ by a phase shift of 1/4 period [11, 12] in order to accommodate a travelling-wave like structure [13, 14]. However, this is not quite the case in high-Reynolds number wakes, where the second POD eigenvalue is found to be smaller than the first one [15, 16].

An example of the symmetric leading mode in case BuInf is shown in figure 4. In case BuInf, the leading modes corresponding to (a_1, a_2) are symmetric about $y = 0$ (as shown in figure 4) and leading modes corresponding to (a_3, a_4) are antisymmetric (not shown). In cases Bu25 and Bu1, the strict symmetry is broken due to the rotation but can still be found visually [1].

For all three cases and in the plane $z/h = 0.25$ as in figure 3, the first four POD modes captures 31%-36% of the variance of ω_z', and the first ten POD modes captures 38%-43%, in the near wake. The far wake shows stronger low-rank behavior that the first four modes contains 42%-64% of the variance, while the first ten contains 53%-78%. The significantly increased low-rank behavior encourages the reduced-order modelling (ROM) of the far wake.

Although the present POD does not involve all velocity components, the vorticity modes have similar shapes and wavenumbers that are shared with the velocity modes (see [17]) and are more suitable for the present problem of a well-defined vortex wake. With respect to convergence of the numerical results, the first 10 POD eigenvalues are well converged using $N_t = 400$, and the first 100 eigenvalues are converged using $N_t = 4000$. To balance efficiency and accuracy, we only analyze the first 10 POD modes and the choice of $N_t = 2000$ is sufficient.

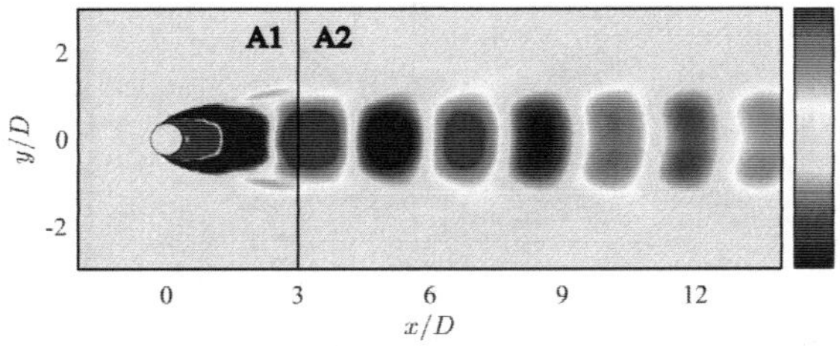

Figure 4: Leading POD modes of ω_z in the near (A1) and the far (A2) wakes pieced together, for plane at $z/h = 0.25$ in case BuInf. Note that the POD problem is solved in A1 and A2 separately, and the near wake mode is multiplied by a constant for better visual comparison.

3.2. Phase-space evolution of POD coefficients

The dynamics of wakes past bluff bodies of various cross-sections can be explained by a fixed point (mean flow) and a limit cycle around it (periodic shedding), after the first Hopf bifurcation [12, 18]. The limit cycle can be characterized by the periodic orbits of the first two coefficients of proper orthogonal decomposition (POD) modes, in wakes at Reynolds number ranging from laminar to turbulent regimes [15, 11, 12, 16, 17]. Following the framework proposed by [12], it was found in [17] that the coherent motions in the turbulent wake of a wall mounted cylinder can be well approximated by the subspace spanned by the first four POD modes. Thus, reduced order models (ROMs) of the present near and far wake are attractive. First, by comparing the leading modes measured in the far wake, it becomes possible to infer the information (such as the Rossby number which depends on the dimension of the wake generator), in an inverse problem. On the other hand, in simulations of far wakes, the time step is restricted by the fine grid near the body for a turbulence-resolving simulation and, since a substantial size of domain ($O(100D)$) is needed to reach the far wake, the computational cost becomes quite large [19, 20]. The combination of ROM (such as a Galerkin system) and broadband turbulence as a substitute inflow data generator that replaces the time-resolved near wake is appealing.

In this section, we consider the framework of Galerkin projection and it's applicability in the present flow. The time-resolved snapshots are projected on the POD modes to obtain the expansion coefficient in (11):

$$a_i(t) = \frac{(\boldsymbol{q}(\boldsymbol{x},t), \boldsymbol{\phi}_i(\boldsymbol{x}))_{\boldsymbol{W}}}{(\boldsymbol{\phi}_i(\boldsymbol{x}), \boldsymbol{\phi}_i(\boldsymbol{x}))_{\boldsymbol{W}}}. \tag{12}$$

From eqn. (11) it can also be shown that

$$\langle a_i a_j \rangle = \delta_{ij}\lambda_i, \tag{13}$$

i.e., the expected value of $\langle a_i^2 \rangle$ is λ_i. Assume the evolution of $a_i(t)$ takes the form of

$$a_i(t) = \sqrt{2\lambda_i}\cos(t), \tag{14}$$

where the factor $\sqrt{2}$ is the ratio of the maximum to the variance of a sinusoidal signal, such that the expectation of the maximum of $a_i(t)/\sqrt{2\lambda_i}$ is unity. This normalization is used for figure 5.

The trajectories of the leading pairs of POD coefficients (a_1, a_2) and (a_3, a_4) are plotted in figure 5, which are reminiscent of Lissajous curves. The trajectories of (a_1, a_i) where a_i $(i > 2)$ is any higher-rank POD coefficient are Lissajous curves with different frequency and phase relations, similar to [11, 17] and are not shown. The first few pairs (a_{2i-1}, a_{2i}) are found to be quasi periodic in laminar and turbulent bluff body wakes [15, 11, 16, 17]. Even though each

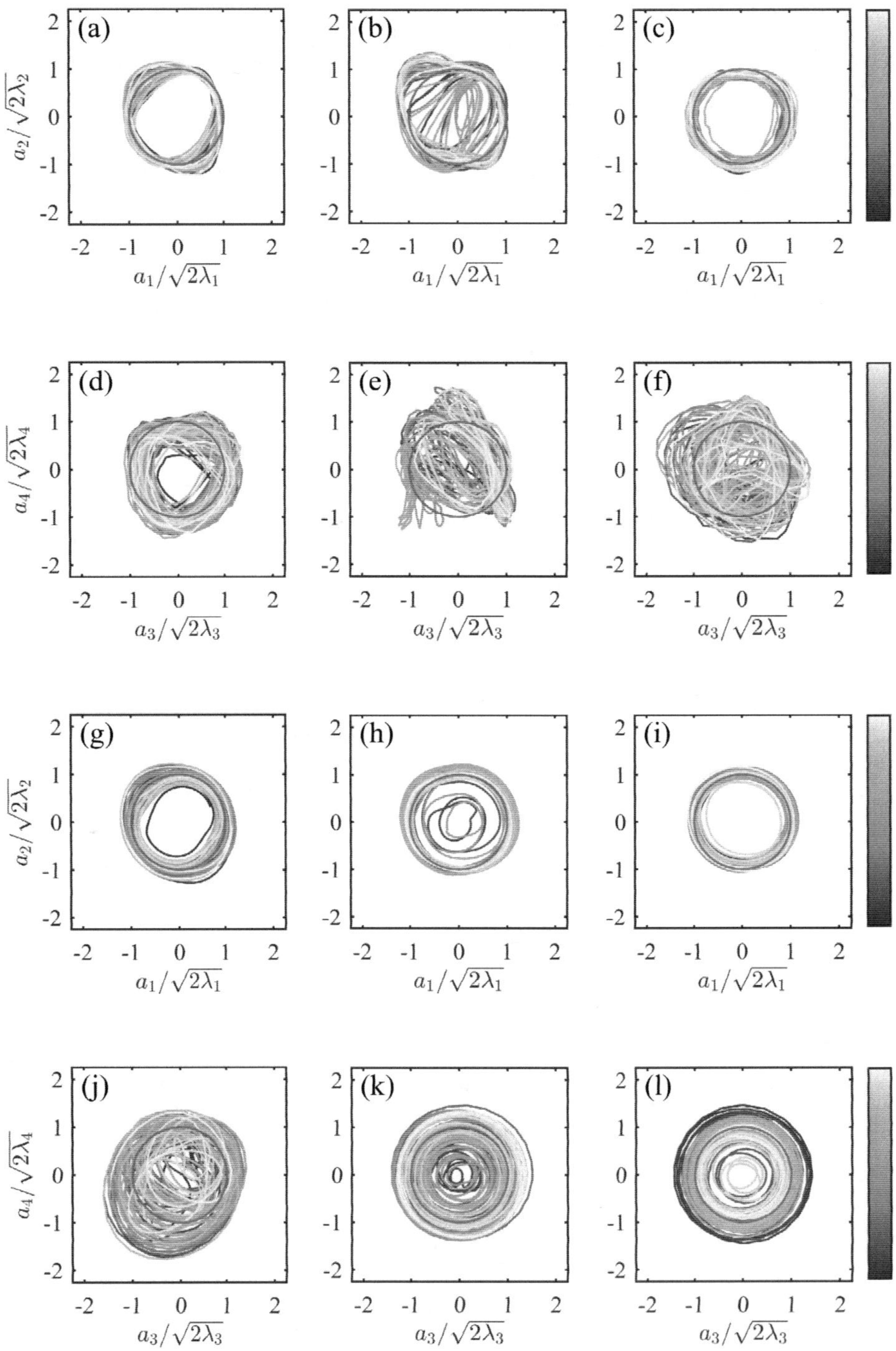

Figure 5: Trajectories of the pairs of POD coefficient (a_1, a_2) and (a_3, a_4). Left: BuInf; middle: Bu25, Right: Bu1. First two rows: near wake; last two rows: far wake. Plane locates at $z/h = 0.25$ and color denotes time (in convective units) that spans more than 30 VS cycles. The signs of a_2 and a_4 are chosen to make all the trajectories counter-clockwise. A red unit circle is also plotted as a reference.

pair of POD modes is statistically uncorrelated by construction, the members of the pair are not independent.

In figure 5, most of the trajectories are orbital while some of them are space-filling. In general, the near wake (figure 5, top two rows) is less ordered than the far wake (figure 5, bottom two rows). As the coherent VS structures emerge and dominate, the phase trajectories of both pairs of (a_1, a_2) and (a_3, a_4) approach circular orbits. Comparatively, the phase space of (a_3, a_4) is more filled than that of (a_1, a_2) for all cases, corresponding to a greater extent of variance in higher-rank modes. For the pair (a_3, a_4), although the orbits are closer to being circular going from the near to the far wake, there is still a large amplitude variation $a_3/\sqrt{2\lambda_3}, a_4/\sqrt{\lambda_4}$ ranging from 0 to about 1.5 in figure 5.

Given the nearly circular orbits of (a_1, a_2) in figure 5, and the fact that leading POD modes occur in pairs, we define the amplitude and phase of the first pair of POD modes that represent the VS motion as

$$\theta = \arctan(\frac{a_2/\sqrt{2\lambda_2}}{a_1/\sqrt{2\lambda_1}}) \tag{15}$$

$$r = \sqrt{(\frac{a_1}{\sqrt{2\lambda_1}})^2 + (\frac{a_2}{\sqrt{2\lambda_2}})^2}, \tag{16}$$

such that $(a_1/\sqrt{2\lambda_1}, a_2/\sqrt{2\lambda_2}) = re^{i\theta}$. This POD phase is found to agree well with the phase identified from the pressure signals [21]. Both θ and r will be analyzed in detail in a later section.

3.3. Inter-plane correlation between POD coefficients

With the aim of understanding the inter-plane connection that could be helpful in the modelling such flows, we study the correlations between POD coefficients and amplitudes. As seen in the previous section, the first two POD coefficients (a_1 and a_2) are not independent, and represent the same dynamics of VS motion. Hence, we only analyze the inter-plane correlation of a_1 in this section. The correlation of a_1 at two different heights is defined as

$$C_1(\Delta t; z_1, z_2) = \frac{\langle a_1(t; z_1) a_1(t + \Delta t; z_2)\rangle}{\langle a_1(t; z_1)^2\rangle^{1/2} \langle a_1(t; z_2)^2\rangle^{1/2}}, \tag{17}$$

where z_1, z_2 are the distance from the bottom wall. Since $C_1(\Delta t) \neq C_1(-\Delta t)$ in general, we require the order of $z_1 < z_2$. Moreover, the value of $C_1(0; z_1, z_2)$ can be used to evaluate the phase lag between two planes, assuming that both are evolving at the same frequency. Assume two harmonic signals in $[0, 2\pi]$

$$q_1(t) = \sin t,\ q_2(t) = \sin(t + \psi) \tag{18}$$

share the same frequency but only differ by a phase shift. Their correlation at $\Delta t = 0$ is

$$C_q = \frac{\langle q_1 q_2\rangle}{\langle q_1^2\rangle^{1/2} \langle q_2^2\rangle^{1/2}} = \frac{\langle \frac{1}{2}\cos\psi - \frac{1}{2}\cos(2t + \psi)\rangle}{\frac{1}{\sqrt{2}}\frac{1}{\sqrt{2}}} = \cos\psi, \tag{19}$$

which only depend on the phase difference (ψ) between these two signals. Here $\langle\cdot\rangle$ is interpreted as $1/(2\pi)\int_0^{2\pi} dt$. If q_1 and q_2 are in phase ($\psi = 0$) then $C_q(0) = 1$ is true.

Figure 6 compares the correlation (17) between the near and the far wake. The pairs of planes are taken from $z/h = 0.12, 0.25, 0.50, 0.75$. The correlation coefficients are harmonic, with the period being close to the VS period ($T_{\rm VS} \approx 4D/U_\infty$), but the maximum can vary. A harmonic correlation function with a maximum close to unity indicates that the POD modes of different planes oscillate at the same frequency but only with a phase shift. This is the case in figure 6.

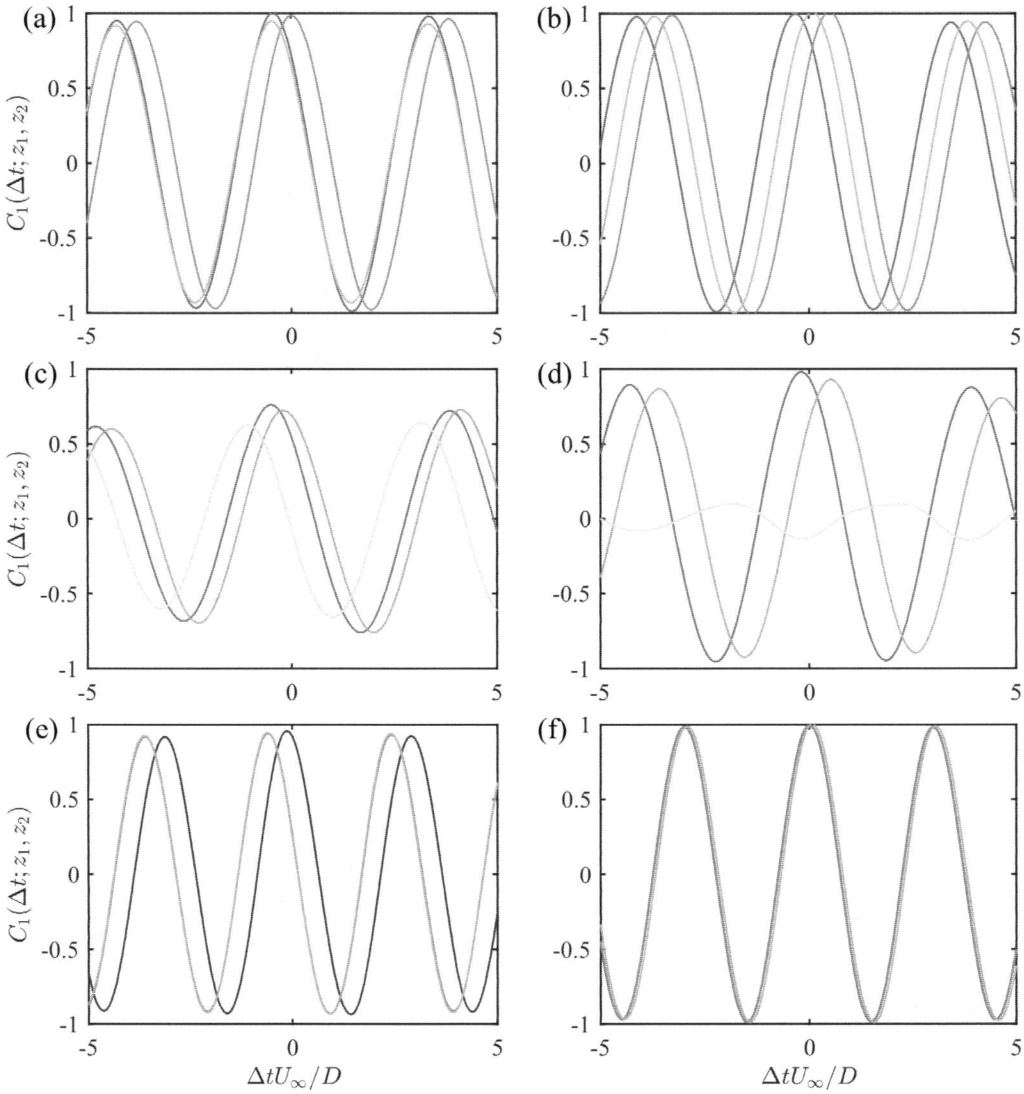

Figure 6: POD correlation coefficients $C_1(\Delta t; z_1, z_2)$ of the first mode at two different planes. (a,c,e) show the near wake (A1) and (b,d,f) show the far wake (A2). (a,b) BuInf; (c,d) Bu25; (e,f) Bu1. Colors from dark to light indicate pairs of planes at $(z_1/h, z_2/h) = (0.12, 0.25), (0.12, 0.50), (0.12, 0.75)$.

For case BuInf, both the correlations in the near and the far wake achieve the maximum close to unity, indicating that a constant VS frequency is maintained both in the near and the far wake but the phase lag between different planes has change a bit.

The correlations in case Bu1 are similar to BuInf in the near wake, both in terms of phase lag and maximum. However, the far wake in case Bu1 shows exactly the same phase in the four planes examined, as implied by $C_1(0; z_1, z_2) = 1$ and according to (19). This is in agreement with the vertically aligned vortex 'columns' shown in figure 2(c) where the VS in each height are in phase. The simpler vertical structures in the strongly rotating case Bu1 implies the possibility of neglecting the inter-plane phase difference of VS when modelling is considered.

Contrary to BuInf and Bu1, the inter-plane correlation in case Bu25 is larger in the far wake than in the near wake, except for the farthest pair $z/h = (0.12, 0.75)$. This increased degree of order is consistent with the phase portraits of (a_1, a_2) and (a_3, a_4) (middle column of figure 5) which are more regular in the far wake. Additionally, as shown in figure 6(d), the first POD mode at $z/h = 0.75$ in case Bu25 is hardly correlated to that at $z/h = 0.12$. This doesn't indicate a complete loss of vertical coherence at $z/h = 0.75$ but is a result of mixed frequencies in POD modes. A spectral POD analysis which separates structures that oscillate at different frequencies was performed in [1] and the dominant mode in that plane is still the VS mode at the VS frequency, and the three-dimensional coherent VS structures are found to extend vertically beyond $z/h = 0.75$. In all, the difference observed in Bu25 implies a non-monotonic dependence of wake coherence on rotation, which needs to be accounted for in any modeling strategies.

From figure 6, it can be inferred that the correlation coefficients can generally achieve a value of nearly unity for a certain time lag Δt between two planes separated at $z_1/h = 0.12$ and $z_2/h = 0.75$, suggesting that the flow remains coherent in this range. Note that the uppermost and lowermost planes ($z/h = 0.12$ and $z/h = 0.75$) are chosen to be the rough boundaries of the core of the vortex wake - neither too close to the top of the hill to be influenced by internal gravity waves nor in the bottom boundary layer where the wake is disturbed. This gives some hope that the core of the wake - isolated from the obstacle as well as the bottom wall - can be described with a ROM.

In all cases, the maximum location of $C_1(\Delta t) = 1$ nearest to $\Delta t = 0$ tend to be negative in the near wake. Combining with (17) and the fact that $z_1 < z_2$, it implies that the upper planes lead in the phase, consistent with figure 2 and the global modes in [1]. The phase of each plane will be further analyzed in the next section.

3.4. Analysis of the phase and the amplitude of the VS mode

In order to model these wakes by assuming periodic base flow in the form of periodic VS motions in each plane with a phase shift between planes, one needs to know whether constant VS frequency and phase lag are maintained throughout the evolution. Also, the variability of the amplitude r of the periodic leading mode $a_1\phi_1 + a_2\phi_2$ is important to modelling. Besides, as shown in figure 5, the variation of the amplitude of the leading modes during their evolution is not negligible. The correlation between amplitudes at two offset planes would be of interest as it gives an idea as to how much the knowledge in one plane can be used to infer another and whether this system can be further reduced. In this section, the phase θ and amplitude r of the orbital motion of (a_1, a_2), defined in (15), will be analyzed.

Figure 7(a) shows the phase angle of the leading pair defined in (15) in the far wake as a continuous function of time. The angle is allowed to exceed 2π as more cycles elapse. The slopes give an estimate of the averaged VS Strouhal number $St_{\rm VS} \approx 0.26$ in all three cases, and the VS frequency is quite constant during the evolution, for all cases (excluding $z/h = 0.75$ in Bu25 that will be discussed below).

For case Bu25 (figure 7c), the phase angle in plane $z/h = 0.75$ varies significantly from others, and leads to a frequency almost twice of the shedding frequency. This is due to the fact that in this case the f-k_x relation of the POD modes is not linear, or in other words, there are mixed frequencies. Here f and k_x are the dominant frequency and streamwise wavenumber of a POD mode, respectively. In cylinder wakes [11, 12] or the present hill wake [1] the relation between f and k_x is quite linear - a pair of POD modes will have a single characteristic streamwise wavenumber and a single frequency of associated coefficients. For example, the first pair has a wavelength $k_x = 2\pi/\lambda_x$ (λ_x is the average separation between same-sign vortices) and evolves at $f_{\rm VS}$, and the second pair has a wavelength $2k_x$ and evolves at $2f_{\rm VS}$, and so forth. This is found to be true in the present wake (except for $z/h = 0.75$, case Bu25) for at least the best-resolved first 6 POD modes. This linear relation results in the same spatial modes if the spectral POD (SPOD) is carried out at the first 3 frequencies - $f_{\rm VS}, 2f_{\rm VS}, 3f_{\rm VS}$. In case Bu25, where the first

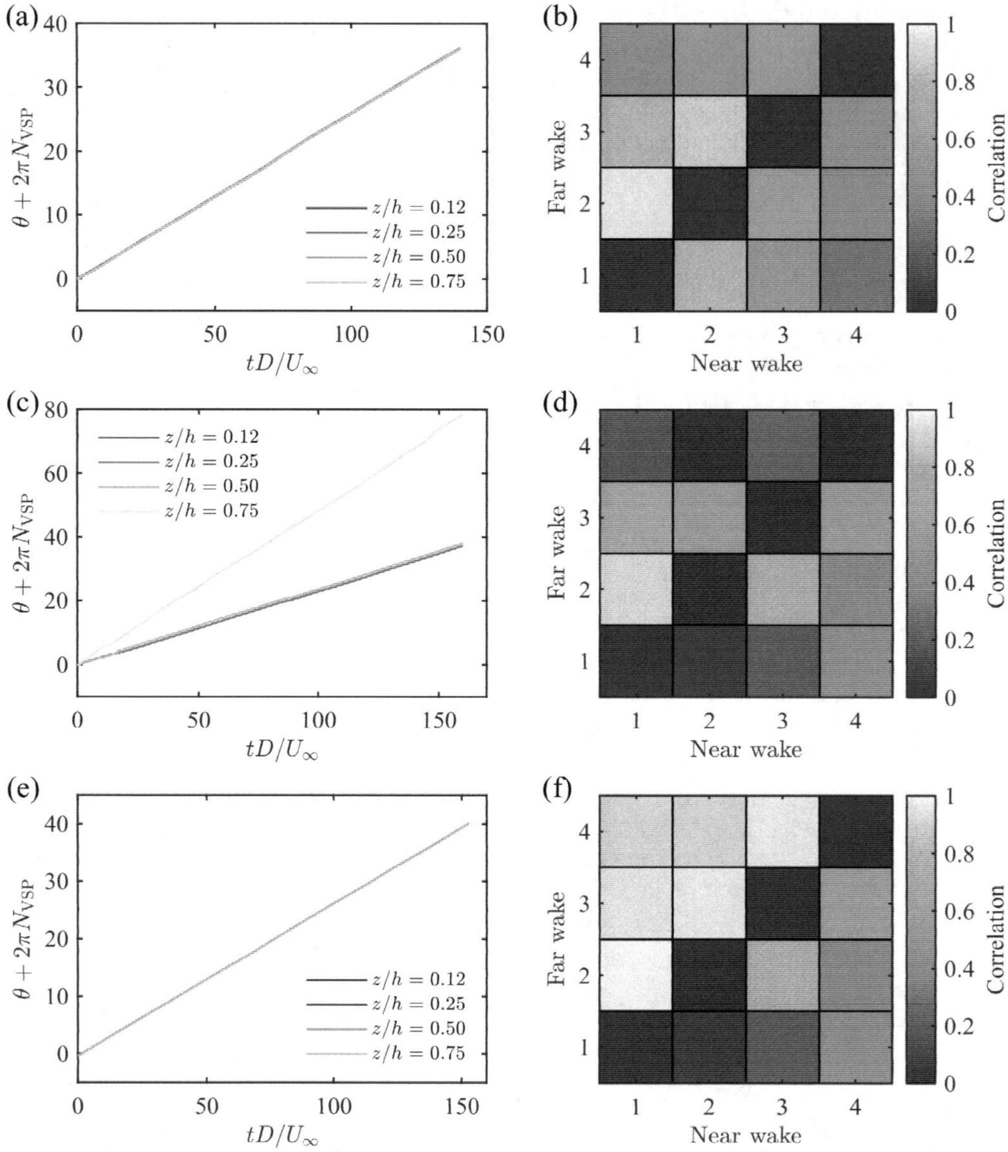

Figure 7: (a,c,e) The POD phase angle θ as a function of time, in the far wake. The phases are shifted to be continuous by adding integer multiples of the VS period (VSP). (b,d,f) Coherence map in the near wake and in the far wake. Cases Buinf (a,b), Bu25 (c,d), and Bu1 (e,f). Color represents correlation coefficient $C_r(\Delta t_{\max}; z_1, z_2)$ between two planes, where $z_1 < z_2$ and $\Delta t_{\max}$ is the optimal time lag between two planes that maximizes the correlation coefficient. The elements below the diagonal line are for the near wake, while those above are for the far wake. Elements on the diagonal line representing self-correlation which are expected to be unity are removed for clarity. Numbers from 1 to 4 represent planes at $z/h = 0.12, 0.25, 0.50, 0.75$.

4 POD modes all have the same wavelength (not shown) and similar shapes with a tendency to symmetry about $y = 0$ (not shown), hence the second harmonic ($2f_{\rm VS}$) is aliased to the projection coefficient of the first pair of POD modes. The preceding argument explains the anomalous slope in figure 7(c) and the low amplitude of the correlation function in figure 6 in the $z/h = 0.75$ plane. A SPOD analysis in [1] which is advantageous over POD in this special situation ascertains that the dominant frequency of this plane is still the same as other planes ($f_{\rm VS}$).

The correlations between the amplitude of the first pair POD modes (defined in (17)) at two different planes are compiled into correlation maps in figure 7(b,d,f). A unit amplitude of r is regarded as a baseline of the model periodic flow $(a_1, a_2) = re^{i\theta}$ and is subtracted. We denote the deviation of r from unity as $r' = r - 1$ and similar to (17), the correlation between the values of r' in two different planes is denoted as $C_r(\Delta t_{\max}; z_1, z_2)$. The optimal time lag $\Delta t_{\max}$ that maximizes $C_r(\Delta t; z_1, z_2)$ differs among cases. To take advantage of the symmetry with respect to the commutation of z_1 and z_2, C_r in the near wake and the far wake are plotted in the lower and upper triangles of the matrices in figure 7, respectively, for better comparison. The matrices are colored according to the magnitudes of correlation coefficients. Here we use 1 to 4 to refer to planes from bottom to top.

In case BuInf, it can be seen that planes 1 and 2 have the strongest correlation in the near wake, while this correlation is strengthened in the far wake, accompanied by the increased correlation between planes 1, 3 and 2, 3. This also implies an emergence of order during the evolution. No matter in the near or in the far wake, the topmost plane 4 at $z/h = 0.75$ is not well correlated with others. In case Bu25, there is only one strong correlation in the near and in the far wake, respectively. And all planes are mutually less correlated for Bu25 compared to the other two cases. In case Bu1, there is only one pair of planes that have a correlation greater than 0.6 in the near wake. However, when the flow enters the far wake, the magnitudes show that all planes become correlated mutually, suggesting the best vertical coherence in this case and significant simplification in its modelling.

4. Summary and conclusions

Geophysical wakes visually exhibit coherent eddies in their horizontal motion (coherent vertical vorticity ω_z) as is well known from satellite and other images of cloud cover and ocean color. Less known is the systematic influence of density stratification and rotation on the dynamics of these wake eddies. Thus motivated, we conduct POD analysis of a time-resolved wake dataset. The specific problem is a steady current past a conical obstacle - a hill or seamount - with 30° slope and $Re = U_\infty D/\nu = 10\,000$. Unstratified flow past a conical obstacle does not exhibit coherent ω_z eddies. The stratification, which has to be sufficiently large for flow to be primarily around the obstacle than over it, is chosen so that the height-based Froude number is $Fr = U_\infty/Nh = 0.15$. System rotation is systematically increased from the non-rotating case to values corresponding to Rossby number $Ro = U_\infty/fD = -0.75$ and $Ro = -0.15$. The negative value of Ro follows from the chosen Southern Hemisphere location with negative (clockwise) rotation and Coriolis frequency (f_c). Associated with the change of rotation rate at fixed stratification, the Burger number $Bu = (Ro/Fr)^2$ (a measure of the relative strength of stratification to rotation) takes the values of ∞ (case BuInf), 25 (case Bu25) and 1 (case Bu1) corresponding to $Ro = \infty$, -0.75 and -0.15, respectively.

POD is the tool employed here to diagnose coherent dynamics in the simulated wakes with a focus on the effect of system rotation. POD is conducted on the vertical vorticity at selected horizontal planes ($z/h = 0.12$, 0.25. 0.5 and 0.75) chosen to encompass the wake core, away from the bottom boundary layer and also from the upper region with internal gravity waves. The leading members of the set of POD coefficients $\{a_i(t)\}$ are investigated and their cross-correlation between planes at different elevations is computed to reveal inter-plane coherence.

The near wake ($x/D < 3$) has turbulence and also interacts with internal waves. Therefore, the near and far wakes are analyzed separately.

Vortex shedding at a specific frequency and advection of the vortices by a constant speed would lead to a limit cycle in the trajectory of $a_2(t)$ plotted against $a_1(t)$. The expectation that an increase of rotation rate (decreasing $|Ro|$ and decreasing Bu) would lead to a *systematic* increase of coherence as measured by POD is *not* borne out. The overarching conclusion of the POD analysis is that the dependence is more complex as summarized and discussed below.

For the near wake, trajectories in $a_1(t), a_2(t)$ coordinates show that there is a *decrease* in coherence relative to the baseline non-rotating case when $|Ro|$ decreases to 0.75 (case Bu25) followed by an increase when $|Ro|$ decreases to 0.15 (case Bu1). In particular, the trajectories in case Bu25 (see figure 5b) are space-filling rather than repeating on an orbital path. Since $Ro = -0.75$ is in the submesoscale regime, the newly-formed vortices can be unstable to submesoscale instabilities that are known to occur at $Ro = O(1)$ and cause deviations from a defined orbit. For the far wake, it is the cases $Bu = 25, 1$ with rotation that display more ordered trajectories while the non-rotating case (BuInf) show more variance.

The inter-plane cross-correlation coefficient $C_1(\Delta t; z_1, z_2)$ is also computed to diagnose correlation across planes at different elevations. The characteristic vertical scale is U/N which, when nondimensionalized by h becomes $Fr = U_\infty/Nh$ and takes the value of 0.15 in all cases. The correlation between the VS modes in different horizontal planes can stay correlated with a vertical separation several times greater than the stratification length scale for overturn motions (U/N), as can be seen in figure 6. The vertical coherence and temporal periodicity of vortex motions encourages the ROM of such wakes. The simplest model of the leading POD coefficients that is consistent with vortex shedding is a limit cycle with frequency that is constant across z_1 and z_2. Such a model would lead to $C_1(\Delta t; z_1, z_2)$ taking the form of a harmonic function with unity amplitude. Indeed, the non-rotating, strongly-stratified case BuInf and the strongly-rotating, strongly-stratified case Bu1 do conform to the simple model in both the near and the far wake. It is intermediately strong rotation that breaks this model. At $Ro = -0.75$ (Bu25), the behavior is that of a harmonic function but with amplitude reduced from 1 to about 0.6 in the near wake. The amplitude recovers to near unity in the far wake except for the ($z_1 = 0.12, z_2 = 0.75$) pair which includes planes at the upper and lower edges of the wake core.

A correlation map (figure 7, right column) summarizes inter-plane correlation. It is evident that as the flow progresses from the near to the far wake there is a strong increase in mutual correlation, i.e., a significant increase in order. We conclude that ROM could provide a possible surrogate for the coherent part of wake dynamics and also as part of the inflow generation procedure for more complex problems, e.g., the interaction of wake vortices with downstream features such as broad-band roughness, other isolated topographic elements, a different stratification, etc. A caveat is that, at the large Re of geophysical flows, additional modeling would be needed for the incoherent part and its potential feedback to the low-order model. Furthermore, it is quite possible that, at even stronger stratification and values of Fr less than 0.15, the inter-plane coherence of wake eddies would be disrupted [22].

Acknowledgements

The authors would like to acknowledge the financial support from ONR grant N00014-22-1-2024. This work was also supported by the European Research Council under the advanced grant ERC-AdG-101018287 ('CausT') during the stay of J. L. in the fifth Madrid turbulence summer school, for which the hospitality by the Fluid Dynamics Group at Universidad Politécnica de Madrid is gratefully acknowledged. We especially thank Scott Dawson and Barbara Lopez-Doriga from Illinois Institute of Technology for reviewing the draft of this paper.

References

[1] Liu J, Puthan P and Sarkar S 2023 *to appear on arXiv*
[2] Nicoud F and Ducros F 1999 *Flow, Turbulence and Combustion* **62** 183–200
[3] Balaras E 2004 *Computers & Fluids* **33** 375–404
[4] Yang J and Balaras E 2006 *Journal of Computational Physics* **215** 12–40
[5] Puthan P, Jalali M, Ortiz-Tarin J L, Chongsiripinyo K, Pawlak G and Sarkar S 2020 *Ocean Model.* **149** 101611
[6] Garcia-Villalba M, Li N, Rodi W and Leschziner M 2009 *J. Fluid Mech.* **627** 55–96
[7] Lumley J L 1967 *Atmospheric turbulence and radio wave propagation* 166–178
[8] Lumley J L 1970 *Stochastic tools in turbulence* (Academic Press)
[9] Holmes P, Lumley J L, Berkooz G and Rowley C W 2012 *Turbulence, Coherent Structures, Dynamical Systems and Symmetry* 2nd ed (Cambridge university press)
[10] Sirovich L 1987 *Quarterly of applied mathematics* **45** 561–571
[11] Ma X and Karniadakis G E 2002 *J. Fluid Mech.* **458** 181–190
[12] Noack B R, Afanasiev K, Morzyński M, Tadmor G and Thiele F 2003 *J. Fluid Mech.* **497** 335–363
[13] Rempfer D and Fasel H F 1994 *J. Fluid Mech.* **260** 351–375
[14] Taira K, Brunton S L, Dawson S T, Rowley C W, Colonius T, McKeon B J, Schmidt O T, Gordeyev S, Theofilis V and Ukeiley L S 2017 *AIAA J.* **55** 4013–4041
[15] Ma X, Karamanos G S and Karniadakis G 2000 *J. Fluid Mech.* **410** 29–65
[16] Oudheusden B v, Scarano F, Hinsberg N v and Watt D 2005 *Experiments in Fluids* **39** 86–98
[17] Bourgeois J, Noack B and Martinuzzi R 2013 *J. Fluid Mech.* **736** 316–350
[18] Barkley D 2006 *Europhys. Lett.)* **75** 750
[19] VanDine A, Chongsiripinyo K and Sarkar S 2018 *Computers & Fluids* **171** 41–52
[20] Ortiz-Tarin J, Nidhan S and Sarkar S 2021 *J. Fluid Mech.* **918** A30
[21] Perrin R, Braza M, Cid E, Cazin S, Barthet A, Sevrain A, Mockett C and Thiele F 2007 *Exp. Fluids* **43** 341–355
[22] Perfect B, Kumar N and Riley J 2018 *Geophysical Research Letters* **45** 9098–9105

Energy transfer in the formation and evolution of energy-containing motions in turbulent boundary layers

Antonio Matas, Kevin Liu, Ezhilsabareesh Kannadasan, Callum Atkinson and Julio Soria

Laboratory for Turbulence Research in Aerospace & Combustion (LTRAC), Department of Mechanical and Aerospace Engineering, Monash University, Victoria 3800, Australia

E-mail: antonio.matassalva@monash.edu

Abstract. This paper examines using direct numerical simulation (DNS) the recovery of the energy-containing scales of motion (energy-containing motions) when these are removed at the inflow boundary condition of a flat plate zero-pressure gradient turbulent boundary layer (ZPG-TBL). The energy-containing motions are identified from the spanwise spectral energy density of the interscale turbulent transport. Analysis of the resulting spanwise spectral energy densitiy of velocity and vorticity suggests that near-wall streaks are the initial energy-containing motions formed with quasi-streamwise vortices only developing further downstream. This result is in agreement with the previous study on turbulent channel flows by Kannadasan *et al.* [1].

1. Introduction

In wall-bounded flows, turbulence is often interpreted in the frame of the attached-eddy model (AEM), which conceptualises the flow as a superposition of energy-containing eddying motions spanning a wide range of sizes across the boundary layer. The original model, confined to the so called logarithmic region (log-region), builds on the hypothesis that the flow in this region is dominated by a superposition of self-similar energy-containing motions whose flow field is attached to the wall and whose geometry scales with distance from their center to the wall [2, 3]. However, although the existence of such attached-eddies is endorsed by a growing number of numerical and experimental studies, the precise details about which motions constitute each attached-eddy (*i.e., their structure*) and their dynamical origin have proven a challenging task [4, 3]. Ultimately, extending the AEM beyond the log-region or, more broadly, formulating a model that describes the eddy motions of wall-turbulence as a set of representative motions would be desirable to facilitate the understanding and ability to make predictions for these flows. A recurrent approach to clarify such aspects has consisted in conducting numerical experiments that simulate isolated length scales in the flow, achieved by considering a subset of the flow, restricting the range of wavenumbers, or a combination of both strategies [4]. These experiments provided evidence supporting that energy-containing motions complete a self-sustaining regeneration cycle in the near-wall [5, 6, 7] and in the log [8, 9] and outer regions [10], suggesting the possibility that their generation at each length scale may rely on an autonomous process independent of the rest of the flow. Recent work [11, 12] suggested that the statistical features of the self-sustaining energy-containing motions at a given spanwise length scale are consistent with Townsend's attached-eddies and that a single attached-eddy would be composed of two distinct elements, one of which would be a long streaky motion reaching the near-wall region and the other a vortex packet. Hwang [11] further inferred that the smallest attached-eddy would correspond to a buffer-layer streak with quasi-streamwise vortices aligned to it, whereas the largest one to a very-large-scale motion (VLSM) with large-scale motions (LSMs) aligned to it. Hence, these findings derive into a description of the eddy structure of wall-turbulence by incorporating all known coherent structures into the framework of the attached-eddy theory [11]. However, despite the body of numerical work quantifying the nature of the energy-containing motions of wall-turbulence, the specific details of how these motions originate remain unclear. With this in mind, a direct numerical simulation (DNS) of a flat plate zero-pressure-gradient turbulent boundary layer (ZPG-TBL) in which the energy-containing motions are removed at the inflow boundary has been undertaken to investigate this aspect. This numerical experiment therefore aims to unveil mechanisms that might otherwise remain hidden in a canonical ZPG-TBL flow, and to evidence them by examining the role that different length scales play on the recovery of such motions within the flow. Given that the energy-containing motions of wall-turbulence can be characterised by their spanwise wavelength [8, 11], and leveraging the spanwise statistical homogeneity of boundary layers, this study makes use of the evolution equation of turbulent kinetic energy (TKE) for each spanwise Fourier mode to identify and study these motions. In particular, we associate the bulk of the energy-containing motions with a distinctive connected (y, λ_z)-region within the spanwise spectral energy density of the interscale turbulent transport that is characterized by negative values. This region in the (y, λ_z)-plane is representative of spanwise scales of motion that, on the average, supply most energy to other spanwise scales of motion and are therefore defined as *the energy-containing motions*. Given that the numerical method employed in the DNS of the ZPG-TBL is spectral in the spanwise coordinate, it enables the precise elimination of the velocity fluctuations within the (y, λ_z)-region that represent the energy-containing motions at the inflow boundary of the ZPG-TBL. This study therefore focuses on analysing the statistical behaviour of the flow when the bulk of the energy-containing motions is removed at a flow cross-plane, as opposed to a structural approach where individual coherent motions are identified and suppressed in order to study their genesis and dynamics. The remainder of this paper is organized as follows. The details of the analysis method and of the numerical experiments are presented in §2. The streamwise evolution of the mean velocity and Reynolds stresses and of the the spectral energy densities of velocity and streamwise vorticity are examined in §3. Finally, the conclusions and ongoing work are discussed in §4.

2. Methods

The ZPG-TBL is computed in a parallelepiped rectangular domain defined within a Cartesian coordinate system with the streamwise, wall-normal, and spanwise directions denoted by x, y and z. The corresponding instantaneous fluid velocities are denoted by $\widetilde{u}$, $\widetilde{v}$, and $\widetilde{w}$, respectively. In index notation, these are indicated by x_i and $\widetilde{u_i}$ with $i = \{1, 2, 3\}$, respectively. The wall on which the ZPG-TBL develops is located at $y = 0$. The subscript $\cdot|_x$ is used to express the streamwise coordinate value at which a variable is evaluated. The ensemble average $\langle\cdot\rangle$ denotes averaging over the spanwise homogeneous direction and time. Mean values are denoted by capital letters, $U_i = \langle\widetilde{u_i}\rangle$, while fluctuations with respect to these averages by lower-cases, $u_i = \widetilde{u_i} - U_i$. The notation $\widehat{\cdot}$ denotes Fourier expansion coefficient in the spanwise direction, for instance $u(t, x, y, z) = \sum_k \widehat{u}(t, x, y, k_z)e^{ik_z z}$, with k_z the spanwise wavenumber with associated spanwise wavelength $\lambda_z = 2\pi/k_z$. The one-dimensional spanwise spectra are premultiplied by k_z to express energy density per $\log k$ on the log-scale plots. The terms in the transport equation are likewise premultiplied by y. The superscript '+' denotes quantities in viscous units that are non-dimensionalised appropriately with the kinematic viscosity ν and the streamwise-dependent friction velocity u_τ, taken at the evaluated x location unless otherwise specified and defined as $u_\tau \equiv \sqrt{\tau_w/\rho}$, where τ_w is the mean wall shear stress and ρ is the constant fluid density. The free-stream velocity is denoted by U_∞ and the 99% boundary layer thickness by δ. The Reynolds number based on the momentum thickness, θ, is defined as $Re_\theta = U_\infty \theta/\nu$.

2.1. Numerical code and experiments

As mentioned above, the turbulent boundary layer flow is directly simulated numerically in a parallelepiped domain over a smooth no-slip wall with spanwise periodicity and streamwise non-periodic inflow and outflow. The TBL DNS code solves the primitive-variable formulation of the Navier–Stokes equations for incompressible fluids, which are integrated using a fractional step method [13, 14] on a x-y staggered grid using semi-implicit low-storage third-order Runge-Kutta time-integration [15]. Fourth order compact spatial discretization [16] is employed for the convective and viscous terms in the streamwise and wall-normal directions. A second order discretization for the pressure is used in the directions perpendicular to the span, which is spectral. Dealiasing using the 2/3 rule is employed in the spanwise periodic direction. A complete detailed description of the numerical methodology and the code can be found in [15, 17].

The setup for the present DNS consists of two concatenated domains that run synchronously. The first domain, BL_1, runs a reference ZPG-TBL. This domain generates its own inflow by rescaling the velocity fluctuations of a downstream cross-plane located at $x = X_r$ using a method based on Lund *et al.* [18], with the mean inflow velocity fixed to a prescribed profile to ensure a constant mean inflow mass flux [17]. The velocity field at the cross-plane of this domain located at $x = X_t$, and termed the transfer plane, is employed to generate the turbulent inflow condition of the second domain, BL_2. In the current simulation, X_t is taken such that $Re_{\theta|X_t} = 2500$. BL_2 simulates a ZPG-TBL in which the energy-containing motions are removed at the inflow boundary condition, as described below, and their generation and downstream development are studied. The inflow of BL_2 is obtained by transferring at each time step the velocity field of the transfer plane with the energy-containing motions filtered out, as detailed in §2.2. The mean pressure gradient is controlled and kept small by imposing a streamwise dependent wall-normal velocity, constant in time, at the upper boundary [17]. Convective boundary conditions are applied at the outflow of both domains. Minor adjustments to the outflow boundary conditions are allowed to enforce mass conservation [15].

Each boundary layer is simulated within a computational domain of dimensions L_x, L_y and L_z. Both boundary layers share the same L_y and L_z and only L_x varies. L_x of BL_2 matches the distance from X_t to the outlet of BL_1. To minimize free-stream fluctuations, L_y is selected to be 2.4 times larger than the maximum value of the boundary layer thickness within the DNS, which, in this case, occurs at the outflow of BL_1. The corresponding collocated grid sizes are N_x, N_y and N_z respectively. The grid spacing is kept constant in the streamwise direction, Δx^+, and the spanwise direction, Δz^+. A non-uniform grid is employed in the wall-normal direction, with the smallest grid spacing, Δy^+_{min}, located near the wall. DNSs are run with a constant time-step size, Δt^+. The preliminary statistics presented here have been compiled over a time interval T after the initial transients were washed out. T is expressed in eddy-turnover time units u_τ/δ, with u_τ and δ taken at the BL_1 transfer plane.

Table 1. ZPG-TBL DNS parameters.

Case	Re_θ	$(L_x, L_y, L_z)/\delta$	Δx^+, Δy^+_{min}, Δz^+	N_x, N_y, N_z	Tu_τ/δ
BL_1	$290 - 4000$	$101.64, 4.82, 5.4\pi$	$11, 0.28, 5.51$	$8193, 315, 2730$	8.9
BL_2	$2500 - 3750$	$48.38, 4.82, 5.4\pi$	$11, 0.28, 5.51$	$3900, 315, 2730$	8.9

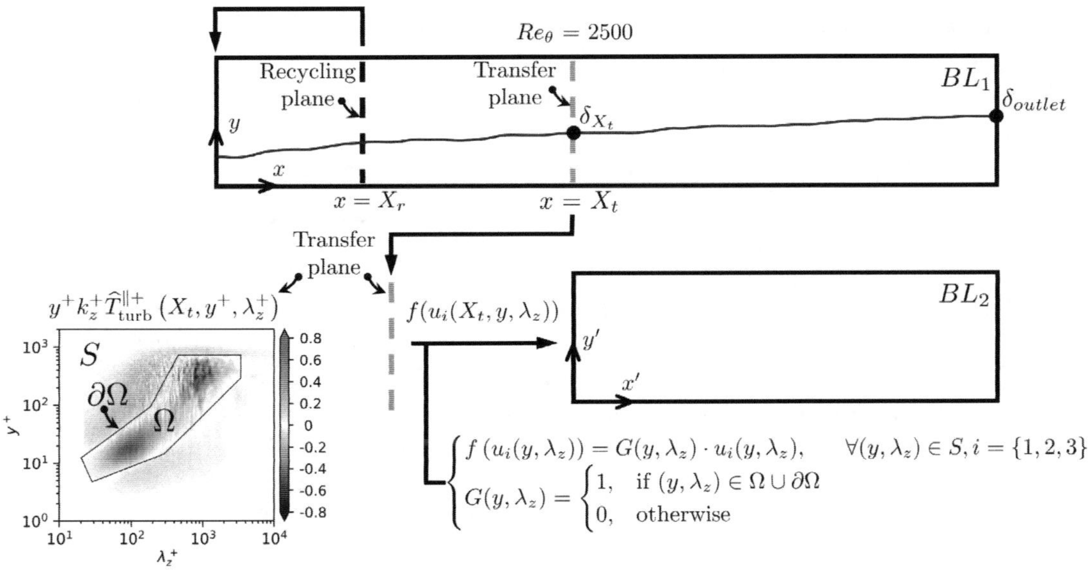

Figure 1. Sketch of the DNS set-up and of the filtering strategy employed for the removal of the energy-containing motions at the inflow boundary of the BL_2 ZPG-TBL DNS.

2.2. Formulation

To identify the energy-containing motions of the turbulence and study the energy transfers associated with their recovery when these are removed at the inflow boundary condition of a ZPG-TBL DNS, the evolution equation of TKE for each spanwise Fourier mode is considered. This equation can be derived by Fourier-transforming the trace of the equation of the two-point correlations of the velocity fluctuations [19, 20] or, equivalently, by starting from the Navier–Stokes equation for the Fourier modes of the velocity and pressure fluctuations [21, 22, 23]. We adapt the formulation in Mizuno [22] for channel flows to ZPG-TBLs, and express the evolution equation for the TKE spectrum, $\widetilde{e} = \left(|\widehat{u}|^2 + |\widehat{v}|^2 + |\widehat{w}|^2\right)/2$, as:

$$\frac{\partial\overline{\widehat{e}(x,k_z)}}{\partial t} = \underbrace{\mathrm{Re}\left\{\overline{-\widehat{u}^*\widehat{v}\frac{dU}{dy}}\right\}}_{\widehat{P}(x,y,k_z)} + \underbrace{-\nu\overline{\frac{\partial\widehat{u}_i}{\partial x_j}\frac{\partial\widehat{u}_i^*}{\partial x_j}}}_{\widehat{\epsilon}(x,y,k_z)} + \underbrace{\mathrm{Re}\left\{\overline{-\widehat{u_i}^*\frac{\partial}{\partial x_j}\widehat{(u_iu_j)}}\right\}}_{\widehat{T}_{turb}(x,y,k_z)}$$
$$+ \underbrace{\mathrm{Re}\left\{\frac{d}{dy}\left(-\frac{\overline{\widehat{p}\widehat{v}^*}}{\rho}\right)\right\}}_{\widehat{T}_p(x,y,k_z)} + \underbrace{\nu\frac{d^2\overline{\widehat{e}}}{dy^2}}_{\widehat{T}_\nu(x,y,k_z)} \tag{1}$$

with $(\partial/\partial x_1, \partial/\partial x_2, \partial/\partial x_3) = (\partial/\partial x, \partial/\partial y, ik_z)$ and summation implied over repeated indices. The overline $\overline{\square}$ indicates time averaging, the superscript * represents complex conjugation and $\mathrm{Re}\{\square\}$ means the real part of the complex number. The left hand side of the equation represents the rate of change of TKE of each spanwise Fourier mode. This term vanishes if the flow is statistically steady. The terms on the right-hand side are known as the rate of: (i) turbulence production ($\widehat{P}$), (ii) viscous dissipation ($\widehat{\epsilon}$), (iii) turbulent transport ($\widehat{T}_{turb}$), (iv) pressure transport ($\widehat{T}_p$) and (v) viscous transport ($\widehat{T}_\nu$). The budget equation of the TKE in physical space is obtained by summing the terms of the equation over all wavenumbers. Notice that the only term in the right hand side involving an interaction between spanwise wavenumbers is the non-linear turbulent transport term. This term can be further decomposed into two terms [22]:

$$\widehat{T}_{turb}(x,y,k_z) = \underbrace{\mathrm{Re}\left\{\overline{-\frac{1}{2}\left(\widehat{u_i^*}\frac{d}{dy}(u_iv)\right)}\right\}}_{\widehat{T}^{\perp}_{turb}(x,y,k_z)} + \underbrace{\mathrm{Re}\left\{\overline{\widehat{u_i}^*\frac{\partial}{\partial x_j}\widehat{(u_iu_j)}} - \overline{\frac{1}{2}\left(\widehat{u_i^*}\frac{d}{dy}(u_iv)\right)}\right\}}_{\widehat{T}^{\parallel}_{turb}(x,y,k_z)} \tag{2}$$

The first term in the right hand side of (2) yields the turbulent energy transport in the wall-normal direction at constant spanwise wavenumber. The second term in the expression yields the turbulent energy transport between spanwise wavenumbers at constant y, *i.e., the interscale turbulent transport.* Notice that this term accounts for the redistribution of turbulent energy among the various scales of motion and that does not make a contribution to the total energy budget:

$$\sum_k \widehat{T}^{\parallel}_{turb}(x,y,k_z) = 0 \tag{3}$$

It should be noticed that the decomposition of the turbulent transport term is not unique. The adopted decomposition follows Mizuno [22] and is consistent with other studies in the literature [24, 20].

2.3. *Identification and removal of the energy-containing motions*

The energy-containing motions of the turbulence are understood as those fluid motions that carry most of the kinetic energy and which, through a cascading process, feed energy to the rest of fluid motions. Since in wall-bounded turbulence the energy-containing motions can be characterised by their spanwise length scales [8, 11], the TKE equation for each spanwise Fourier mode provides an effective framework to identify and study these motions. In particular, it is reasonable to argue that the bulk of the energy-containing motions would leave a distinct imprint on the energy spectrum of $\widehat{T}^{\parallel}_{turb}$, in the form of a discernible region of negative energy. That is, that region would be, by definition, mostly representative of the scales of motion that provide most energy to other scales. Figure 2 presents the premultiplied spectral energy density of $\widehat{T}^{\parallel}_{turb}$ at the transfer plane. A contour enclosing the region characterized by negative values in the spectrum has been approximated and is represented in the figure by the solid black curve. The inflow boundary condition of BL_2 is generated by transferring at each time-step the velocity field at the transfer plane to the inlet of BL_2 with the energy-containing motions removed. This filtering is accomplished by setting to zero the velocity fluctuations with coordinate (y, λ_z) located within the approximated contour. Since the DNS code employed is spectral in the spanwise direction, this filtering operation is easily implemented. Notice that this filtering operation does not modify the spanwise zeroth Fourier mode and consequently the mean mass flow rate is not affected by this operation.

3. Results

The effect of removing the energy-containing motions at the inflow boundary condition of a ZPG-TBL is first studied by examining the evolution of the wall-normal profiles of the mean velocity and Reynolds stresses. The premultipled spanwise spectral energy density of the velocity and of the streamwise vorticity are subsequently analysed.

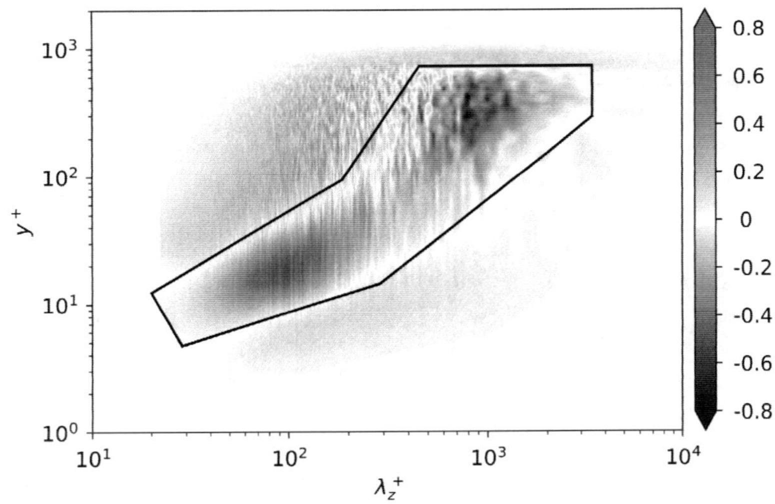

Figure 2. $y^+k_z^+\widehat{T}_{turb}^{||+}(y^+,\lambda_z^+)$ at $x = X_t$. Contour levels are -0.8(0.1)0.8 times the absolute maximum of the spectrum.

3.1. Statistical wall-normal velocity profiles

Figure 3, compares the wall-normal profiles of the mean velocity, the root-mean-squared (rms) velocity fluctuations, and the Reynolds shear stress between of BL_2 and BL_1 at various streamwise locations. The coordinate $\tilde{x}$ defines the streamwise location where quantities are being compared based on the distance from this x-station to the transfer plane when referring to BL_1, and to the domain's inlet when referring to BL_2. This distance is given in terms of the boundary layer thickness at the transfer plane, $\delta_{|X_t}$. The streamwise locations shown in the figure were chosen by examining the evolution of the absolute and relative differences of the mean and fluctuating velocities at fixed y^+ between BL_1 and BL_2 (not shown), and selecting the streamwise locations which best represent the evolution of these quantities. The interest in identifying the wall-normal region along the streamwise direction that is affected by the removal of the energy-containing motions at the inflow of BL_2, motivates the normalization of BL_2 quantities using the friction velocity of BL_1 taken at the evaluated streamwise location $x = \tilde{x}$.

Figure 3 (*a-c*) compares the mean velocity profile of the two domains. Since the filtering operation that removes the energy-containing motions at the inflow of BL_2 does not affect the zeroth spanwise Fourier mode, the mean velocity profile of BL_2 is equal to that of BL_1 at $\tilde{x} = 0\delta_{|X_t}$. As the flow evolves downstream, a small reduction appears in the profile of BL_2 at the vicinity of wall, while the rest of the profile remains to practical effects unperturbed. As the flow develops downstream, this disturbance increases in magnitude and gradually propagates away from the wall, eventually reaching the free-stream, while a gradual recovery builds from the wall upwards. Downstream of $\tilde{x} = 20\delta_{|X_t}$, the wall-normal profile up to approximately the end of the wake region gradually recovers. At the streamwise end of the computational domain of BL_2, the deviation of the corresponding mean velocity profile of BL_2 from that of BL_1 is within 10%. Although BL_2 requires a longer streamwise domain to recover the mean velocity profile of BL_1, it should be noted that most of the profile recovery occurs within the first $20\delta_{X_t}$. During the remaining $26\delta_{X_t}$ up to the streamwise end of BL_2, the recuperation proceeds at a considerably slower pace. Hence, it is reasonable to expect that the streamwise computational domain is large enough to capture the relevant processes that lead the recovery of the energy-containing motions within BL_2. The evolution of the mean velocity profile of BL_2 is better understood by examining it together with the evolution of the Reynolds stresses.

Figure 3 (d) shows that removing the energy-containing motions at the inflow of BL_2 results in a significant reduction, approximately by half, of the rms value of streamwise velocity fluctuations from the buffer layer to the free-stream region. In the neighborhood of station $\tilde{x} = 0.5\delta_{|X_t}$, the rms of the streamwise fluctuations in the near-wall reach its maximum level of reduction. Subsequently, between stations $\tilde{x} = 0.5 - 1.75\delta_{|X_t}$, the profile of the rms of the streamwise fluctuations with $y^+ < 90$ begins to recover and a near-wall peak becomes observable. Within this streamwise range, the profile of the rms of the wall-normal velocity fluctuations (Figure 3 (g)) continues to decrease in the log and wake regions while the rms profile of the spanwise velocity fluctuations (Figure 3 (j)) presents no significant evolution. Downstream of $\tilde{x} = 1.75\delta_{|X_t}$, the near-wall peak of the rms of the streamwise fluctuations grows in magnitude (Figures 3 (e)) and the rms profile of the transversal, *i.e., wall-normal and spanwise*, velocity fluctuations begins to recover from the wall upwards displaying similar trends (Figures 3 (h, k)). At the streamwise end of BL_2 (Figures 3 (f, i, l)), the deviation of the rms velocity profiles of BL_2 from those of BL_1, for each point of the profiles, is within a 10%. Figure 3 (m) shows that at the inflow of BL_2, the Reynolds shear stress has been mostly removed. By the time the flow reaches $\tilde{x} = 1.75\delta_{|X_t}$, the profile below approximately the midpoint of the log-region shows recovery, while a slight reduction is observed on the above half of the log-region, with the rest of the profile remaining relatively unchanged. At the x-station, the profile of the root-mean-square of the streamwise fluctuations exhibits a near-wall peak and the profile of the rms of the wall-normal fluctuations begins to recover, resulting in the gradual recovery of the Reynolds shear stress.

The removal of the energy-containing motions at the inflow boundary of BL_2 results in a reduction of the Reynolds stresses at this x-station. As the flow evolves downstream, the mean velocity profile adjusts to the remaining Reynolds shear stress which results in a reduction of the profile. This reduction originates at the vicinity of the wall, which is where the wall-normal gradient of the Reynolds shear stress is larger. As the flow progresses downstream, this reduction propagates upwards from the wall, accompanied by a gradual recovery which builds upwards from the wall. The reduction of the mean velocity profile and the recovery of the streamwise velocity fluctuations imply that energy from mean flow is gradually transferred to recover the streamwise velocity fluctuations through the production term. The inter-component energy redistribution from the streamwise to the transversal velocity components does not become apparent until the streamwise velocity fluctuations within the buffer layer are partially recovered, which also translates in a recovery of the Reynolds shear stress. Turbulence in the flow therefore recovers. At the streamwise end of the BL_2 computational domain, the deviation of each point of the profiles of the mean and rms velocities of BL_2 is within a 10% of those of BL_1. Most of the streamwise evolution of the profiles is captured within the first $20\delta_{X_t}$, which is followed by a slower recovery that continues until the end of the domain, which is consistent with the observed evolution of the mean velocity profile.

3.2. Spectral energy density of velocity and vorticity

In this section, the streamwise evolution of the k_z premultiplied spanwise spectral energy density (k_zE) of the velocity and streamwise vorticity as a function of the spanwise wavelenght λ_z and of the wall normal distance y are examined. Quantities corresponding to BL_1 have been streamwise averaged over a distance equivalent to one boundary layer thickness, measured at the outflow and centered at the location where quantities are given. No averaging is performed for quantities corresponding to BL_2 due to their rapid streamwise variation, particularly within the initial $20\delta_{X_t}$ of the domain. At each streamwise location, wall-scaled quantities employ the friction velocity of the boundary layer case at which they are evaluated. All figures in this section follow the same contour-lines and colour-levels arrangement. The contour-lines represent the

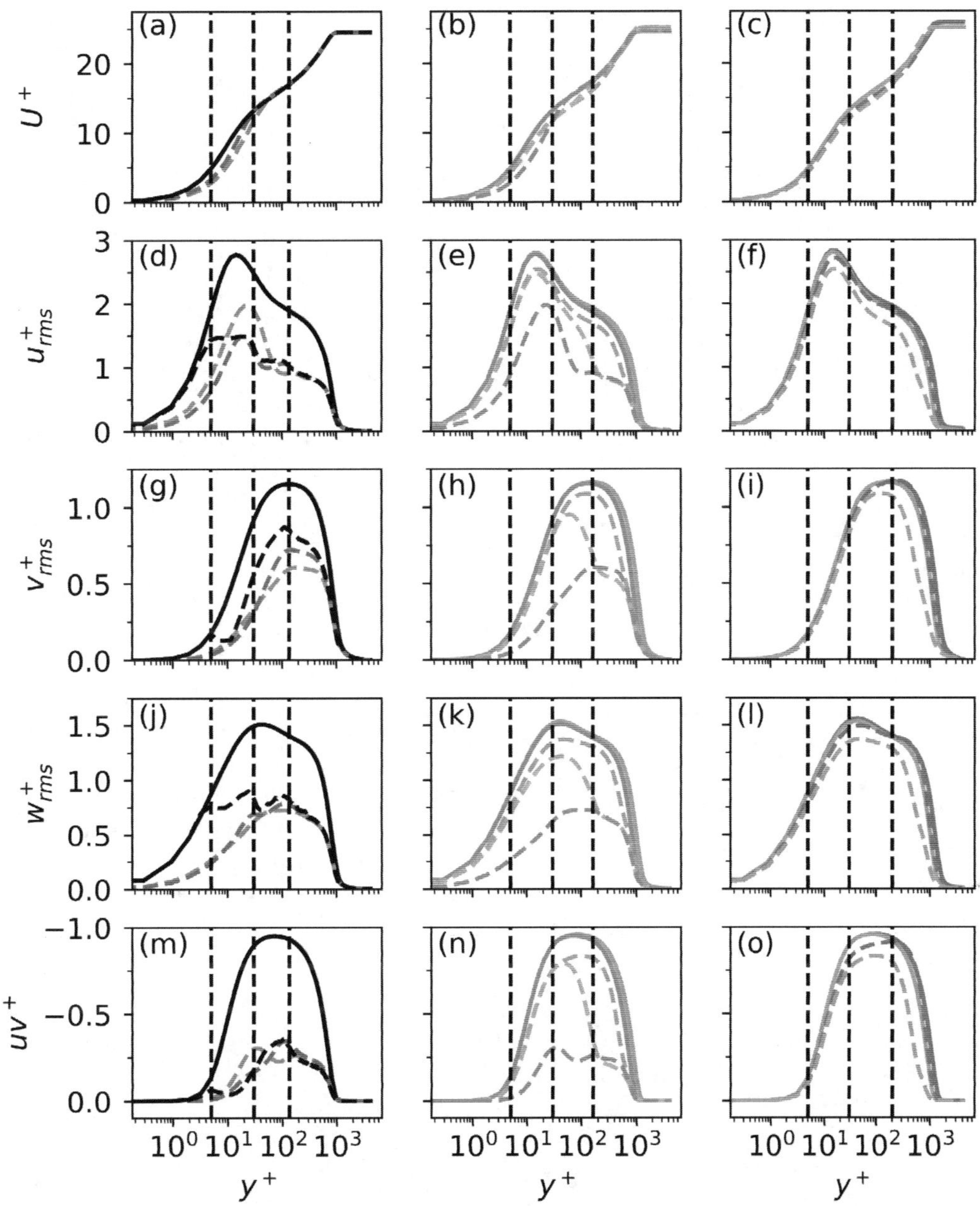

Figure 3. Statistical velocity profiles. (*a-c*) Mean velocity profile. Root-mean-squared of the: (*d-f*) streamwise, (*g-i*) wall-normal (*j-l*) and spanwise velocity fluctuations. (*m-n*) Reynolds shear stress. BL1 (—); BL2 (– –);. Streamwise locations $\tilde{x}/\delta_{X_t}$: 0 (—), 0.5 (—), 1.75 (—), 6 (—), 20 (—) and 46 (—). Vertical dashed black lines indicate, from left to right, the upper boundary of the viscous, buffer and log layers of BL_1, respectively.

(0.1, 0.2, 0.4, 0.6, 0.8) contours of the maximum value of the energy spectrum of the specified quantity within BL_1 at the given streamwise location. The spectral energy densities of quantities of BL_2 require significantly more samples than the corresponding BL_1 quantities to reach the same statistical significance and uncertainty, thus they are presented as a coloured-flood-contour

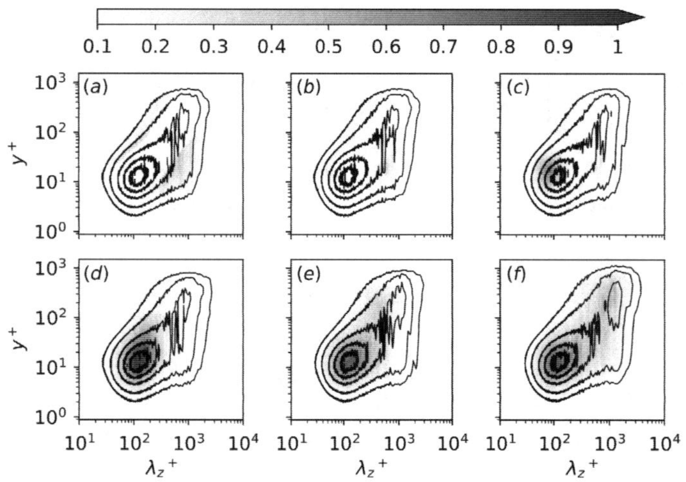

Figure 4. $k_z E_{uu}(\lambda_z^+, y^+)$ at streamwise locations $\tilde{x}/\delta_{X_t}$: (a) 0, (b) 0.5, (c) 1.75, (d) 6, (e) 20 and (f) 46. The spectrum of BL_1 is illustrated by the black countour lines, which are $(0.1, 0.2, 0.4, 0.6, 0.8)$ times the maximum of the spectrum at the given location. The colormap visualizes the spectrum of BL_2, with contour levels 0.1 to 1 times the maximum of the spectrum of BL_1 at the transverse plane X_t.

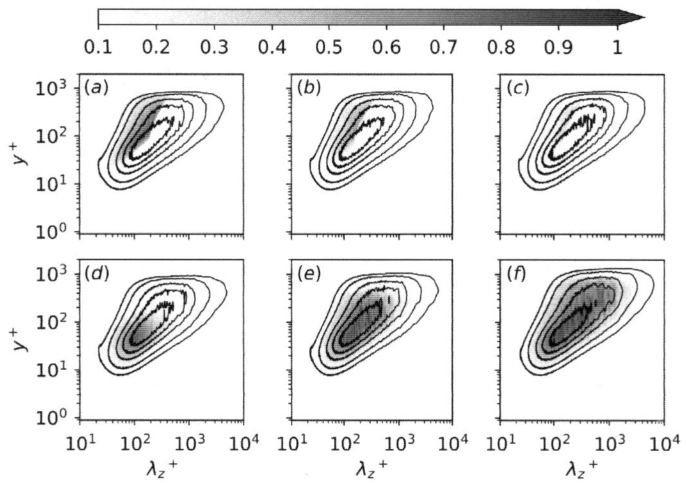

Figure 5. $k_z E_{vv}(\lambda_z^+, y^+)$ at streamwise locations $\tilde{x}/\delta_{X_t}$: (a) 0, (b) 0.5, (c) 1.75, (d) 6, (e) 20 and (f) 46. The spectrum of BL_1 is illustrated by the black countour lines, which are $(0.1, 0.2, 0.4, 0.6, 0.8)$ times the maximum of the spectrum at the given location. The colormap visualizes the spectrum of BL_2, with contour levels 0.1 to 1 times the maximum of the spectrum of BL_1 at the transverse plane X_t.

plot instead of contour lines, with values normalized by the maximum of the spectrum of the specified quantity taken at the transfer plane. Figure 4 shows the streamwise evolution of the spectral energy density of the streamwise velocity fluctuations ($k_z E_{uu}$). The suppression of the energy-containing motions at the inflow of BL_2 results in the removal of the near-wall peak, located at $\lambda_z^+ \approx 100$ and $y^+ \approx 15$ for BL_1. This near-wall peak is attributed to the near-wall cycle of streaks of streamwise velocity and quasi-streamwise vortices, and its location is known to be roughly independent of the Reynolds number [25]. As the flow progresses from the inflow to $x = 6\delta_{|X_t}$, the energy in the spectrum of BL_2 builds around a buffer layer peak, that gradually shifts its location as the flow evolves, while it reduces in the rest of the spectrum. Downstream of $x = 6\delta_{|X_t}$, the spectral energy gradually propagates towards regions located further away from the wall and into larger spanwise wavelengths. Figures 5 and 6 show $k_z E_{vv}$ and $k_z E_{ww}$, respectively. From the inflow of BL_2 to $x = 1.75\delta_{|X_t}$, both $k_z E_{vv}$ and $k_z E_{ww}$ spectra present an energy reduction. Once the $k_z E_{uu}$ spectrum of BL_2 begins to reveal a near-wall peak, $k_z E_{vv}$ and $k_z E_{ww}$ build energy gradually from the wall-upwards as the flow moves downstream through the redistribution of streamwise turbulent energy. A spectral analysis of the pressure-strain needs to be performed for a more in depth investigation of the transfers of energy between velocity components.

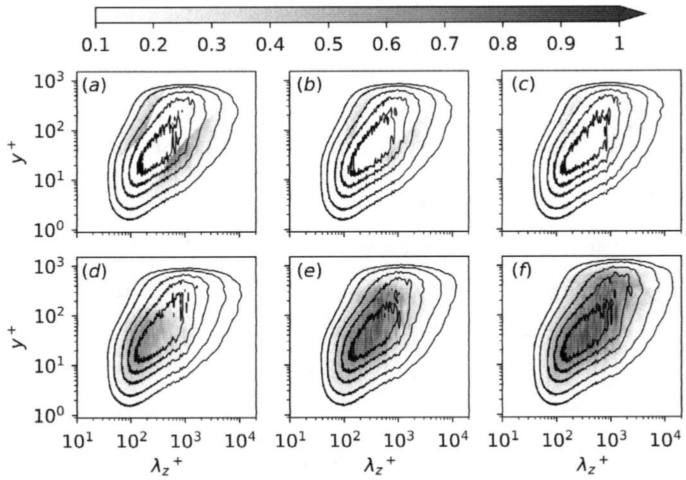

Figure 6. $k_z E_{ww}(\lambda_z^+, y^+)$ at streamwise locations $\tilde{x}/\delta_{X_t}$: (a) 0, (b) 0.5, (c) 1.75, (d) 6, (e) 20 and (f) 46. The spectrum of BL_1 is illustrated by the black countour lines, which are $(0.1, 0.2, 0.4, 0.6, 0.8)$ times the maximum of the spectrum at the given location. The colormap visualizes the spectrum of BL_2, with contour levels 0.1 to 1 times the maximum of the spectrum of BL_1 at the transverse plane X_t.

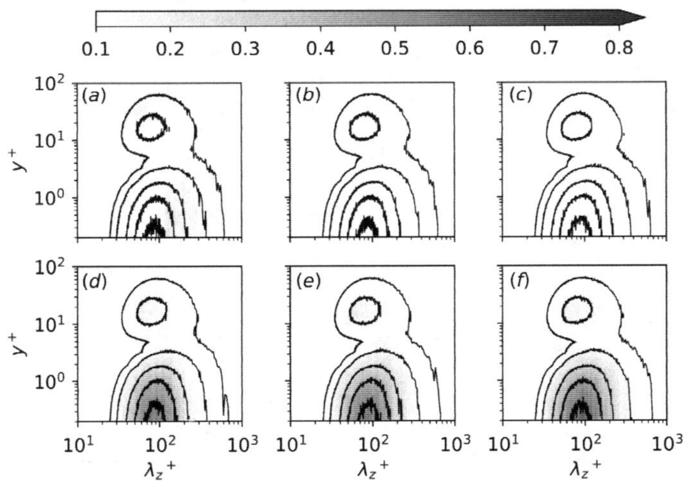

Figure 7. $k_z E_{\omega_x \omega_x}(\lambda_z^+, y^+)$ at streamwise locations $\tilde{x}/\delta_{X_t}$: (a) 0, (b) 0.5, (c) 1.75, (d) 6, (e) 20 and (f) 46. The spectrum of BL_1 is illustrated by the black countour lines, which are $(0.1, 0.2, 0.4, 0.6, 0.8)$ times the maximum of the spectrum at the given location. The colormap visualizes the spectrum of BL_2, with contour levels 0.1 to 1 times the maximum of the spectrum of BL_1 at the transverse plane X_t.

The recovery of the near-wall cycle of BL_2 can be further examined by analysing the evolution of the energy spectral density of the streamwise vorticity ($k_z E_{\omega_x \omega_x}$), presented in Figure 7. At $x = 0\delta_{|X_t}$, the $k_z E_{\omega_x \omega_x}$ spectrum of BL_1 exhibits two peaks. The peak within the buffer layer is attributed to streamwise vortices, while the peak closer to the wall arises due to vortices created by the no-slip condition at the wall [26]. In the spectrum of BL_2, both of these peaks have been removed by the filtering operation that removes the energy-containing motions, and only traces of energy of the streamwise vorticity are detected. No noticeable energy in the spectrum of BL_2 seem to be present for locations $x = 0.5 - 1.75\delta_{|X_t}$. It should be noted that at $x = 1.75\delta_{|X_t}$, the $k_z E_{uu}$ spectrum of BL_2 shows the presence of a peak in the near-wall. This presence of a near-wall peak in the $k_z E_{uu}$ spectrum and the absence of energy in the near-wall of the spectrum of $k_z E_{\omega_x \omega_x}$, suggest the presence of near-wall streaks and the absence of near-wall quasi-streamwise vortices. At $x = 6\delta_{|X_t}$, the $k_z E_{\omega_x \omega_x}$ spectrum of BL_2 exhibits energy peaks at the same locations as those detected in the spectrum of BL_1. In particular, the buffer-layer peak in the spectrum suggests the presence of the quasi-streamwise vortices that are involved in the near-wall cycle. This would imply that a near-wall cycle was initiated between $x = 1.75 - 6\delta_{|X_t}$, and that the near-wall quasi-streamwise vortices observed are being generated as a result of the breakdown of the near-wall streaks. Thus, the formation of near-wall streaks seems to be the driving process in the recovery of the energy-containing motions, or at least in the near-wall,

Journal of Physics: Conference Series **2753** (2024) 012008 doi:10.1088/1742-6596/2753/1/012008

when these are removed at an upstream flow cross-plane. However, the formation of near-wall streaks with the apparent absence of streamwise vorticity remains unclear, and likely requires looking beyond time averaged quantities. The current observations indicate that the near-wall region recovers faster than the outer region. This is consistent with the shorter time-scales of the turbulence near the wall. However, based on the presented results, it is not possible to draw conclusions regarding any potential dependency of the recovery of the energy-containing motions of the outer region on those on the near-wall region. Current research suggests that the energy-containing motions in the outer region are self-sustaining [27, 12]. However, it is not clear whether the re-established near-wall energy-containing motions have a role in the initial resurgence of the outer region energy-containing motions.

4. Discussion

This study has investigated the recovery of the energy-containing motions when they are removed at the inflow boundary of a flat plate ZPG-TBL DNS. The bulk of the energy-containing motions have been identified with a distinctive connected (y, λ_z)-region within the spanwise spectral energy density of the interscale turbulent transport characterized by negative values. The removal of these motions at the inflow of BL_2 results in a significant reduction of the Reynolds stresses at this x-station. As the flow evolves downstream, the mean velocity profile adjusts to the remaining Reynolds shear stress, resulting in a reduction of the profile, initially occurring at the vicinity of the wall. This profile reduction propagates from the wall upwards and is followed by a gradual recovery, which also builds from upwards from the wall. Simultaneously, the streamwise velocity fluctuations gradually recover. The reduction of the mean velocity profile and the recovery of the streamwise velocity fluctuations imply that energy from mean flow is gradually transferred to recover the streamwise velocity fluctuations through the production term. On the other hand, the intercomponent energy redistribution of energy from the streamwise to the transversal velocity components is negligible until the streamwise velocity fluctuations within the buffer layer have commenced the recovery towards their statistically stationary ZPG-TBL state. The turbulence in the flow therefore recovers as the flow evolves downstream.

The suppression of the energy-containing motions at the inflow of BL_2 results in the removal of most of the energy in the $k_z E_{uu}$ and $k_z E_{\omega_x \omega_x}$ spectra, and in particular in the removal of the respective buffer layer energy peaks. As the flow evolves downstream, the recovery of the buffer layer peak in the $k_z E_{uu}$ spectrum and the absence of it in the $k_z E_{\omega_x \omega_x}$ spectrum until further downstream, implies that near-wall streaks are the primary energy-containing motion formed, with quasi-streamwise vortices only developing further downstream. This suggests that the formation of the near-wall streaks may be the primary process driving the recovery of the energy-containing motions, or at least in the near-wall region, when the energy-containing motions are removed at an upstream cross-plane. Nevertheless, it is not clear how the near-wall streaks are formed with the apparent absence of streamwise vorticity in the flow. This matter necessitates further analysis, likely requiring looking beyond time averaged quantities. As the flow evolves further downstream, energy in the spectrum of the velocities gradually propagates towards regions located further away from the wall and into larger spanwise wavelengths. The above observation indicates that the near-wall region recovers faster than the outer region, consistent with the shorter time-scales of the turbulence near the wall. However, based on the presented results, it is not possible to draw conclusions regarding a potential dependency, if any, of the recovery of the energy-containing motions in the outer region on those in the near-wall region. Whilst current research suggests that the energy-containing motions in the outer region are self-sustaining [10, 12], whether the re-established near-wall energy-containing motions have a role in the initial resurgence of the energy-containing motions in the outer region is not clear and requiring further investigation.

The results presented in this study find agreement with the previous study by

Kannadasan *et al.* [1] on turbulent channel flows. One of the differences of the present study has been in only suppressing the velocity fluctuations at the (y, λ_z)-locations corresponding to the energy-donor region identified in the spectrum of the interscale turbulent transport. This allowed the conservation of scales in the near-wall that receive energy, and therefore not identified as energy-containing motions, that otherwise would have been removed.

Ongoing work concerns the spectral analysis of the Reynolds stress transport equation and the study of the scale-to-scale interactions developing in the flow, which will be reported in subsequent papers.

Acknowledgments
The support of the Australian Research Council (ARC) and the computational resources provided through NCMAS by the Australian National Computational Infrastructure and the Pawsey Supercomputing Centre for this work is gratefully acknowledged. This work was supported in part by the European Research Council under the CausT grant ERC-ADG-101018287. We are grateful to Professor Ricardo Garcia-Mayoral for his careful review of the manuscript and insightful comments.

References

[1] Kannadasan E, Atkinson C and Soria J 2023 *J. Fluid Mech.* **955** R1
[2] Townsend A A 1976 *The Structure of Turbulent Shear Flow* (Cambridge Univ. Press.)
[3] Marusic I and Monty J P 2019 *Annu. Rev. Fluid Mech.* **51** 49–74
[4] Jiménez J and Lozano-Durán A 2016 Coherent structures in wall-bounded turbulence *Progress in Wall Turbulence 2* (Springer Int. Pub.) pp 37–46
[5] Jimenez J and Moin P 1991 *J. Fluid Mech.* **225** 213–240
[6] Hamilton J, Kim J and Waleffe F 1995 *J. Fluid Mech.* **287** 317–348
[7] Jimenez J and Pinelli A 1999 *J. Fluid Mech.* **389** 335–359
[8] Flores O and Jiménez J 2010 *Phys. Fluids* **22**
[9] Hwang Y and Cossu C 2011 *Phys. Fluids* **23**
[10] Hwang Y and Cossu C 2010 *Phys. Rev. Lett.* **105** 044505
[11] Hwang Y 2015 *J. Fluid Mech.* **767** 254–289
[12] Cossu C and Hwang Y 2017 *Philos Trans A Math Phys Eng Sci* **375** 20160088
[13] Harlow F H and Welch J E 1965 *Phys. Fluids* **8** 2182–2189
[14] Perot J B 1993 *J. Comput. Phys.* **108** 51–58
[15] Simens M P, Jiménez J, Hoyas S and Mizuno Y 2009 *J. Comput. Phys.* **228** 4128–4231
[16] Lele S K 1992 *J. Comput. Phys.* **103** 16–42
[17] Sillero J A, Jiménez J and Moser R D 2013 *Phys. Fluids* **25** 105–102
[18] Lund T S, Wu X and Squires K D 1998 *J. Comput. Phys.* ISSN 0021-9991
[19] Lumley J L 1964 *Phys. Fluids* **7** 190–196
[20] Lee M and Moser R 2019 *J. Fluid Mech.* **860** 886–938
[21] Bolotnov I, Lahey R, Drew D, Jansen K and Oberai A 2010 *Comput. Fluids.* **39** 640–655
[22] Mizuno Y 2016 *J. Fluid Mech.* **805** 171–187
[23] Cho M, Hwang Y and Choi H 2018 *J. Fluid Mech.* **854** 474–504
[24] Mansour N N, Kim J and Moin P 1988 *J. Fluid Mech.* **194** 15–44
[25] Jiménez J 2013 *Phys. Fluids* **25** 101302
[26] Jeong J, Hussain F, Schoppa W and Kim J 1997 *J. Fluid Mech.* **332** 185–214
[27] Hwang Y and Bengana Y 2016 *J. Fluid Mech.* **795** 708–738

Journal of Physics: Conference Series **2753** (2024) 012009 doi:10.1088/1742-6596/2753/1/012009

Predictability of decay events in transitional wall-bounded flows

Daniel Morón[1], Alberto Vela-Martín[2] and Marc Avila[1,3]

[1] University of Bremen, Center of Applied Space Technology and Microgravity (ZARM), Am Fallturm 2, 28359 Bremen, Germany.

[2] Department of Aerospace Engineering, Universidad Carlos III de Madrid, Avda. de la Universidad 30, 28911 Leganés, Madrid, Spain.

[3] University of Bremen, MAPEX Center for Materials and Processes, Am Biologischen Garten 2, 28359 Bremen, Germany.

E-mail: daniel.moron@zarm.uni-bremen.de

Abstract. Extreme events in turbulent flows are in general linked with sudden changes in the dynamics of the system. Despite the increasing interest in modelling these events, their predictability remains largely unexplored. Here we focus on the case of sudden turbulence/chaos decay in pipe flow and in a reduced-order model of shear flows (Moehlis, Faisst and Eckhardt, 2004). We study the predictability of decay events with massive ensembles of simulations. For this purpose we introduce a new pseudo-spectral GPU code to perform fast simulations of turbulent pipe flow and describe a methodology to characterize predictability of decay events using the results of the ensembles of simulations. We report the decrease of predictability as one goes back in time with respect to a decay event, and show that this decrease does not need to be monotonic.

1. Introduction

Despite lacking a concrete definition, extreme events in dynamical systems are identified according mainly to two factors. First, they are linked to high amplitude phenomena, which dominate the behaviour of a dynamical system for a certain time span. Second, they take place at infrequent times [1]. They are mostly studied in the atmospheric sciences, where the predictability of events such as tornadoes or hurricanes is of obvious interest. Particularly now, in the context of climate change, they are expected to have more severe consequences and to occur more often [2].

Extreme events are also found in simpler dynamical/complex systems [1]. A notable example is turbulence in shear flows, where extreme events have been linked with the self-sustaining process of turbulence in shear flows [3], or to sudden increases of the shear at the wall [4]. Due to the effect extreme events have on the dynamics of turbulence in general, there is an increasing interest on studying their predictability and causes, in order to better model and predict them [5].

Extreme events are also found in transitional shear flows, particularly in the dynamics of localized turbulent structures (like turbulent puffs) in linearly stable flows like pipe flow [6]. The behaviour of these structures is governed by the Reynolds number Re alone. In the case of pipe flow, $Re = \frac{UD}{\nu}$, where U is the bulk velocity, D the pipe diameter and ν the kinematic viscosity. At $1500 \lesssim Re \lesssim 2040$ turbulent puffs can be triggered in the flow due to sufficiently large perturbations, but decay after a certain time [7]. The statistics of these decay events follow an exponential distribution, see fig. 1, with a mean lifetime τ that scales super-exponentially with Re $\tau \propto \exp(\exp(aRe + b))$ [8]. Here a and b are two fitting constants. This super-exponential scaling has been derived with the use of extreme value theory and linked to extreme events in the flow [9, 10]; see a recent review about the transition to turbulence in pipe flow in [11]. Decay events, whose statistics follow an exponential distribution, are not limited to pipe flow, and are also found in other shear flows such as channel [12] and Couette [13, 14] flows, or reduced models of shear flows like the MFE model (Moehlis, J., Faisst, H., & Eckhardt, B. (2004)) [15, 16].

Here we look for methods to characterize predictability of decay in pipe flow and the MFE model. Assessing predictability is of paramount importance, as any predictive model must also include its predictability limits, which depend on the initial data uncertainty [17]. In what follows, we include a short survey of studies related with predictability of decay in these two systems, and methods to characterize predictability.

Góme *et al.* [18] looked at the most probable paths to extreme events of localized turbulent patches in channel flow, using a rare event algorithm. For the MFE model, recent studies explore the use of data-driven techniques to make predictions of the model [19, 20, 21]. Lellep *et al.* [22] identified states of the MFE model that are related with decay events using an explainable artificial intelligence method. The method is able to predict decay events at relatively short times before the event itself. As one goes back in time with respect to the decay event, the method performs worse until it cannot make any accurate prediction. This is rooted in the predictability limit of the MFE model, that, as far as we know, has not been characterized yet. This will be done in this paper.

Predictability has historically been linked with the Lyapunov exponents of the chaotic attractor [23, 24, 25, 26]. They measure the rate at which trajectories, that are initially separated by a small δ in phase-space, separate from each other due to the chaotic behaviour of the dynamical system. The expression "Lyapunov exponent" corresponds to the ergodic average of the separation rate for asymptotically small δ over the attractor. Lyapunov exponents can also be computed locally in phase space, what is known as "effective Lyapunov exponents". It is known that (effective and ergodic) Lyapunov exponents are not good indicators of predictability in two situations: in the case of non-infinitesimal perturbations (not small enough δ) and in the case of systems with several characteristic time scales [24], like turbulence. Therefore, other studies have opted for more fundamental ways of characterizing predictability. One option is to look at the evolution of a set of initial states of the system, localized in a region of phase space. This corresponds to the study of the evolution of the probability distribution function of possible states of the system [27], conditioned to be initialized at certain locations of phase space. The evolution of the probability density function of a set of initial conditions, can be shown to follow the Liouville conservation equation [27], which, in theory, has an analytical solution. For practical purposes, however, it is currently unfeasible to define this equation for high dimensional systems like turbulence, let alone integrate it. Recently Jiménez [28] was able to integrate the probability density function of a reduced representation of a turbulent channel flow using the Perron-Frobenius operator. His study relies on physically inspired and educated guesses, that radically reduce the number of dimensions of the representation of the system, making it feasible to study the evolution of the corresponding probability density function. However, without the correct guesses, and such good knowledge of the system of interest, it is difficult to replicate this analysis for other turbulent flows.

In this paper we assess predictability with massive ensembles of simulations. We use the results of the ensembles of simulations to approximate the probability distribution function of possible future states of the system. It is now feasible to perform this huge number of simulations thanks to the use of Graphics Processing Units GPUs, that allow for fast codes. The purpose

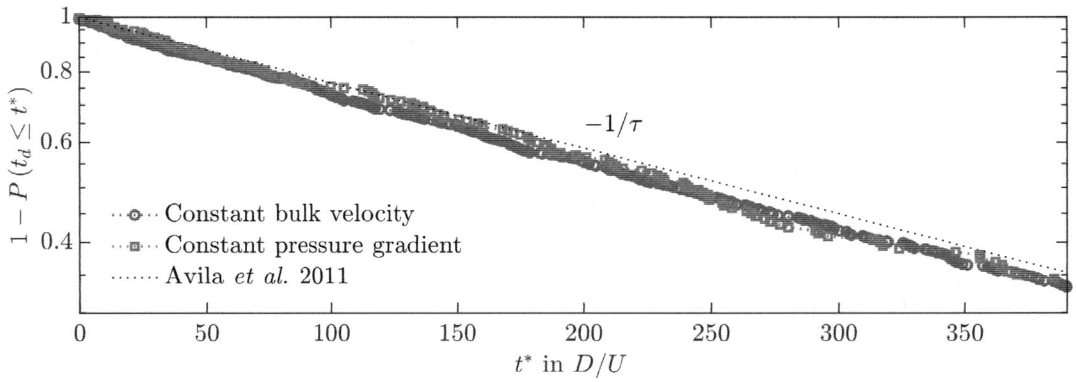

Figure 1: Lifetimes statistics of puffs in DNS using the new `nsPipe`-GPU code for $L_x = 50D$ long pipes at $Re = 1850$. The vertical axis represents one minus the cumulative distribution function of puff decay after a time step size t^*. In blue (red) and circles (squares) cases with a constant bulk velocity (constant pressure gradient). The dotted line corresponds to the value of the exponential distribution proposed by Avila *et al.* [8], that fits the experimental data of Hof *et al.* [7]. Here $1 - P_q \approx \exp(-t^*/\tau)$, where $\tau = \exp(\exp(0.005556 \cdot Re - 8.499))$.

of this paper is two-fold. First, we describe a new GPU version of our `nsPipe` code, that allows us to perform fast pipe flow DNS. Then, using results produced with this new code, together with results of the MFE model, we define the methods we use to characterize the predictability of these two systems. At the end of the paper we discuss our preliminary findings.

The manuscript is organized as follows. In §2 and §3, we describe the GPU codes we use to perform DNS of pipe flow and the MFE model, respectively. In §4, we discuss the way we assess the predictability of decay events. In §5, we show the main results of the analysis and in §6, we draw some conclusions.

2. Puffs in pipe flow

We consider the flow of a viscous Newtonian fluid with constant properties in a straight smooth rigid pipe of circular cross-section. The flow is assumed to be incompressible and governed by the dimensionless Navier–Stokes equations (NSE)

$$\frac{\partial \mathbf{u}}{\partial t} + (\mathbf{u} \cdot \nabla)\,\mathbf{u} = -\nabla p + \frac{1}{Re}\nabla^2 \mathbf{u} + f_p(t) \cdot \mathbf{e}_x \quad \text{and} \quad \nabla \cdot \mathbf{u} = 0. \tag{1}$$

Here, $\mathbf{u}$ is the fluid velocity, p the pressure, $\mathbf{e}_x$ the unit vector in the axial direction and $f_p(t)$ the pressure gradient that drives the flow. The latter can either be constant, or time-dependent. In the second case, $f_p(t)$ is adjusted at each time step in order to force a constant bulk velocity. All variables are rendered dimensionless using the pipe diameter (D), the bulk velocity (U) of the corresponding (parabolic) laminar profile, and the fluid density (ρ_f). The equations are formulated in cylindrical coordinates (r, θ, x), where the velocity field has three components $\mathbf{u} = (u_r, u_\theta, u_x)$ in the radial, azimuthal and axial directions, respectively.

2.1. Direct Numerical Simulations (DNS)

We solve the NSE numerically using a newly developed C-CUDA version of the pseudo-spectral `nsPipe` code. Find a detailed description of this new version, `nsPipe`-GPU, later in this section.

We perform DNS in a $L_x = 50D$ long pipe domain, at a constant $Re = 1850$. We initialize the simulations with the Hagen-Poiseuille laminar profile in the whole domain. On top of the parabolic profile, first we superpose a pair of localized streamwise vortices defined as:

$$u'_r = A_p \left(1 - r^2\right)^2 \exp\left[-10 \sin^2\left(\pi x / L_x\right)\right] \sin\left(\theta\right), \tag{2}$$

$$u'_\theta = A_p \left[\left(1 - r^2\right)^2 - 4r^2\left(1 - r^2\right)\right] \exp\left[-10 \sin^2\left(\pi x / L_x\right)\right] \cos\left(\theta\right), \tag{3}$$

where $A_p = 0.6$. This is a common perturbation used in the literature to trigger a single puff in DNS of pipe flow [11]. Second, we introduce, for each degree of freedom, an additive white Gaussian noise, with magnitude $|\epsilon| \leq 1e - 2$. These perturbations trigger a single turbulent puff in the flow domain.

We use $N_r = 48$ radial points and $M_\theta = 96$ azimuthal and $M_x = 768$ axial physical points after de-aliasing. The maximum shear Reynolds number measured is of $Re_\tau \approx 70$ which results in a grid spacing in viscous units of $0.06 \lesssim \Delta r^+ \lesssim 2.2$, $D\Delta\theta^+/2 \approx 4.5$ and $\Delta x^+ \approx 9$. The time step size is set equal to $\Delta t = 0.0025D/U$. The grid has been optimized to have the minimal number of physical points needed to keep a sufficient spatial resolution. We are confident that our grid choice does not have an impact on our results, as the lifetime distribution of puffs agrees well with experiments, see fig. 1.

As a heuristic threshold we assume that puffs decay when the volume-averaged cross section kinetic energy is smaller than:

$$\frac{1}{V} \iiint_V \left(u_r^2 + u_\theta^2\right) \mathrm{d}V = \left\langle u_r^2 + u_\theta^2 \right\rangle_{r,\theta,x} \leq 1e - 7. \tag{4}$$

The new `nsPipe`-GPU version implements the same numerical schemes of the original `Fortran` version; see a detailed description of the original code in López *et al.* [29] and references therein. Here we describe the main capabilities of the code, together with its architecture, performance and validation.

The equations of motion are discretized using a Fourier-Galerkin ansatz in the azimuthal (θ) and axial (x) directions. In the radial (inhomogeneous) direction, high-order finite differences (FD) are used. The user can select the stencil-length of the FD scheme, being the default one a length of 7. The code assumes periodic boundary conditions in the azimuthal and axial directions. It also considers a non-homogeneously distributed grid spacing in the radial direction, having more points clustered close to the pipe wall than close to the pipe center-line. The NSE are integrated forward in time with a predictor-corrector method. The divergence free condition is enforced at each time step after iterating in the corrector step. The boundary conditions at the wall are imposed using an influence matrix method, following `openpipeflow` [30].

There are two options to drive the flow in the `nsPipe`-GPU code. In the default one, we consider a time-varying driving force. In this case, during the iterations of the corrector step, f_p is adjusted so the bulk velocity remains constant and equal to U. In the other version of the code, the driving force is fixed to a constant. The constant value is set so the bulk velocity in the turbulent case is close to U. Unless stated otherwise, the simulations are performed with a fixed bulk velocity (time-dependent driving force).

All the relevant parameters (number of Fourier modes, radial points, time step size, Reynolds number, etc.) can be written in the `head.h` file before the code is compiled and run.

2.2. *Code architecture and performance*

In order to run the `nsPipe`-GPU code, a CUDA-capable GPU device with CUDA compute capability 2.0 (or higher), support for double-precision arithmetic and `NVIDIA`'s CUDA toolkit are required. The GPU code runs in a massively parallel setup with thousands of GPU threads

Journal of Physics: Conference Series **2753** (2024) 012009 doi:10.1088/1742-6596/2753/1/012009

Table 1: Performance of the GPU code compared with the performance of the CPU code, in ms per time step. The code has been tested for two different grids, described in the three first columns with the total number of physical radial N_r, azimuthal M_θ and axial M_x points after de-aliasing. The fourth column denotes the total memory required to run the code in the GPU. The rest of the columns denote the computing time of performing a time step in ms, averaged over 1000 time steps. The fifth column corresponds to the GPU code and the rest to the CPU code. The latter was run using different number of cores. The GPU used was an `A100 80GB`. The CPU code was run with different numbers of `Xeon 8360Y` processors. A single processor has 36 cores.

N_r	M_θ	M_x	Memory GPU	GPU	2×36	4×36	8×36	2×36
48	96	768	2GB	8.65	40.15	22.37	14.16	8.92
96	240	5760	30GB	397.45	2302.66	1230.55	649.68	372.28

and highly efficient memory management. It relies on CUDA kernels for linear algebra and Fast-Fourier Transforms (`cuFFT`). An in-house (LU) algorithm is implemented to invert the one-dimensional (radial) Helmholtz and Poisson problems that result from the discretization of the velocity and pressure equations, respectively.

This GPU version runs on single GPU devices, although we plan to further develop it for hybrid MPI-GPU applications. Currently the maximum size of the case that can be run by the code is limited by the memory of the GPU.

The performance of the code was tested by comparing the computing time per time step of the GPU code with the computing time of the CPU code for different computation grids and number of cores. See the main results of this comparison in table 1. We observe an outstanding speed-up of the GPU code compared with the original `nsPipe` code run in 72, 144 or 288 cores. Only by running the CPU `nsPipe` code in 576 or more cores, we observe a similar cost per time step between the two. In sum, the GPU code is much more efficient than the CPU code.

2.3. Code validation

We perform lifetime statistics of puffs at $Re = 1850$. We use the two versions of the GPU code, one with constant and another with a time-varying driving force of the flow.

See in figure 1 the lifetime distribution of puff decay using the two versions of the `nsPipe`-GPU code. The simulations are capped to a maximum run time of 400 advective time units, D/U. The two distributions are compared with the exponential distribution inferred by Hof *et al.* [7], experimentally. As seen in the figure, the distributions produced by the code are close to the experimental value. Note that the statistics we obtain are independent of the type of driving we choose.

3. Low-dimensional model of shear flow

The MFE model originally proposed by Moehlis *et al.* [15] describes the evolution of a set of Fourier modes that approximate the fluid flow between two free-slip walls subjected to a sinusoidal body force, known as Waleffe flow. In particular, the model considers the amplitudes $a_j(t)$ of nine modes $\mathbf{v}_j$, that have a direct interpretation in terms of the flow close to the wall. The mode $j = 1$ represents the mean velocity profile, in case $a_1 \equiv 1$ the flow is laminar. The mode $j = 2$ represents velocity streaks; $j = 3$ downstream vortexes; $j = 4$ and $j = 5$ the spanwise flows; $j = 6$ and $j = 7$ two wall normal vortex flows; and $j = 9$ represents a correction to the mean velocity profile. The mode $j = 8$ does not have a clear interpretation, but represents a three-dimensional mode that interacts with the others.

All the quantities are normalized by the channel half height $h/2$ and the laminar velocity U_0 at the walls. The non-dimensional velocity field is:

$$\mathbf{u}(\mathbf{x},t) = \sum_{j=1}^{9} a_j(t)\,\mathbf{v}_j(\mathbf{x}), \tag{5}$$

where $\mathbf{x}$ is the non-dimensional position of the flow in Cartesian coordinates, and t the time in $h/2U_0$ units.

By plugging this discretization in the Navier–Stokes equations, and after some simplifications and assumptions, the evolution equation of each amplitude a_j is governed by an ODE with the form:

$$\frac{\mathrm{d}a_j}{\mathrm{d}t} = L_j a_j + N_j(\mathbf{a}) + c_j, \tag{6}$$

where, L_j and $N_j(\mathbf{a})$ are linear and nonlinear operators acting on the vector $\mathbf{a} = (a_1, a_2, ...)^T$; and c_j model constants. The MFE model depends on the Reynolds number, in this case defined as $Re = \frac{U_0 h}{2\nu}$, and the spanwise/streamwise domain sizes L_z and L_x. We fix $Re = 400$, $L_x = 2\pi h$ and $L_z = \pi h$. At these flow parameters, and when it is initially sufficiently perturbed, the MFE model shows a transient chaotic dynamic. Specifically, chaos in the MFE model decays at random times that also follow an exponential distribution as in the case of puffs in pipe flow, fig. 1. We set as a heuristic threshold that a decay event corresponds to the case where:

$$a_1 \geq 0.995. \tag{7}$$

We integrate the MFE model forward in time using a low-storage explicit 4th order Runge–Kutta method, with a constant time step size $\Delta t = 0.05h/2U_0$. We implemented this integration in a C-CUDA code where massive ensembles of MFE model can be simultaneously integrated. In the code, each GPU thread integrates its particular MFE model independently of the rest of threads. We initialize the MFE model simulations with a random distribution of amplitudes a_j, whose energy is set equal to:

$$\sum_{j=1}^{9} a_j^2(t=0) = 0.3^2. \tag{8}$$

4. Method to assess predictability

In this work we assess the predictability of puff and MFE model decay in three steps.

4.1. Step 1: base trajectories

We run $i = 1, ..., N_b$ pipe (or MFE model) base trajectory simulations that have a decay event at a given time T_d^i. These decay events are determined according to heuristic thresholds: eq. (4) in the case of puffs and eq. (7) in the case of the MFE model. See in figure 2 two base trajectories as examples of puff and MFE model decay events, respectively.

4.2. Step 2: sampling the base trajectory

We save the instantaneous state of each base trajectory at several N_t times t_j^i, where $j = 1, 2, ..., N_t$, before the decay event $t_j^i < T_d^i$. The predictability of each of these instantaneous states will be assessed. One must select which time steps to save before the decay event. The first time step in the past is $T_d^i - t_1^i$, and the rest of sampling times are $t_j^i < t_1^i$. Here we save the states of the system at points separated by a constant time step size $\Delta t_j = t_{j-1} - t_j$. We set different values of N_t, t_1 and Δt_j depending on the system considered (pipe flow or MFE model), see table 2.

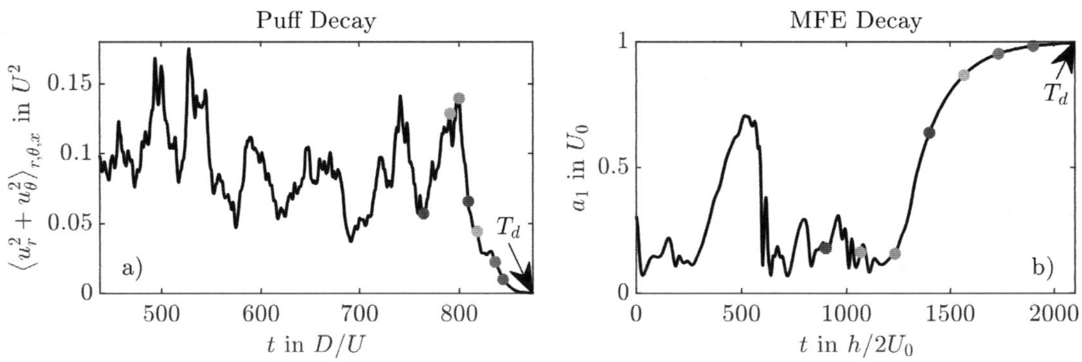

Figure 2: Exemplary base trajectories of puff (left) and MFE model (right) with a decay event at $t \equiv T_d$. In the case of puffs we plot the volume averaged cross section kinetic energy. In the case of the MFE model, the coefficient of the mean profile a_1. With points we denote the positions of the trajectory t_j where we launch ensembles of simulations. We define the decay events according to our heuristic thresholds: eq. (4) in the case of puffs and eq. (7) in the case of the MFE model.

Table 2: Default values of the parameters used in this study.

Parameter	Puff	MFE	Description
N_b	1	14000	Number of base trajectories
N_t	10	400	Number of samples of the trajectory
N_e	200	1000	Number of simulations per ensemble
$T_d - t_1$	40 D/U	100 $h/2U_0$	First time step considered for the sampling
Δt_j	8 D/U	5 $h/2U_0$	Time step size between consecutive samples
ϵ_0	1e-2 U	1e-4 U_0	Magnitude of the noise of ensemble simulations

4.3. Step 3: run ensembles of simulations

For each of the N_b base trajectories and each of the N_t instantaneous states, we perform massive ensembles of simulations. Each ensemble has N_e simulations, and all of them are initialized with the corresponding instantaneous field at t_j^i of one of the N_b base trajectories, plus an additive white Gaussian noise with magnitude ϵ_0. In the case of puffs the noise is added in the whole domain, and in each velocity component. In the case of the MFE model, the noise is added to all coefficients a_j.

In total we perform $N_b \times N_t \times N_e$ individual simulations for both systems. We use the ensembles of simulations to either gather statistics of decay events or to compute ensemble-averaged effective (local in phase-space) Lyapunov exponents λ.

See in table 2 the default parameters we use for the two systems studied here. In the case of the MFE model we perform ensembles of simulations at additional ϵ_0, as described below.

5. Results

Here we show the main results of the two set-ups, puff decay and MFE model decay. As gathered in table 2, for the case of puff decay we only consider one base trajectory, the one shown in fig. 2a, sampled at 10 time steps. In the case of the MFE model, unless stated otherwise, we consider

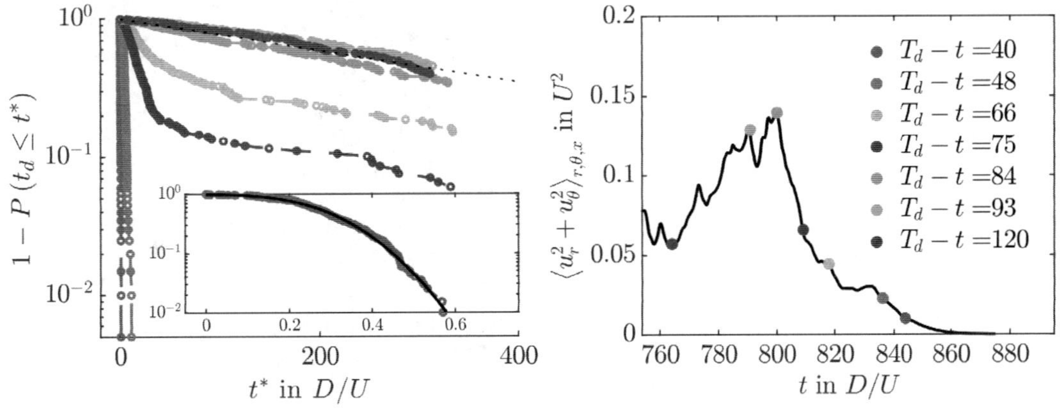

Figure 3: To the left: one minus the cumulative distribution function of puff decay after a time span t^* after initialization, for ensembles of simulations initiated close to different puff states before decay. With a dotted line, the expected distribution, $1 - P_q \approx \exp(-1/\tau)$. The auxiliary plot represents a zoom in on the left-most cumulative distribution. The black line is the cumulative distribution of a fitted normal distribution $f = 0.5 - 0.5\text{erf}\left[b\left(t^* - c\right)\right]$. Here b and c are fitted parameters and erf stands for the error function. To the right: a zoom in the base trajectory shown in figure 2a. The points denote some instantaneous times t_j^i where the ensembles are launched. The colors of the dots correspond to the colors of the distributions shown in the left plot.

$N_b = 14000$ base trajectories, sampled at $N_t = 400$ time steps summing up to a grand total of $N_b \times N_t \times N_e = 5.6e9$ individual MFE model simulations.

5.1. *Conditional statistics*

See in figure 3 (left) conditioned puff decay statistics. Each line corresponds to a different ensemble, initialized close to an instantaneous state of the base trajectory, at different t_j^i before the decay event at T_d^i. Here t^* represents the lifetime of puffs in the ensembles after initialization and t_d the time of decay. For the sake of comparison, the distributions are horizontally shifted so $t^* = 0$ corresponds to the earliest decay event observed in each ensemble. Although it is not shown here, we observe the same behaviour in the MFE model. Note how, the distribution of puff decay tends to be clustered close to $t^* = 0$, if the puff used to initialize the ensemble is close to the base trajectory decay event. We hypothesize that, puffs so close to the decay event are on the laminar side of the edge and can thus be already considered as relaminarized. Note that, the edge state is a manifold that separates locally the phase space in two regions: in one part trajectories decay at short times, in the other, they exhibit chaotic dynamics before decay [18]. In this region of phase space, all trajectories close to the base one will also decay shortly after or before the base trajectory. How long they will take to do so depends on the particular direction of the white Gaussian noise used to initialize the simulation. That is the reason why the distribution of lifetimes, for these cases, follows a Gaussian instead of the expected exponential one, thereby confirming our hypothesis.

As one uses puffs farther back in time with respect to the decay event as initial condition, the resultant survival distribution tends to look more and more similar to the expected exponential distribution, see fig. 1.

In this paper we use these conditional lifetime distributions to characterize the predictability

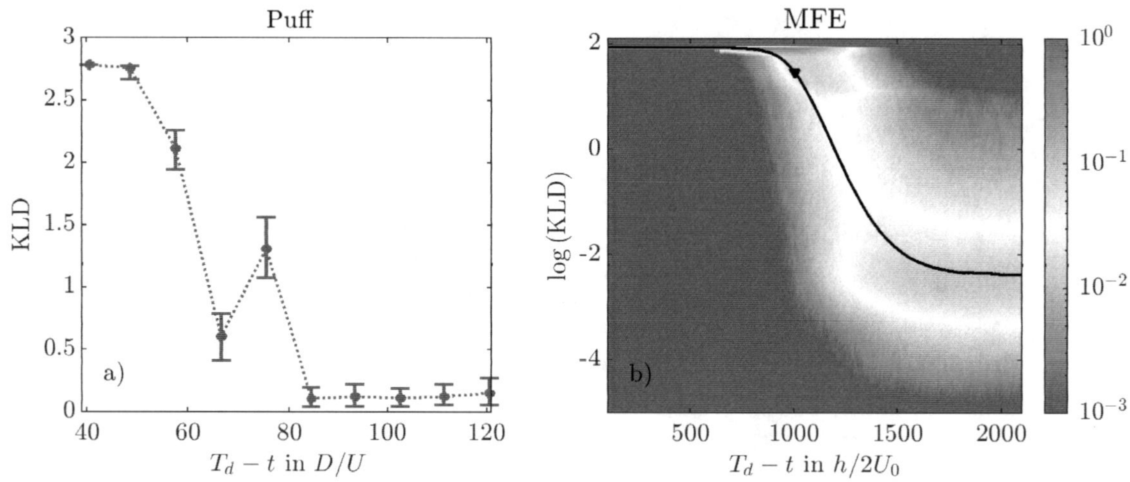

Figure 4: Kullback-Leibler divergence as a function of time for puff (left) and MFE model (right) decay events. The positive horizontal axis represents back in time with respect to a decay event T_d. In the case of the puff (a, left) we present results for only one base trajectory. The error-bars denote the uncertainty of computing the KLD using $N_e = 200$ discrete data points according to a bootstrapping analysis. In the case of the MFE model (b, right) the KLD is presented in logarithm form and the results correspond to $N_b = 14000$ different base trajectories. The colormap represents a normalized probability distribution function of the KLD at each time step. In a black line, see the mean KLD.

of each puff state. We compute the predictability as the difference between these conditional probability distributions (conditioned to be initialized close to a given state), and the expected exponential one (according to experiments).

5.2. Kullback-Leibler divergence

The Kullback-Leibler divergence (KLD) measures the difference between two probability distributions p and p_q, and is given as:

$$\mathrm{KLD} = \sum_x p(x) \log\left(\frac{p(x)}{p_q(x)}\right). \tag{9}$$

Note that the KLD is minimum when $p \equiv p_q$, then $KLD \to 0$. In our case, p is the normalized probability distribution function $p = \frac{\mathrm{d}P(t^*)}{\mathrm{d}t^*}$ of the conditional statistics of decay shown in fig. 3. The normalized probability distribution $p_q = \frac{\mathrm{d}P_q(t^*)}{\mathrm{d}t^*}$ corresponds to the expected exponential lifetime distribution, see fig. 1. The units of the KLD are arbitrary, and they are different between the puff and MFE model cases. For both cases, the bigger the KLD is, the more different the two distributions are, and therefore, the more predictable the members of the ensemble are.

See in figure 4a the corresponding KLD of the base trajectory shown in figure 2a. The positive horizontal axis here, and in the rest of the figures, corresponds to back in time with respect to the decay event at T_d^i. Close to the decay event, the instantaneous state of the puff is highly predictable and the KLD is maximum. At this point in time there is no uncertainty on the value of the KLD, as, even after performing a bootstrapping analysis, the value of the KLD is almost unaffected. This means that, all the members of the ensemble decay in less than $\approx 40D/U$ advective time units. Therefore, for the selected ϵ_0, the decay of this base trajectory can be predicted with full accuracy at $T_d - t \lesssim 40D/U$ advective time units before the event.

As one goes further back in time, the predictability decreases until it reaches a minimum at $T_d - t \approx 80$ advective time units before the decay event. For this base trajectory the decay predictability is completely lost in less than 40 advective time units. Also, for this particular base trajectory, predictability decreases in a non-monotonic way, meaning that, as the base trajectory approaches the decay event, it visits regions of phase space with increasing and decaying predictabilities.

In figure 4b we show the evolution of the KLD for $N_b = 14000$ different MFE model base trajectories. We also compute statistics of the KLD conditioned to the particular time t_j^i before the decay event at T_d^i. With colors we denote a normalized probability distribution function at each time step. Red means high probability and dark blue zero probability that a base trajectory has a particular value of the KLD. The black line represents the mean KLD. Close to the decay event, all trajectories have maximum predictability, as seen by the narrow probability distribution of KLD at $T_d - t \to 0$. This is the case until $T_d - t \approx 700h/2U_0$. This means that, for the selected heuristic threshold, and initial noise ϵ_0, the decay of the MFE model can be predicted with full accuracy only at $T_d - t \lesssim 700h/2U_0$ time units before the event. As one goes back in time, the predictability decreases. In general, the predictability of the MFE model is lost in $\approx 800h/2U_0$ time units. However, as denoted by the probability distribution function, at the same time step before the decay event, some trajectories can have predictabilities that differ by several orders of magnitude from one another.

5.3. *An additional metric of predictability*

Here we propose an additional metric of predictability. Note that the conditional distributions, see fig. 3, have a shape that is always somewhere in between two limiting cases. One is the exponential distribution, that has the smallest slope possible. The other is the Gaussian-like distributions, that has a quadratic decay with time, and that is usually observed for ensembles initialized with instantaneous states close to the decay.

Let $t^*_{P=P_i}$ denote the time at which the conditional probability $P\left(t^*_{P=P_i}\right) = P_i$, being P_i a desired probability, and t^e the time the exponential distribution reaches $P_q\left(t^e\right) = P_{q,i}$. One can define an indicator:

$$\mathrm{I} = 1 - \frac{t^*_{P=P_i}}{t^e}, \tag{10}$$

that measures the difference between the slopes of the conditional probability and the exponential one. Note that, if the conditional probability is equal to the exponential one $\mathrm{I} \equiv 0$. In any other case $0 < \mathrm{I} < 1$. If $\mathrm{I} \approx 1$ the ensemble, and therefore the instantaneous state, are highly predictable.

We set a target probability $P_i = 0.99$ and look at what values of $t^*_{P=P_i}$ the distributions reach this value. If the available distribution data ends before this value is reached, we assume that the rest of the distribution has a slope equal to the exponential one $\propto -\tau^{-1}$. This assumption comes from a simple observation of figure 3. There, the distributions with the slowest slope tend to follow the exponential distribution.

See in figure 5 the results of this new indicator. By comparing this figure with fig. 4, it is clear that the new metric almost perfectly reproduces the behaviour of predictability obtained with the KLD. As an advantage, and different to the KLD, this metric has well defined and interpretable values.

5.4. *The effect of the noise magnitude*

In this section we compare results of ensembles of the MFE model using four different $\epsilon_0 = 1e-3, 5e-4, 1e-4, 5e-5$. For each ϵ_0 we perform ensembles of $N_e = 1000$ members, of $N_b = 2000$ base trajectories sampled at $N_t = 400$ time steps.

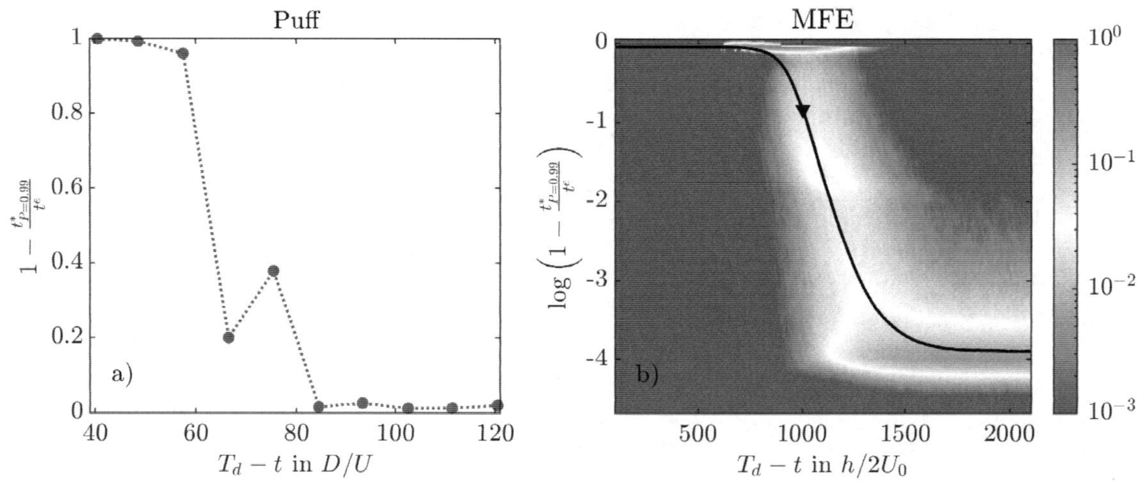

Figure 5: New metric of predictability, see §5.3, eq. (10), for puffs (a,left) and the MFE model (b, right). We consider one base trajectory for the case of puffs, and $N_b = 14000$ MFE model trajectories, se table 2. The color-map denotes the normalized probability distribution of the metric at each time step.

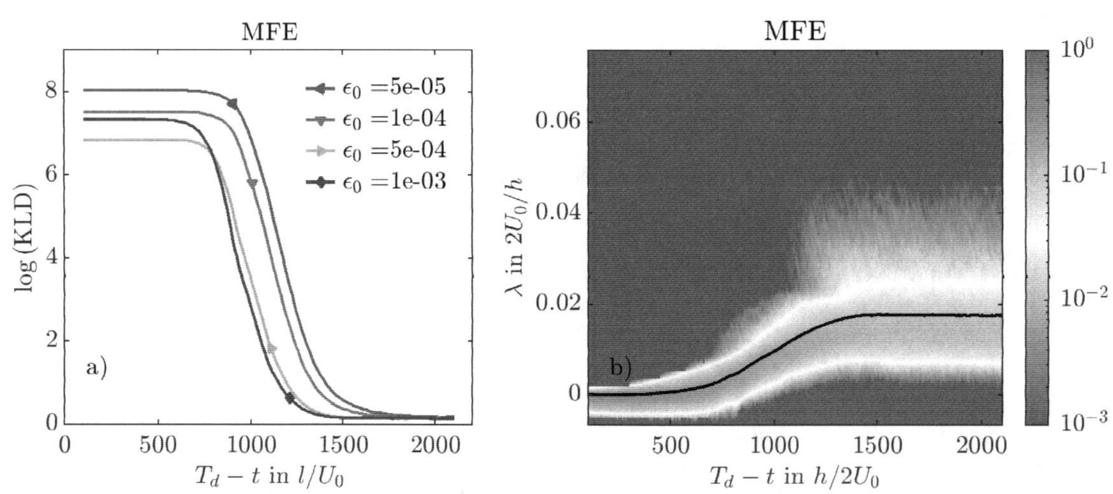

Figure 6: To the left, mean KLD of different ensembles of the MFE model, with $N_e = 1000$, $N_b = 2000$ and $N_t = 400$, initialized with different noise magnitudes ϵ_0. To the right, the resultant local Lyapunov exponents of ensembles of the MFE model with $N_e = 1000$, $N_b = 2000$ and $N_t = 400$. The colormap denotes a normalised probability distribution function at each time step.

See in figure 6a the mean KLD of all the cases at different ϵ_0. Note that all the cases considered here, saturate at similar low KLD values, as one goes back in time. Also, they all have a similar slope of predictability loss. The time at which the predictability starts to decrease for all the cases is a function of ϵ_0. This is expected. As one increases the initial magnitude of the noise, the trajectories initially separate faster and result in less predictable distributions.

All the KLD shown here saturate at different maximum KLD. This has to do with the methods we use to compute the KLD but also has an interesting explanation. Close to a decay event, the conditional distribution becomes more similar to a Gaussian distribution, as seen in figure 3. The standard deviation of this Gaussian distribution depends on the initial noise ϵ_0. If this noise is small, the Gaussian tends to have a smaller variability. Gaussians with a smaller variability are less similar to the exponential distribution, and result in a higher KLD. We expect a similar behaviour in the case of puffs, if we would change the initial noise size ϵ_0.

5.5. The effective Lyapunov exponent

Given a state $\mathbf{x_0}$ of a system, one can define a second state $\mathbf{x_1} = \mathbf{x_0} + \delta\mathbf{x}$. Provided that $\delta\mathbf{x}$ is small enough in magnitude, as time marches, the trajectories that result from the two initial conditions will initially diverge exponentially. The rate at which they do so is what is known as the effective Lyapunov exponent of the state $\mathbf{x_0}$ [24]. One can repeat this, not for just one, but for different $\delta\mathbf{x}$, and obtain an average effective Lyapunov exponent. In our case, we consider as initial states $\mathbf{x_0}$ the instantaneous states of our MFE model for $N_b = 2000$ base trajectories, sampled at $N_t = 400$ times. For each of these instantaneous states, we perform $N_e = 1000$ ensembles of simulations initialized with a small $\epsilon_0 = 1e-7$. Note that, in order to compute the effective Lyapunov exponents, we need that the trajectories initially diverge in an exponential way, and this will only be the case if ϵ_0 is small enough. We then compute the ensemble-average effective Lyapunov exponents as the ensemble average rate at which all the trajectories of the ensemble diverge from the base one.

See in figure 6b the effective Lyapunov exponents computed for the MFE model. The colormap denotes a probability distribution function of the effective Lyapunov exponent at each time step. Note that, close to the decay event, $\lambda \lesssim 0$, meaning that neighbour trajectories tend to come closer to one another. Farther back in time, the mean effective Lyapunov exponent increases, and comes closer to the expected value of $\lambda \approx 0.0244\,(2U_0/h)$ [19]. By comparing this color-map with figure 4b, it is clear that the effective Lyapunov exponents approximate the predictability of the instantaneous MFE model. However, close to decay, λ shows much more uncertainty than the KLD. Below, we compare both metrics in terms of predictability.

5.6. The relation between predictability metrics

In this paper we have proposed three methods to characterize the predictability of puffs and MFE model decay. Here we compare these three methods. In figure 7 we show statistics of the correlation coefficient between the KLD and either the effective Lyapunov exponent or the new metric of predictability described in §5.3. Here we only show results of the MFE model for $N_b = 2000$ base trajectories and $\epsilon_0 = 1e-4$.

We observe that the new indicator, see eq. (10), almost perfectly correlates with the KLD for all the cases considered here, fig. 7. This means that, in order to assess the predictability of an ensemble of simulations of transient chaos, one can use any of the two metrics, and obtain the same information.

The effective Lyapunov exponents and the KLD correlate reasonably well, fig. 7, but we observe a far larger standard deviation of the correlation coefficient between these two than for the case of KLD and the new predictability indicator. This means that the effective Lyapunov exponent can give some idea of the predictability of certain regions of phase space. However, and even though the MFE model does not have a broad spectrum of time scales, λ is not an

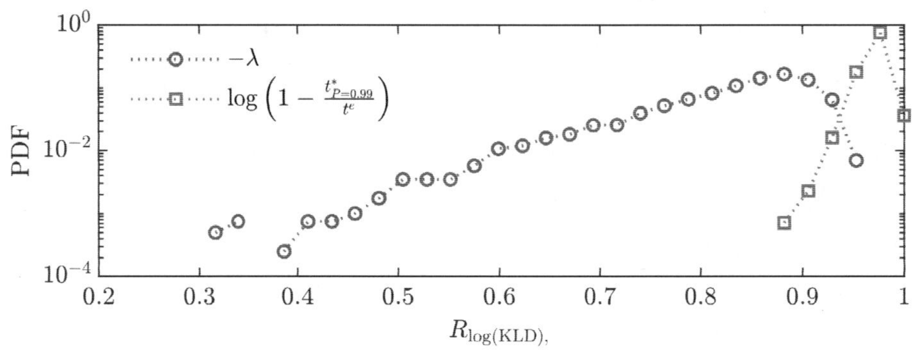

Figure 7: Normalized probability distribution functions of the correlation coefficient between the logarithm of KLD and either the minus effective Lyapunov exponents λ, or the logarithm of the new indicator of predictability, eq. (10), described in §5.3. The results correspond to $N_b = 2000$ base trajectories sampled at $N_t = 400$ sampled times of the MFE model.

accurate measure of predictability [24]. Puffs have a broader range of time scales, and therefore, the correlation is expected to be even worse in this case.

6. Discussion

In this paper we have studied the predictability of turbulence/chaos decay events in wall-bounded flows. We have considered two different systems: a reduced-order model of shear flows, the MFE model, and the case of puffs in pipe flow. As we need massive ensembles of simulations to assess the predictability of decay events, we here present for the first time a new C-CUDA code version of the open-source `nsPipe` code of pipe flow. We show how the new code outperforms the original CPU version in speed and memory-efficiency.

In the paper we describe the methods we use to assess predictability of decay. The main idea is to perform massive ensembles of simulations initiated close to states of the system that have a known decay event in the future. By studying the decay statistics of the ensembles, we can assess the predictability of regions of phase space in the vicinity of these states. We assess the predictability using different metrics. In general, we find that, as one goes back in time with respect to a decay event, predictability decreases. Interestingly, predictability does not always decrease monotonically. We also find that, at the same time before decay, some trajectories have radically different predictabilities compared to others.

In this paper we explore some details of the proposed methods, using only the MFE model. We find that, the main results regarding predictability of the model, are independent to the value of the initial noise of the simulations, as long as the initial noise ϵ_0 stays in a certain range. We also compare different metrics to assess predictability. We show how, a newly defined metric to assess predictability perfectly correlates with the Kullback-Leibler divergence, with the advantage of having a better interpretability. We also compare the KLD with the effective Lyapunov exponent. We report how, the local Lyapunov exponent can approximate the predictability of the MFE model, but it is not very accurate. We expect to find a worse correlation in the case of turbulent puffs.

As future steps of the analysis we plan to report on predictability characteristics of not only one but multiple puff base trajectories. We also plan to classify instantaneous states of puffs and MFE model according to their predictability. We believe that, by identifying what makes an instantaneous state more predictable, we can find the causes behind decay events.

Acknowledgements
Dr. Markus Rampp (Max Planck Computing and Data Facility, MPCDF) conducted the performance studies of the GPU code comparing it with the CPU code. His assistance is gratefully acknowledged. The attendants of the Fifth Madrid Turbulence Workshop are also acknowledged for interesting and fruitful discussions. This work was supported in part by the European Research Council under the Caust grant ERC-AdG-101018287, and are here gratefully acknowledged.

References

[1] Stewart M, Carleton W C and Groucutt H S 2022 *Frontiers in Earth Science* **10** 786829
[2] Sillmann J, Thorarinsdottir T, Keenlyside N, Schaller N, Alexander L V, Hegerl G, Seneviratne S I, Vautard R, Zhang X and Zwiers F W 2017 *Weather and climate extremes* **18** 65–74
[3] Jiménez J 2013 *Physics of Fluids* **25**
[4] Guerrero B, Lambert M F and Chin R C 2020 *Journal of Fluid Mechanics* **904** A18
[5] Hack M P and Schmidt O T 2021 *Journal of Fluid Mechanics* **907** A9
[6] Eckhardt B, Schneider T M, Hof B and Westerweel J 2007 *Annu. Rev. Fluid Mech.* **39** 447–468
[7] Hof B, De Lozar A, Kuik D J and Westerweel J 2008 *Physical review letters* **101** 214501
[8] Avila K, Moxey D, De Lozar A, Avila M, Barkley D and Hof B 2011 *Science* **333** 192–196
[9] Goldenfeld N and Shih H Y 2017 *Journal of Statistical Physics* **167** 575–594
[10] Nemoto T and Alexakis A 2021 *Journal of Fluid Mechanics* **912** A38
[11] Avila M, Barkley D and Hof B 2023 *Annual Review of Fluid Mechanics* **55** 575–602
[12] Gomé S, Tuckerman L S and Barkley D 2020 *Physical Review Fluids* **5** 083905
[13] Bottin S and Chaté H 1998 *The European Physical Journal B-Condensed Matter and Complex Systems* **6** 143–155
[14] Shi L, Avila M and Hof B 2013 *Physical review letters* **110** 204502
[15] Moehlis J, Faisst H and Eckhardt B 2004 *New Journal of Physics* **6** 56
[16] Moehlis J, Faisst H and Eckhardt B 2005 *SIAM Journal on Applied Dynamical Systems* **4** 352–376
[17] Palmer T and Hagedorn R 2006 *Predictability of weather and climate* (Cambridge University Press)
[18] Gomé S, Tuckerman L S and Barkley D 2022 *Philosophical Transactions of the Royal Society A* **380** 20210036
[19] Srinivasan P A, Guastoni L, Azizpour H, Schlatter P and Vinuesa R 2019 *Physical Review Fluids* **4** 054603
[20] Fox A J, Constante-Amores C R and Graham M D 2023 *Physical Review Fluids* **8** 094401
[21] Martínez-Sánchez Á, López E, Le Clainche S, Lozano-Durán A, Srivastava A and Vinuesa R 2023 *Journal of Fluid Mechanics* **967** A1
[22] Lellep M, Prexl J, Eckhardt B and Linkmann M 2022 *Journal of Fluid Mechanics* **942** A2
[23] Lorenz E N 1965 *Tellus* **17** 321–333
[24] Aurell E, Boffetta G, Crisanti A, Paladin G and Vulpiani A 1996 *Physical Review E* **53** 2337
[25] Boffetta G, Cencini M, Falcioni M and Vulpiani A 2002 *Physics reports* **356** 367–474
[26] Ziehmann C, Smith L A and Kurths J 2000 *Physics Letters A* **271** 237–251
[27] Palmer T N 2000 *Reports on progress in Physics* **63** 71
[28] Jiménez J 2023 *Journal of Fluid Mechanics* **968** A10
[29] López J M, Feldmann D, Rampp M, Vela-Martín A, Shi L and Avila M 2020 *SoftwareX* **11** 100395
[30] Willis A P 2017 *SoftwareX* **6** 124–127

The role of nonlinear interactions in the onset of drag increase in flow over riblets

B Viggiano[1], C J Camobreco[2], J Wong[2], M Luhar[3], R García-Mayoral[4], D Chung[2] and D Gayme[5]

[1]Génie Mécanique, Polytechnique Montréal, Montréal, QC H3T 1J4, Canada
[2]Department of Mechanical Engineering, University of Melbourne, Victoria 3010, Australia
[3]Department of Aerospace and Mechanical Engineering, University of Southern California, Los Angeles, CA 90089, USA
[4]Department of Engineering, University of Cambridge, Cambridge CB2 1PZ, UK
[5]Department of Mechanical Engineering, Johns Hopkins University, Baltimore, MD 21218, USA

E-mail: bianca.viggiano@polymtl.ca

Abstract. Characterizing the mechanisms that contribute to the onset of drag increase over micro-grooves (riblets) as the spacing increases is critical to design strategies for riblet-based drag reduction. This study decomposes the roughness function to investigate different mechanisms associated with the breakdown of drag reduction as riblet spacing is increased. We obtain the roughness function through direct numerical simulations (DNS) in a minimal channel and restricted nonlinear (RNL) models. Both the traditional RNL decomposition and an augmented RNL (ARNL) model that includes additional nonlinear interactions are employed as computationally tractable, reduced order representations of the flow field. RNL and ARNL results are compared to those of DNS in minimal channels to investigate the role of the different scale-dependent nonlinear interactions contributing to the roughness function. A comparison of the co-spectra arising from the minimal channel DNS with that from RNL and ARNL simulations indicates that general trends are captured by both reduced order models. However, the additional nonlinearity introduced in the ARNL model produces closer correspondence in the observed structural features of the DNS results. In particular, the ARNL better captures the signatures of the dispersive flow and the texture-coherent fluctuations. There is also a noticeable improvement observed in the profiles of the added stress contributions obtained with the ARNL model versus the RNL model.

1. Introduction

Transport efficiency losses in a range of aviation applications have been attributed to skin-friction drag [1, 2, 3]. Mitigation of this drag can be achieved through the inclusion of small-scale spanwise-varying surface elevations known as riblets [4]. If sized correctly, riblet lined surfaces in low Reynolds number flows can reduce skin-friction drag by nearly ∼10% compared to a smooth wall [4, 5, 6]. The potential to exploit this behavior in passive flow control has motivated detailed studies of the effect of riblets both experimentally [5, 6] and numerically [7, 8, 9]. For flows over small riblets, the turbulence remains smooth-wall-like, albeit with a wall-normal displacement [10, 11, 12]. Here small refers to riblets with a square-root of groove area $\ell_g^+ \lesssim 10$ [4], and the '+' superscript denotes scaling by the wall shear stress τ_w and kinematic viscosity of the fluid ν. As the spacing is increased, the riblet-induced drag reduction lessens and eventually an increase in drag is seen in flows over larger riblets. An in-depth understanding of this so-called breakdown of riblet-induced drag reduction is required to robustly deploy riblets.

Efforts to understand and further define the mechanisms underlying the breakdown of riblet-induced drag reduction have focused on contributions to the roughness function. The roughness function $\Delta U^+ \equiv U_S^+ - U_R^+$ quantifies the shift between the mean streamwise velocity profile of the flow over smooth and riblet surfaces, denoted by the subscripts S and R, respectively. García-Mayoral *et al.* (2011) [13] decomposed the change in skin-friction coefficient (related to ΔU^+) into contributions from the slip velocity and the effect of extra Reynolds stresses. Contributions to the roughness function ΔU^+ have also been studied for rough and permeable surfaces [14, 15]. Endrikat *et al.* (2021) [9] recently performed a detailed campaign of numerical simulations for multiple riblet geometries and spacings to assess the contribution of Kelvin–Helmholtz-like (spanwise) rollers to ΔU^+. They also assessed their role in the breakdown of the drag-reduction. Although this role was previously noted in blade riblet geometries [4], the results in [9] showed that spanwise rollers only formed for certain riblet geometries and only accounted for a portion of the drag increase.

In smaller riblets, the effects of extra Reynolds stresses, e.g. as considered in García-Mayoral *et al.* (2011) [13], are negligible. The roughness function is then directly related to the virtual origins of the mean velocity ℓ_U^+ and of the turbulence ℓ_T^+, i.e. $\Delta U^+ = -(\ell_U^+ - \ell_T^+)$ [10, 11], where ℓ_U^+ and ℓ_T^+ are both measured from the riblet crests and defined as positive toward the groove floors. Accurately measuring ΔU^+ and the virtual origins ℓ_U^+ and ℓ_T^+ in a self-consistent manner is more challenging for larger riblets (at sizes when the drag reduction breaks down). First, the turbulence is no longer smooth-wall-like, and the resulting difference in Reynolds shear stresses (between smooth and riblet walls) contributes to the roughness function, i.e. $\Delta U^+ = -(\ell_U^+ - \ell_T^+) + \Delta U_{u'v'}^+$, where, e.g. $\Delta U_{u'v'}^+ = \int \left[-\langle u'_S v'_S\rangle_{x,z,t}^+ + \langle u'_R v'_R\rangle_{x,z,t}^+\right] \, dy^+$. Here, subscripts S or R on velocity fluctuations u' and v' respectively refer to smooth or riblet lined walls, and angled brackets indicate averaging over the directions specified in the subscript. Second, although the uncertainties in ΔU^+ and ℓ_U^+ are relatively small for riblets with sizes $10 \lesssim \ell_g^+ \lesssim 20$, cf. [12], uncertainties in the turbulence origin ℓ_T^+ and Reynolds stress contribution $\Delta U_{u'v'}^+$ become large (given that both are related). While this issue does not prevent measuring an approximate friction velocity that characterizes the shifted but otherwise similar log-layer of the flow above rough walls and canopies (as related to either a zero-plane-displacement or penetration depth [16, 17, 18, 19, 20]), it hampers understanding of the drag-increasing mechanisms.

Measuring ΔU^+ requires both a clearly defined log-layer and a log-layer slope (von Kármán constant) matched with that of a smooth wall. Thus, the thickness of the boundary layer (set by, e.g. the channel height) must be much larger than the height of the roughness sublayer [21]. Outer-layer similarity must also be maintained both in the wake and the log-layer [20]. This condition is typically satisfied when the characteristic lengths of the outer scales are much larger than the characteristic lengths of the roughness. In a numerical setting, the log-layer measured roughness function ΔU^+ can be efficiently obtained by constraining the dimensions of the periodic box and by presuming outer-layer similarity. These minimal log-layer channels have been shown to capture both near-wall and log-layer turbulence over a smooth wall, up to a given critical wall-normal location within the outer-layer similar region [22]. For rough walls, this minimal channel technique was similarly shown to correctly predict the velocity shift ΔU^+ [23] and spectrally truncated turbulent quantities [24], although the technique is unable to predict the bulk velocity directly. Furthermore, one of the drag-increasing contributions to ΔU^+, specifically, that associated with spanwise rollers, is reproduced by minimal channels [25]. While even smaller minimal buffer-layer channels can capture individual coherent structures (e.g. a single streak) [26], these smaller domains do not encompass the logarithmic region required to simultaneously measure ΔU^+ and the mechanisms contributing to it. Thus, contributions to the roughness function ΔU^+ for riblets up to and just beyond drag-breakdown will be assessed herein with minimal log-layer channels.

The fine resolution required for DNS of flows over riblets has also motivated the use of more computationally efficient reduced order flow representations. In particular, the correlation between skin-friction drag [16, 27] and the coherent structures in wall-bounded turbulence motivates a coherent structure based representation, such as the RNL model [28]. This modeling framework has been shown to reproduce salient features of low and moderate Reynolds number wall-bounded turbulence [29, 30, 28], as well as riblets and non-homogeneous rough surfaces [31] at vastly reduced computational expense. This model decomposes the flow into a large scale streamwise constant mean and small-scale perturbations about that mean, and restricts nonlinear interactions to those contributing to the mean flow. This full representation of the cross-plane motions associated with the secondary flows contributing to $\Delta U^+_{u'v'}$, suggests that an RNL model may be well suited to computationally efficient studies of the associated flow mechanisms. We also introduce an augmented RNL (ARNL) model, which permits intermediate scales to interact nonlinearly with the RNL mean flow but maintains order reduction by dynamically constraining inter-scale interactions [32]. The use of the two models enables the evaluation of the role of the additional nonlinearity in the ARNL model in the mechanisms of interest.

In this work, we aim to study the flow mechanisms that are related to the $\Delta U^+_{u'v'}$ contribution to the roughness function (ΔU^+), and thereby characterize the drag increase generated by larger (post-optimal) riblets. However, mechanisms that appear related to the $\Delta U^+_{u'v'}$ contribution may be misinterpreted if the virtual origin for turbulence ℓ_T^+ (or zero-plane-displacement) is incorrectly identified. Thus, our first aim is to assess the sensitivity of $\Delta U^+_{u'v'}$ to ℓ_T^+, and to further investigate whether portions of the $\Delta U^+_{u'v'}$ contribution can be rendered insensitive to ℓ_T^+, through analysis of minimal channel DNS data. Thereafter, our second aim is to use the RNL and ARNL models to assess whether a reduced set of nonlinear interactions can recover $\Delta U^+_{u'v'}$, while similarly assessing their sensitivity to ℓ_T^+ within these reduced order settings.

The remainder of this paper is organized as follows. We first present the mathematical formulation of the reduced order models in Section 2. The setup of the minimal channel DNS, as well as the RNL and ARNL models is described in Section 3. This is followed by a description of the decomposition of the roughness function in Section 4. The resulting stress contributions are investigated in Section 5. Our final conclusions and outlook for this work are provided in Section 6.

2. Restricted nonlinear models

The RNL and ARNL equations are formed by first decomposing the total velocity and pressure fields, respectively $\boldsymbol{u}(x,y,z,t)$ and $\boldsymbol{p}(x,y,z,t)$, into a mean and perturbations about that mean, i.e., $\boldsymbol{u}(x,y,z,t) = \boldsymbol{U}^\star(x,y,z,t) + \boldsymbol{u}^\star(x,y,z,t)$ and $\boldsymbol{p}(x,y,z,t) = \boldsymbol{P}^\star(x,y,z,t) + \boldsymbol{p}^\star(x,y,z,t)$ Here the mean of the (A)RNL velocity and pressure fields are defined through a spatial filter $\boldsymbol{U}^\star(x,y,z,t) = [\boldsymbol{u}(x,y,z,t)]_l$ and $\boldsymbol{P}^\star(x,y,z,t) = [\boldsymbol{p}(x,y,z,t)]_l$ that limits the contributions to a specified set of streamwise Fourier modes $k_x \in \mathcal{L}$. The perturbations about the mean of the velocity and pressure field are respectively denoted $\boldsymbol{u}^\star(x,y,z,t)$ and $\boldsymbol{p}^\star(x,y,z,t)$. These perturbation fields are also restricted using a small-scale spectral filter, denoted as $[\cdot]_s$, which specifies the support of the perturbation dynamics to the set of modes $k_x \in \mathcal{S}$. The RNL and the ARNL dynamics are given by $\nabla \cdot \boldsymbol{U}^\star = \nabla \cdot \boldsymbol{u}^\star = 0$, and the momentum equations,

$$\partial_t \boldsymbol{U}^\star + [\boldsymbol{U}^\star \cdot \nabla \boldsymbol{U}^\star]_l + \nabla \boldsymbol{P}^\star/\rho - \nu \nabla^2 \boldsymbol{U}^\star = -[\boldsymbol{u}^\star \cdot \nabla \boldsymbol{u}^\star]_l, \tag{1a}$$

$$\partial_t \boldsymbol{u}^\star + [\boldsymbol{U}^\star \cdot \nabla \boldsymbol{u}^\star]_s + [\boldsymbol{u}^\star \cdot \nabla \boldsymbol{U}^\star]_s + \nabla \boldsymbol{p}^\star/\rho - \nu \nabla^2 \boldsymbol{u}^\star = 0. \tag{1b}$$

These models restrict the dynamics by omitting nonlinear interactions between perturbations that do not contribute to the large scale modes $k_x \in \mathcal{L}$. In the RNL model, the single large-scale streamwise mode is $k_x = 0$, i.e. $\boldsymbol{U}^\star(y,z,t) = \langle \boldsymbol{u} \rangle_x$ where $\langle \cdot \rangle_x$ denotes streamwise averaging. The RNL small scales, $k_x \neq 0 \in \mathcal{S}$, are restricted to modes in the peak range of the outer-layer surrogate dissipation spectra, see e.g. [30] for further details.

In ARNL the set of large-scale modes $\mathcal{L}$ is expanded to include one or more intermediate scales $k_x \neq 0$ that interact non-linearly with the other *large-scale* modes (i.e. $[\boldsymbol{U}^\star \cdot \nabla \boldsymbol{U}^\star]_l$ in Eq. 1a). In what follows, we denote the large and intermediate $k_x \in \mathcal{L}$ modes in the ARNL model as K_x to differentiate them from the small-scale modes supporting the perturbation dynamics $k_x \in \mathcal{S}$. Similar to the GQL model [33], the additional large-scale modes $K_x \neq 0$, denoted as $K_x^{(n)}$, interact with the small-scale non-zero modes, e.g. $k_x^{(1)}, k_x^{(2)} \in \mathcal{S}$, in a consistent manner, such that for a given single large-scale non-zero mode, $n = 1$, i.e. $K_x^{(1)}$, $k_x^{(2)} = K_x^{(1)} + k_x^{(1)}$, see [32] for further details. As in the RNL model, the ARNL small scales are taken from the peak region of the surrogate dissipation spectrum. We note that this selection of large and small scales modes differs from the GQL modeling approach, where a cutoff defines the respective small and large scales [33].

3. Simulation setup

The direct numerical simulations (DNS) are performed with the solver Cliff, by Cascade Technologies Inc. [34, 35]. The finite-volume node-based solver is second-order accurate, and solves the full Navier–Stokes equations over a body-fitted grid. All DNS employ minimal channels, where the domain size and simulation times are guided by [23, 36]. Minimal channel domain lengths of $L_x^+ \approx 1000$ and spans $L_z^+ \approx 250$ ensure a critical height up to which turbulence remains realistic of $y_c^+ \approx 100$. Simulations times are accordingly increased to ensure the simulation time L_t scaled by the largest realistic eddy (of size y_c) is at least $L_t u_\tau / y_c \approx 300$. The channel half-heights measured from the riblet mean height are set to $\delta^+ = 395$. For further details, see Table 2 or [9, 12]. DNS data for six blade riblet sizes are investigated, two cases from [9] and four from [12]. All blade riblets have a height to spacing ratio of 0.5 and thickness to spacing ratio of 0.2 (Table 1). Three cases correspond to riblets below the optimal groove area [4], and three above. All except the largest case are drag reducing.

A pseudo-spectral, open-source code (LESGO) is used to solve the RNL and ARNL Eqs. (1a) and (1b). The derivatives in the streamwise and spanwise domain are evaluated using Fourier transforms. Central second-order finite-differences are applied for the wall-normal direction. A hyperbolic tangent stretched coordinate system is used to ensure enough grid points near the roughness elements [37]. A Crank–Nicolson scheme is also implemented in the wall-normal domain for time advancement of the diffusive terms. An Adams–Bashforth scheme is applied for the remaining terms. The simplified dynamics associated with the proposed models allow the nonlinear term to be computed as a convolution in Fourier space (k_x, y, z, t) instead of switching to physical space, saving computational costs [38].

The RNL and ARNL simulations are carried out in a half-channel domain with stress-free $(\partial u/\partial y = \partial w/\partial y = 0)$ and zero permeability boundary conditions imposed at the top and bottom of the calculation domain, respectively. Periodic boundary conditions are applied in the streamwise and spanwise directions. No-slip conditions are imposed at the bottom wall. The riblet surface is defined using the immersed boundary method of [39]. In particular, inside the riblets, a force is applied such that the velocity is set to zero for all time steps. Outside of the riblets, the force is set to zero and therefore the velocity is unchanged. The implementation of the immersed boundary method to the RNL and ARNL representations is described in detail in Minnick (2022) [31].

Three blade riblet sizes are considered for the both RNL and ARNL cases, the riblet geometries and simulation parameters for all cases are provided in Table 1. Smooth wall

Table 1. Characteristics of blade geometries considered, where s^+ is the riblet spacing, and the groove area $A_g^+ = \ell_g^{+2}$. DNS cases are minimal channels; those marked with a ‡ superscript are from [9], and those with a † superscript are from [12]. All riblets cross-sections have $h/s = 1/2$ and $b/s = 1/5$.

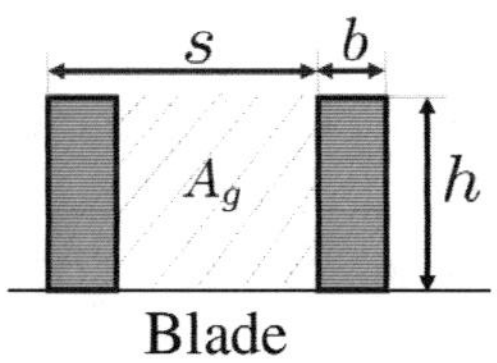

Case	s^+	ℓ_g^+	Case	s^+	ℓ_g^+
BL20$_{\text{RNL}}$	20	12.6	BL08$^{\dagger}_{\text{DNS}}$	7.8	4.9
BL30$_{\text{RNL}}$	30	19.0	BL12$^{\dagger}_{\text{DNS}}$	11.9	7.5
BL40$_{\text{RNL}}$	40	25.3	BL16$^{\dagger}_{\text{DNS}}$	15.6	9.9
BL20$_{\text{ARNL}}$	20	12.6	BL20$^{\ddagger}_{\text{DNS}}$	20.3	12.8
BL30$_{\text{ARNL}}$	30	19.0	BL25$^{\dagger}_{\text{DNS}}$	25.0	15.8
BL40$_{\text{ARNL}}$	40	25.3	BL33$^{\ddagger}_{\text{DNS}}$	33.3	21.1

Table 2. Description of the computational domain and grid resolution for the minimal channel DNS and RNL (including ARNL) simulations, and wavenumbers retained in the RNL and ARNL. L_x^+ and L_z^+ are the streamwise and spanwise domain lengths, respectively, while δ^+ is the domain height measured from the riblet mean height. The resolution in the streamwise, wall-normal and spanwise directions are $\Delta x^+ = u_\tau \Delta x/\nu$, $\Delta y^+ = u_\tau \Delta y/\nu$ (at the riblet tips, and at the channel centreline) and $\Delta z^+ = u_\tau \Delta z/\nu$ (at the riblet tips, for the smallest and largest riblets), respectively.

Case	L_x^+	δ^+	L_z^+	Δx^+	Δy^+	Δz^+
DNS	1002–1027	395	250–264	6	0.30–7.00	0.20–0.83
RNL	-	395	1200	-	0.62–3.41	1
ARNL	-	395	1200	-	0.62–3.41	1

simulations of the RNL and ARNL models with the same resolution and Reynolds number are also performed. Note, the same resolution is employed because this data is used to initialize the RNL and ARNL riblet simulations. For the DNS, the smooth-wall case has similar streamwise and wall-normal resolution, but approximately four times larger spanwise resolution. The details of the computational grids for the DNS and RNL simulations are provided in Table 2. A summary of the streamwise wavenumbers used in the reduced-order models is included in Table 3, where the non-dimensional wavenumber is defined as $k_x\delta = 2\pi\delta/L_x \times (n)$, in which δ is the half-channel height, L_x is the streamwise domain length[1] and n is a non-negative integer. RNL simulations include the streamwise mean, $K_x = 0$ while ARNL cases also include a second large-scale mode, $K_x^{(1)}$, which corresponds to a large-scale wavelength, Λ_x/δ, equal to 0.898, which is the nearest in value to the half-channel height δ. As described above, the small-scale modes supporting the RNL dynamics are associated with wavelengths in the region associated with outer-layer peak of the surrogate dissipation spectrum. While the ARNL small-scales are also constrained to this region, they differ from those in the RNL model to ensure that $k_x^{(2)} = K_x^{(1)} + k_x^{(1)}$.

[1] L_x is used here to define the wavenumber although there is no physical grid used in the reduced order simulations.

Table 3. The wavenumbers and corresponding wavelengths retained in the RNL and ARNL.

Case	$K_x\delta$	$\approx \Lambda_x/\delta$	$k_x\delta$	$\approx \lambda_x^+$
RNL	0	∞	15.5, 16, 16.5	160.1, 155.1, 150.4
ARNL	0, 7	∞, 0.898	12.5, 19.5	198.5, 127.3

4. Texture-coherent contribution to the roughness function

As discussed in Section 1, the roughness function is decomposed as $\Delta U^+ = -(\ell_U^+ - \ell_T^+) + \Delta U^+_{u'v'}$, where, for small riblets, wall turbulence behaves like that over a smooth wall down to the inner viscous sublayer [12], i.e. $\Delta U^+_{u'v'} \approxeq 0$, indicating that the drag increase is only due to a change in virtual origin. For larger riblets, the inner layer changes, where the smooth-wall-like viscous sublayer is replaced by a riblet sublayer such that i.e. $\Delta U^+_{u'v'} \neq 0$. One approach to understand this change is to isolate its texture-coherent contributions from the texture-incoherent contributions [40, 41]. A standard approach to extract the texture-coherent part of $\Delta U^+_{u'v'}$ is to time- and streamwise-average the velocity components, which is generally referred to as the dispersive flow [42]. However, for other non-smooth surfaces, signs of texture-coherence in the remaining fluctuations (instantaneous minus dispersive flow) can still be seen at wavelengths both greater than and less than the roughness spacing, identified as the sidebands of modulation [40, 41, 43]. Thus, a more general class of texture-coherent flows ($\lambda_x^+ \leq \infty$, $\lambda_z^+ \lesssim s^+$), beyond dispersive flows ($\lambda_x^+ \to \infty$, $\lambda_z^+ = s^+, s^+/2, s^+/3, \ldots$), will be considered when evaluating texture-coherent contributions to $\Delta U^+_{u'v'}$.

Measuring both ΔU^+ and $\Delta U^+_{u'v'}$ relies on identifying ℓ_T^+. In this work we choose to identify ℓ_T^+ as the wall-normal shift minimizing the sum-squared error in the Reynolds stress profiles, i.e. minimizing $-\langle u'_S v'_S\rangle^+_{x,z,t}(y^+) + \langle u'_R v'_R\rangle^+_{x,z,t}(y_R^+ = y^+ - \ell_T^+)$ at matched y^+ in the vicinity of the maximum-Reynolds-stress gradient. Herein, this range is chosen as $4 \lesssim y^+ \lesssim 6$ [12]. The reference riblet wall-normal coordinate y_R^+ is zero at the riblet crests. This fitting method is only appropriate when the flow is smooth-wall-like (for small riblets), and does not guarantee sensible values of ℓ_T^+ for larger riblets. For small riblets, this method also agrees well with virtual origins defined by profiles of the root-mean-square of the velocity fluctuations [44].

With a turbulence origin identified, this leaves the issue of evaluating the contribution from the difference in the Reynolds shear stress,

$$\Delta U^+_{u'v'} = -\int_{\ell_T^+}^{y_m^+} \langle u'_S v'_S\rangle^+_{x,z,t}(y^+)\,\mathrm{d}y^+ + \int_{\ell_T^+}^{y_m^+} \langle u'_R v'_R\rangle^+_{x,z,t}(y_R^+ = y^+ - \ell_T^+)\,\mathrm{d}y^+, \qquad (2)$$

where $y^+ = 0$ is at the smooth wall and $y^+ = y_m^+$ is a wall-normal height within the log-layer, above the riblet sublayer [12]. With flow changes occurring only in the riblet sublayer, i.e. with outer-layer similarity, the difference in Reynolds stress beyond the upper integration limit is zero at matched friction Reynolds number $\delta_S^+ = \delta_R'^+$ ($\delta_R'^+$ is the channel height measured from $y_R^+ = -\ell_T^+$), or is nearly zero with sufficiently thick log layers even if the outer flow is mismatched. Eq. (2) depends on ℓ_T^+ in three ways. First, as the wall-normal coordinates of the Reynolds stress profiles are not matched, but differ by ℓ_T^+. If ℓ_T^+ is uncertain, this can translate to leading order changes in $\Delta U^+_{u'v'}$ for larger riblets. Second, through the lower integration bound, which for riblets starts at the crests ($y_R^+ = 0$), and for the smooth wall starts at ℓ_T^+ above the wall. This is less of an issue than matching wall-normal coordinates, as the smooth-wall Reynolds stresses are small near the wall. Third, through the friction velocity u_τ that defines the wall units for the riblets (although the uncertainty in ℓ_T^+ does not greatly alter u_τ). Overall, these uncertainties present a challenge to interpreting the drag-increasing mechanisms that set the minimum of ΔU^+ and hence the optimal riblet size. This motivates further subdividing $\Delta U^+_{u'v'}$ into contributions from coherent flow structures [9, 13, 42], e.g. from typical secondary (dispersive) flows both above and within the riblets grooves, or from spanwise-coherent rollers above the crests. Assessing the sensitivity of some of these contributions to $\Delta U^+_{u'v'}$ is the first aim of analysing the DNS data.

The first issue is to separate out the dispersive stress contributions due to secondary flows. Defining ΔU^+ based on the mean momentum balance $\partial_y U^+ - \langle u'v' \rangle^+_{x,z,t} = 1 - y/\delta'_R$ [12] above the riblet tips requires, for consistency, that fluctuations are defined by

$$u(x, y, z, t) = \langle u \rangle_{x,z,t}(y) + u'(x, y, z, t), \tag{3}$$

and similarly for v and w. Herein, $u(x, y, z, t)$ is the instantaneous velocity, $\langle u \rangle_{x,z,t}(y)$ is the horizontal plane- and time-averaged velocity, henceforth referred to as the spanwise mean flow, and $u'(x, y, z, t)$ is the fluctuation about this spanwise mean (denoted by $'$). To then separate out the dispersive contribution $\tilde{u}$, a fluctuation about the riblet-average u'' is defined, and u' re-expressed as

$$u'(x, y, z, t) = \tilde{u}(y, z) + u''(x, y, z, t). \tag{4}$$

where $\tilde{u}(y, z) = \langle u \rangle_{x,r,t}(y, z) - \langle u \rangle_{x,z,t}(y)$, following the notation of e.g. [9, 40, 41]. Substituting Eq. (4) into Eq. (3) recovers $u(x, y, z, t) = \langle u \rangle_{x,r,t}(y, z) + u''(x, y, z, t)$, highlighting that u'' represents a fluctuation about the riblet-averaged flow $\langle u \rangle_{x,r,t}$ (the flow within each riblet groove, averaged across all the riblet periods, i.e. averaging across s^+-width segments of the domain). From the decomposition in Eq. (4), the product of fluctuations about the spanwise mean can be re-expressed as

$$\langle u'v' \rangle_{x,z,t} = \langle [\tilde{u} + u''][\tilde{v} + v''] \rangle_{x,z,t}, = \langle u''v'' \rangle_{x,z,t} + \langle \tilde{u}\tilde{v} \rangle_{x,z,t}. \tag{5}$$

Thus, the dispersive contribution $\Delta U^+_D = \int_{\ell^+_T}^{y^+_m} \langle \tilde{u}_R \tilde{v}_R \rangle^+_{x,z,t} (y^+_R = y^+ - \ell^+_T)\, \mathrm{d}y^+$ has been identified, where $\Delta U^+_{u'v'} = \Delta U^+_D + \Delta U^+_{u''v''}$. Note that ΔU^+_D is almost unaffected by ambiguity in ℓ^+_T, apart from the change in friction velocity u_τ (the lower limit of the integral is always at the fixed riblet crest, and there is no dispersive smooth-wall equivalent to cancel with).

The preceding discussion to obtain ΔU^+_D indicates that contributions to $\Delta U^+_{u'v'}$ that do not have smooth-wall counterparts, such as those that are coherent with the texture, are resilient to ambiguity in ℓ^+_T. However, the dispersive contributions alone do not constitute all of the texture-coherent contributions. This is because amplitude modulation of the s^+-periodic dispersive flow by the overlying near-wall turbulence (e.g. quasi-streamwise vortices centered around $\lambda^+_z \approx 50$ [45]) gives rise to spectral content at wavelengths above and below s^+ (and also at wavelengths above and below the harmonics $s^+/2, s^+/3, \dots$) [40, 41]. These λ^+_z-sidebands that flank s^+ are visible in Figure 1(a), at $\lambda^+_x \approx 150$ especially, and in spectra of flow over other textures [46, 47]. Provided sufficient scale separation exists between the texture s^+ and the near-wall turbulence, a simple approach is to deem all Reynolds stress contributions with spanwise wavelengths $\lambda^+_z \lesssim 1.25 s^+$ texture coherent ($\Delta U^+_{TCu'v'}$). As illustrated in Figure 1(a), this cut-off includes the modulations sidebands as well as the dispersive flow at $\lambda^+_x \to \infty$. For riblets at larger spanwise spacings s^+ for which scale separation is not sufficient, this texture-coherent cut-off includes near-wall turbulence which may remain texture-incoherent. To minimize this issue, the smooth-wall Reynolds stresses are integrated below the same $\lambda^+_z \lesssim 1.25 s^+$ cut-off for each riblet, i.e. $\Delta U^+_{TCu'v'} = \int_{\ell^+_T}^{y^+_m} -\langle u'_S v'_S \rangle^+_{x,z,t}|_{\lambda^+_z \lesssim 1.25 s^+} (y^+) \mathrm{d}y^+ + \int_{\ell^+_T}^{y^+_m} \langle u'_R v'_R \rangle^+_{x,z,t}|_{\lambda^+_z \lesssim 1.25 s^+} (y^+_R = y^+ - \ell^+_T) \mathrm{d}y^+$.

The isolation of both dispersive and more general texture-coherent contributions in the analysis of the RNL and ARNL models will further highlight the capabilities and limitations

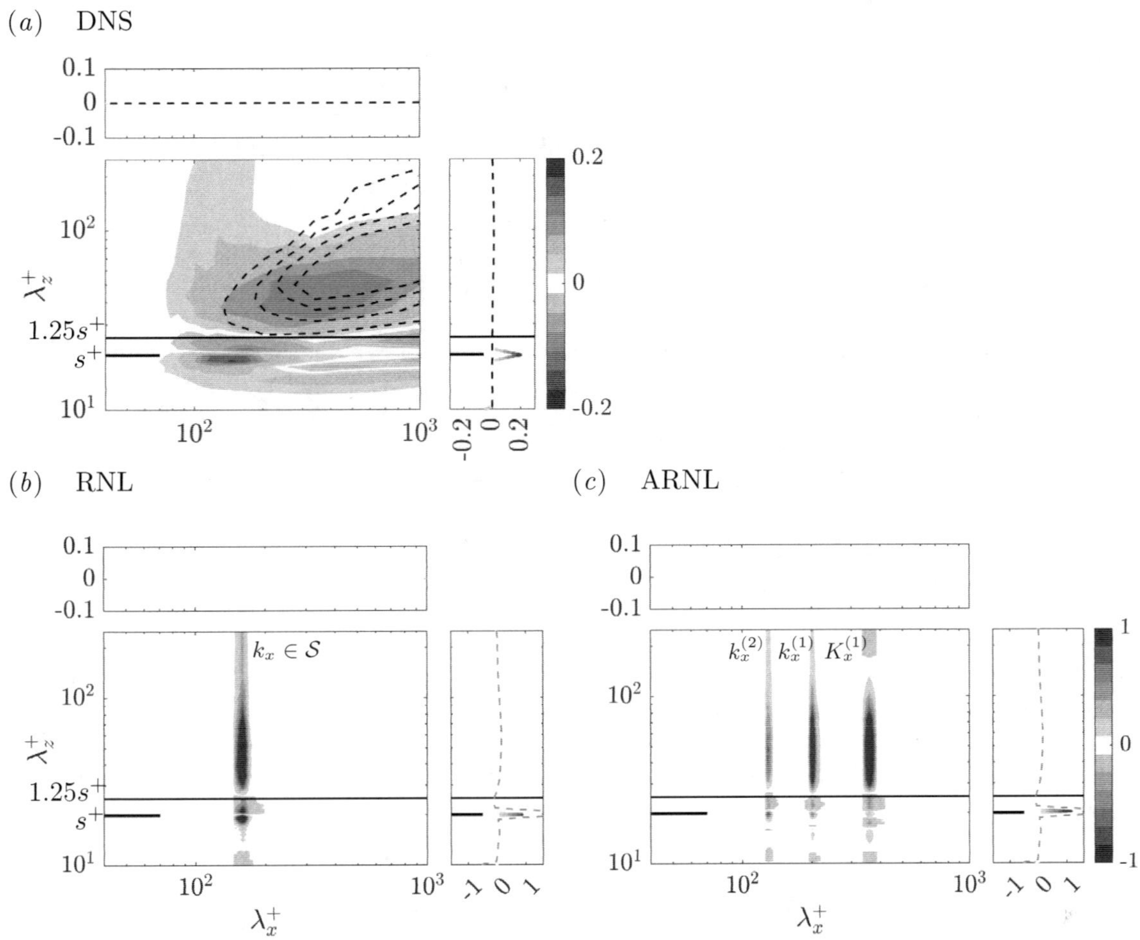

Figure 1. Premultiplied Reynolds stress co-spectra $\langle u'v' \rangle^+$ for $\ell_g^+ \approx 12.8$ blade riblets at $y_R^+ \approx 1$ (colored contours), normalized by the integral of the Reynolds stresses at this y_R^+. (*a*) Minimal log-layer channel DNS. (*b*) RNL model. (*c*) ARNL model. For the DNS in (*a*), the smooth wall spectra are included in dashed black contours at $y^+ \approx 1 + \ell_T^+$ (extrapolated model ℓ_T^+), plotting every second level. For the RNL and ARNL in (*b*, *c*), the dashed gray curves are the sum of the DNS Fourier coefficients for all modes with $\lambda_x^+ \gtrsim 314$. For clarity, premultiplied Reynolds stresses with absolute values < 0.01 have been removed. The RNL and ARNL model show spectra until $\lambda_z^+ = 250$ for consistency with the truncated minimal channel DNS.

of these reduced order models. The RNL model only allows nonlinear interactions between streamwise Fourier modes that contribute to the mean flow, $K_x^{(0)}$, while the ARNL model includes additional triadic contributions to the non-zero 'large-scale' mode, $K_x^{(1)}$. The comparison of $\Delta U_{u'v'}^+$ and $\Delta U_{u''v''}^+$ from both RNL and ARNL thus allows investigation of whether these additional nonlinear improve fidelity of the representation of dispersive stresses due to the riblets.

Journal of Physics: Conference Series **2753** (2024) 012010 doi:10.1088/1742-6596/2753/1/012010

5. Discussion

5.1. Reynolds stress co-spectra

Figure 1(a) shows the premultiplied Reynolds stress co-spectra computed from the minimal channel DNS. Included in this figure is the $1.25s^+$ demarcation of spectral space for just post-optimal riblets ($\ell_g^+ \approx 12.8$, $s^+ \approx 20.3$), where all Reynolds stress contributions in the region $\lambda_z^+ \lesssim 1.25s^+ \approx 25$ are approximated as texture coherent (i.e. are included in $\Delta U^+_{TCu'v'}$), which includes the dispersive contribution ΔU^+_D. Note that the smooth wall contours (dashed lines) remain above the $1.25s^+$ cut-off for this s^+, and so provide almost no cancellation with the texture-coherent Reynolds stresses in the vicinity of s^+, thus improving the robustness of the $\Delta U^+_{TCu'v'}$ measurement. The sidebar of Figure 1(a) also highlights the signature of dispersive flows in the $\lambda_x^+ \to \infty$ and $\lambda_z^+ \leq s^+$ modes, which increase drag via $\Delta U^+_{TCu'v'}$ and are less sensitive to differences in ℓ_T^+. This dispersive signature at $\lambda_x^+ \to \infty$ and $\lambda_z^+ \leq s^+$ vanishes when instead defining fluctuations about the riblet average, and so is not measured in $\Delta U^+_{TCu''v''}$, recalling Eq. (5), and as considered in Section 5.2.

The co-spectra of the RNL model and the ARNL model output are provided in Figures 1(b, c) for a similar riblet spacing of $s^+ \approx 20$. Although the DNS, RNL, and ARNL spectra integrate to similar Reynolds stress values, the RNL and ARNL spectral densities display higher magnitudes because the non-zero streamwise wavenumber support of the dynamics comprises narrow spectral regions. Both models are able to capture the signature of the dispersive stresses in the $K_x = 0$ mode, as observed in the sidebar of Figure 1(b,c), although the DNS estimate here, which includes $\lambda_x^+ \gtrsim 314$, exceeds that of the RNL. Texture-coherent flow signatures in the $\langle u'v' \rangle$ co-spectra are also observed near $s^+ \approx 20$ in the limited RNL and ARNL modes, Figures 1(b,c). For the texture-coherent flows in particular, the signs of the Reynolds stress signatures from RNL and ARNL models match those of the DNS (negative just above s^+, positive just below s^+), for all resolved streamwise scales. These sidebands, cf. [40], suggest that the restricted modes and nonlinear interactions included in both RNL and ARNL are sufficient to capture the modulation effect. We note the inclusion of the intermediate mode at $K_x\delta = 7$ in the ARNL model enables closer comparison with the DNS. The improvement of the ARNL model is evidenced by the redistribution of $\langle u'v' \rangle$ across the three λ_x^+ values, as is similarly observed in the DNS.

5.2. Drag breakdown

Before investigating the decomposition of ΔU^+, the sensitivity of ΔU^+ to the selection of the virtual origin is evaluated by comparing results obtained using an ℓ_T^+ value based on the fitting described in Section 4 and one computed using an extrapolated model (to be described later). The analysis focuses on riblet spacings associated with drag-reduction breakdown. The first approach, as described in Section 4, measures ℓ_T^+ by minimizing the difference between the smooth and riblet DNS Reynolds stress profiles, with the resulting ΔU^+ values depicted in Figure 2(a), ■ markers. The second approach instead employs an extrapolated value for ℓ_T^+, extrapolated from smaller riblet sizes, and is shown in Figure 2(b), ■ markers. As an example, considering the results at $\ell_g^+ \approx 21$, ΔU^+ dropped from ≈ 0.7 to ≈ 0.3 when switching from the first-approach $\ell_T^+ \approx 5.7$, Figure 2(a), to the second-approach $\ell_T^+ \approx 2.4$, Figure 2(b). Note that the mean velocity profiles are normalized by the u_τ consistent with the chosen ℓ_T^+, before the mean flow offsets ΔU^+ are measured above the roughness sublayer ($50 \lesssim y^+ \lesssim 100$ for the minimal channel DNS). The friction velocities consistent with the fitted and extrapolated model ℓ_T^+ differed by $\approx 0.4\%$, where a 1% change in u_τ typically leads to an $\approx 10\%$ change in ΔU^+. Thus, addressing ambiguity in ℓ_T^+ is more important than addressing ambiguity in u_τ, with the ambiguous ℓ_T^+ having noticeably impacted ΔU^+ for post-optimal riblets, at least for $10 \lesssim \ell_g^+ \lesssim 20$. For reference, the extrapolated model ℓ_T^+ values are extrapolated from the virtual origins obtained for small ($\ell_g^+ \lesssim 10$) riblets [12]. These assume $\ell_T^+ \approx h_\perp^+ + m_T {\ell_g^+}^2$, where $h_\perp^+$

is the riblet perpendicular protrusion height [6, 48], and where m_T is the gradient of the ℓ_T/ℓ_g versus ℓ_g^+ curve for small riblets. The viscous model [12] employed to predict this ℓ_T/ℓ_g versus ℓ_g^+ curve maintains good agreement with DNS-fitted ℓ_T^+ values for small riblets ($\ell_g^+ \lesssim 10$), and the DNS-fitted ℓ_T^+ values for small riblets could equally have been extrapolated directly.

Even having shown that the ΔU^+ measurements are impacted by ambiguity in ℓ_T^+, the overall aim remains unchanged: to assess the texture-coherent contributions $\Delta U^+_{TCu'v'}$ which may remain resilient to ambiguity in ℓ_T^+. Thus, the drag breakdown, excluding the highly sensitive texture-incoherent contributions ($\Delta U^+_{TIu'v'}$), becomes $\Delta U^+ \approx -(\ell_U^+ - \ell_T^+) + \Delta U^+_{TCu'v'}$. As shown in Figure 2($a$), the smooth-wall-like and texture coherent-contributions $-(\ell_U^+ - \ell_T^+) + \Delta U^+_{TCu'v'}$ do not recover ΔU^+ with the fitted ℓ_T^+. Thus, a large texture-incoherent Reynolds stress contribution $\Delta U^+_{TIu'v'}$, which is drag reducing, would be required for the mean-momentum budget to balance (note that due to sensitivity to ℓ_T^+, $\Delta U^+_{TIu'v'}$ is not computed separately). However, when using the extrapolated model ℓ_T^+, as shown in Figure 2(b), the texture-incoherent contribution $\Delta U^+_{TIu'v'}$ is now drag increasing, although much smaller in magnitude. Thus, without a method to unambiguously identify ℓ_T^+, no firm statements can be made about some of the drag-reducing or -increasing mechanisms leading to the eventual drag breakdown of riblets. While the Reynolds stress contribution $\Delta U^+_{u'v'}$ has typically been thought to be drag increasing, e.g. [6, 13, 42, 24], explanations have also been put forward that larger riblets may pin the quasi-streamwise vortices, stabilizing the near-wall streaks and thereby reducing drag [49, 50]. For these larger riblets, whether it is appropriate to define ℓ_T^+ is also debatable when the flow is no longer smooth-wall-like. However, ℓ_T^+ should still be retained to allow for a gradual continuation into the larger-riblet regime in which outer-layer similar turbulence presumably remains, otherwise its complete exclusion from the $\Delta U^+ = -(\ell_U^+ - \ell_T^+) + \Delta U^+_{u'v'}$ decomposition could artificially inflate the relevance and meaning of $\Delta U^+_{u'v'}$.

The magnitudes of the dispersive and texture-coherent contributions for the DNS are now considered, as is their resilience to ambiguity in ℓ_T^+. The dispersive contributions are represented by the difference between the —— and —— curves in Figures 2(a, b), this difference only affected by u_τ and not by ℓ_T^+ directly. Similarly, the texture-coherent contributions are represented by the difference between the —— and —— curves in Figures 2(a, b). The texture-coherent contributions are weakly sensitive to ℓ_T^+ due to a small cancellation with some smooth wall Reynolds stress contributions below $\lambda_z \lesssim 1.25s^+$, but overall the differences between the —— and —— curves when comparing Figures 2(a, b) are minimal. When taking the extrapolated model ℓ_T^+ values as a more sensible extrapolation for larger riblets, the total smooth-wall-like and texture-coherent contributions in Figure 2(b), i.e. $-(\ell_U^+ - \ell_T^+) + \Delta U^+_{TCu'v'}$ (—— curve), come closer to recovering the measured ΔU^+ (—— curve). Thus, texture incoherent contributions may not be greatly important for just post-optimal riblets, so long as the correct virtual origin (which may be close to the model extrapolation) can be identified.

Figures 2(c, d) show the same results as in Figures 2(a, b) based on the RNL simulations, alongside a comparison to the mean flow offsets measured from DNS (■). Again, the drag decomposition is performed based on fitted ℓ_T^+ values as well as the extrapolated model ℓ_T^+ values, as described above for the DNS in the respective discussion of Figures 2(a, b). Here, there are similar trends are to those observed in the DNS decomposition of the roughness function based on the fitted ℓ_T^+ value, presented in Figure 2(a). More specifically, good agreement is observed for all methods for the smallest riblets investigated, and, as the spacing increases, the measured ΔU^+ is not fully recovered. However, the magnitude of the decomposition is underrepresented by the RNL model. Given this method of calculating the turbulence virtual origin provides an underestimation of the added stresses and in turn, the curves actually follow closely the offset for the fitted ℓ_T^+ values. This behavior is not equivalent to that observed for the DNS, nor is the over estimation of stresses based on the extrapolated model ℓ_T^+ values. Although the DNS data also

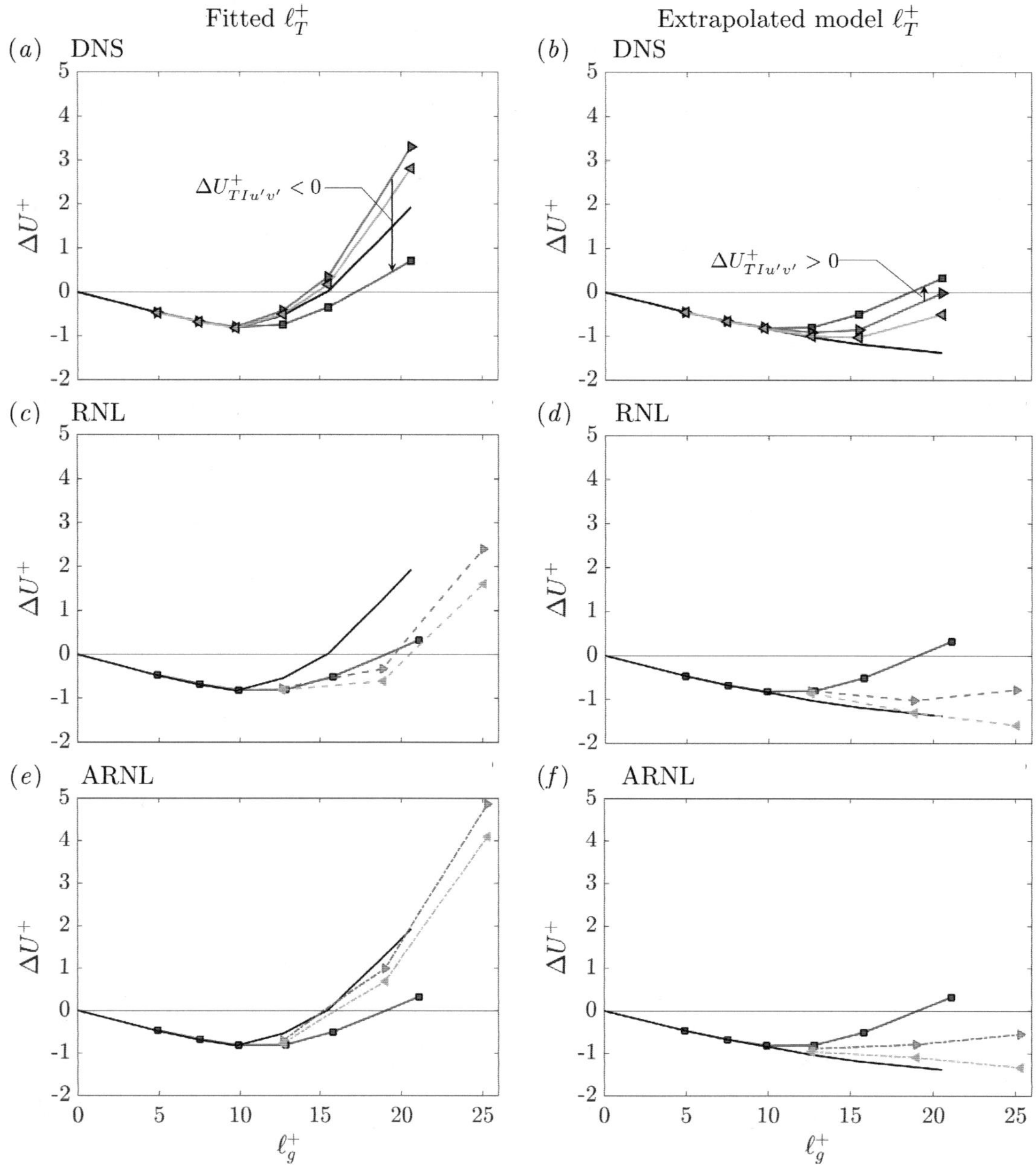

Figure 2. (a) Comparing the DNS-measured ΔU^+ to decompositions of ΔU^+ based on fitted virtual origins ℓ_T^+. For $\ell_g^+ \lesssim 10$, $\Delta U^+ \approxeq -(\ell_U^+ - \ell_T^+)$. For larger ℓ_g^+, the non-zero $\Delta U^+_{TIu'v'}$ is sensitive to ℓ_T^+. ■: ΔU^+ (log-layer measurement), —: $(-\ell_U^+ - \ell_T^+)$, ▶: $-(\ell_U^+ - \ell_T^+) + \Delta U^+_{TCu'v'}$, ◀ : $-(\ell_U^+ - \ell_T^+) + \Delta U^+_{TCu''v''}$. ($b$) As in ($a$), except with extrapolated model ℓ_T^+ values [12]. For the decomposition to balance, $\Delta U^+_{TIu'v'}$ would now be drag increasing, rather than drag reducing. (c) The same decompositions as for the DNS except with the RNL results, with the RNL-fitted ℓ_T^+, compared to the DNS-measured ΔU^+. (d) As in (c), except with the extrapolated model ℓ_T^+ values [12]. (e) The same decompositions as for the DNS except with the ARNL results, with the ARNL-fitted ℓ_T^+, compared to the DNS-measured ΔU^+. (f) As in (e), except with the extrapolated model ℓ_T^+ values. RNL and ARNL curves are denoted with dashed and dashed-dotted lines, respectively. For the reduced order models the markers are specified as ■: ΔU^+ (log-layer measurement - DNS), ▶: $-(\ell_U^+ - \ell_T^+) + \Delta U^+_{TCu'v'}$, ◀ : $-(\ell_U^+ - \ell_T^+) + \Delta U^+_{TCu''v''}$.

indicates that $\Delta U^+_{TIu'v'}$ may be drag increasing for the fitted ℓ^+_T case, the RNL overestimates its effect. These discrepancies could in part be due to the inability of the RNL model to capture all of the nonlinearities of the flow over the larger riblets, indicating that the missing nonlinearity may play an important role the mechanisms associated with the texture incoherent contribution $\Delta U^+_{TIu'v'}$.

We next investigate whether capturing additional nonlinearity improves these estimates by introducing the specific intermediate modes to interact nonlinearly in the ARNL model described in Section 2. Figures 2(e, f) present the same decomposition of the roughness function ΔU^+ for the fitted ℓ^+_T and extrapolated model ℓ^+_T values based on ARNL simulations, along with the DNS measured mean flow offsets (■). Figure 2(e) shows curves which are very similar to those observed in the DNS decomposition of the roughness function. Here, good agreement is not only observed in the just post-optimal spacings but also during the onset of the drag increase regime. However, the curves over-estimate the drag increase and $\Delta U^+_{TIu'v'} < 0$. The extrapolated model ℓ^+_T curves, Figure 2(f), present over-estimations of $\Delta U^+_{TIu'v'}$, similar to those present for the RNL model but less extreme in comparison to the DNS curves. As the ARNL framework allows additional nonlinearity to be added it would be of interest to probe whether one or more additional intermediate scales further improves the model fidelity.

6. Conclusions and Outlook

The onset of drag increase in turbulent flow over riblets is investigated through the decomposition of the roughness function. The nonlinear interactions are highlighted through comparisons of the decomposed stresses resulting from minimal channel log-layer DNS, the RNL and ARNL models. From analysis of the DNS dataset, there are three conclusions. First, for riblets of sizes $10 \lesssim \ell^+_g \lesssim 20$, sensitivity in the measured ΔU^+ due to ambiguity in the turbulence origin ℓ^+_T is non-trivial (for larger ℓ^+_g, the magnitude of ΔU^+ is much larger, and this is less of an issue). Second, for riblets of sizes $\ell^+_g \gtrsim 10$, ambiguity in ℓ^+_T can alter whether Reynolds stress contributions appear as drag reducing or as drag increasing. Finally, decomposing the Reynolds stress contributions by extracting the texture-coherent parts can alleviate sensitivity issues, with $\Delta U^+_{TCu'v'}$ even coming close to accounting for all of $\Delta U^+_{u'v'}$ for post-optimal riblets for one sensibly identified set of ℓ^+_T values.

The 2D co-spectra of the spanwise- and riblet-averaged stress from the minimal channel DNS data and both the ARNL and RNL models present similar trends. The dispersive signature is observed in all the model outputs. Again better agreement is observed between the DNS and the ARNL model versus the RNL model co-spectra, although the differences are less pronounced than those observed in the roughness function decomposition. All reduced order models show similar structures to those of the DNS.

The stress contribution decomposition is not well captured in the RNL model but with the addition of the intermediate scale of the ARNL reduced order model, similar trends to those observed in the DNS are recovered. It is of note that the ARNL model uses the same number of non-zero wavenumbers as the RNL model, but captures the dynamics observed in the DNS with higher accuracy due to its construction and the allowed nonlinear interactions. Therefore this additional fidelity does not pose a significant increase in computational cost.

This project will continue to assess the capabilities of the reduced order models in capturing the nonlinearities observed in the DNS. One method to better understand the abilities of the RNL and ARNL models is to isolate the contributions of individual streamwise modes for the two models to better understand their specific contributions to the spectra and to the added stresses. The results will help to understand how the intermediate mode in the ARNL model interacts with the small-scale modes (and how that compares to the RNL model), and in turn, its role in the representation of the dynamics above the riblets.

Acknowledgements
This work was partially supported by the European Research Council (Caust grant ERC-AdG-101018287) and the US National Science Foundation (grant no. 1652244). Simulations employed the Advanced Research Computing at Hopkins (ARCH) core facility (rockfish.jhu.edu), which is supported by the National Science Foundation (OAC 1920103). C. C., J. W. and D. C. gratefully acknowledge the support of the Air Force Office of Scientific Research under award number FA2386-21-1-4018 (program managers: D. J. Newell and R. W. Carr, AOARD). We also thank D. R. Smith (EOARD) and G. L. Abate (AFOSR).

References

[1] Bushnell D 1983 Turbulent drag reduction for external flows *21st Aerospace Sciences Meeting* p 227
[2] Viswanath P R 2002 *Prog. Aerosp. Sci.* **38** 571–600
[3] Spalart P R and McLean J D 2011 *Phil. Trans. R. Soc. A* **369** 1556–1569
[4] García-Mayoral R and Jiménez J 2011 *Phil. Trans. R. Soc. A* **369** 1412–1427
[5] Walsh M 1982 Turbulent boundary layer drag reduction using riblets *20th Aerospace Sciences Meeting* p 169
[6] Bechert D W, Bruse M, Hage W, Van Der Hoeven J G T and Hoppe G 1997 *J. Fluid Mech.* **338** 59–87
[7] Choi H, Moin P and Kim J 1993 *J. Fluid Mech.* **255** 503–539
[8] Garcia-Mayoral R and Jiménez J 2012 *Phys. Fluids* **24** 105101
[9] Endrikat S, Modesti D, García-Mayoral R, Hutchins N and Chung D 2021 *J. Fluid Mech.* **913** A37
[10] Luchini P 1996 Reducing the turbulent skin friction *Computational methods in applied sciences '96 (Paris, 9-13 September 1996)* ed Désidéri J A, Hirsch C, Le Tallec P, Oñate E, Pandolfi M, Périaux J and Stein E (John Wiley, Chichester United Kingdom) pp 465–470
[11] Ibrahim J I, Gómez-de-Segura G, Chung D and García-Mayoral R 2021 *J. Fluid Mech.* **915** A56
[12] Wong J, Camobreco C J, García-Mayoral R, Hutchins N and Chung D 2024 *J. Fluid Mech.* **978** A18
[13] García-Mayoral R and Jiménez J 2011 *J. Fluid Mech.* **678** 317–347
[14] MacDonald M, Chan L, Chung D, Hutchins N and Ooi A 2016 *J. Fluid Mech.* **804** 130–161
[15] Gómez-de-Segura G and García-Mayoral R 2019 *J. Fluid Mech.* **875** 124–172
[16] Raupach M R, Antonia R A and Rajagopalan S 1991 *Appl. Mech. Rev.* **44** 1–25
[17] Perry A E, Schofield W H and Joubert P N 1969 *J. Fluid Mech.* **37** 383–413
[18] Nikora V I, Koll K, Mclean S R, Ditrich A and Aberle J 2002 Zero-plane displacement for rough-bed open-channel flows *Fluvial Hydraulics River Flow* pp 83–92
[19] Luhar M, Rominger J and Nepf H 2008 *Environ. Fluid Mech.* **8** 423–439
[20] Chen Z and García-Mayoral R 2023 *J. Fluid Mech.* **973** A31
[21] Jiménez J 2004 *Annu. Rev. Fluid Mech.* **36** 173–196
[22] Flores O and Jiménez J 2010 *Phys. Fluids* **22** 071704
[23] Chung D, Chan L, MacDonald M, Hutchins N and Ooi A 2015 *J. Fluid Mech.* **773** 418–431
[24] Endrikat S, Newton R, Modesti D, García-Mayoral R, Hutchins N and Chung D 2022 *J. Fluid Mech.* **952** A27
[25] Endrikat S, Modesti D, MacDonald M, García-Mayoral R, Hutchins N and Chung D 2021 *Flow Turbul. Combust.* **107** 1–29
[26] Jiménez J and Moin P 1991 *J. Fluid Mech.* **225** 213–240
[27] Castro I P 2007 *J. Fluid Mech.* **585** 469–485
[28] Gayme D F and Minnick B A 2019 *Phys. Rev. Fluids* **4** 110505
[29] Bretheim J U, Meneveau C and Gayme D F 2015 *Phys. Fluids* **27** 011702
[30] Minnick B A and Gayme D F 2019 Characterizing energy transfer in restricted nonlinear wall-bounded turbulence *International Symposium on Turbulence and Shear Flow Phenomena (TSFP-11)*
[31] Minnick B A 2022 *A Restricted Nonlinear Approach to Momentum and Scalar Mixing in Wall-Turbulence* Ph.D. thesis Johns Hopkins University
[32] Minnick B A, Viggiano B and Gayme D F 2023 Augmented Restricted Nonlinear (ARNL) Model for High Reynolds Number Wall-Turbulence *Progress in Turbulence X: Proceedings of the iTi Conference on Turbulence 2023* (Springer)
[33] Marston J B, Chini G P and Tobias S M 2016 *Phys. Rev. Lett.* **116** 214501
[34] Ham F, Mattsson K and Iaccarino G 2006 Accurate and stable finite volume operators for unstructured flow solvers *CTR Annu. Res. Briefs* (Center for Turbulence Research) pp 243–261
[35] Ham F, Mattsson K, Iaccarino G and Moin P 2007 Towards time-stable and accurate LES on unstructured grids *Complex Effects in Large Eddy Simulations. Lecture Notes in Computational Science and Engineering, vol. 56* ed Kassinos S C, Langer C A, Iaccarino G and Moin P (Springer, Berin Heidelberg) pp 235–249

[36] MacDonald M, Chung D, Hutchins N, Chan L, Ooi A and García-Mayoral R 2017 *J. Fluid Mech.* **816** 5—42
[37] Jelly T O, Jung S Y and Zaki T A 2014 *Phys. Fluids* **26** 095102
[38] Bretheim J U, Meneveau C and Gayme D F 2018 *J. Turbul.* **19** 141–166
[39] Peskin C S 2002 *Acta Numer.* **11** 479–517
[40] Abderrahaman-Elena N, Fairhall C T and García-Mayoral R 2019 *J. Fluid Mech.* **865** 1042–1071
[41] Xie W 2022 *The Effect of Texture Granularity on Turbulence* Ph.D. thesis University of Cambridge
[42] Modesti D, Endrikat S, Hutchins N and Chung D 2021 *J. Fluid Mech.* **917** A55
[43] Fairhall C T, Abderrahaman-Elena N and García-Mayoral R 2019 *J. Fluid Mech.* **861** 88–118
[44] Gómez-de-Segura G and García-Mayoral R 2020 *Int. J. Heat Fluid Flow* **86** 108675
[45] Jiménez J, Del Álamo J C and Flores O 2004 *J. Fluid Mech.* **505** 179–199
[46] Fairhall C T and García-Mayoral R 2018 *Flow Turbul. Combust.* **100** 961–978
[47] Sharma A and García-Mayoral R 2020 *J. Fluid Mech.* **888** A2
[48] Luchini P, Manzo F and Pozzi A 1991 *J. Fluid Mech.* **228** 87–109
[49] Goldstein D, Handler R and Sirovich L 1995 *J. Fluid Mech.* **302** 333–376
[50] Goldstein D B and Tuan T C 1998 *J. Fluid Mech.* **363** 115–151

Journal of Physics: Conference Series **2753** (2024) 012011 doi:10.1088/1742-6596/2753/1/012011

Temporal vortex tracking in isotropic turbulence

Afonso A. Ghira[1], Alberto Vela-Martin[2] and Carlos B. da Silva[1]

[1] LAETA, IDMEC, Instituto Superior Técnico, Universidade de Lisboa, Lisboa, Portugal
[2] Universidad Carlos III de Marid, Aerospace Engineering Department, Madrid, Spain

E-mail: afonso.ghira@tecnico.ulisboa.pt

Abstract. An algorithm to allow the temporal vortex tracking of the intense vorticity structures is introduced and applied to direct numerical simulations (DNS) of homogeneous isotropic turbulence (HIT). The algorithm allows the identification of the individual life events of the worms, including events of splits into several structures and of merging between different structures. Results obtained from a relatively small DNS, with Taylor based Reynolds number equal to $Re_\lambda = 54$, for a simulation lasting more than 6 integral time scales, shows that the complement of the CDF of the life time of the IVS is well approximated by an exponential distribution function. The mean life time of the IVS which is approximately $3.6\tau_\eta$, where τ_η is the Kolmogorov time, and the birth rate of these structures is close to $4.5/\tau_\eta$ (number of new structures per Kolmogorov time). A population model is built to predict the number of existing structures at any time, which gives a number of existing IVS equal to ≈ 16, which agrees well with the results from the DNS.

1. Introduction

It is well know that intense vorticity structures (IVS) or 'worms' abound in virtually all turbulent flows. These structure are important for the study of the viscous dissipation mechanism and to internal intermittency [1], and their study is also motivated by their role in the transport, mixing and diffusing at the small scales of motion [2, 3].

The IVS are typically defined as regions of particularly strong concentrated vorticity and are often mentioned as having a life time which is big compared with the characteristic time scale of the flow [4]. Several methods have been employed to identify these structures as discussed in [5, 2]. These include the use of iso-surfaces of vorticity, which has been mostly used in homogeneous and isotropic turbulence [6, 7, 8, 9], and quantities derived from the velocity gradient invariants [2, 10, 11]. However, the criteria to define the IVS that we will use here is the one proposed by [12, 13] who defined these structures as the regions of particularly strong vorticity, defined by flow points with the highest enstrophy representing 1% of the total, [12, 13].

Although the IVS have been primarily studied in isotropic turbulence *e.g.* [6, 7, 8, 9, 14, 15, 16, 12, 5, 13, 17, 18] other flow types have also been studied such as mixing layers [11], channel flows [10] and jets [19, 3]. It turns out that many of the characteristics of the IVS observed in isotropic turbulence, such as the size of the vortex core radius, are also observed in other flow types.

For instance, it has been shown that the vortex core radius of the worms is of the order of the Kolmogorov micro-scale $\langle R_{ivs}\rangle/\eta \approx 4-5$, while the mean tangential velocity is close to the root-mean-square velocity $\langle U_{ivs}\rangle/u' \approx 1$, both in isotropic turbulence [13, 17] and also in mixing layers [11], channel flows [10]. Recently, Ghira *et al.* [17] have shown that the length of the IVS also scales with the Kolmogorov time, with $\langle L_{ivs}\rangle/\eta \approx 60$, in isotropic turbulence.

Despite the large number of these studies, a detailed temporal analysis of the life events and life times of these structures is still lacking. In the present work we develop an algorithm to allow the temporal vortex tracking of the intense vorticity structures and we apply this to a single simulation of isotropic turbulence with relatively small Reynolds number as illustration. Larger simulations using the same procedure will be reported elsewhere. The algorithm allows the identification of the individual life events of the worms, including events of 'splits' into several structures and of 'merging' between different structures. This allows us to rigorously determine the life times of the IVS in isotropic turbulence.

This paper is organised as follows. The next section (Section II) describes the DNS used in the present work. Section III describes the temporal tracking algorithm to compute the IVS lives, while Section IV analyses the results obtained in the present work. The work ends in section V with an overview of the main results and conclusions.

2. Direct numerical simulations of homogeneous isotropic turbulence

For the purposes of this work we employ a single simulation of forced Homogeneous Isotropic Turbulence (HIT) obtained by solving the incompressible Navier-Stokes equations in a triple periodic domain using a pseudo-spectral solver. We use the Alvelius' forcing scheme [20] and let the turbulence evolve until statistically steady conditions are reached. Statistical stationarity is obtained by imposing a power input through the imposed forcing that balances the viscous dissipation rate, $P = \epsilon$, where $P = \overline{f_i u_i}$ is the power of the input forcing and $\epsilon = 2\nu\overline{s_{ij}s_{ij}}$, is the viscous dissipation rate, while $s_{ij} = (\partial u_i/\partial x_j + \partial u_j/\partial x_i)/2$ is the rate-of-strain tensor.

Table 1 summarises the physical and computational parameters of this simulation in the statistically stationary conditions used in the subsequent analysis. The simulation consists on a small HIT case using $N^3 = 128^3$ collocation points, for a Taylor based Reynolds number of $Re_\lambda = 54$ and a resolution of $k_{\max}\eta = 2.0$. The analysis is done using a total of $N_f = 1400$ instantaneous fields and lasts for close to 7 integral time scales and more then 50 Kolmogorov time scales.

Figure 1 shows the evolution of the kinetic energy and dissipation rate normalised by their initial values at statistically steady conditions, where the snapshots analyzed by the Wormtracker were taken. The figure shows the typical variability present in similar (statistically stationary) simulations of HIT. As expected the viscous dissipation exhibits proportionally greater variations than the kinetic energy, however the instantaneous value of ϵ never exceeds 10% its mean value during the simulated times.

Table 1. Physical and computational parameters of the DNS: Number of collocation points (N^3); Taylor micro-scale based Reynolds number (Re_λ); Kinematic viscosity (ν); Root-mean-square of the velocity fluctuations ($u' = \sqrt{u'^2}$); Taylor micro-scale (λ); Longitudinal integral scale (L_{11}); Viscous dissipation rate (ε); Kolmogorov's micro-scale (η); Maximum effective wave number normalised by the Kolmogorov micro-scale ($k_{\max}\eta$); Number of analyzed instantaneous fields (N_f); Kolmogorov's timescale (τ_η); Integral time scale ($\tau_L = L_{11}/u'$); Final simulation time normalized by the Kolmogorov's timescale (t_f/τ_L).

N^3	Re_λ	ν	u'	λ	L_{11}	ϵ	η	$k_{\max}\eta$	N_f	τ_η	τ_L	t_f/τ_L
128^3	54	0.0370	2.90	0.68	1.48	10.01	0.0474	2.0	1400	0.061	0.51	6.70

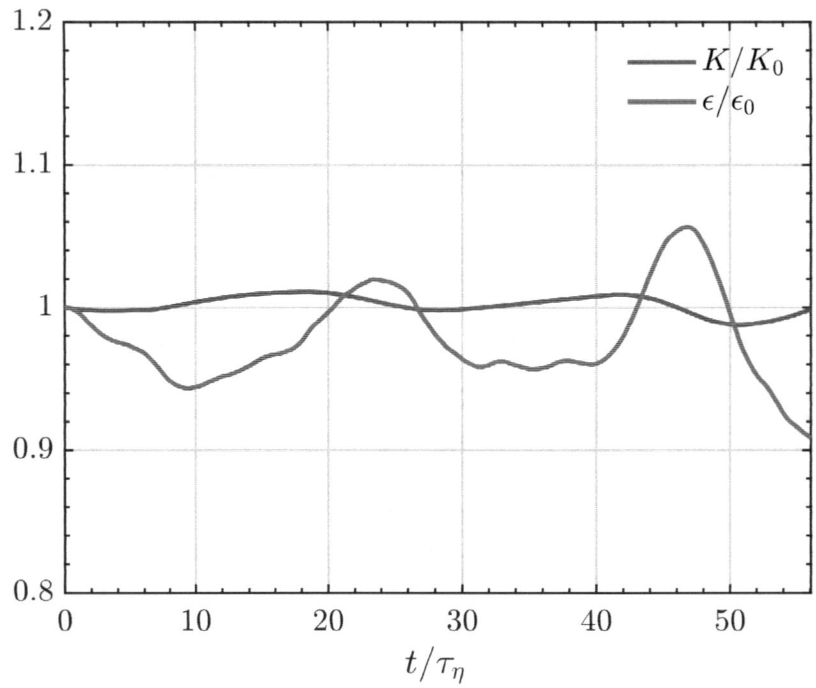

Figure 1. Evolution of the kinetic energy (K) and dissipation rate (ϵ) normalized by their initial values K_0 and ϵ_0, respectively. During this time window the Wormtracker analyzes the instantaneous fields at each timestep of temporal integration.

3. Temporal Tracking Algorithm

3.1. Wormtracker

As described in the introduction the intense vorticity structures (IVS) are defined by flow points (from the DNS grid) with the highest vorticity magnitude covering 1% of the total flow domain. The algorithm used to detect and quantify these structures ('wormtracker') is based in the one from Jiménez *et al.* [12, 13]), with some improvements already described in Ghira *et al.* [17]. The only aspects of the wormtracker that are relevant to the present work are related to the detection, and automatic assignment/numbering of these structures. Other aspects of the wormtracker, such as the computation of the associated kinematic and geometrical aspects (*e.g* the IVS' tangential velocity and vortex radius), are detailed in Ghira *et al.* [17].

In order to detect these structures an histogram of vorticity magnitude is computed for each instantaneous field of the simulation (N_f fields in total) in order to obtain the reference vorticity magnitude threshold ω_{ivs} corresponding to the 1% highest vorticity magnitude. Any point for which the local vorticity magnitude $\omega = (\omega_i\omega_i)^{1/2}$ (ω_i is the vorticity vector) is greater than $\omega > \omega_{ivs}$ is assigned to an individual structure. The workmtracker uses the direction of the local vorticity vector of each individual structure to build the full axis of the structure and all the IVS are numbered, starting (number one) with the structure with the strongest vorticity, to the structure with smallest vorticity. As in [12, 13, 17] structures with less than 20 axis points (from the ones in the DNS grid that have been assigned to an IVS axis) are discarded from the statistical sample, however, unlike in previous works, the wormtracker algorithm runs together with the Navier-Stokes solver so that there is no need to store and analyse each one of the individual instantaneous fields in subsequent post-processing, *i.e.* the wormtracking is done during the actual simulation. This proved to be essential to allow the time tracking of each one of the individual structures during very long times, which is essential to make a proper temporal analysis of the IVS.

3.2. Connectivity

In order to connect structures of two contiguous snapshots, an algorithm was developed based on spacial overlapping principles.

Given a snapshot at time t_s, all identified structures are assigned an index/number, with the number **1** being assigned to the structure having the most intense vorticity point, followed by **2** for the second most intense structure, and so on, until the last, less intense vortex axis has been numbered. Next, using a scalar variable, we assign the same index/number to this scalar in all mesh points within a distance of one Δx (mesh spacing) from each point along the filament in every direction. The result is a scalar variable identifying in space the position of a filament and its representative cloud. For instance, the scalar field will display the value of **1** to all the points of the cloud surrounding the most intense axis (the axis containing the most intense vorticity point). This choice is based on the assumption that a filament is not expected to move away, appreciably, from its cloud within one timestep Δt due to the Courrant Number restriction.

For the next snapshot, at time $t_s + \Delta t$, a new set of structures is identified and a new indexing/axis numbering takes place, again starting from the axis of the most intense vortex structure. However, by overlapping the new structures with respect to the scalar field of clouds from the previous snapshot, we are able to connect and refer back in time this new indexing. In other words, we can assign a global index/numbering that identifies unambiguously each structure from its 'birth' to its 'death' and relate it to the local indexing as function of time using all the existing/instantaneous snapshots from a given simulation.

This procedure allows us to have a spacial recognition about the positions of the structures in subsequent timesteps. Nevertheless, we still need to build some criteria to make comparisons. We start by defining the **overlap operator** O_{ij}, which represents (in %) for a given cloud i how much of a worm j belongs to it (where only the points belonging to some cloud are taken into account). For instance, consider a given worm j which has n points along its filament, and suppose that from these n axis points, a number of x points belong to cloud a, y axis points to cloud b and z axis points do not overlap with any other existing cloud ($n = x + y + z$). In this case the the matrix elements of the overlap operators O_{aj} and O_{bj} are defined by $O_{aj} = x/(n-z)$ and $O_{bj} = y/(n-z)$, respectively. We then define **Complete Superposition** when $O_{ij} > 90\%$ and **Incomplete Superposition** if $10\% \le O_{ij} \le 90\%$, where by superposition we mean that worm j is overlapping cloud i.

Using these definitions and metrics we can now look for the most simple events in the temporal evolution of a worm, namely, *i)* the continuation of the life of the worm (**Same Worm**), *ii)* **Pure Splitting**, *iii)* **Pure Merging**, *iv)* **Death**, and *v)* **Birth**. Same Worm events occurs whenever there is a Complete Superposition of a cloud by one structure. A Pure Splitting means that a cloud has more than one structure with Complete Superpositions. A pure merging happens when more than one cloud merge into one structure with Incomplete Superpositions on them and, at the same time, all the clouds involved must satisfy the same merging criteria. A Death is detected if one cloud does not have any structure overlapping. Finally, a Birth occurs because a structure is not overlapping any existing cloud. These criteria are summarised in Table 2.

As mentioned before, the above connectivity algorithm allows one to build a global worm index using every one of the local indexes obtained for each time step/instant of the simulation. With this procedure all statistics taken at each instant of time can be traced back to each specific individual structure, as a time evolving entity. The most basic entity arising from the connectivity algorithm is the **Lifetime Map**, which is shown in Figure 2 for the present simulation, whose parameters have been presented in Table 1.

This figure shows a zoom of the lifetime map showing the history of several worms. Every

Table 2. Classification criteria based on the number of structures contributing to a specific range of values of the overlap operator O_{ij}. This table should be interpreted with respect to each one of the existing clouds i. Note that a cloud with given merging criteria will only be considered as merged if all the involved clouds satisfy the same criteria.

Complete superposition ($O_{ij} > 90\%$)	Incomplete superposition ($10\% \leq O_{ij} \leq 90\%$)	Classification
$= 1$	$= 0$	Same Worm
> 1	$= 0$	Pure Splitting
≥ 1	> 0	–
$= 0$	> 0	Pure Merging (all clouds must satisfy)
$= 0$	$= 0$	Death

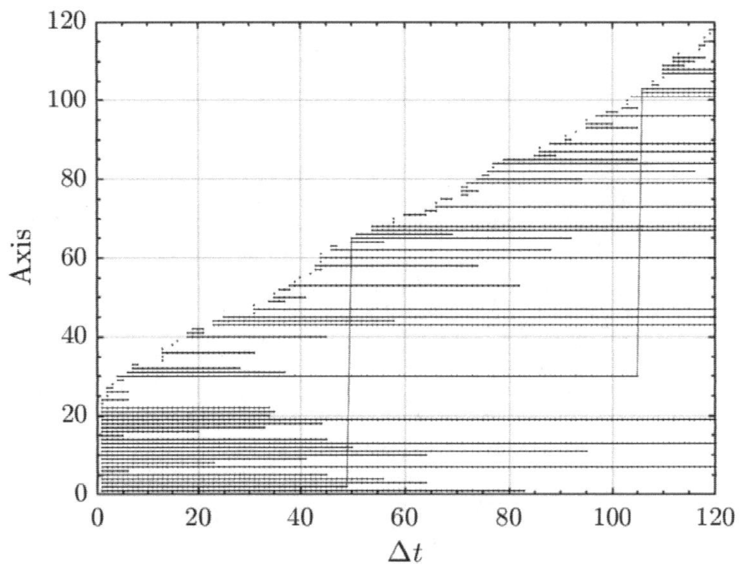

Figure 2. Zoom of the lifetime map showing the history of several worms. Every horizontal line represents a worm that is 'alive'. The vertical axis shows the global index/number assigned to each structure while the horizontal axis represents the elapsed time in timestep units. The red lines show the connection among branches arising from splitting events.

horizontal line represents a worm that is 'alive' at some time interval. The vertical axis shows the global index/number assigned to each structure while the horizontal axis represents the elapsed time in timestep units. The red lines illustrate what happens when a structure splits. Note that a new global index is assigned for each new branch arising from the structures that have split. The zoom is from the first $120\Delta t$ of the simulation (the total analysis contains $1400\Delta t$). Note that the present analysis starts from the statistically stationary state, which explains the existence of about ≈ 20 axis already at $\Delta t = 0$.

The first detected splitting event is shown in Figure 3, in exactly the two (contiguous) times when that event was recorded. There we can see (Figure 3(a)) that before splitting a filament

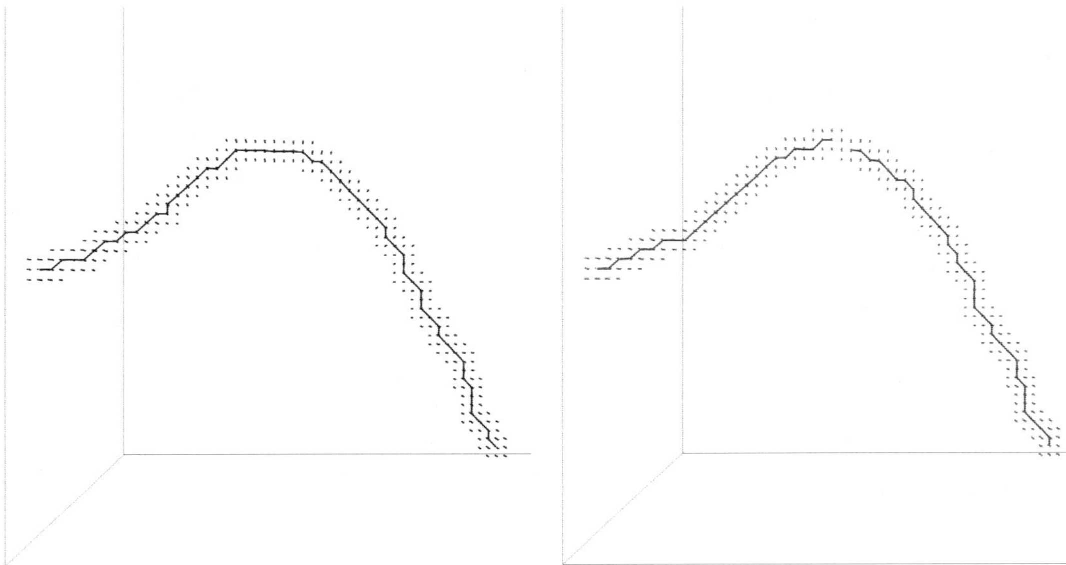

(a) (b)

Figure 3. Visualisation of the first splitting, showing (a) the structure undergoing that splitting and (b) the resulting branches for one IVS located at one of the corners of the computational domain (the total size of the computational domain is much larger). Note that in (a) the cloud is also shown for the respective filament. That cloud is compared with the positions occupied by the two structures appearing in (b), thus identifying this chain of events. In (b) we also see the two clouds surrounding each new filament. The process of cloud reconstruction takes place at the end of each snapshot analysis for further time tracking.

exists with the corresponding cloud. In the very next instant (Figure 3(b)) we see two different structures instead. The connection is made by comparing those two structures with the cloud presented in Figure 3(a). By the end of the second timestep, the scalar variable with the cloud information is reconstructed to allow the continuation of the analysis in the next time step.

4. Results

In this section we will use the temporal vortex tracking algorithm described to estimate the lifetime and the population of the IVS. Analysis of the other types of events defined above, as well as the computation of these quantities at higher Reynolds numbers will be reported elsewhere.

4.1. The life time of the intense vorticity structures

From the data gathered, we can compute a simple estimation for the Cumulative Distribution Function (CDF) of the worms' lifetime, $P\left(T/\tau_\eta < t/\tau_\eta\right)$, where T is the lifetime random variable, t is the measured lifetime and τ_η the Kolmogorov's timescale. In this process we discard the

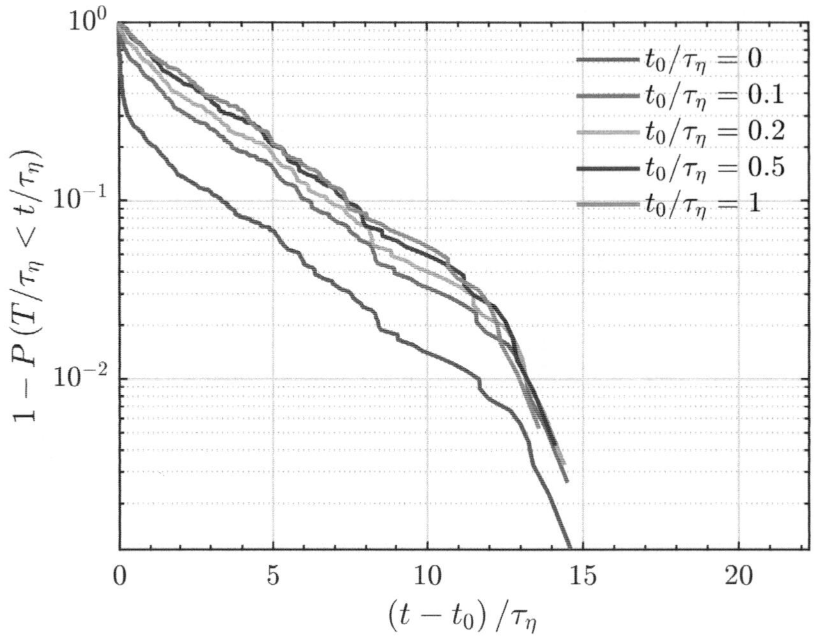

Figure 4. Complementary probability of the CDF of the worms' lifetimes, where data for lifetimes smaller than a given value t_0 have been removed. T is the random variable describing the lifetime t and τ_η is the Kolmogorov's timescale. In the CDF for $t_0/\tau_\eta = 0$ all the lifetimes have been used.

contribution from structures that are 'alive' both at $t = 0$ and $t = t_f$ (final time), because we do not know their past and future history, and their inclusion would affect the estimate. This is justified because the simulation used here was run for a very long time, with more than 6 integral time scales (see table 1).

Since the CDF is computed for each known (measured) value of the lifetime data the use of thresholds in the detection method described above may lead to some numerical artifacts in the results. We expect to see effects from these artifacts especially for very small timescales. Therefore, we investigate their effect by recomputing the CDF eliminating structures living less than a given time t_0. Figure 4 shows 5 CDFs obtained by removing from the statistical sample lifetimes smaller than times between $0 \leq t_0/\tau_\eta \leq 5$. In the CDF for $t_0/\tau_\eta = 0$ no removal was done *i.e.* all the lifetimes have been used (note that in this figure the vertical axis represents not the CDF but the complementary probability of the CDF).

The figure shows that using different cutoffs t_0/τ_η affects the CDFs shape, however all CDFs converge to virtually the same function for $t_0 = 0.5\tau_\eta$. We therefore proceed with our analysis by eliminating all the structures displaying lifetimes smaller than $t_0/\tau_\eta = 0.5$, as it is likely that events lasting less than half a Kolmogorov time are included in artificial numerical 'noise'. This is done with figure 5 which displays the CDF for $t_0 \leq 1\tau_\eta$, which we label as the 'true' probability function for a structure to live more than a certain time t. Noting that this function approximates well to a line with a constant slope, we conclude that the CDF is well approximated by an exponential function. This function, whose parameters are indicated in the enclosed label, is also added in the figure. From this we can now estimate the mean life time of the IVS by computing the reciprocal of 0.32 (from the fitted curve) and summing it to the cutoff chosen value. We obtain $\tau/\tau_\eta = 3.61$ for the average life time of these structures. This value is much

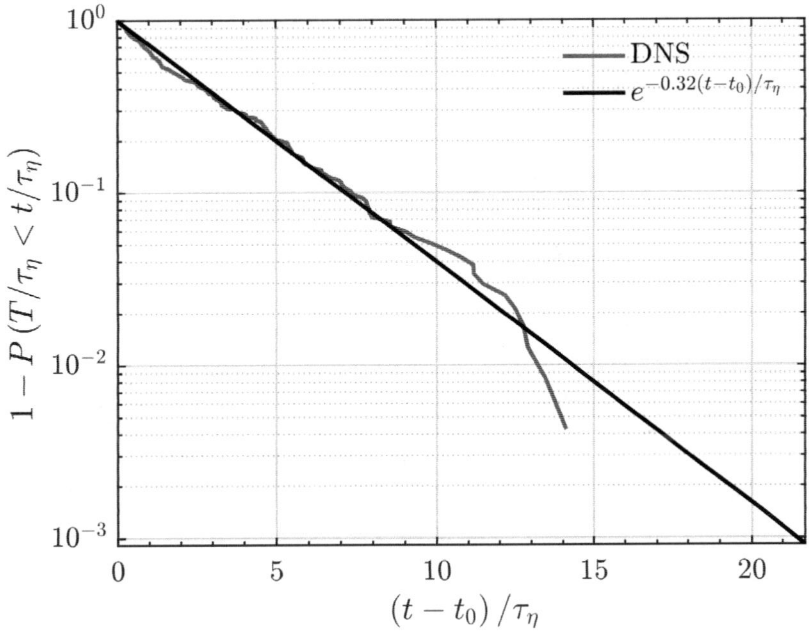

Figure 5. CDF of the worms' lifetimes. T is the random variable describing the lifetime t and τ_η is the Kolmogorov's timescale. t_0 is the removed small time intervale that has been eliminated from the data.

smaller than the integral time scale usually claimed as being the typical life time of the IVS [21]. Our explanation, consubstantiated in several animations, is that even though many of these long life times structures do exit (and maybe these are the structures that we retain in our mind from movies and visualisations) they represent relatively rare events.

4.2. The population of the intense vorticity structures

Using the life time data available from the simulation it is possible to build a population model, to estimate the number of existing 'alive' IVS, as well as the number of Births and Deaths of these structures for a given simulated time. We postulate that the population model follows a well know law used in particle physics. In short, we assume that the number of individuals since a given time $N(t)$ is governed by the following differential equation,

$$\frac{dN}{dt} = -\frac{N}{\tau} + \frac{dN_B}{dt}, \tag{1}$$

where τ is the mean lifetime of the individuals (worms) and N_B the accumulated number of births, (thus dN_B/dt represents their birth rate). Assuming a constant value of this rate $dN_B/dt = C$ we write,

$$\frac{dN}{dt} = -\frac{N}{\tau} + C, \tag{2}$$

where C is the **birth rate constant**. The solution of this equation is,

$$N(t) = N_0 e^{-t/\tau} + C\tau\left[1 - e^{-t/\tau}\right], \tag{3}$$

where $N_0 = N(0)$ is the initial number of individuals/worms, at the initial time when the time tracking starts. With the filtered data used in our analysis, where we have discarded the structures at $t = 0$ (see above), we have $N_0 = 0$ and therefore we can estimate the birth rate constant C by fitting our experimental curve of accumulated number of births to the model $N_B = Ct$. Moreover, for sufficiently long simulating times, the model predicts a saturated number of individuals $N(\infty) = C\tau$ existing at a given (long) simulated time. Table 3 summarises all the results, including the mean lifetime estimated in the as previous section. We see that for this relatively small simulation we have about ≈ 16 worms alive at any given instant, and the normalised birth rate constant is $C\tau_\eta \approx 4.5$. These values are well supported from the DNS data.

Table 3. Estimates from DNS data and from the population model: Number of collocation points (N^3); Taylor micro-scale based Reynolds number (Re_λ); Mean lifetime normalized by the Kolmogorov's timescale (τ/τ_η); Birth rate normalized by the Kolmogorov's timescale ($C\tau_\eta$); Number of individuals predicted by the population model at saturated conditions ($N(\infty) = C\tau$).

N^3	Re_λ	τ/τ_η	$C\tau_\eta$	$N(\infty) = C\tau$
128^3	54	3.61	4.54	16

Finally, we again use the lifetime map data to compute a series of statistics such as the number of births and deaths of worms, and the number merges and splitting events. Table 4 gives those numbers for the final simulated time. The number of existing structures at $t = 0$ and $t = t_f$ is also shown. Due to statistically steady conditions of our simulation, we expect the number of IVS to remain approximately constant and thus comparable to what our model predicts at saturated conditions. This is indeed the case as shown in the table. We start/end the analysis with approximately the same number of structures, 24 and 19, respectively. During the simulations we had approximately 900 births and deaths, and the number of splits (48) is slightly larger than the number of merger (16).

Table 4. Summary of some statistics for the data: Number of colocation points (N^3); Taylor micro-scale based Reynolds number (Re_λ); Initial number of individuals ($N(0)$); Final number of individuals ($N(t_f)$); Final accumulated number of births ($N_B(t_f)$); Final accumulated number of deaths ($N_D(t_f)$); Final accumulated number of merges ($N_M(t_f)$); Final accumulated number of splittings ($N_S(t_f)$).

N^3	Re_λ	$N(0)$	$N(t_f)$	$N_B(t_f)$	$N_D(t_f)$	$N_M(t_f)$	$N_S(t_f)$
128^3	54	24	19	924	929	16	48

5. Conclusions

Intense vorticity structures (IVS) or 'worms' are known to populate virtually all turbulent flows and a detailed temporal analysis of the life events and life times of these structures is still lacking. In the present work an algorithm to allow the temporal vortex tracking of the intense vorticity structures is introduced and applied to direct numerical simulations (DNS) of homogeneous isotropic turbulence (HIT).

The algorithm allows the identification of the individual life events of the worms, including events of splits into several structures and of merging between different structures. Results obtained from a relatively small DNS, with Taylor based Reynolds number equal to $Re_\lambda = 54$, for a simulation lasting more than 6 integral time scales, shows that the complement of the CDF of the life time of the IVS is well approximated by an exponential function.

This allows one to compute the mean life time of the IVS which is approximately $3.6\tau_\eta$, where τ_η is the Kolmogorov time, while the birth rate of these structures is close to $4.5/\tau_\eta$ (number of new structures per Kolmogorov time). A population model is built to predict the number of existing structures at any time instant giving a number of existing IVS equal to 16, that agrees well with the results from the DNS.

A more involved discussion of these events, including more advanced analysis methods and much higher Reynolds numbers will be published elsewhere.

Acknowledgments
This work was totally performed in one month during the Fifth Madrid Summer Workshop, funded by the European Research Council under the Caust grant ERCAdG-101018287. A. A. Ghira ackowledges Fundação para a Ciência e Tecnologia (FCT) for the scholarship 2021.08206.BD.

References

[1] Ishihara T, Kaneda Y, Yokokawa M, Itakura K and Uno A 2007 *J. Fluid Mech.* **592** 335–366
[2] Kida S and Miura H 1998 *Eur. J. Mech. B/Fluids* **17** 471–489
[3] da Silva C B, dos Reis R J N and Pereira J C F 2011 *J. Fluid Mech.* **685** 165–190
[4] Davidson P A 2015 *Turbulence, an introduction for scientists and engineers (second edition)* (Oxford University Press)
[5] Cadot O, Douady S and Couder Y 1995 *Phys. Fluids* **7** 630–646
[6] Siggia E D 1981 *J. Fluid Mech.* **107** 375–406
[7] Kerr R 1985 *J. Fluid Mech.* **153** 31–305
[8] She Z S, Jackson E and Orszag S 1991 *Proc. Roy. Soc. London A* **434** 101
[9] Vincent A and Meneguzzi M 1991 *J. Fluid Mech.* **225** 1–20
[10] Kang S J, Tanahashi M and Miyauchi T 2008 *Journal of Turbulence* **8(52)** 1–19
[11] Tanahashi M, Iwase S and Miyauchi T 2001 *Journal of Turbulence* **2(6)** 1–17
[12] Jiménez J, Wray A, Saffman P and Rogallo R 1993 *J. Fluid Mech.* **255** 65–90
[13] Jiménez J and Wray A 1998 *J. Fluid Mech.* **373** 255–285
[14] Ruetsch G R and Maxey M R 1991 *Phys. Fluids* **3** 1587–1597
[15] Ruetsch G R and Maxey M R 1992 *Phys. Fluids* **4** 2747–2760
[16] Vincent A and Meneguzzi M 1994 *J. Fluid Mech.* **258** 245–254
[17] Ghira A A, Elsinga G and da Silva C B 2022 *Phys. Review Fluids* **7** 104605
[18] Aligolzadeh F, Holzner M and Dawson J R 2023 *J. Fluid Mech.* **A5** 973
[19] Ganapathisubramani B, Lakshminarasimhan K and Clemens N T 2008 *J. Fluid Mech.* **598** 141–175
[20] Alvelius K 1999 *Phys. Fluids* **11(7)** 1880–1889
[21] Verzico R and Jiménez J 1999 *J. Fluid Mech.* **394** 261–279

Coherent structure tracking of the second Mack mode in transitional hypersonic boundary layers

Riwan Hammachi[1,2], **Estelle Piot**[1], **Hugues Deniau**[1], **Jean-Philippe Brazier**[1], **Guillaume Daviller**[2], **José I. Cardesa**[1]
[1]ONERA/DMPE, Université de Toulouse, Toulouse, F-31055 , France
[2]CERFACS, 42 avenue Gaspard Coriolis, Toulouse, 31057, France

E-mail: riwan.hammachi@onera.fr

Abstract. The natural transition of hypersonic boundary layers (HBLs) is often expressed in terms of discrete modes and their linear stability. A frequent interpretation revolves around fast and slow acoustic modes interacting in the vicinity of the vortical/entropic branches of the continuous spectrum found from stability analyses. Yet several transition scenarios are contingent upon factors such as the spectral content of the free-stream disturbances, or the interactions between the discrete modes within the boundary layer and the free-stream disturbances near the leading edge which can be decomposed into vortical, acoustic and entropic nature based on the fluid-thermodynamic (FT) components. Yet the interpretations of linear stability applied to discrete modes can lead to semantic conflicts with the terminology of FT components. To clarify the current description of the processes involved, this study chooses an approach aimed at characterizing the dynamics of the second Mack mode in transitional HBLs through coherent structure tracking. The method involves decomposing the flow perturbations into acoustic, vortical and entropic content, and following their associated coherent structures over time. For this purpose, direct numerical simulations are carried out to investigate the dynamics of the second Mack mode instability in two-dimensional HBLs, considering a flow at Mach 6 over a cooled and an insulated wall. It is found that vortical structures coexist at different heights along the wall surface, forming alternating sign doublets around the critical layer and above the relative sonic line. These structures are found to merge in the region of maximum second Mack mode instability.

1. Introduction

The pursuit of hypersonic-capable vehicles represents a complex technological challenge and a profound scientific endeavor. Among the myriad hurdles, the comprehension and mastery of hypersonic aerodynamics stands as paramount. At these high velocities, the transition to turbulence in the near-wall flow introduces a substantial increase in surface skin friction and significant wall heating. Consequently, unraveling the transition mechanisms within the hypersonic boundary layer (HBL) takes on pivotal importance in shaping the design of high-performance hypersonic vehicles. According to our current understanding of the physics governing this phenomenon, transition is a complex process that varies depending on numerous parameters such as environmental conditions, mean flow properties, and vehicle geometry. In the vicinity of hypersonic vehicles, the boundary layer is susceptible to various disturbances that can impact its dynamics, ultimately leading to the onset of turbulence [1]. In light of the consensus that atmospheric conditions during hypersonic flight correspond to a low-disturbance

environment, the pathway toward turbulence follows the "natural transition" scenario which predominantly hinges on the modal growth of the relevant instabilities [2]. In the initial phase, the freetream disturbances enter the boundary layer and excite its wave modes through the receptivity process. Subsequently, unstable eigenmodes (when present) undergo exponential growth up to certain amplitudes past which secondary instabilities, nonlinear interactions and three-dimensional effects enter the picture, leading ultimately to breakdown and turbulence.
It is important to note that in the hypersonic regime, multiple instability mechanisms coexist within the boundary layer. This is in contrast with two-dimensional (2D) low-speed boundary layers, which typically exhibit only the viscous Tollmien–Schlichting (T–S) wave instability [3]. A specific characteristic of a 2D hypersonic boundary layer lies in the existence of a family of inviscid unstable modes: the Mack modes. In the terminology of Mack [4], they are denoted as first, second and higher modes. In his study, Mack revealed the influence of wall temperature on their stability. Wall cooling destabilizes the second and higher mode instabilities while stabilizing the first one [5]. Moreover, an increase in the Mach number at the boundary layer edge (M_e) produces similar effects. In the case of insulated 2D boundary layers at hypersonic speeds, starting from about $M_e > 4$, the transition is primarily governed by second-mode waves [6]. The Mack second, third, and higher modes are convectively-amplified waveguide normal modes arising from an inviscid instability which is characterized by trapped acoustic waves [7, 8, 9]. The existence of these waves is sustained by the existence of a relative sonic line at a wall height y_s, where $\bar{M}(y_s) = \frac{\bar{u}(y_s)-c}{\bar{a}(y_s)} = -1$ [10, 5]. Here, $\bar{M}$, $\bar{a}$, and c are the relative Mach number, the speed of sound and the disturbance propagation speed, respectively. The presence of such a line implies that relative to the flow speed, disturbances travel subsonically above it and supersonically below it. Hence the lower part of the boundary layer behaves as an acoustic waveguide, where acoustic rays are reflected off the wall and turn around near the relative sonic line towards the wall. Above this turning point at $y = y_s$, the disturbance generates a distinctive "rope-like" wave pattern, as observed by numerous researchers both experimentally and numerically [11, 12]. These structures are centered around $\bar{M}(y_c) = 0$, where y_c denotes the critical layer height. A schematic view in Figure 1, adapted from [13], concisely summarizes previous authors explanations on the behavior of the second Mack mode.

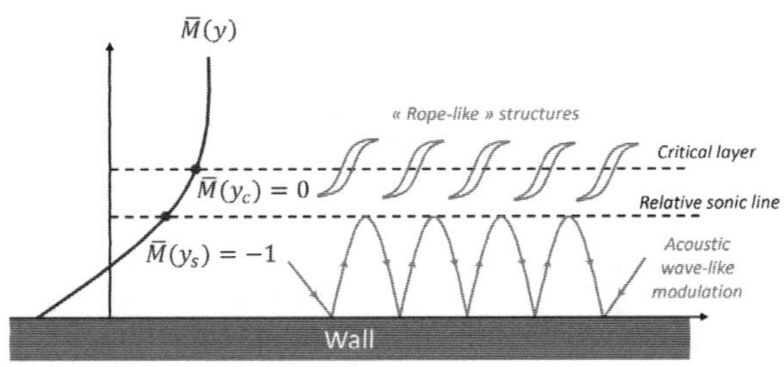

Figure 1: Schematic view of the Mack second mode in a high-speed boundary layer, adapted from Knisely *et al.* [13]

The term "acoustic" has been resorted to in order to classify the physical nature of the second and higher Mack modes. It is common practice in this field to sort the various boundary layer modes into categories that link them in one way or another to a type of physical process or quantity. Examples include entropy modes, vorticity modes, fast or slow acoustic modes. This habit feeds from results provided by linear stability analysis, a tool that has significantly shaped our understanding of hypersonic boundary layer transition mechanisms. The spectrum obtained from such analysis and the structure of the normal modes have led to a refinement in the terminology used to describe the unstable modes in the hypersonic boundary layer. Yet the nomenclature of Mack's first and second modes is not particularly descriptive and can be limiting, both in qualitatively describing the transition mechanisms associated with these instabilities and in interpreting the physical aspects of these phenomena. In fact, it turns out that the first and second modes are not truly distinct modes. Notably, the works of Fedorov and Khokhlov [14] have revealed the continuity of the spectrum between Mack's so-called first and second modes. Fedorov and Tumin [15] subsequently introduced a new terminology based on the behavior of truly distinct modes in the linear stability spectrum. Their new terminology aimed to provide an alternative interpretation of the physical and mathematical description of hypersonic boundary layer instabilities, introducing modes S and F1 alongside continuous modes already identified previously: entropy, vorticity, fast and slow acoustic. For sake of simplicity, we follow the notation introduced by Fedorov [9] and label mode F1 as F.

Near the leading edge, modes S and F initiate in the slow (S) and the fast (F) acoustic part of the spectrum, respectively. Their initial phase velocity close to that of acoustic waves promotes the receptivity of these two modes to acoustic disturbances in the flow. Moving downstream, the phase velocity of mode F decreases to a point where it may synchronize with the vortical and entropic modes of the continuous spectrum. At this point, mode F becomes susceptible to free-stream temperature spottiness and turbulence [16]. Further downstream, if the tendency of mode F to become slower and S to become faster is unaltered, a second synchronization point occurs. Both synchronization points impact the stability of modes S and F. Many studies employing direct numerical simulations have verified these observations [17, 18]. Figure 2 illustrates the different scenarios leading to the growth of Mack modes in relation to the receptivity mechanisms of the flow disturbances. It is important to highlight that the evolution of the S and F modes, their possible interaction and their stability are complex functions of the flow conditions. Depending on the Mach number and the wall temperature, the stability of the S and F branches can switch, or the synchronization point between S and F modes can vanish [15].

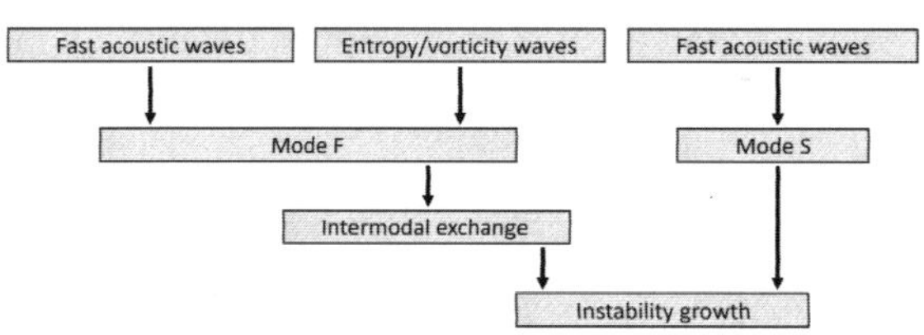

Figure 2: Scenarios of Mack second mode evolution for a high-speed boundary layer on a flat plate, adapted from Fedorov *et al.* [9]

While linear stability analysis have shed light on the role of the boundary layer modes in the transition process, the method is far from complete when it comes to physical interpretation. Its shortcomings motivated extensions or alternative approaches. Kuehl [19] proposed a thermoacoustic analysis on the second mode using an inviscid, Lagrangian, cycle-averaged disturbance acoustic energy formulation, leading to the hypothesis that the physical origins of Mack's second mode are in the thermoacoustic Reynolds stress. Tian *et al.* [20] performed a stability analyses based on the rates of change of perturbations associated to the Mack second mode. They found that the variation in fluctuating internal energy is maintained by the advection of perturbed thermal energy near the critical layer (where the flow propagates at the velocity of the disturbance) and the fluctuation in dilatation near the wall. Additionally, streamwise velocity perturbations were attributed to the simultaneous influence of momentum transfer from the wall-normal velocity fluctuation and the streamwise gradient of the pressure perturbation near the wall. Chen *et al.* [21] conducted an analysis of the phase of each term in the linear stability equation to identify the significant physical sources of boundary layer instabilities. Noteworthy among their observations is the role of wall cooling, which destabilizes the second mode by establishing a pronounced region of wall-normal internal energy transport beneath the second generalized inflection point (GIP).

Unnikrishnan *et al.* [22] caution against semantic misinterpretations made in the literature associated with interpreting the nature of stability modes in terms of acoustic, vortical, or entropic content through their trajectory in the eigenspectra, which prompts the questions: can these three contributions be isolated in the flow? Can the dynamics of these contributions be studied in order to elucidate their true role in the transition process? A first step towards this involves being able to carry out an energy decomposition where each component can be assigned a type of contribution: acoustic, vortical, or entropic. Such a classification for small energy disturbances in a uniform flow can be inferred from the work of Chu [23], yet its limitations have given rise to several studies on hypersonic flows which rely on the momentum potential theory (MPT) decomposition by Doak [24] in order to individually analyse the different contributions [25, 22, 26, 27, 28]. Based on Doak's MPT, Unnikrishnan *et al.* [22] were able to study the evolution of fluid-thermodynamic (FT) components associated with the Mack second mode instability, revealing many interesting results. In particular, they showed that vortical components are the most prominent in the Mack second mode instability, followed by thermal and acoustic components. They also showed that the thermal component is the most significant source term for the growth of acoustic perturbations, an effect which becomes pronounced under wall cooling conditions. Their results highlight that the vortical structures manifest as a series of rope-shaped cells across the GIP, and that both the acoustic and thermal components undergo a change in structural shape downstream of the mode synchronization exhibiting a "trapped" structure. The acoustic component displays alternating monopole arrays between the wall and the critical layer, while the thermal component is confined to the region between the wall and a line straddling the GIP. The MPT-based FT decomposition was also fruitfully used by Long *et al.* [26] to study the "sound radiation" caused by the supersonic mode [29], and the influence of a porous coating on this instability. All these studies seem to converge on the significant role of the GIP and the critical layer in energy amplification mechanisms, as well as on the utility of the Doak decomposition in studying the unstable modes in hypersonic boundary layers.

In the present study, we report on our progress towards exploring the dynamics of the second-mode instability through FT components. The study begins with a linear stability analysis to identify the distinct linear modes within hypersonic 2D flat plate boundary layers under different wall temperature conditions. Following this, forced direct numerical simulations (DNS) are conducted, introducing linear harmonic disturbances in the boundary layer that excite the normal modes predicted by the stability analyses. We then employ the tracking method pioneered by Lozano-Durán and Jiménez [30], and apply it to a temporal sequence of the fluctuating vorticity. This preliminary step towards our final goal allows us to track coherent structures over time, providing both qualitative and quantitative insights into their temporal behaviour, their lifetimes, their geometrical characterization and their advection velocities in a hypersonic boundary layer. Finally, the Doak decomposition is applied to the DNS results in order to compute, for individual flow fields, the FT components of the perturbations. Comparisons are drawn between the components observed over insulated and cooled wall conditions.

2. Simulation conditions

We consider two cases of two-dimensional HBLs over a flat plate. One where the wall temperature T_w is equal to the recovery temperature T_r (adiabatic case), another where $T_w/T_r = 0.145$ with $T_w = 300$ K. The free-stream Mach and unit Reynolds numbers are given by:

$$M_\infty = \frac{U_\infty}{a_\infty} = 6, \tag{1}$$

$$Re_\infty = \frac{\rho_\infty U_\infty}{\mu_\infty} = 2 \cdot 10^6 \text{ m}^{-1}, \tag{2}$$

where ρ_∞, U_∞, and μ_∞ are the density, velocity, and dynamic viscosity of the free-stream, respectively. $a_\infty = \sqrt{\gamma R T_\infty}$ is the speed of sound based on the free-stream temperature T_∞. The flow is considered to be a calorically perfect gas, with specific heat ratio $\gamma = 1.4$ and gas constant $R = 287.05$ J/(kg · K). The dynamic viscosity μ is chosen to follow Sutherland's law, and the heat conductivity κ is related to μ through a constant Prandtl number $Pr = 0.72$.

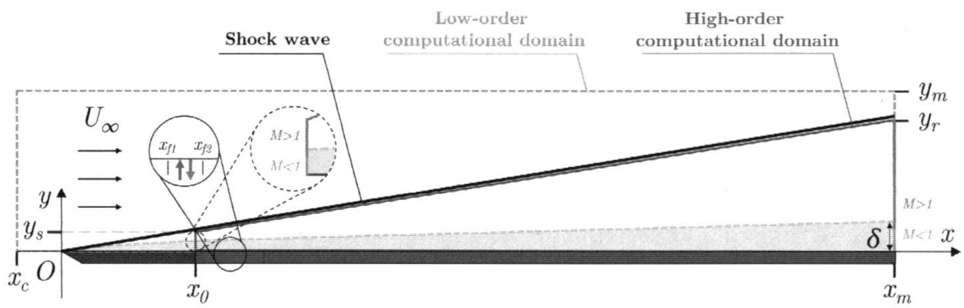

Figure 3: Sketch of the physical model and computational domains.

To initialize the forced simulations, a precursor simulation is run with the finite-volume flow solver CEDRE [31]. This first simulation computes the steady base flow over the flat plate with a second-order finite volume spatial discretization, along with a first-order implicit temporal scheme. The computational domain of the steady base flow is referred to as the "low-order" region in Figure 3. Subsequently, the steady flow field is applied as both the initial condition and as the inlet and outlet boundary conditions within a subdomain where we perform the high-fidelity simulations, denoted as the "high-order computational domain" in Figure 3. The details of the domain geometry for this study are summarized in Table 1. We used bilinear interpolation to project the low-order solution onto the high-order domain. The high-fidelity simulations are run with JAGUAR [32, 33, 34, 35], a high-order spectral difference flow solver collaboratively developed by CERFACS and ONERA. At first, JAGUAR is run without any forcing applied to it for long enough to evacuate initial transients. The latter are due to the fact that the two solvers have different truncation errors. A steady-state is then achieved with JAGUAR, which is in excellent agreement with the low-order base flow solution given by CEDRE. It should be noted that in the case of the cold flat plate, the wall temperature is set to $T_w = 300\ K$ for both CEDRE and JAGUAR. However, for the adiabatic wall, an isothermal wall boundary condition is used in JAGUAR to enforce the wall temperature to $T_w = T_r(x)$, where $T_r(x)$ is the adiabatic wall temperature obtained from CEDRE when enforcing an adiabatic boundary condition.

Once a newly converged steady solution is achieved with Jaguar, it serves as the initial steady state for perturbed unsteady simulations with the same solver. To initiate the Mack second mode instability in the boundary layer, a wall blowing-suction actuator is employed, located within the region between x_{f1} and x_{f2}, as depicted in Figure 3. The blowing/suction velocity follows a sinusoidal space-time dependency given by:

$$v'(t) = A_f \sin\left(2\pi \frac{x - x_{f2}}{x_{f1} - x_{f2}}\right) \sin(\omega t), \qquad x_{f1} \le x \le x_{f2} \tag{3}$$

The value of the frequency f is chosen to be 98 kHz while the forcing amplitude $A_f = 0.1$ m/s is set to trigger a linear response of the Mack second mode instability towards the outlet of the high-order computational domain. The forcing frequency comes from the outcome of a linear stability analysis of our base flow with the MAMOUT code [36], which indicates this frequency to be linearly unstable and generating most of its amplification close to x_m. The forcing amplitude results from preliminary tests that confirmed the linear response of the Navier-Stokes solution for this amplitude value. The Jaguar simulations use a constant number of cells in the wall-normal direction set to 180 and 1505 cells in the streamwise direction. Given that the solution vector within each cell is expressed through a p-order polynomial, which yields $p+1$ internal degrees of freedom within each cell along each spatial direction, the total number of degrees of freedom along each direction can be obtained by multiplying these values by $p+1$. The computational grid used for solving the flow is refined near the wall, with a minimum cell size of approximately $\Delta y \simeq 25$ μm in the wall-normal direction and a constant streamwise spacing of $\Delta x \simeq 0.85$ mm. With this resolution, there are approximately 80 degrees of freedom within a wavelength of Mack's second mode in the streamwise direction. This oversampling is chosen to ensure that coherent structures are defined by a sufficient number of computational cells in the subsequent coherent structure tracking step.

Variable	Value (m)	Variable	Value (m)
x_c	-0.3	x_m	1.5
x_0	0.2	y_s	0.03
x_{f1}	0.2225	y_r	0.235
x_{f2}	0.2375	y_m	0.3425

Table 1: Details on the sizing of the domain geometry in Figure 3.

3. Methodology and Mathematical Description

3.1. Direct Numerical Simulations

The physical problem under investigation is governed by the non-reacting Navier–Stokes equations in 2D generalized coordinates (ξ, η), which consists of the mass conservation equation, the two-component momentum conservation equations, and the energy equation :

$$\frac{\partial \mathbf{Q}}{\partial t} + \frac{\partial \mathbf{E}}{\partial \xi} + \frac{\partial \mathbf{F}}{\partial \eta} = 0 \tag{4}$$

Here, the vector $\mathbf{Q}$ represents the array of conservative variables and $\mathbf{E}$ and $\mathbf{F}$ are flux vectors in the ξ and η directions, respectively. These vectors are defined in relation to the corresponding vectors $\mathbf{Q}_c$, $\mathbf{E}_c$, and $\mathbf{F}_c$ in a cartesian coordinate system.

$$\mathbf{Q} = J\mathbf{Q}_c \tag{5}$$

$$\mathbf{E} = J\left(\mathbf{E}_c\frac{\partial \xi}{\partial x} + \mathbf{F}_c\frac{\partial \xi}{\partial y}\right) \tag{6}$$

$$\mathbf{F} = J\left(\mathbf{E}_c\frac{\partial \eta}{\partial x} + \mathbf{F}_c\frac{\partial \eta}{\partial y}\right) \tag{7}$$

The symbol J denotes the determinant of the transformation jacobian $J = |\partial(x,y)/\partial(\xi,\eta)|$. The vector of conservative variables and the flux vectors can be expressed in the physical coordinate system as:

$$\mathbf{Q}_c = \begin{pmatrix} \rho \\ \rho u \\ \rho v \\ \rho e \end{pmatrix} \tag{8}$$

$$\mathbf{E}_c = \begin{pmatrix} \rho u \\ \rho u^2 + p - \tau_{xx} \\ \rho uv - \tau_{xy} \\ u(\rho e + p) - (u\tau_{xx} + v\tau_{yx} + \kappa\partial_x T) \end{pmatrix} \tag{9}$$

$$\mathbf{F}_c = \begin{pmatrix} \rho v \\ \rho uv - \tau_{xy} \\ \rho v^2 + p - \tau_{yy} \\ v(\rho e + p) - (u\tau_{xy} + v\tau_{yy} + \kappa\partial_y T) \end{pmatrix}, \tag{10}$$

where ρ is the fluid density, u and v are the velocity components along physical coordinates x and y, respectively, p is the static pressure, T the temperature, $e = p/\rho(\gamma-1) + (u^2+v^2)/2$ is the specific total energy and κ the heat conductivity. The components of the viscous stress tensor are labeled τ_{ij}.
The governing equations are solved with Jaguar employing a spectral difference scheme for the spatial discretization. This method relies on a polynomial representation of the solution vector and the flux components inside each cell. The scheme's order $p+1$ is adjustable, with its accuracy level determined by the degree p of the interpolating polynomial. For this study, a fourth-order spectral difference scheme is used while the temporal integration is achieved using a fourth-order five-stage Runge-Kutta time-stepping scheme.

Due to complex nature of the hypersonic boundary layer flow, which encompasses both supersonic ($M > 1$) and subsonic ($M < 1$) regions (see Figure 3), distinct boundary treatments are applied at the inflow and outflow boundaries. Non-reflecting boundary conditions based on Navier–Stokes characteristic boundary conditions [37, 33] are set in the subsonic inflow and outflow regions. For the super/hypersonic inflow ones, Dirichlet conditions are imposed for u, v, p and T, while for the outflow the extrapolated solution from the inner domain is used to define the boundary condition.

3.2. Momentum potential decomposition

Doak's momentum potential theory (MPT) enables the decomposition of momentum density ρu_i into a linear superposition of rotational, acoustic, and thermal components [24]. Following the Helmholtz decomposition, one can first split ρu_i into a rotational (solenoidal) component $\mathbf{B}$ and an irrotational component expressed as $\nabla\psi$:

$$\rho\mathbf{u} = \mathbf{B} - \nabla\psi, \quad \nabla\cdot\mathbf{B} = 0. \tag{11}$$

Assuming the flow to be statistically time-stationary, each instantaneous flow quantity consists of a mean $(\bar{\cdot})$ part and a fluctuation part $(\cdot')$. In this context, the mean scalar potential, denoted as $\overline{\psi}$, is both irrotational and solenoidal and can thus be assumed to be zero, leading to:

$$\overline{\rho\mathbf{u}} = \overline{\mathbf{B}}, \quad (\rho\mathbf{u})' = \mathbf{B}' - \nabla\psi'. \tag{12}$$

The fluctuating solenoidal $\mathbf{B}'$ field represents the hydrodynamic component associated with the momentum density field. The continuity equation can then be re-expressed as follows:

$$\frac{\partial \rho'}{\partial t} + \nabla \cdot (\rho \mathbf{u})' = 0, \quad \nabla \cdot (\overline{\rho \mathbf{u}}) = 0. \tag{13}$$

By inserting equation (12) in (13), a Poisson equation is derived for the fluctuating scalar potential field:

$$\nabla^2 \psi' = \frac{\partial \rho'}{\partial t}. \tag{14}$$

The irrotational field ψ' can now be expressed as the linear superposition of an isentropic (acoustic) ψ'_A and isobaric (thermal, or entropic) ψ'_T contribution as follows:

$$\psi' = \psi'_A + \psi'_T, \quad \nabla^2 \psi'_A = \left(\frac{1}{a^2}\right) \frac{\partial p'}{\partial t}, \quad \nabla^2 \psi'_T = \left(\frac{\partial \rho}{\partial S}\right) \frac{\partial S'}{\partial t}. \tag{15}$$

The fluctuating momentum density is therefore depicted as a combination of a vortical component $\mathbf{B}'$, an acoustic component $-\nabla \psi'_A$, and a thermal component $-\nabla \psi'_T$. Here, $a = \sqrt{\left(\frac{\partial p}{\partial \rho}\right)_S}$ represents the local speed of sound, and S denotes the entropy of the fluid.

In this study, the Poisson equations of the total irrotational component fluctuation ψ' and the acoustic component ψ'_A are solved. Following this, ψ'_T is determined by substitution. On the outer boundaries, thermal fluctuations are considered to be negligible, while strictly vortical disturbances are assumed (and introduced) within the forcing strip. As a result, identical boundary conditions are enforced for both Poisson problems. The Dirichlet boundary conditions for solving the Poisson problems are formulated by integrating along the boundaries, following the procedure detailed in [38, 39, 22]. The Poisson equations are solved using the open-source library Bcube written in the Julia language, employing finite elements (FE) with third-order Lagrange polynomials. The consistency in order across the different interpolations (SD-based Jaguar and FE-based Bcube) was favored with respect to other lower order, finite-volume based Poisson solvers. The impact of the different interpolating schemes on the resulting ψ' fields has not been studied further. The time derivatives appearing on the right-hand sides of the Poisson equations are computed in Jaguar through an upstream fourth-order finite difference scheme.

4. Identification of coherent structures

In the present work, the tracking method considered is based on the work by Lozano-Durán *et al.* [30]. This section presents a definition of coherent structures and outlines the chosen procedure for their identification and extraction.

It is important to recall that the disturbances of interest to us are of the harmonic type. Hence, their characteristic size can be expressed in terms of their wave length L_λ. For the time tracking of such structures of size roughly equal to $L_\lambda/2$ in length, we estimate that within one forcing period T_f a sampling frequency of T_f/n is sufficient for time-tracking when $n > 2$. We set $n = 18$ as a conservative choice, since the structures may be smaller depending on the threshold. To be more precise, we conduct our analysis choosing a total of $N_s = 365$ flow field snapshots that span approximately 20.6 T_f in time. These snapshots are stored at regular intervals of $\Delta t_s = 5.76 \cdot 10^{-7}$ s, during which an object traveling at the phase speed of the mode disturbance covers a distance equivalent to *ca.* 4.5 cells with our mesh resolution. Then, we classify a structure as a cluster of spatially adjacent cells, or pixels, inside which a specific quantity of interest Q exceeds a chosen threshold. More specifically, the structures are identified based on satisfying the condition:

$$Q(x, y, t) > \alpha_p H(x). \tag{16}$$

Here, α_p represents a threshold derived from percolation analysis [40, 41, 27], and $H(x)$ is a shape function that takes into account the streamwise evolution of the perturbation and is defined as:

$$H(x) = \max_{x,y,t}(Q) \frac{\ln(Q'_w)}{\lambda \max\limits_{x,y,t}(\ln(Q'_w))}, \tag{17}$$

where λ a constant set to 1000 and the subscript $\cdot_w$ refers to the wall values. Given that the dynamics of the second Mack mode are confined to a specific region within the boundary layer, any structure with a barycentre position above twice the boundary layer thickness away from the wall will be discarded. Similarly, any structure with an area l^2 below a defined threshold, set to $l = 2.2 \cdot 10^{-3}$ m, will be categorized as noise and consequently not recognized as a valid structure. In comparison, note that the boundary layer thickness span the range of values from 3 to 8 mm. This area criterion resulted from an analysis of the distribution of the coherent structure sizes.

5. Preliminary results and discussion

5.1. Steady state

The u-velocity and temperature profiles obtained from the DNS are plotted at the end of the computational domain in Figure 4. The normalized wall-normal coordinate is determined using the local boundary-layer thickness, which is based on the total enthalpy criterion $h_0(\delta_{h0}) = C_h h_\infty$ [42], with $C_h = 1.008$. Over the insulated wall, the developed boundary layer is thicker, approximately $\delta_{h0} \simeq 14$ mm, in contrast to $\delta_{h0} \simeq 8$ mm over the cold wall. Unnikrishnan *et al.* [43] noted that second-mode instabilities tend to exhibit maximum amplitudes near high-gradient regions. Consequently, amplification dynamics are intensified closer to the wall under colder surface conditions. Moreover, as can be seen from the right-most panels of Figure 4, the insulated configuration displays a GIP, while wall cooling suppresses the GIP. Similar behavior has been observed in [44, 43].

5.2. Linear stability analysis

A spatial local linear stability analysis (see Appendix A) is performed on the HBL profiles for the frequency $f = 98$ kHz. The resulting phase speeds and spatial growth rates associated with F and S modes are plotted in Figure 5. Horizontal dashed lines in the left plots correspond to the continuous branches of the eigenspectra, including the fast acoustic ($c_r/u_e = 1 + 1/M_e$), entropy/vorticity ($c_r/u_e = 1$), and slow acoustic branches ($c_r/u_e = 1 - 1/M_e$). The dashed vertical lines correspond to positions where specific intermodal interaction phenomena can occur due to phase speed synchronisation between two modes. For both the adiabatic and cooled walls, modes F and S appear to originate from the fast and slow acoustic branches near the leading edge, respectively. Subsequently, tracking the trajectory of mode F reveals an intersection with the vorticity/entropy branches. This coincides with the point where the amplification rate of mode S starts moving sharply towards the unstable values (negative α_i) . Further downstream, the phase velocities of modes F and S synchronize for the adiabatic case. At this juncture, mode S seems to reach its maximum instability. Conversely, under the cold wall condition, this synchronization between the two discrete modes is absent, as observed in [15]. Establishing a cause-and-effect relationship between the intermodal interaction phenomena in the eigenspectra and the stability observations of the modes proves to be difficult. Despite this, insights from a range of analytical and numerical receptivity studies [45, 18, 46] indicate that the synchronization phenomena, characterized by phase velocity and eigenfunction interactions between different

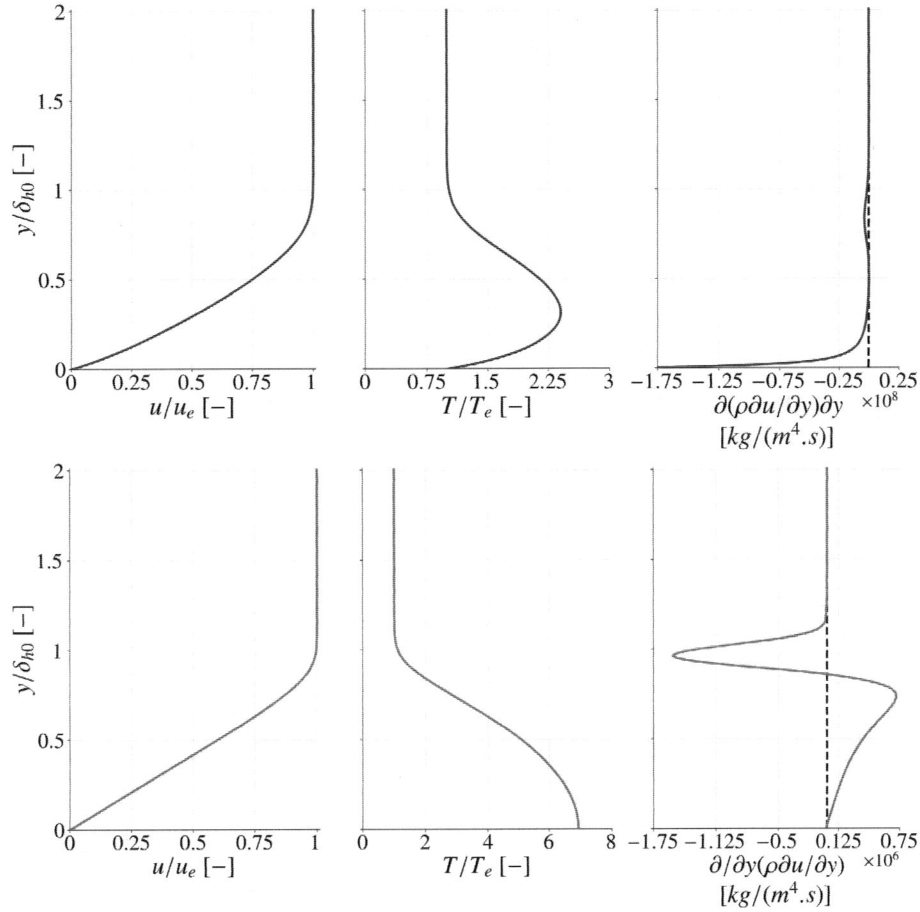

Figure 4: Streamwise velocity profile (left), temperature profile (center), and indicator function for the existence of a GIP (right). Where the indicator function crosses the zero value (vertical dashed line), a GIP exists. All profiles taken from $x = 1.5$ m for the cold wall (top) and the insulated case (bottom).

modes, imply energy transfer processes between the modes involved. Therefore, we expect the synchronization patterns (or lack thereof) observed through the LST results presented above to be reflected in the FT-based analysis that will be presented later. Indeed, it is the scope of our work to study in detail these interactions.
The total amplification rate of an unstable wave is quantified by the N-factor, representing the overall exponential growth of an unstable perturbation. It is defined in relation to the final amplitude A at location x and the initial amplitude A_0 at the critical point x_{cr}. The exact definition is given by:

$$N(x, f) = \ln\left(\frac{A}{A_0}\right) = \int_{x_{cr}}^{x} -\alpha_i(x, f)\, dx. \quad (18)$$

The N-factor curves associated to the Mack second mode (mode S here) are presented in Figure 6. The amplification of the second Mack mode is more pronounced on the cold wall compared to the adiabatic case, in line with the fact that cooling destabilizes this mode [47]. Moreover, in the adiabatic case, the maximum amplification of this instability occurs more

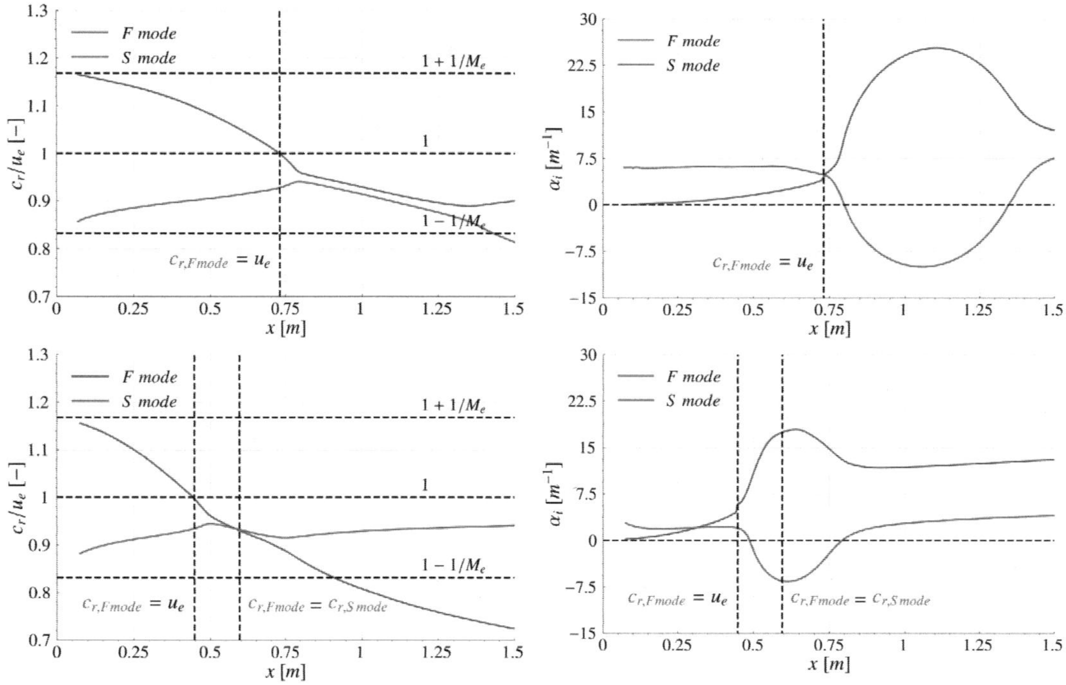

Figure 5: Phase velocities (left) and spatial growth rates (right) of discrete modes F and S at a frequency $f = 98$ kHz computed by local LST for the cold wall (top) and the insulated case (bottom).

upstream. This can be explained by the fact that the adiabatic boundary layer is thicker, and that the frequency of the instability is inversely proportional to the thickness δ [48]. Hence at equal frequency, the amplification peak is shifted upstream in the adiabatic case.

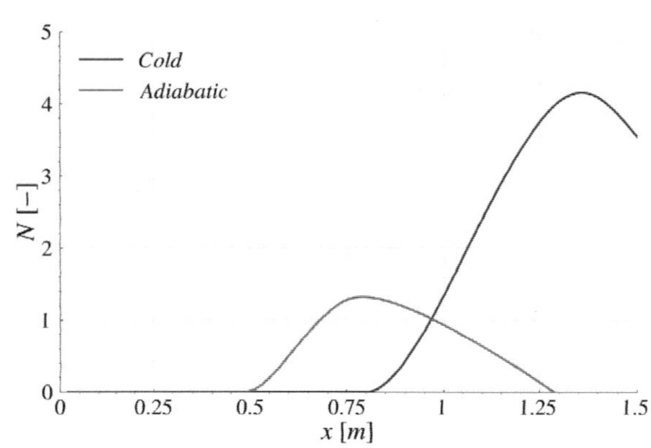

Figure 6: N-factor curves computed by local LST for the cold wall and the insulated case.

5.3. Coherent structures tracking based on fluctuating vorticity

In this section, the results of the time-tracking of structures associated with the fluctuations of vorticity over the cold wall are discussed.

5.3.1. Percolation Analysis In order to determine the threshold parameter α_p in equation 16, we conduct a percolation analysis by systematically varying α_p over a range of values. Of particular interest is the evolution of both the number and the area of structures that are identified at the various thresholds. The methodology, as employed in previous studies [41, 30, 49], is described in great detail in [27]. Within the context of our study, the structures based on the fluctuating vorticity are studied in a first instance as a surrogate for the solenoidal component of the Doak decomposition. We note in passing that we have preliminary knowledge suggesting that the perturbations will organize themselves following a pattern of aligned sign-alternating doublets. Hence, we plan to distinguish between structures with positive and negative vorticity perturbations by treating them as distinct entities during the clustering process. A fluctuating vorticity structure is therefore defined as a clustered set of spatially-connected pixels satisfying the following conditions:

$$\begin{aligned} \nabla \times \mathbf{u}' &< -\alpha_p H(x), \\ \nabla \times \mathbf{u}' &> \alpha_p H(x). \end{aligned} \tag{19}$$

The results of the percolation analysis are presented in Figure 7 with the ratio of the area of the largest structure A_{max} divided by the total area A_{tot} occupied by all the structures, and the total number of structures N_s normalized by the largest number of identified structures $N_{s,max}$ over all the values of α_p. Logically, at a low value of α_p, a relatively small number of structures are detected, covering a broad surface area. By increasing the value of α_p, these large structures will break down into several smaller but more numerous structures. This process is referred to as the percolation crisis. If the value of α_p is further increased, the ratio A_{max}/A_{tot} tends to increase slowly while the number of objects continues to decrease. We opt for a threshold value of $\alpha_p = 0.4$, located in the range where the area ratio remains roughly constant and the phenomenon of object splitting is still important.

It is worth noting some discrepancies in the percolation curves obtained in this study compared to those commonly reported in the literature. Indeed, typically at a low value of α_p, a large number of pixels are clustered into a single large structure, causing the area curve to start from the value of 1, and the ratio of the number of structures to be very low. Moreover, the area ratio eventually tends to 0 as α_p becomes large, as with a very high threshold, no structures are detected. In our case study, the distinctive feature lies in the separate consideration of structures exhibiting positive and negative vorticity fluctuations. Consequently, in regions where perturbations manifest as sign-alternating doublets, the structures are inherently distinct, irrespective of α_p, justifying the observed evolution of the curves in Figure 7.

5.3.2. Branch classification The evolution of the structures over their lifetime can be organized into branches. Additionally, these branches can be classified into several groups based on their topology within a temporal graph that gathers them according to the procedure described by Lozano-Durán and Jiménez [30]. A branch designated as "primary" is one associated with a structure that can unequivocally be identified as an individual entity throughout its entire lifespan. Secondary branches, on the other hand, encompass two classifications: "incoming" for instances where the associated object merges with another to form a new entity, and "outgoing" for scenarios involving the splitting of an object. The fourth category, called "connectors," consists of branches that originate from the splitting of a structure and conclude by merging into another. Table 2 summarizes classification results for our coherent structures based on fluctuating vorticity. We observe a predominance of primary branches. However, this result is challenging to interpret because the saved temporal sequence does not cover a sufficiently long time interval for the perturbations injected at the initial time to reach the end of the computational domain. Indeed, the duration of the temporal sequence is $0.4\tau_c$, where τ_c represents the advection time defined as the ratio of the average flow velocity outside

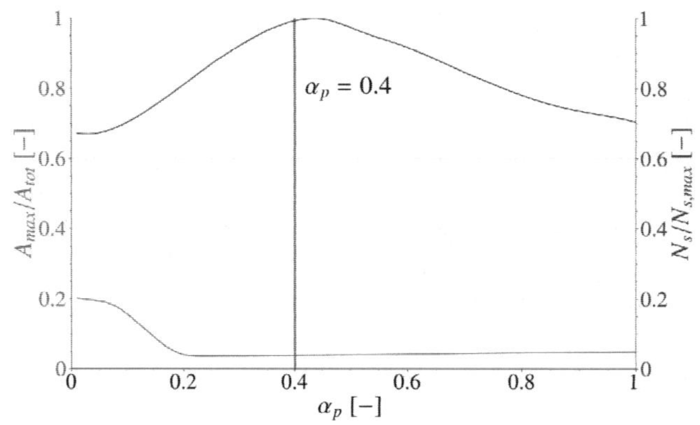

Figure 7: Percolation analysis curves: the solid red line represents the ratio between the area of the largest structure, A_{max}, and the total volume occupied by structures, A_{tot}. The solid blue line depicts the ratio between the number of structures, N_s, and the largest number of identified structures, $N_{s,max}$.

of the boundary layer to the distance between the blowing/suction slot and the end of the computational domain. Therefore, it is not possible to determine the lifespan of a branch from its birth to its death when it lives for longer than $0.4\tau_c$. However, it is noteworthy that outgoing branches are so underrepresented compared to the other branch types within the dynamics of the vorticity fluctuation structures. Additionally, the incoming branches appear to be relatively significant in terms of number and occupied area, showing that mergers occur very frequently. Overall, however, we conclude that branch classification requires a longer time series to show definite results and shed light on the merging/splitting processes. Future work will carry out coherent structure tracking based on MPT components over a longer time series.

Branch type	$N_b/N_{b,tot}$	$A_b/A_{b,tot}$
Primary	0.498	0.447
Incoming	0.255	0.408
Outgoing	0.043	0.030
Connectors	0.204	0.115

Table 2: Branch number fraction and branch area fraction displayed for each branch type.

5.3.3. Vertical evolution Geometrical properties are derived from the temporal evolution along the identified branches. This section investigates the dynamics of vorticity fluctuation structures along the wall-normal direction by analyzing geometrical parameters, including the minimum and maximum y coordinates along the streamwise direction (y_{min} and y_{max}, respectively). Additionally, the study encompasses the structures' heights, represented by $L_y = y_{max} - y_{min}$, and the y-coordinate y_{cg} of the structures' barycentre. Figure 8 illustrates the spatial evolution of these geometric properties of the structures. Dashed horizontal lines indicate, from top to bottom, the boundary layer edge, the critical layer, and the relative sonic line. Additionally, solid vertical lines mark positions where key phenomena in the structures' behavior are observed, which will be clarified in the discussion that follows. A color code is also employed to categorize structures with similar geometrical properties. Again, those properties will become apparent below.

Observations reveal that slightly downstream from the perturbation injection slit, structures attached to the wall coexist with other structures situated slightly above. Figure 9 provides a snapshot view of the vorticity fluctuation contours zoomed in on this area, providing qualitative insights into the behavior of the structures. The presence of elongated structures of vorticity fluctuations with opposite signs overlaid is noticeable, indicating the presence of a shear layer associated with the imposition of a blowing/suction condition at the wall. After $x = 0.32$ m, a point marked by the first vertical line, the structures appear to organize into different patterns. Two groups of structures are distinguishable in Figure 8: a first group of objects (in green) attached to the wall and extending up to the relative sonic line, and a second group of objects (in blue) between the sonic line and the boundary layer edge. It is also evident from the y_{cg} locations (see Figure 8) that this second group of objects organizes around the critical layer. This spatial arrangement is clearly depicted in Figure 9, which provides a zoomed-in view of this area. The presence of sign-alternating doublets is observed, with two regions of opposite vorticity perturbation: one trapped between the wall and the relative sonic line, and the other around the critical layer. It is noted that the behavior of vorticity fluctuation structures varies according to the sign of the fluctuation. Structures exhibiting negative fluctuations display a more significant size near the wall compared to their counterparts with positive fluctuations. Conversely, this trend is reversed for the alternate set of structures situated above the relative sonic line. From the position $x = 0.83$ m (marked by the second vertical line), corresponding to the point at which the S mode of linear stability analysis becomes unstable (see Figure 5), a change in the behavior of these structures becomes observable. The structures aligned with the critical layer become thicker over time, mainly because they extend downward while remaining aligned with the critical layer (see Figure 8). This phenomenon reaches a critical point at $x = 1.05$ m, marked by the third vertical line, corresponding to the point where the S mode is most unstable according to the LST (see Figure 5). At this point, the upper structures appear to merge with objects attached to the wall, and the new structures obtained then extend from the wall to the boundary layer edge (in yellow). In Figure 9, it is observed that this merging occurs between regions of vorticity fluctuation with the same sign but located at two different heights. These observations regarding the behavior of vorticity fluctuation structures seem to indicate the generation of a shear layer that results from applying a perturbation to the wall in the form of a blowing/suction source. This layer eventually gives rise to sign-alternating doublets exhibiting trapped wave behavior between the wall and the relative sonic line. Additionally, a zone of vorticity perturbation organizes around the critical layer. Finally, we observe that the positions marking changes in the stability of the S mode from the LST play a crucial role in the dynamics of vorticity fluctuation structures.

5.3.4. Advection velocities The streamwise advection velocity of the structures is computed based on the temporal displacement of the structures' streamwise barycenter position, $x_{\rm cg}$. In Figure 10, we present a comparative analysis of the advection velocities of vorticity fluctuation structures and the phase speeds of the discrete modes F and S obtained from LST analysis. Interestingly, in the proximity of the excitation point, the structures seem to be advected at the bulk flow velocity u_e. Shortly after, their advection velocity follows a trend similar to that of mode F, including the presence of multiple peaks. These could be indicative of the superposition of multiple modes associated with the blowing/suction forcing, which excites mode F, mode S, acoustic waves, and entropy/vorticity waves simultaneously just downstream of the actuator [50]. What can be stated based on our current results is that this region of varying convection speed also corresponds to significant topological changes in the structure's shape as can be seen

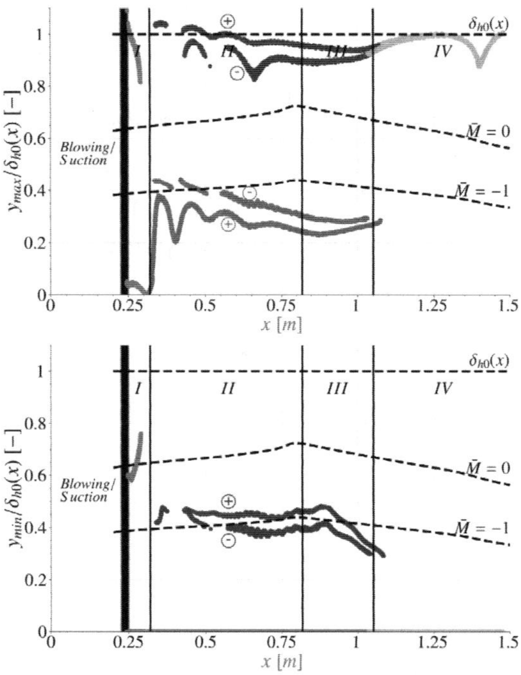

Figure 8: Spatial evolution of the geometrical features related to the vertical extent and position of vortical coherent structures: maximum wall-normal coordinate (top left), minimum wall-normal coordinate (bottom left), height (top right), wall-normal barycenter position (bottom right). The vertical solid lines represent the delimitation of region of interest I, II, III and IV. The dashed horizontal lines correspond to the boundary layer edge, the critical layer, and the relative sonic line, from top to bottom, respectively. Structures within each region are represented by distinct color: • for region I, • for regions II and III attached to the wall, • for regions II and III not attached to the wall, and • for region IV. Additionally, ⊕ and ⊖ denote structures associated with positive and negative fluctuations, respectively, where the distinction is relevant.

on Figure 8.

Further downstream, the structures are advected with a velocity close to the phase velocity of mode S. This transition appears to occur around $x = 0.83$ m, corresponding to the point where the S mode becomes unstable according to linear stability analysis (see Figure 5), and also coincides with a change in the vertical evolution behavior of the structures (Figure 8). Such behavior appears to be explained by the stability of different modes present in the boundary layer. Near the actuator, the modes of the fast acoustic branch and the vorticity/entropy branch are excited in addition to the F and S modes, as mentioned earlier. Further downstream, the F mode is relatively less stable (growth rate close to 0), along with certain hydrodynamic and entropic modes. The perturbations propagating into the boundary layer are then dominated by the latter. Finally, the S mode becomes unstable and eventually prevails. A comparison of the wall-normal perturbation amplitudes obtained from DNS and the eigenfunctions of the discrete modes F and S obtained by LST at two different locations, $x = 0.55$ m and $x = 1.35$ m, is presented in Figure 11. The consistency between the wall-normal perturbation profiles and the eigenfunctions of mode F at $x = 0.55$ m suggests that the fluctuations are dominated by this mode in this region. At $x = 1.35$ m, the trend reverses, and the perturbation profiles align closely with the eigenfunctions of mode S. These results support the analysis conducted

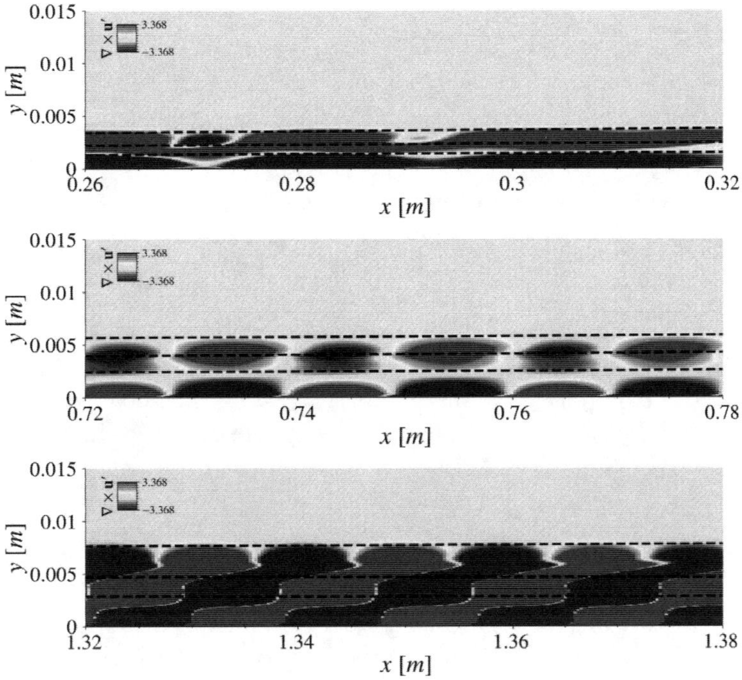

Figure 9: Instantaneous magnified view of vorticity fluctuation in three zones of interest I, II and IV. The black dashed lines correspond to the boundary layer edge, the critical layer, and the relative sonic line, from top to bottom, respectively.

on the comparison of the phase velocities of the two discrete modes from LST and the advection velocity of the fluctuating vorticity structures.

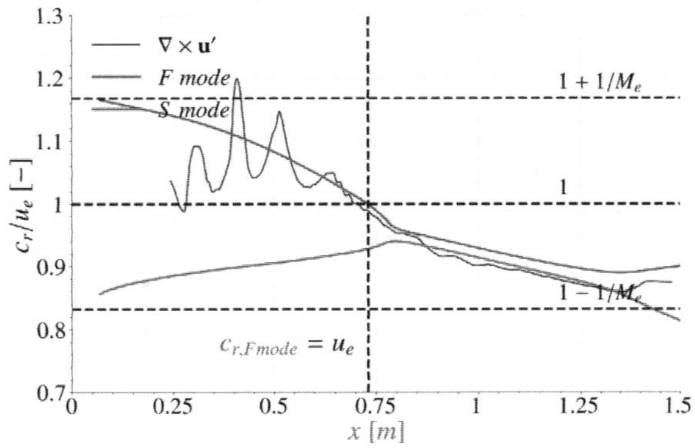

Figure 10: Phase speed velocity of the modes present in the cooled-wall boundary layer obtained through LST, and mean advection velocity of the vorticity structures found through coherent structure tracking.

5.4. Fluid-thermodynamic decomposition

We apply Doak's MPT decomposition to instantaneous flow fields of fluctuating quantities extracted from the DNS (cooled and adiabatic walls). Figure 12 presents snapshots of the

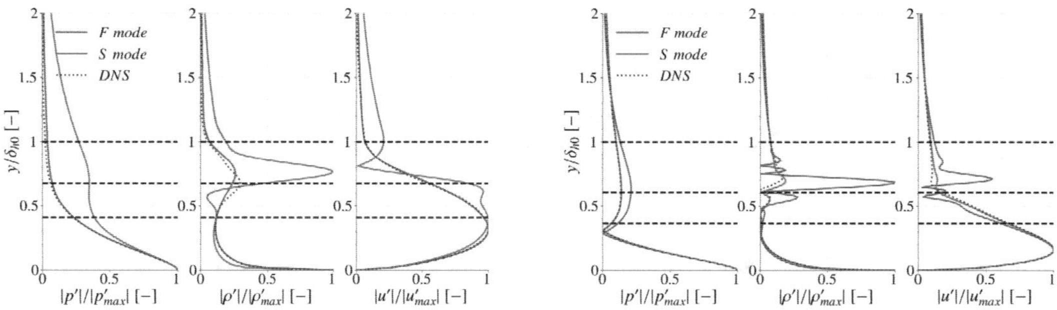

Figure 11: Wall-normal perturbation amplitudes from DNS and F and S mode eigenfunctions obtained by LST at $x = 0.55$ m (left) and $x = 1.35$ m (right). The black dashed lines correspond to the boundary layer edge, the critical layer, and the relative sonic line, from top to bottom, respectively.

magnitudes of the instantaneous vortical ($||B||$), acoustic ($||\nabla\psi_A||$), and thermal ($||\nabla\psi_T||$) fluctuations over the cold plate surface, shown for the entire computational domain. It is clear from the contours that within the region where the second Mack mode experiences maximum amplification (approximately $x = 1.35$ m as indicated by the LST), all three FT components undergo amplification within the boundary layer, while outside of the boundary layer the vortical and the acoustic components are the most intense. The three components maintain a relatively low intensity until reaching the amplification region of the second Mack mode. The vortical and acoustic components feature radiation outward from the boundary layer, guided by the oblique shocklet induced by the blowing/suction source term. Simultaneously, this shocklet appears to generate vortical fluctuations. In contrast, the thermal component is concentrated mainly within the boundary layer, and displays no radiation tendency away from the boundary layer. This observation aligns well with findings from previous studies [22, 26]. Similar observations were made for the insulated flat plate case (not shown for brevity).

Figure 13 allows us to be more quantitative, illustrating the magnitudes of the FT components at the wall surface for both the cold wall and the insulated wall cases. The first thing to note is that the magnitudes are significantly lower over the adiabatic wall. This trend follows from the fact that perturbations are able to grow significantly more in the cooled wall case, as shown by the N-factor curves in Figure 6. Recalling from equation 18 that the amplitude grows as e^N, differences in FT-component magnitudes of one order of magnitude can be expected and are indeed observed.

The region of growth leading to the maximum magnitudes in these components is clearly aligned with the location where the second Mack mode predicted by LST is most unstable in both cases (see Figure 5). Moreover, the acoustic component is the most intense of the three for both wall temperatures, followed by the thermal component, while the vortical component experiences the mildest amplification. In the case of the cold wall, the acoustic component appears to be much more predominant with respect to the other two components in the second Mack mode amplification region. This observation is consistent with the destabilizing influence of wall cooling and with the acoustic nature of this mode. The agreement observed between our findings and the prior research conducted by Unnikrishnan *et al.* [22] and Long *et al.* [26] validates our implementation of the Doak decomposition on flow fields of hypersonic boundary layers.

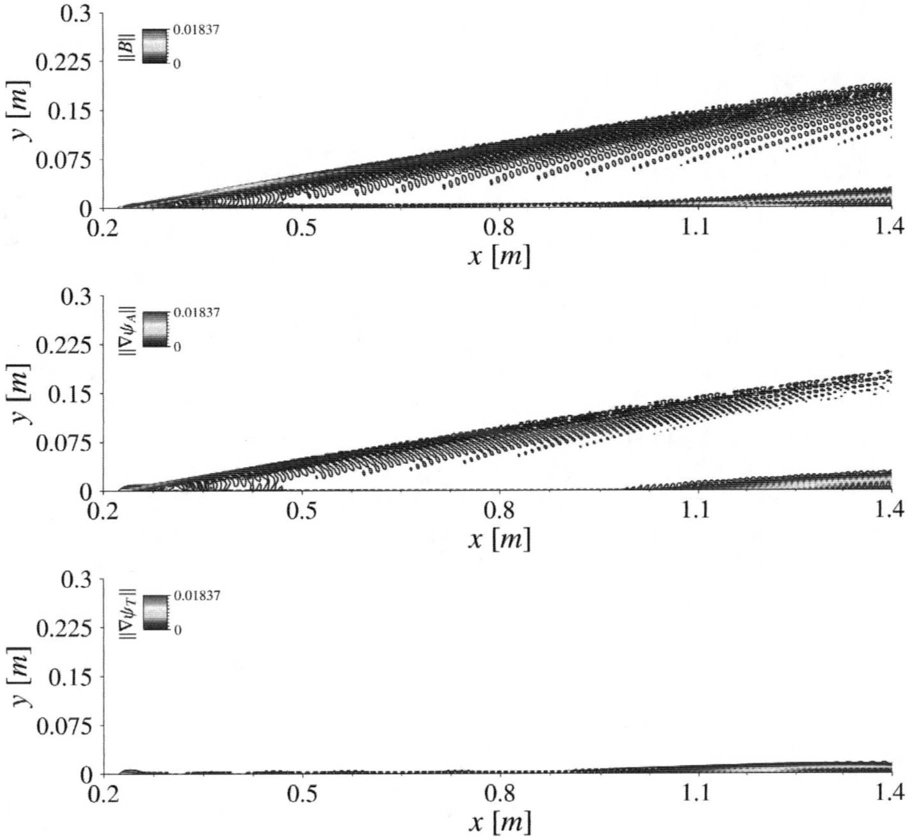

Figure 12: Instantaneous view of magnitudes of vortical (top), acoustic (center) and thermal (bottom) perturbations over the cold plate surface.

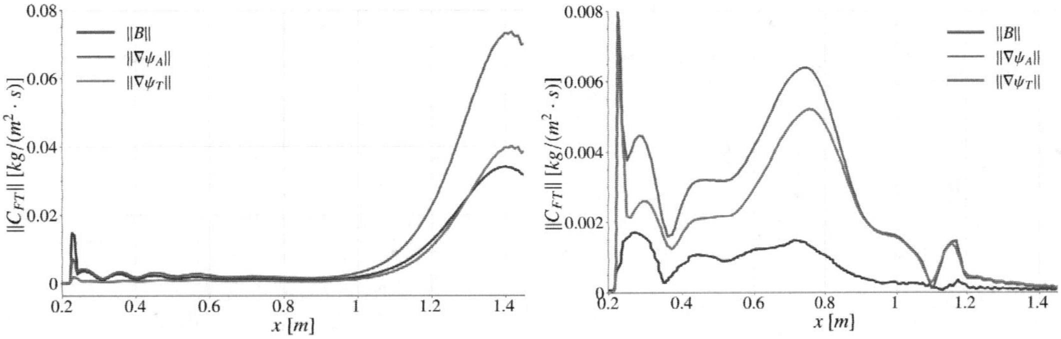

Figure 13: Magnitudes of the fluid-thermodynamic components at $y = 0$ in the case of the cold wall (left) and the adiabatic wall (right).

6. Conclusion

In this study, we have delved into the dynamics of high-speed boundary layer flows over insulated and cooled walls, exploring the link between stability analysis and the properties of coherent structures of vorticity found in the boundary layer. This study is an initial exploration on how coherent structure tracking methods and the FT component decomposition of the flow field can uncover the mechanisms underlying the second Mack mode instability.

A local linear stability analysis lead to the expected distinct behavior in the branching of discrete modes for adiabatic and cooled wall conditions, with an absence of synchronization in the cooled wall case. Coherent structure tracking based on fluctuating vorticity provided a qualitative examination of the spatial and temporal evolution of structures within the boundary layer. The spatial evolution of coherent structures aligns with changes in the linear stability of the discrete modes, underscoring the interplay between coherent structures and the underlying instability mechanisms in high-speed boundary layers. A noteworthy result stems from the evolution of the vertical dimension and location of the structures, revealing the coexistence of vorticity structures at different heights along the wall that feature alternating sign doublets during mode F predominance. These structures organize around the critical layer, and above the relative sonic line. Structure mergers at a critical point aligned with the maximum instability of the S mode has also been revealed, the dynamical implications of which need to be further investigated.

A classification of the coherent structure dynamics into branches within graphs was attempted. It revealed that for this time-periodic flow, a longer time sequence is necessary in order for the objects that appear during the simulation to have the time to be washed out of the domain.

The MPT-based FT decomposition introduced by Doak has been implemented as a post-processing tool to our boundary layer data, which is the preliminary step towards time-tracking of coherent structures based on this quantity. Individual snapshots confirm the insights into the different energetic modes within the boundary layer that have been observed previously. Here, we have linked the quantitative trends in the acoustic and thermal components of the FT decomposition with the amplification region of the second Mack mode instability. The acoustic component, particularly pronounced in the cold wall case, emphasizes the destabilizing influence of wall cooling. Future work will involve carrying out the time-tracking used in this study for vortical structures but applied to FT-based objects instead, and on the basis of a data set spanning a longer time series.

Acknowledgments

This work was supported in part by the European Research Council under the Caust grant ERC-AdG-101018287. The authors express their gratitude to the developers of Bcube at ONERA/DMPE, namely Maxime Bouyges, Ghislain Blanchard, and Lokman Bennani, for their valuable assistance in configuring and managing the solution of the Poisson problems.

Appendix A. Linear Stability Theory

The Linear Stability Theory (LST) equations are derived from the governing Navier-Stokes equations. Initially, LST assumes that the instantaneous flow variables $\mathbf{q} = (p, u, v, w, T)^T$ can be decomposed into mean and fluctuating components, represented as $\bar{\mathbf{q}}$ and $\mathbf{q}'$, respectively, where $\mathbf{q} = \bar{\mathbf{q}} + \mathbf{q}'$. In cases of natural transition, the disturbances are assumed to be very small, which allows the linearization of the Navier-Stokes equations, considering that quadratic and higher-order terms are negligible. The parallel or quasi-parallel flow assumption is applied, *i.e.* the mean flow is considered as predominantly dependent on the wall-normal direction y. Under these conditions, normal mode solutions are sought in the form:

$$\mathbf{q}'(x,y,t) = \hat{\mathbf{q}}(y) \exp\left[i(\alpha x + \beta z - \omega t)\right] + c.c. \tag{A.1}$$

Here, α and β represent the x and z components of the wavenumber vector $\mathbf{k}$, while ω is the frequency, and $\hat{\mathbf{q}}(y)$ represents eigenfunctions that describe the mode structure within the boundary layer. By introducing the normal modes into the linearized Navier-Stokes equations, a one-dimensional differential eigenvalue problem for the modal functions is obtained:

$$\left(\mathbf{A}\frac{d}{dy^2} + \mathbf{B}\frac{d}{dy} + \mathbf{C}\right)\hat{\mathbf{q}} = 0, \tag{A.2}$$

Here, $\mathbf{A}$, $\mathbf{B}$, and $\mathbf{C}$ are complex square matrices of size 5. In the context of a boundary layer flow, the perturbation field can be described by the following boundary conditions:

$$\text{At } \ y = 0, \quad \hat{u} = \hat{v} = \hat{w} = 0, \tag{A.3}$$

$$\hat{T} = 0 \quad \text{for an isothermal wall,} \tag{A.4}$$

$$\frac{d\hat{T}}{dy} = 0 \quad \text{for an adiabatic wall,} \tag{A.5}$$

$$\text{At } \ y \to \infty, \quad \hat{u} = \hat{v} = \hat{w} = \hat{T} \to 0 \tag{A.6}$$

For spatial stability analysis, the frequency ω is real, whereas α is a complex eigenvalue, with $\alpha = \alpha_r + i\alpha_i$. When $\alpha_i < 0$, the flow is unstable, exhibiting a spatial growth rate α_i and a streamwise wavenumber α_r. The local phase velocity is defined as $c_r = \omega/\alpha_r$. Additionally, in the context of the Mack second mode, which primarily exhibits two-dimensional dynamics, β is assumed to be 0 in this study. The relationship for the eigenvalues, often referred to as the dispersion relation, can be expressed as:

$$\alpha = \mathbf{\Omega}(\omega). \tag{A.7}$$

The evaluation of the dispersion relation for a boundary layer profile is a fundamental aspect of stability theory. The eigenvalues, along with their corresponding eigenfunctions, provide a complete description of the normal modes for a given boundary layer profile. In this study, these computations are carried out using the in-house LST solver, MAMOUT [36], which numerically solves the dispersion relation for the HBL profiles extracted from the precursor solution (with the CEDRE finite-volume solver). This is achieved through a high-ordre discretization scheme, as detailed in [51]. It should be mentioned that a convergence analysis with respect to the base flow resolution was conducted to assess the accuracy of our stability results.

References

[1] Saric W S, Reed H L and Kerschen E J 2002 Boundary-layer receptivity to freestream disturbances *Annu. Rev. Fluid Mech.* **34** 291–319
[2] Morkovin M V 1994 Transition in open flow systems a reassessment *Bull. Am. Phys. Soc.* **39** 1882
[3] Goldstein M E 1983 The evolution of Tollmien–Schlichting waves near a leading edge *J. Fluid Mech.* **127** 59–81
[4] Mack L M 1969 *Boundary-layer stability theory* (Jet Propulsion Laboratory)
[5] Mack L M 1984 Boundary-layer linear stability theory *AGARD Rep.* **709** 1–3
[6] Özgen S and Kırcalı S A 2008 Linear stability analysis in compressible, flat-plate boundary-layers, *Theor. Comput. Fluid Dyn.* **22** 1–20
[7] Mack L M 1990 On the inviscid acoustic-mode instability of supersonic shear flows: Part 1: Two-dimensional waves, *Theor. Comput. Fluid Dyn.* **2** 97–123
[8] Malmuth N, Fedorov A, Shalaev V, Cole J, Hites M, Williams D, and Khokhlov A 1998 *Problems in high speed flow prediction relevant to control Proc. Int. Conf. 2nd AIAA, Theoretical Fluid Mechanics Meeting (Albuquerque)* (AIAA Paper 2695)

[9] Fedorov A 2011 Transition and stability of high-speed boundary layers *Annu. Rev. Fluid Mech.* **43** 79–95
[10] Mack L M 1975 *A numerical method for the prediction of high-speed boundary-layer transition using linear theory Proc. Int. Conf. Analyses Requiring Advanced Computers (Hampton)* Part 1
[11] Demetriades A 1974 *Hypersonic viscous flow over a slender cone. III-Laminar instability and transition Proc. Int. Conf. 7th Fluid and Plasma Dynamics Conference (Palo Alto)* (AIAA Paper 535)
[12] Egorov I V, Fedorov A V, and Soudakov V G 2006 Direct numerical simulation of disturbances generated by periodic suction-blowing in a hypersonic boundary layer *Theor. Comput. Fluid Dyn.* **20** 41–54
[13] Knisely C P and Zhong X 2019 Sound radiation by supersonic unstable modes in hypersonic blunt cone boundary layers. II. Direct numerical simulation *Phys. Fluids* **31**
[14] Fedorov A V and Khokhlov A P 2002 Receptivity of hypersonic boundary layer to wall disturbances *Theor. Comput. Fluid Dyn.* **15** 231–254
[15] Fedorov A and Tumin A 2011 High-speed boundary-layer instability: old terminology and a new framework *AIAA J.* **49** 1647–1657
[16] Fedorov A and Tumin A 2003 Initial-value problem for hypersonic boundary-layer flows *AIAA J.* **41** 379–389
[17] Ma Y and Zhong X 2003 Receptivity of a supersonic boundary layer over a flat plate. Part 1. Wave structures and interactions *J. Fluid Mech.* **488** 31–78
[18] Ma Y and Zhong X 2003 Receptivity of a supersonic boundary layer over a flat plate. Part 2. Receptivity to free-stream sound *J. Fluid Mech.* **488** 79–121
[19] Kuehl J J 2018 Thermoacoustic interpretation of second-mode instability *AIAA J.* **56** 3585–3592
[20] Tian X and Wen C 2021 Growth mechanisms of second-mode instability in hypersonic boundary layers *J. Fluid Mech.* **908** R4
[21] Chen Y, Guo P and Wen C 2023 A unified explanation of energy growth sources for unstable modes in flat-plate boundary layers *J. Fluid Mech.* **972** A5
[22] Unnikrishnan S and Gaitonde D V 2019 Interactions between vortical, acoustic and thermal components during hypersonic transition *J. Fluid Mech.* **868** 611–647
[23] Chu B-T 1965 On the energy transfer to small disturbances in fluid flow (Part I) *Acta Mech.* **1** 215–234
[24] Doak P E 1989 Momentum potential theory of energy flux carried by momentum fluctuations *J. Sound Vib.* **131** 67–90
[25] Jahanbakhshi R and Zaki T A 2021 Optimal heat flux for delaying transition to turbulence in a high-speed boundary layer *J. Fluid Mech.* **916** A46
[26] Long T, Dong Y, Zhao R and Wen C 2021 Mechanism of stabilization of porous coatings on unstable supersonic mode in hypersonic boundary layers *Phys. Fluids* **33**
[27] Atzori M, Vinuesa R, Lozano-Durán A and Schlatter P 2018 Characterization of turbulent coherent structures in square duct flow *J. Phys.: Conf. Ser.* **1001** 012008
[28] Tumuklu O, Theofilis V, and Levin D A 2018 On the unsteadiness of shock–laminar boundary layer interactions of hypersonic flows over a double cone *Phys. Fluids* **30**
[29] Chuvakhov P V and Fedorov A V 2016 Spontaneous radiation of sound by instability of a highly cooled hypersonic boundary layer *J. Fluid Mech.* **805** 188–206
[30] Lozano-Durán A and Jiménez J 2014 Time-resolved evolution of coherent structures in turbulent channels: characterization of eddies and cascades *J. Fluid Mech.* **759** 432–471
[31] Refloch A, Courbet B, Murrone A, Villedieu P, Laurent C, Gilbank P, Troyes J, Tessé L, Chaineray G, Dargaud J B, and others 2011 CEDRE software *Aerospace Lab* **2** 1–10
[32] Cassagne A, Boussuge J-F, Villedieu N, Puigt G, d'Ast I and Genot A 2015 *JAGUAR: a new CFD code dedicated to massively parallel high-order LES computations on complex geometry Proc. Int. Conf. 50th 3AF International Conference on Applied Aerodynamics (Toulouse)*
[33] Fièvet R, Deniau H, Piot E 2020 Strong compact formalism for characteristic boundary conditions with discontinuous spectral methods *J. Comput. Phys.* **408** 109276
[34] Veilleux A, Puigt G, Deniau H, Daviller G 2022 A stable Spectral Difference approach for computations with triangular and hybrid grids up to the 6th order of accuracy *J. Comput. Phys.* **449** 110774
[35] Veilleux A, Puigt G, Deniau H, Daviller G 2022 Stable Spectral Difference Approach Using Raviart-Thomas Elements for 3D Computations on Tetrahedral Grids *J. Sci. Comput.* **91** 7
[36] Brazier J-P and Perraud J 2022 *Laminar-Turbulent Transition Modelling in High-Speed Flows Proc. Int. Conf. 2nd International Conference on High-Speed Vehicle Science Technology (Bruges)*
[37] Poinsot T J and Lele S K 1992 Boundary conditions for direct simulations of compressible viscous flows *J. Comput. Phys.* **101** 104–129
[38] Daviller G, Comte P, and Jordan P 2009 Flow decomposition for the study of source mechanisms (AIAA Paper 3305)
[39] Jordan P, Daviller G, and Comte P 2013 Doak's momentum potential theory of energy flux used to study a solenoidal wavepacket *J. Sound Vib.* **332** 3924–3936

[40] Moisy F and Jiménez J 2004 Geometry and clustering of intense structures in isotropic turbulence *J. Fluid Mech.* **513** 111–133

[41] Del Alamo JC, Jiménez J, Zandonade P and Moser RD 2006 Self-similar vortex clusters in the turbulent logarithmic region *J. Fluid Mech.* **561** 329–358

[42] Hao Z, Yan C, Zhou L and Qin Y 2017 Development of a boundary layer parameters identification method for transition prediction with complex grids *J. Aerosp. Eng.* **231** 2068–2084

[43] Unnikrishnan S and Gaitonde DV 2021 Instabilities and transition in cooled wall hypersonic boundary layers *J. Fluid Mech.* **915** A26

[44] Zhang S, Liu J, and Luo J 2016 Effect of wall-cooling on Mack-mode instability in high speed flat-plate boundary layers *Appl. Math. Mech.*, **37**, 1219–1230

[45] Fedorov AV and Khokhlov AP 2001 Prehistory of instability in a hypersonic boundary layer *Theor. Comput. Fluid Dyn.* **14** 359–375

[46] Ma Y and Zhong X 2005 Receptivity of a supersonic boundary layer over a flat plate. Part 3. Effects of different types of free-stream disturbances *J. Fluid Mech.* **532** 63–109

[47] Mack L M 1975 Linear stability theory and the problem of supersonic boundary-layer transition *AIAA J.* **13** 278–289

[48] Stetson K, Thompson E, Donaldson J and Siler L 1989 *Laminar boundary layer stability experiments on a cone at Mach 8. V-Tests with a cooled model, Proc. Int. Conf. 20th Fluid Dynamics, Plasma Dynamics and Lasers Conference (Buffalo)* (AIAA Paper 6.1989-1895)

[49] Osawa K and Jiménez J 2018 Intense structures of different momentum fluxes in turbulent channels *Phys. Rev. Fluids* **3** 084603

[50] Wang X, Zhong X and Ma Y 2011 Response of a hypersonic boundary layer to wall blowing-suction *AIAA J.* **49** 1336–1353

[51] Malik M R 1990 Numerical methods for hypersonic boundary layer stability *J. Comput. Phys.* **86** 376–413

Periodic bursting cycles on the edge to square duct turbulence

Markus Scherer[1,*], Markus Uhlmann[1,†], and Genta Kawahara[2,‡]

[1]Institute for Hydromechanics, Karlsruhe Institute of Technology, 76131 Karlsruhe, Germany
[2]Graduate School of Engineering Science, Osaka University, 1-3 Machikaneyama, Toyonaka, Osaka 560-8531, Japan

E-mail: *markus.scherer@kit.edu, †markus.uhlmann@kit.edu, ‡genta.kawahara.es@osaka-u.ac.jp

Abstract. The long-time behaviour of trajectories embedded in the edge manifold between the laminar and turbulent basins of attraction is investigated in order to identify simple invariant solutions to the Navier-Stokes equations in straight ducts with square cross-section. With the aid of the iterative bisection procedure of Toh and Itano (*J. Fluid Mech.*, vol. 481, 2003, pp. 67-76), we detect three travelling waves that represent attracting 'edge states' within the separatrix and saddles w.r.t. the full state space. All three travelling waves are presumably connected to solutions known in the community. In other situations, the edge states to which trajectories are attracted seem to be invariant periodic cycles that feature a rich dynamics: Along each cycle, the flow oscillates periodically between two flow states, in each of which streamwise vorticity is predominantly residing near one set of parallel walls, while the vortical activity associated to the other pair of opposing walls is markedly reduced. This 'switching' dynamics is accompanied by intermittent bursting-like high-dissipation excursions. While we cannot rigorously prove that the identified periodic cycle is a periodic orbit rather than a finite-precision approximation to a heteroclinic cycle, sound arguments for the former supposition are given. Interestingly, the different phases of the periodic cycle qualitatively resemble a similar 'switching' behaviour observed in (chaotic) marginal turbulence, which suggests that the found periodic cycles could help understanding the characteristic dynamics of duct turbulence at low Reynolds numbers.

1. Introduction

A profound understanding of turbulent fluid motion in straight pipes is of fundamental importance for the design of many industrial processes. Nowadays, much is known about the characteristics of turbulence in pipes with circular cross-section in the transitional and fully-developed regime [1, 2]. Less well understood is, however, the more complicated situation of a duct with square cross-section: In contrast to 'canonical' plane channel and circular pipe flows, the presence of four impermeable bounding walls break the lateral homogeneity of the system and induce a mean secondary flow of Prandtl's second kind [3]. Even though of comparably low amplitude (about $\mathcal{O}(1\%)$ of the bulk velocity u_b), this turbulence-driven secondary motion is well-known to markedly alter the momentum transport and heat transfer normal to the mean flow direction [4]. Previous experiments [5, 6] and direct numerical simulations (DNS) [7–9] therefore investigated the characteristics of the mean secondary flow field, the spatial distribution of the Reynolds stress tensor components, the role of the individual terms in the vorticity transport equation as well as the interplay between secondary currents and thermal convection

[10, 11]. Uhlmann et al. [12] and Pinelli et al. [13], on the other hand, analysed the origin of the mean secondary flow phenomena at the level of individual coherent structures including streamwise velocity streaks and quasi-streamwise vortices, which are elementary ingredients of the self-sustaining regeneration cycle of near-wall turbulence [14–16]. In particular, it was shown that the spatial constraints in form of the four bounding no-slip walls restrict the mobility of individual quasi-streamwise vortices [17] and thus cause the appearance of the characteristic cross-sectional pattern of the mean streamwise vorticity. From a dynamical systems perspective on turbulence [18], the coherent structures embedded in the chaotic turbulent attractor represent low-dimensional, spatiotemporally organised patterns that carry a relevant portion of energy and/or contribute to its dissipation [19]. Within this framework, chaotic turbulence dynamics might be interpreted as a 'pinballing' between individual unstable low-dimensional invariant sets including equilibria (EQ) and periodic orbits (PO) within a high-dimensional state space [20, 21] – a picture that has proven extraordinarily useful in furthering our understanding of sustained turbulence [22, 23] and the transition to turbulence [2, 24]. In square duct flows, several families of travelling waves (TW), which are equilibrium solutions to the Navier-Stokes equations when observed in an appropriate co-moving frame of reference, have been found in the past: The solutions detected by Wedin et al. [25] and Okino et al. [26] can be classified as 'four-vortex states' in that two dominant pairs of vortices residing on two opposite walls characterise the streamwise-averaged secondary flow field. Okino and Nagata [27] later extended their database by a 'two-vortex state' TW which features just a single pair of counter-rotating mean secondary flow vortices at one of the four solid walls. The family of TWs presented by Uhlmann et al. [28], in turn, reveals an 'eight-vortex state' mean secondary flow pattern that possesses a striking similarity to the characteristic long-time averaged secondary flow field of chaotic turbulent trajectories. The 'four-vortex state' of Okino et al. [26], on the other hand, resembles time averages over intermediate intervals of $\mathcal{O}(100)$ bulk time units length at marginal bulk Reynolds numbers $Re_b \equiv u_b H/\nu \approx 1100$, where u_b, ν and H denote the bulk velocity, kinematic viscosity and half the side-length of the square cross-section, respectively (cf. § 2). Uhlmann et al. [12] showed that under such marginal conditions, the cross-section scaled in wall units is too small to accommodate a full turbulent regeneration cycle [14] on each of the four walls, so that the turbulent flow activity is constrained to only two walls facing each other. Note that analogous observations are reported in minimal closed channels, where turbulence activity is often constrainted to one of the two walls [29]. Which pair of parallel walls is currently 'active', though, changes intermittently, so that in the long-time average, the characteristic 'eight-vortex pattern' is recovered even for marginal conditions. While the TW solution of Okino et al. [26] might indeed represent a flow state visited by turbulent trajectories at marginal conditions, the understanding of the dynamics of this 'switching' behaviour remains incomplete. Of particular interest is therefore the identification of time-dependent invariant sets such as – in the simplest case – individual (relative) periodic orbits or groups of saddle TWs that are interconnected by heteroclinic orbits, often termed *heteroclinic cycles* [cf. 30, p. 46ff and 31, p. 631ff].

In the current study, we have isolated several time-recurrent trajectories within the 'edge manifold' (that is, the manifold separating the laminar and turbulent basins of attraction) using the iterative bisection procedure ('edge-tracking') of Toh and Itano [32] (in the following referred to as TI03). In the context of rectangular duct flows, only two previous studies are known to the authors that have performed edge-tracking in their analysis: Biau and Bottaro [33] used it to underline the similarity between the secondary flow patterns of states embedded in the edge manifold and their solutions obtained by an optimisation procedure, without making further efforts to identify invariant solutions along the edge. Okino [34, in Japanese], on the other hand, performed an edge-tracking study for square duct flow in domains of different streamwise period $L_x/H \in \{2\pi, 4\pi, 8\pi\}$. For the three streamwise domain lengths, the edge state was found to be a 'domain-filling' TW solution, a chaotic relative attractor and a spatially localised puff-like TW

Journal of Physics: Conference Series **2753** (2024) 012013 doi:10.1088/1742-6596/2753/1/012013

solution similar to that found by Avila et al. [35] in a circular pipe, respectively. However, to the best of the authors' knowledge, no time-dependent invariant solution in rectangular duct flow was known prior to the current study, so that we consider the current results to be of significant interest for the understanding of the turbulent dynamics at marginally turbulent conditions.

The manuscript is organised as follows: In § 2, the physical system under consideration as well as the admissible symmetries of the governing equations are described. The different numerical methods that are required to perform an edge-tracking analysis and to later converge simple invariant sets are presented in § 3. The simulation results obtained with the aid of these techniques are discussed in § 4. The manuscript closes with a summary of our findings and a discussion of the 'nature' of the isolated time-recurrent invariant sets in § 5. Specifically, sound arguments will be presented which imply that the trajectories shadow a (relative) periodic orbit rather than a heteroclinic cycle.

2. Theoretical framework

2.1. Flow configuration

In the remainder of this work, we analyse the flow of an incompressible Newtonian fluid in a streamwise periodic duct of square cross-section with side lengths $2H$ and no-slip conditions imposed along the four impermeable walls. The origin of a Cartesian coordinate system is placed in the centre of the cross-section, with its basis vectors pointing in the streamwise (x) and the two cross-stream, wall-normal directions (y, z). The velocity field is denoted by $\boldsymbol{u}(\boldsymbol{x}, t) = (u, v, w)^T$, with u, v and w indicating fluid motion along the x-, y- and z-direction, respectively. The corresponding vorticity field is introduced as $\boldsymbol{\omega}(\boldsymbol{x}, t) \equiv \nabla \times \boldsymbol{u} \equiv (\omega_x, \omega_y, \omega_z)^T$. The flow is driven by a time-dependent streamwise pressure gradient that is adjusted at each time step to ensure a constant mass flow rate Q and thus bulk velocity $u_b \equiv Q/(4H^2)$. Wall shear stress $\tau_w(t)$ and friction velocity $u_\tau(t) = \sqrt{\tau_w(t)/\rho}$ are understood as averages over the duct perimeter. Averaging operators of the form $\langle \bullet \rangle_j$ ($j \in \{x, y, z, t\}$) indicate averaging along one or more spatial directions or time. All simulations performed in the course of this study feature streamwise-periodic domains with fundamental period $L_x = 2\pi H$ (with $\alpha \equiv 2\pi/L_x = 1/H$ denoting the fundamental wavenumber), which is for all considered Reynolds numbers markedly larger than the minimal streamwise domain length $(L_x^+)_{min} \approx 200$ required to sustain turbulence in a square duct [12]. Here, the notation $\bullet^+$ indicates scaling in wall units, that is, non-dimensionalisation using $\langle u_\tau \rangle_t$ and ν. Throughout, time is measured in bulk time units $T_b = H/u_b$.

2.2. Symmetric subspaces

Subject to the given boundary conditions, the Navier-Stokes system allows for a number of continuous and discrete symmetries. The streamwise periodicity of the domain leads to a continuous translational symmetry in x, viz.

$$l_x(s_x): \quad [u, v, w]\,(x, y, z) \quad \rightarrow \quad [u, v, w]\,(x + s_x, y, z) \qquad \forall\, s_x \in \left[-\frac{\pi}{\alpha}, \frac{\pi}{\alpha}\right), \tag{1}$$

that allows for the existence of travelling waves (relative equilibria) and relative periodic orbits, where 'relative' refers to invariance within a co-moving frame of reference. In contrast to pipes with circular cross-section, the boundary conditions in the square duct case break the continuous azimuthal rotational symmetry and reduce it to the discrete dihedral group D_4. For the sake of generality, we restrict ourselves in the following sections to the elements of the dihedral symmetry group $D_2 = \{e, \mathbf{Z}_y, \mathbf{Z}_z, \mathbf{R}_\pi\}$ which is also fulfilled in ducts with rectangular cross-section (with e denoting the identity). The two reflectional symmetries $\mathbf{Z}_y, \mathbf{Z}_z$ and the π-rotational symmetry

$\mathbf{R}_\pi$ contained in D_2 are defined as

$$\left.\begin{array}{rlcl} \mathbf{Z}_y: & [u,v,w](x,y,z) & \rightarrow & [u,-v,w](x,-y,z) \\ \mathbf{Z}_z: & [u,v,w](x,y,z) & \rightarrow & [u,v,-w](x,y,-z) \\ \mathbf{R}_\pi: & [u,v,w](x,y,z) & \rightarrow & [u,-v,-w](x,-y,-z) \end{array}\right\}. \tag{2}$$

The governing equations are thus equivariant under the symmetry group $\Gamma = SO(2)_x \times D_2$. In the current work, the following subgroups of Γ for specific streamwise shifts are of specific interest and therefore merit a separate definition:

$$\left.\begin{array}{rlcl} \mathbf{S}_y: & [u,v,w](x,y,z) & \rightarrow & [u,-v,w]\left(x+\frac{\pi}{\alpha},-y,z\right) \\ \mathbf{S}_z: & [u,v,w](x,y,z) & \rightarrow & [u,v,-w]\left(x+\frac{\pi}{\alpha},y,-z\right) \\ \mathbf{\Omega}_\pi: & [u,v,w](x,y,z) & \rightarrow & [u,-v,-w]\left(x+\frac{\pi}{\alpha},-y,-z\right) \end{array}\right\}. \tag{3}$$

3. Numerical method

3.1. Fluid solver

The Navier-Stokes equations are integrated using a pseudo-spectral method in primitive variable formulation for the spatial derivatives together with a fractional step method and a mixed explicit-implicit time marching scheme [cf. 13 for a detailed overview]. The velocity and pressure fields are expanded in terms of truncated Fourier series in the periodic streamwise direction and Chebyshev polynomials in the two cross-stream directions. Gauss-Chebyshev-Lobatto points are chosen as collocation points in the cross-stream directions, whereas a grid of equispaced nodes is used in the streamwise direction. In the current simulations, fields are expanded by $N_x = 64$ Fourier modes in the streamwise direction and between $N_y = N_z = 65$ and $N_y = N_z = 129$ Chebyshev polynomials in the cross-stream directions, ensuring that the grid spacing fulfills $\Delta x^+ \leq 14$ and $\max(\Delta y^+), \max(\Delta z^+) \leq 3.5$ in all considered cases. In the streamwise homogeneous direction, back and forth transformation between physical and Fourier space is performed based on fast Fourier transforms (FFT), taking into account de-aliasing according to the 2/3-rule. Poisson and Helmholtz problems arising from the numerical discretisation are solved in Fourier space for each streamwise wavenumber separately with the aid of a fast-diagonalisation technique [36], whereas nonlinear contributions are solved in physical space. The time integration scheme combines a semi-implicit treatment of the viscous terms with an explicit three-step low-storage Runge-Kutta method for the remaining ones [37, 38]. Throughout, the time step is chosen small enough to keep the CFL number below 0.3.

3.2. Newton-GMRES hookstep algorithm

To converge nearly recurrent episodes within a DNS trajectory to a truly invariant set such as a TW or a PO, a Newton-Raphson shooting method is used within the high-dimensional state space spanned by the coefficients of the Fourier-Chebyshev-Chebyshev expansion. The linear system arising in this context is approximately solved by a variant of the generalised minimal residual (GMRES) algorithm of Saad and Schultz [39] [cf. 40 for a detailed overview]. Since a straight Newton-GMRES procedure is typically not sufficient for convergence [41], the size of the determined Newton step is afterwards restricted to the region in which the linearisation is a sufficiently good approximation to the functional using the 'hookstep' procedure outlined by Dennis and Schnabel [42]. In the past, this combined Newton-GMRES-hookstep approach has been proven suitable to converge solutions even in high-dimensional state spaces [41, 43]. The general structure of the current implementation follows to a large extent that in the

`Openpipeflow` code of Willis [44], but represents an independently developed code whose core is the DNS solver described in section 3.1. To validate the implementation, we have successfully re-converged one of the upper-branch TW solutions found in Uhlmann et al. [28].

3.3. Edge-tracking

The iterative 'edge-tracking' algorithm of TI03 is a repeated bisection procedure that identifies numerical trajectories within the edge manifold $\mathscr{M}$ that separates the laminar and turbulent basins of attraction [45]. For the ith time interval, a parameter $\beta_i \in \mathbb{R}$ ($\forall i \in \mathbb{N}$) is sought that rescales the perturbation of a given flow state $\boldsymbol{u}_{i-1}(\boldsymbol{x}, t_{i-1})$ w.r.t. the laminar base flow $\boldsymbol{u}_{lam}(y,z) = (u_{lam}, 0, 0)^T$ at the beginning of each iteration interval t_{i-1} as

$$\begin{cases} \boldsymbol{u}_{\beta_i}(\boldsymbol{x}, t_{i-1}) \equiv \boldsymbol{u}_{lam}(y,z) + \beta_i \left(\boldsymbol{u}_{i-1}(\boldsymbol{x}, t_{i-1}) - \boldsymbol{u}_{lam}(y,z)\right) & \text{at } t = t_{i-1} \\ \boldsymbol{u}_{\beta_i}(\boldsymbol{x}, t) \equiv \boldsymbol{u}_{\beta_i}(\boldsymbol{x}, t_{i-1}) + \int\limits_{t'=t_{i-1}}^{t} \partial_t \boldsymbol{u} \, \mathrm{d}t' & \forall t \in (t_{i-1}, t_i) \end{cases} \quad (4)$$

to ensure that the initial condition is located on $\mathscr{M}$. Here, the laminar base flow $\boldsymbol{u}_{lam}(y,z)$ is determined from the Poisson equation $(\partial_{,yy} + \partial_{,zz}) u_{lam} = -\Pi/\nu$, where Π is the constant driving pressure gradient required to maintain a laminar flow with identical mass flow rate Q.

In practice, in each bisection step, a flow trajectory is integrated forward in time according to equation (4) using the DNS code outlined in section 3.1 until it either relaminarises or leaves $\mathscr{M}$ towards the turbulent basin of attraction. The interval in which β_i is expected is then halved with the lower and upper limits representing values that cause the flow to relaminarise and to become fully turbulent, respectively. As an indicator for when a trajectory does no longer reside on the edge manifold $\mathscr{M}$, we monitor the root mean square (r.m.s.) values of the kinetic energy residing in the streamwise-dependent Fourier modes, viz.

$$E_{rms} \equiv \sum_j [E_{3D}]_j^{1/2} \quad \text{with} \quad [E_{3D}]_j = \sum_{i=1}^{N_x} \iint\limits_{y,z=-H}^{H} [\hat{\boldsymbol{u}}_i]_j [\hat{\boldsymbol{u}}_i^*]_j \, \mathrm{d}z \mathrm{d}y \quad \forall j \in \{x, y, z\}, \quad (5)$$

where $\hat{\boldsymbol{u}}_i$ and $\hat{\boldsymbol{u}}_i^*$ denote the ith Fourier coefficient and its complex conjugate, respectively. The notation $[\bullet]_j$ indicates the jth component of a scalar or vector (field). The trajectory is considered as deviated to either of the two basins of attraction when the energy E_{rms}/u_b either drops below $E_{rms}^l/u_b \approx 2 \cdot 10^{-3}$ or, after a rapid energy growth, exceeds $E_{rms}^t/u_b \approx 2 \cdot 10^{-1}$.

Even though each bisection is continued until β_i has been determined up to 14 digits, the exponential growth of the remaining deviation makes an iterative procedure including a re-computation of β_i after $\mathcal{O}(100T_b)$ typically unavoidable [32, 45]. In the current work, the initial condition for the first iteration $i = 1$ is a flow state $\boldsymbol{u}_0(\boldsymbol{x}, t_0)$ randomly chosen from a fully-turbulent trajectory. The first parameter value β_1 is then clearly smaller than unity. After β_i has been determined up to the desired precision, the next iteration $i+1$ is initiated with a flow field taken from the previous edge trajectory $\boldsymbol{u}_i(\boldsymbol{x}, t_i) \equiv \boldsymbol{u}_{\beta_i}(\boldsymbol{x}, t_{i-1} + \delta t)$, where the interval for re-initialisation is with $\delta t \approx 340T_b$ of comparable size as in TI03. Due to the iterative procedure with intermittent restarts, the obtained curve is only piecewise continuous, with discontinuities occurring at times t_i and is thus no solution of the Navier-Stokes equations in a strict sense. For the sake of consistency, we follow TI03 and use the expression 'chain solution' for the resulting object. Even though the 'chain solution' itself is not a continuous solution trajectory of the governing equations, we can assume that a continuous trajectory that lies in $\mathscr{M}$ exists for all times and that for the current choice of δt, the 'chain solution' represents a valid approximation to the latter. The fact that the deviation between two consecutive parameter values at the time interval boundaries is seen to be small (β_i is close to unity $\forall i \geq 2$) gives us some confidence that this is indeed the case.

Table 1. List of travelling wave solutions that are identified as edge states with their respective phase velocity c_x. Note that the TW solutions obey more symmetries than the symmetric subspaces in which the edge trajectories are sought - these 'additional' symmetries are enclosed in round brackets. All solutions have been converged using the Newton-GMRES solver outlined in section 3.2 with $[N_x \times N_y \times N_z] = [64 \times 65 \times 65]$ Fourier-Chebyshev-Chebyshev modes.

Case	Re_b	c_x/u_b	Symmetries	(Suspected) corresponding TW
TW_1	1400	1.61	$\mathbf{R}_\pi$,$\mathbf{S}_y$,$\mathbf{S}_z$,($\mathbf{D}_{yz}$),($\mathbf{D}_{zy}$)	'eight-vortex state', Uhlmann et al. [28]
TW_2	2000	1.74	$\mathbf{Z}_y$,($\mathbf{\Omega}_\pi$),($\mathbf{S}_z$)	'four-vortex state', Wedin et al. [25]
TW_3	3200	1.62	$\mathbf{R}_\pi$,($\mathbf{S}_y$),($\mathbf{S}_z$)	'four-vortex state', Okino [34]

4. Results

In the current study, we have analysed the behaviour of trajectories along the edge manifold in different symmetric subspaces. For some of those systems, we repeated the edge tracking procedure for varying Reynolds numbers to scrutinise the impact of the latter on the type of the identified edge states. In the following two sections, we will discuss some selected cases for which the edge trajectories reveal particularly interesting dynamics in that they are attracted by simple, non-chaotic invariant sets such as TWs or periodic cycles.

4.1. Travelling waves

First, we discuss edge trajectories in three different symmetric subspaces for different Reynolds numbers that are attracted by travelling waves. These latter represent saddles in the full state space, but take the role of purely attracting sets within $\mathscr{M}$. The three considered cases TW_1, TW_2 and TW_3 are listed in table 1, together with the respective Reynolds numbers and symmetries. Figure 1(*a-c*) shows the variation of the r.m.s. kinetic energy E_{rms} as the trajectories approach the relative attractors: After an initial apparently chaotic transient of varying length, all three trajectories enter the basin of attraction of the respective TW. Flow states extracted from the respective trajectories in these 'plateau regimes' represent excellent initial guesses for the Newton-GMRES solver and can be converged to truly invariant sets within few Newton iterations. Figure 1(*d-f*) visualises the spatial structure of the three travelling wave solutions in terms of their (exact) coherent structures: Isosurfaces of the streamwise velocity indicate the position of elongated streaks along the no-slip walls, while quasi-streamwise vortices are identified by regions of intense streamwise vorticity. Further informations including the relative phase speed c_x/u_b and the symmetries that the different 'edge states' fulfill are provided in table 1. In this context, it is noteworthy that all edge trajectories approach invariant sets that are located in even higher-symmetric subspaces than those in which the trajectories are searched for. So, TW_1 reveals in addition to the previously defined symmetry group additional reflectional symmetries w.r.t. the diagonals $y = z$ and $y = -z$, viz.

$$\left.\begin{array}{llll} \mathbf{D}_{yz}: & [u,v,w]\,(x,y,z) & \rightarrow & [u,w,v]\,(x,z,y) \\ \mathbf{D}_{zy}: & [u,v,w]\,(x,y,z) & \rightarrow & [u,-w,-v]\,(x,-z,-y) \end{array}\right\}. \quad (6)$$

While all three TWs are propagating in positive x-direction at comparable phase speeds $c_x/u_b \in [1.6, 1.75]$, the spatial organisation of their flow fields is markedly different: TW_1 shown in figure 1(*a,d*) features low-speed streaks and a staggered arrangement of counter-rotating quasi-streamwise vortices along each of the four walls. In TW_2 and TW_3 (cf. figure 1(*b,e*) and (*c,f*), respectively), on the other hand, marked elongated streaks accompanied by pairs of

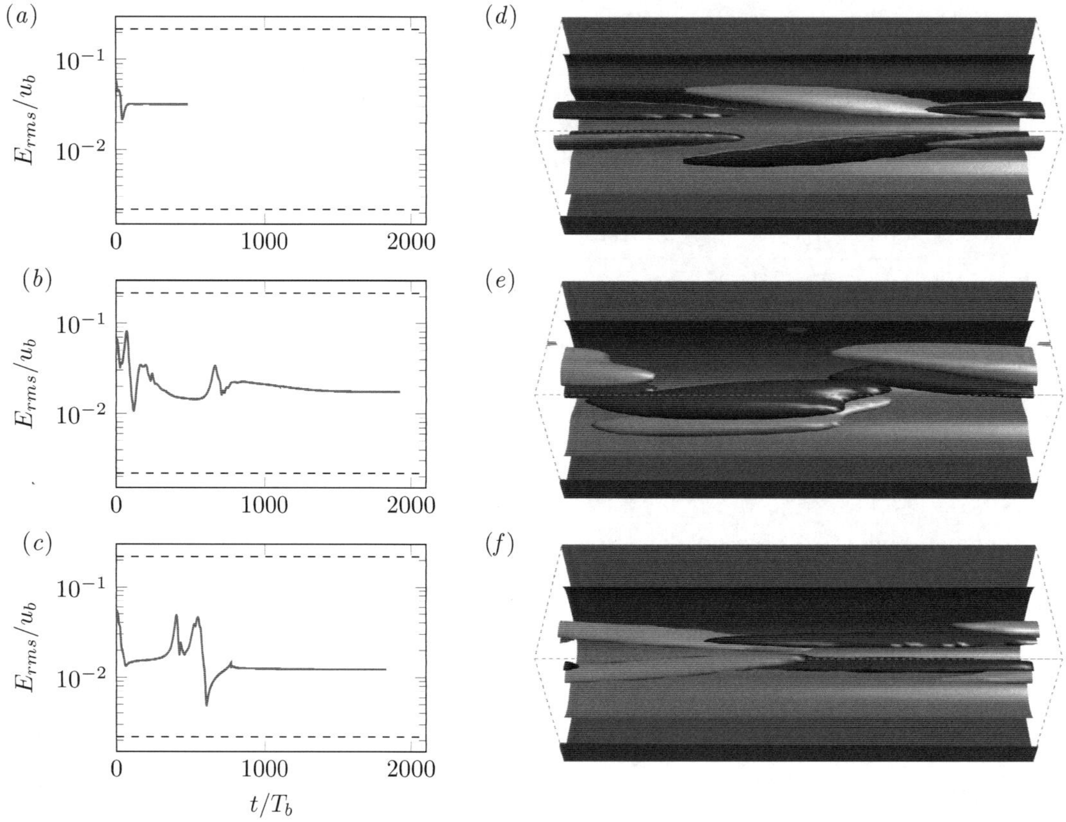

Figure 1. Edge trajectories that are attracted by a travelling wave solution. (*a-c*) Time evolution of the r.m.s. energy signal $E_{rms}(t)$ for varying Reynolds numbers and symmetry groups: (*a*) $Re_b = 1400$, $S = \{e, \mathbf{R}_\pi, \mathbf{S}_y, \mathbf{S}_z\}$, TW_1; (*b*) $Re_b = 2000$, $S = \{e, \mathbf{Z}_y\}$, TW_2; (*c*) $Re_b = 3200$, $S = \{e, \mathbf{R}_\pi\}$, TW_3. See table 1 for the 'additional' symmetries fulfilled by the TW solutions. Horizontal dashed lines indicate the limits above (below) which a trajectory is considered as turbulent (laminar). Shown is only the time up to the flow state that was used as initial condition for the Newton-GMRES iteration. (*d-f*) Three-dimensional visualisations of the converged TW 'edge states' in (*a-c*). Shown are isosurfaces of the streamwise velocity $u = 0.55\max(|u|)$ (green) and isocontours of the streamwise vorticity $\omega_x = \pm 0.55\max(|\omega_x|)$ (clockwise/counter-clockwise rotation indicated by orange/blue color), with mean flow from left to right. For the sake of visualisation, the two frontal walls are not shown and the isocontour of u is visualised for $y \leq -z$ only.

counter-rotating vortices are identified near two of the four walls only. A comparison of the converged TWs with those known in the community reveals that TW_1 belongs to the family of lower-branch 'eight-vortex solutions' of Uhlmann et al. [28] (compare the visualisation in their fig. 5 for identical conditions $Re_b = 1400$ and $\alpha H = 1$). The remaining two travelling wave solutions, TW_2 and TW_3, are assumed to be connected to the solution branches identified by Wedin et al. [25] on the one hand and Okino et al. [46] and Okino [34] on the other hand. In both cases, the mean secondary flow field is a 'four-vortex state' with two pairs of counter-rotating vortices associated with two walls facing each other. But since the current domain size and Reynolds numbers do not coincide with those in these studies, branch continuation in the parameter space spanned by Re_b and αH would be necessary to prove that the converged TWs

Table 2. List of attracting periodic cycles identified as edge states together with the asymmetric chaotic relative attractor. For the cycles, the estimated period T is determined as the value to which the sequences T^k_{min} and T^k_{max} converge for $t \to \infty$. The last column indicates which of the four triangular sectors Ω_i (cf. equation (9)) contribute (at least temporarily) a significant portion of streamwise vorticity to the overall budget.

Case	Re_b	T/T_b	Symmetries	'active' sections
PC2200$_y$	2200	≈ 592	$\mathbf{Z}_y$	$\Omega_W, \Omega_N/\Omega_S$
PC2400$_y$	2400	653.2	$\mathbf{Z}_y$	$\Omega_W, \Omega_N/\Omega_S$
PC3200$_{yz}$	3200	563.8	$\mathbf{Z}_y$,$\mathbf{Z}_z$	$\Omega_W/\Omega_E, \Omega_N/\Omega_S$
CA3200	3200	-	-	all

indeed belong to the same family of solutions. Such a continuation campaign is beyond the scope of the current study and is left for future investigations. Nonetheless, supposing that the solution families are indeed the same, the current results imply that most of the currently known TW solutions in square duct flow take the role of 'edge states' in appropriate symmetric subspaces [compare the discussion for circular pipe flow in 45 and 47]. Note that this is not necessarily to be expected *a priori* as all previously known solution families have been identified by homotopy and back-continuation from either externally forced or internally heated systems [22], without an *a priori* guarantee that these solutions are indeed visited by chaotic flow trajectories.

4.2. Quasi-periodic cycles

We now turn our attention to situations in which the analysed edge trajectories are observed to approach time-dependent invariant sets that seem to be limit cycles in $\mathscr{M}$. Among these cases that are listed in table 2, cases PC2200$_y$ and PC2400$_y$ feature only a single reflectional symmetry $\mathbf{Z}_y$ and differ solely by the Reynolds number, whereas case PC3200$_{yz}$ simultaneously obeys the two reflectional symmetries $\mathbf{Z}_y$ and $\mathbf{Z}_z$. As a reference, table 2 also contains case CA3200, which does not fulfill any symmetry other than the identity $S = \{e\}$ and reveals a chaotic behaviour throughout the observation interval. Before discussing the dynamics in more detail, let us remark that the systems with a single reflectional symmetry $\mathbf{Z}_y$ are seen to approach a TW solution at $Re_b = 2000$, while the edge states detected for higher Reynolds numbers $Re_b \gtrsim 2200$ are all periodic cycles. Clearly, conclusions based on single phase space trajectories have to be drawn with caution, as $\mathscr{M}$ might be of very complex shape and feature a variety of different edge states. Nonetheless, it can be speculated whether the observed (relative) attracting cycles are connected to TW$_2$ by some global bifurcation.

The quasi-periodic dynamics of the 'chain solution' in cases PC2400$_y$ and PC3200$_{yz}$ are clearly visible in the time evolution of the r.m.s. energy signal presented in figure 2(a,c), respectively. For both cases, relatively quiet periods alternate with intermittently occurring bursting events, during which the streamwise-dependent Fourier modes record, temporarily, a significant gain of energy. The length of the individual bursting sequences is visually much shorter than a full recurrence period. At first glance, the fundamental period of case PC2400$_y$ seems to be at least twice as long as the one in case PC3200$_{yz}$, but featuring two energetic peaks of different amplitude per cycle. As we shall see later, both cycles 'oscillate' between two different flow states, in which vorticity is either concentrated near the walls at $y/H \in \{-1, 1\}$ or those at $z/H \in \{-1, 1\}$. The difference is that for case PC3200$_{yz}$, these two states are found to be identical modulo a rotation by $\pi/2$ about the x-axis, so that both states are indistinguishable in terms of the energy signal. In fact, the global kinetic energy evolution is in the pressure-driven

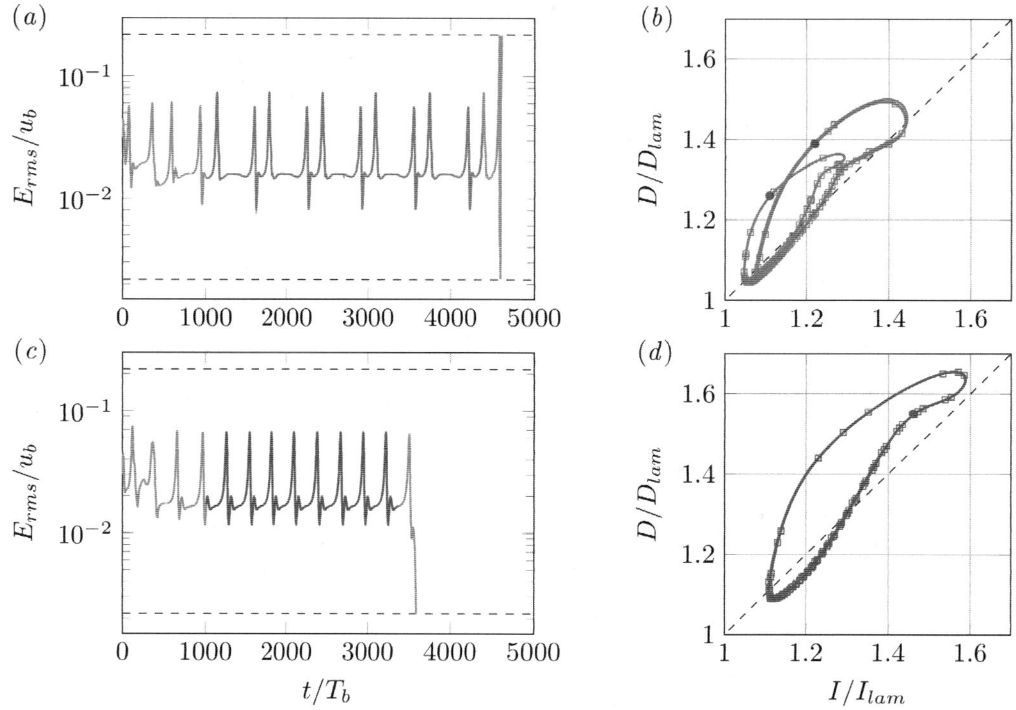

Figure 2. Time evolution of energy and dissipation along the edge trajectories (*a*,*b*) PC2400$_y$ and (*c*,*d*) PC3200$_{yz}$ that are attracted by a periodic cycle. (*a*,*c*) Time evolution of the r.m.s. energy signal $E_{rms}(t)$. The time interval, during which the dynamics along the 'chain solution' can be considered as essentially periodic, is highlighted. (*b*,*d*) Two-dimensional projection of the trajectory dynamics onto a plane spanned by the global dissipation D and energy input rate I. Both quantities are normalised with the respective values in a laminar flow at identical mass flow rate Q, i.e. $D_{lam} = I_{lam}$. Open square symbols are placed along the curves in time intervals of $\approx 25T_b$ length to give an impression of how 'fast' D and I change along the curve. Note that only the time interval highlighted in (*a*,*c*) is shown in the (D,I)-diagram, with blue dots marking when the snapshots shown in figure 4 have been extracted.

square duct flow fully described by a simple balance between energy input I and dissipation D, viz.

$$\frac{\mathrm{d}E}{\mathrm{d}t} = I - D, \tag{7}$$

where

$$E = \frac{1}{2|V|}\int\limits_{\Omega} ||\boldsymbol{u}||^2 \,\mathrm{d}\boldsymbol{x}, \quad I = -\frac{1}{|V|}\int\limits_{\Omega} \partial_x\left(\frac{pu}{\rho}\right) + \Pi_t u \,\mathrm{d}\boldsymbol{x}, \quad D = \frac{1}{|V|}\int\limits_{\Omega} \nu||\boldsymbol{\omega}||^2 \,\mathrm{d}\boldsymbol{x}$$

with $\Omega = [0, 2\pi/\alpha) \times [-H, H] \times [-H, H]$, $V = 8\pi H^2/\alpha$ and the time varying driving streamwise pressure gradient Π_t. In the classical energy input-dissipation diagrams in figure 2(*b*,*d*), the bursting phases in the energy signal manifest themselves as high-dissipation excursions from the more quiescent regions close to the laminar attractor at (D_{lam}, I_{lam}). In the two-dimensional projection onto the (D, I)-plane, the state space trajectories of the 'chain solution' collapse almost perfectly onto two closed cycles. These latter are oriented in such a way that wandering

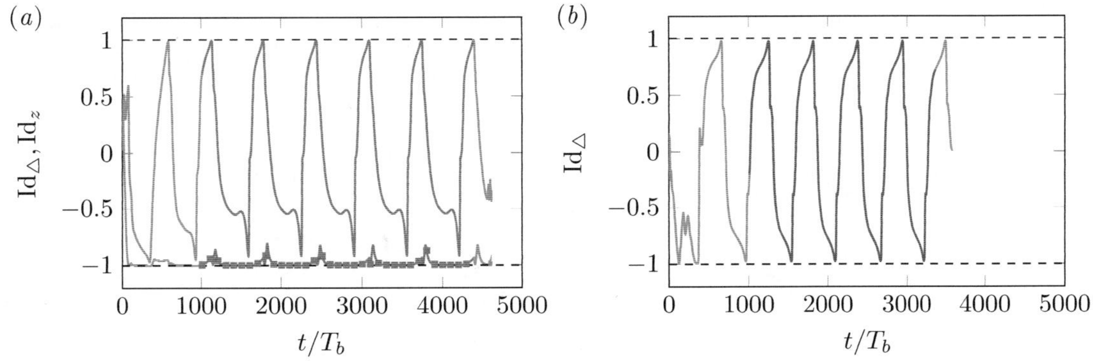

Figure 3. Time evolution of the indicator functions $\mathrm{Id}_\triangle$ (—) and Id_z (-■-) as defined in equations (8) and (10), respectively, for cases (*a*) PC2400$_y$ and (*b*) PC3200$_{yz}$.

along the 'chain solution' as $t \to \infty$ means orbiting along the curve in the (D, I)-plane in clockwise sense. In other words, each bursting phase starts with a rapid increase in dissipation, followed by an intensification of the wall friction and thus the external energy input I with a small time lag of $\mathcal{O}(10^0 T_b)$ length. This characteristic behaviour is consistent with the classical concept of the self-sustaining interaction process between streamwise velocity streaks and quasi-streamwise vortices in the buffer layer of turbulent flows [14, 15] and has been reported for periodic orbits [48] and quasi-periodic cycles [32] in various flow configurations.

In accordance with the earlier discussed qualitatively different evolution of E_{rms} for cases PC2400$_y$ and PC3200$_{yz}$, the (D, I)-projection of the latter case collapses onto a single bean-like loop, whereas PC2400$_y$ follows a more complicated, double loop orbit. Again, this qualitative difference originates in the fact that both trajectories oscillate between two flow states, but in case PC3200$_{yz}$ the two states are identical modulo a rotation by $\pi/2$ in the cross-section, such that the two bursting loops collapse onto a single one in the two-dimensional projection. Before discussing the alternating flow states in more detail, let us come back to the different time scales in the periodic 'chain-solution'. As indicated by the square symbols placed in time intervals of about $\approx 25T_b$ along the curves in figure 2(*b*,*d*), the 'speed' at which the flow orbits along the (D, I)-curve varies significantly in the plane: Both 'chain solutions' spend a considerable fraction of the period in the more quiescent regions at low values of D and I, with I slightly exceeding the values attained by D. On the other hand, they quickly pass the high-dissipation sequences during which $D > I$. Note that this is inasmuch expected as $\langle D\rangle_t = \langle I\rangle_t$ must hold for sufficiently long time intervals, where $\langle\bullet\rangle_t$ indicates long-time averages along a single flow trajectory.

In order to quantify the foreshadowed intermittent 'switching' between two different flow states, we determine a dimensionless indicator function [12]

$$\mathrm{Id}_\triangle \equiv \frac{(S_N + S_S) - (S_W + S_E)}{S_N + S_S + S_W + S_E}, \tag{8}$$

with

$$\left.\begin{aligned} S_i &= \int_{\Omega_i} \langle\omega_x\rangle_x^2 \,\mathrm{d}A \qquad \forall i \in \{W, S, E, N\}, \\ \Omega_S &= \{(y,z)|\; y < z \wedge y < -z\}, \quad \Omega_N = \{(y,z)|\; y > z \wedge y > -z\}, \\ \Omega_W &= \{(y,z)|\; y > z \wedge y < -z\}, \quad \Omega_E = \{(y,z)|\; y < z \wedge y > -z\}, \end{aligned}\right\}. \tag{9}$$

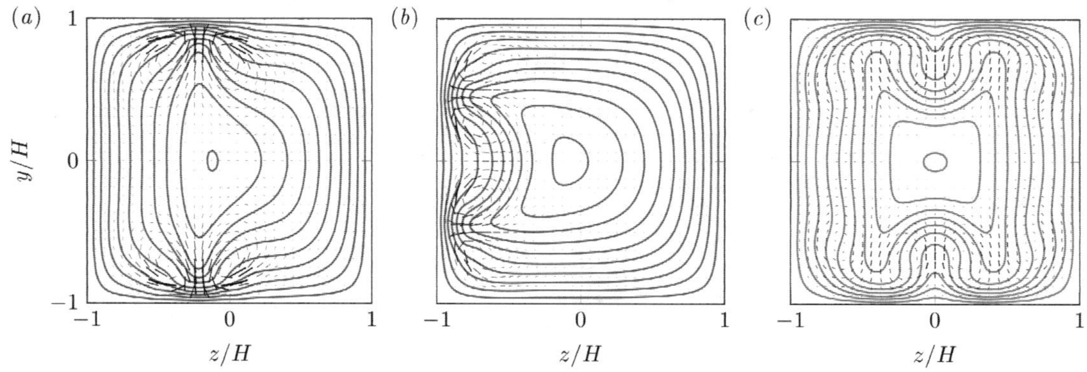

Figure 4. Instantaneous snapshots of the streamwise-averaged primary and secondary flow field $\langle u\rangle_x$ and $(\langle v\rangle_x, \langle w\rangle_x)^T$ for cases (a,b) PC2400$_y$ and (c) PC3200$_{yz}$. Solid lines represent isolevels of $\langle u\rangle_x/u_b$ between 0 and 2, with an increment of 0.2. The intensity and orientation of the secondary flow field $(\langle v\rangle_x/u_b, \langle w\rangle_x/u_b)^T$ is indicated by the vector plot. Note that the arrow scaling differs between (a,b) and (c). The presented snapshots have been extracted from the respective 'chain solutions' at (a) $t = 2440T_b$, $(D/D_{lam}, I/I_{lam}) = (1.39, 1.22)$; (b) $t = 2905T_b$, $(D/D_{lam}, I/I_{lam}) = (1.26, 1.11)$; (c) $t = 2409T_b$ and $t = 2409 \pm 281.9T_b$ (when rotated by $\pi/2$ about the x-axis), $(D/D_{lam}, I/I_{lam}) = (1.55, 1.46)$.

Thus, by definition, $\mathrm{Id}_\triangle$ attains a value close to plus (minus) unity when the enstrophy carried by the streamwise vorticity is predominantly focussed near the walls at $y \in \{-H, H\}$ (the walls at $z \in \{-H, H\}$), respectively. In addition, we introduce the analogously defined indicator functions

$$\mathrm{Id}_y \equiv \frac{S_N - S_S}{S_N + S_S} \quad \text{and} \quad \mathrm{Id}_z \equiv \frac{S_E - S_W}{S_E + S_W}, \tag{10}$$

where Id_y (Id_z) indicates, if close to unity, that vorticity is mainly concentrated near the wall at $y = H$ ($z = H$), or near that at $y = -H$ ($z = -H$) for values close to negative unity.

In figure 3(a,b), $\mathrm{Id}_\triangle$ is evaluated for cases PC2400$_y$ and PC3200$_{yz}$, while Id_y is trivially zero in both cases and Id_z at least in the latter case as a consequence of the enforced symmetries. The curves associated with $\mathrm{Id}_\triangle$ clearly indicate that the dynamics of both systems are characterised by periodic oscillations between two states in which streamwise enstrophy predominantly resides near one pair of parallel walls, while vortical activity is much weaker in the vicinity of the remaining two walls. In contrast to case PC3200$_{yz}$, however, Id_z is close to minus unity over the entire observation interval of case PC2400$_y$, indicating that the wall at $z = -H$ features a much stronger vortical activity than its counterpart at $z = H$. Figure 4 supports this observation by showing the streamwise-averaged flow field for selected snapshots along the 'chain solution' that are representative of either of the two flow states. The extracted snapshots are characterised by a comparably high dissipation (see the blue dots in figure 2(b,d)) and feature between one and two streak-vortex pairs per 'active' wall, with lateral extents $\delta z^+ = \mathcal{O}(10)$ comparable to those typically observed for the quasi-streamwise vortices of the buffer layer in canonical channel [29, 49] and square duct flows [12, 13].

The time signal $\mathrm{Id}_\triangle(t)$ is moreover a good candidate to estimate the fundamental period of the shadowed limit cycles. To this end, we measure the recurrence time of the 'chain solutions' based on the time interval between two consecutive maxima or minima of $\mathrm{Id}_\triangle(t)$ as

$$T^k_{max} = t^{k+1}_{max} - t^k_{max} \quad \text{and} \quad T^k_{min} = t^{k+1}_{min} - t^k_{min}, \tag{11}$$

respectively, where t^k_{max} (t^k_{min}) corresponds to the time at which $\mathrm{Id}_{\triangle}(t)$ attains its kth global maximum (minimum). Note that the thereby estimated value for the cycle period can at most be determined up to a precision of $5\Delta t \approx 0.07T_b$, which is half the interval in which statistics have been stored to disk alongside the simulation runs. The sequence of values attained by T^k_{max} and T^k_{min} for increasing k presented in figure 5 underlines that the estimated period converges within a few cycles to essentially constant values $T \approx 653.2T_b$ (PC2400$_y$) and $T \approx 563.8T_b$ (PC3200$_{yz}$), providing clear evidence that the two edge trajectories indeed approach a limit cycle (within $\mathscr{M}$) with finite period. To quantify the deviation of the 'chain solution' from a truly closed cycle, we evaluate in the following the kinetic energy of the residual flow field

$$\delta \boldsymbol{u}(\boldsymbol{x}, t, \boldsymbol{\delta}, \delta t) = \boldsymbol{u}(\boldsymbol{x}, t) - \boldsymbol{u}(\boldsymbol{x} - \boldsymbol{\delta}, t + \delta t), \tag{12}$$

where $\boldsymbol{\delta} = (s_x, 0, 0)^T$ denotes a spatial shift in the streamwise direction and $\delta t \in \mathbb{R}$ is a shift in time. Defining the kinetic energy contained in the jth velocity component as

$$E_j(\boldsymbol{u}) = \frac{1}{|V|} \int_{\Omega} [\boldsymbol{u}(\boldsymbol{x}, t)]_j^2 \, \mathrm{d}\boldsymbol{x} \quad \text{with} \quad j \in \{x, y, z\}, \tag{13}$$

we can eventually express the relative error in terms of the kinetic energy for the jth velocity component as [32]

$$\hat{E}_j(\boldsymbol{u}, t, \delta t) = \frac{\min_{s_x} E_j(\delta \boldsymbol{u}(\boldsymbol{x}, t, s_x, \delta t))}{E_j(\boldsymbol{u}(\boldsymbol{x}, t))} \quad \text{with} \quad j \in \{x, y, z\}. \tag{14}$$

Note that minimisation over all possible streamwise shifts is necessary, since the phase velocity is *a priori* unknown. In practice, the minimisation is performed over discrete shifts s_x at increasingly fine discretisation in the context of a standard bisection procedure. Using the values to which the series T_{max} and T_{min} converged as guesses for the period T, we have evaluated the relative error $\hat{E}_j(\boldsymbol{u}, t, T)$ for all three velocity components at different times t along the 'chain solution'. Despite the fact that the latter is just an approximation to an actual edge trajectory and notwithstanding the long period that makes an intermediate re-iteration of β_i unavoidable, $\hat{E}_j(\boldsymbol{u}, t, T)$ is for case PC2400$_y$ at most $\mathcal{O}(10^{-5})$ for any of the three velocity components. For case PC3200$_{yz}$, taking into account that the Navier-Stokes equations in square duct flow are equivariant under rotations by $m\pi/2$ ($m \in \mathbb{Z}$) about the x-axis, the deviation is found to be at most $\mathcal{O}(10^{-5})$ for both, the 'full' period $T = 563.8T_b$ and half its value.

5. Discussion and Conclusion

In the previous sections, we have studied the flow in a streamwise periodic duct with square cross-section from a dynamical systems point of view. When artificially forced to stay within the 'edge manifold' $\mathscr{M}$ that separates the laminar and turbulent basins of attraction under additional symmetric constraints, trajectories are found to be attracted by simple invariant sets in phase space, including TWs and periodic cycles. While all TW solutions that take the role of edge states in the considered subspaces are known [25, 28, 34], no comparable periodic limit cycles have been detected before in a square duct. In plane Poiseuille flow, however, a structurally similar phase space object was detected in the pioneering work of TI03 with the same numerical bisection technique as in the current work. TI03 discussed in detail that the observed relatively attracting cycle could be both, a periodic orbit or a heteroclinic cycle. Indeed, it has been shown that structurally stable heteroclinic cycles can exist in systems with symmetry group $S = O(2)$ such as plane Poiseuille flow [50] and there have been some attempts to interpret turbulent bursting events as heteroclinic cycles [51, 52]. However, limited computational resources hindered TI03

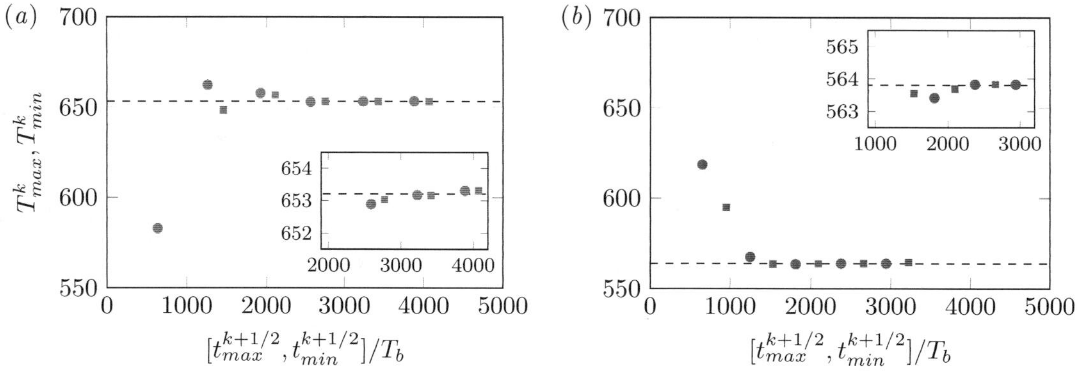

Figure 5. Time evolution of the estimated cycle period determined from the time intervals between two consecutive extrema of $\mathrm{Id}_\triangle$ using equation (11) for cases (*a*) PC2400$_y$ and (*b*) PC3200$_{yz}$. Here, we associate each value of the estimated period with the midpoint of the time interval based on which it has been determined, i.e. $t_\bullet^{k+1/2} = (t_\bullet^k + t_\bullet^{k+1})/2$. Squares and circles represent the kth cycle turn-over time estimated based on the maxima (T_{max}) and minima (T_{min}) of $\mathrm{Id}_\triangle$, respectively. Dashed lines indicate values (*a*) $653.2T_b$ and (*b*) $563.8T_b$, respectively. The insets show close-ups of the same data close to the estimated period length.

from tracking their 'chain solution' around the cyclic orbit for more than one and a half periods and thus from clarifying the type of invariant set that it is shadowing.

For the square duct, the authors are not aware of a comparable statement on the existence of structurally stable heteroclinic cycles under the changed symmetry conditions. Also, the 'chain solution' found by TI03 in the closed channel spends a considerable amount of time near two selected points in the (D, I)-plane that are almost in equilibrium ($I \approx D$), which motivated the speculation that their 'chain solution' is visiting the neighbourhood of two saddle-node travelling waves. In our case, the 'chain solutions' are not seen to stay close to any such typical point with $I \approx D$ for longer time intervals. Instead, we observe a permanent variation of both I and D even during the more quiet time episodes. An attempt to converge a travelling wave solution from nearly recurrent episodes along the 'chain solution' was not successful either. Third, the fact that the period estimated for the 'chain solution' converges to a remarkably robust value after a handful of loops suggests that the shadowed attracting invariant set is indeed a periodic orbit. An attracting heteroclinic cycle has, by definition, an infinite period, but any piecewise continuous, finite-precision approximation to it will possess an 'artificial' period. Though, one should expect that the dynamics of trajectories slow down in the vicinity of a heteroclinic cycle with every loop they spin around and approach the latter [cf. the heteroclinic cycle in $\mathbb{R}^3$ studied by 53] – a behaviour that we do not observe for any of our cases. On the other hand, our observations are in line with the work of Kreilos et al. [54], who successfully isolated a periodic orbit as edge state for an asymptotic suction boundary layer that reveals a structural similarity to the current periodic cycle. The authors provide evidence that the PO emerges from the annihilation of two pairs of fixed points in the context of a saddle-node infinite-period bifurcation, an observation which is of relevance for the ongoing analysis of the presented cases.

Although these arguments provide some evidence that the shadowed relatively attracting cycle is a periodic orbit, none of the above arguments is rigorous enough to conclusively clarify the nature of the attracting invariant set. To close this gap, in the next step one should try to converge the almost recurrent 'chain solutions' to truly periodic orbits with the aid of the Newton-GMRES algorithm. The authors are optimistic that this can be achieved in the near future for the doubly-symmetric case PC3200$_{yz}$, taking into account the excellent quality of the initial guess as well as the fact that the flow seems to repeat itself (modulo a rotation by $\pi/2$ about the x-axis) after half the estimated period $T/2 = 281.9T_b$. In case PC2400$_y$, the situation is more delicate in that the estimated period is too long to be useful as an initial guess for the Newton-GMRES solver, since there is at least one re-iteration of β_i per period necessary to keep the 'chain solution' in the neighbourhood of $\mathscr{M}$. Without further extensions of the numerical approach such as a multipoint shooting technique [e.g. 55] or using an adjointed-based variational Jacobian-free method instead of the Newton-GMRES solver [e.g. 56], there is no hope to converge the current chain solution PC2400$_y$ to a periodic orbit.

But even without the fully-converged time-recurrent invariant set at hand, the current results open up a new dynamical systems-based perspective on the well-known 'switching' behaviour in square duct flow [12]: Fully-chaotic turbulent trajectories are expected to shadow periodic sets like the ones investigated in the current study for a certain period of their lifetime. During those episodes, they follow a similar dynamics as the invariant sets by which they are attracted and 'switch' from one 'two-wall' state to the other. Note that all detected edge states are purely attracting only within the edge manifold, but are inherently unstable solutions w.r.t. the full phase space. All chaotic trajectories will therefore eventually leave their neighbourhood again and, after a while, will possibly approach and shadow another comparable 'switching' cycle and so on and so forth. In this context, it should be emphasised that the 'switching' between (at least) two alternating flow states is no exclusive feature of the considered symmetric subspaces, but a comparable, though chaotic oscillating dynamics is also visible for edge trajectories in the asymmetric case CA3200, which further underlines the relevance of the current findings for the general, chaotic turbulent case. Further efforts will be made to assess the usefulness of this qualitative phase-space picture of square duct dynamics.

Acknowledgements

This work was supported in part by the European Research Council under the Caust grant ERC-AdG-101018287. In this regard, we want to thank Prof. Javier Jiménez and Dr. Kosuke Osawa for their kind hospitality during the fifth Madrid turbulence summer workshop. We also thank Daniel Morón, Prof. Marc Avila and Prof. Lennaert van Veen for very insightful discussions and their valuable comments on an earlier draft of this manuscript. The simulations were performed on the supercomputer bwUniCluster funded by the Ministry of Science, Research and the Arts Baden-Württemberg and the Universities of the State of Baden-Württemberg. The computer resources, technical expertise and assistance provided by the staff are gratefully acknowledged.

References

[1] Smits A J, McKeon B J and Marusic I 2011 *Ann. Rev. Fluid Mech.* **43** 353–375
[2] Avila M, Barkley D and Hof B 2023 *Ann. Rev. Fluid Mech.* **55** 575–602
[3] Bradshaw P 1987 *Ann. Rev. Fluid Mech.* **19** 53–74
[4] Demuren A O and Rodi W 1984 *J. Fluid Mech.* **140** 189–222
[5] Brundrett E and Baines W D 1964 *J. Fluid Mech.* **19** 375–394
[6] Gessner F B 1973 *J. Fluid Mech.* **58** 1–25
[7] Gavrilakis S 1992 *J. Fluid Mech.* **244** 101–129
[8] Huser A and Biringen S 1993 *J. Fluid Mech.* **257** 65–95
[9] Modesti D, Pirozzoli S, Orlandi P and Grasso F 2018 *J. Fluid Mech.* **847** R1
[10] Sekimoto A, Kawahara G, Sekiyama K, Uhlmann M and Pinelli A 2011 *Phys. Fluids* **23** 075103
[11] Modesti D and Pirozzoli S 2022 *J. Fluid Mech.* **941** A16
[12] Uhlmann M, Pinelli A, Kawahara G and Sekimoto A 2007 *J. Fluid Mech.* **588** 153–162

[13] Pinelli A, Uhlmann M, Sekimoto A and Kawahara G 2010 *J. Fluid Mech.* **644** 107–122

[14] Hamilton J M, Kim J and Waleffe F 1995 *J. Fluid Mech.* **287** 317–348

[15] Waleffe F 1997 *Phys. Fluids* **9** 883–900

[16] Schoppa W and Hussain F 2002 *J. Fluid Mech.* **453** 57–108

[17] Kawahara G and Kamada E 2000 *Trans. Japan Soc. Mech. Eng. B* **66**(643) 89–97

[18] Hopf E 1948 *Comm. Pure Appl. Math.* **1** 303–322

[19] Jiménez J 1987 *CTR Annual Research Briefs* 323–324

[20] Waleffe F 2002 Exact coherent structures and their instabilities: toward a dynamical-system theory of shear turbulence *Proc. Int. Symp. on "Dynamics and Statistics of Coherent Structures in Turbulence: Roles of Elementary Vortices"* ed Kida S (National Center of Sciences, Tokyo, Japan) pp 115–128

[21] Gibson J F, Halcrow J and Cvitanović P 2008 *J. Fluid Mech.* **611** 107–130

[22] Kawahara G, Uhlmann M and van Veen L 2012 *Ann. Rev. Fluid Mech.* **44** 203–225

[23] Graham M D and Floryan D 2021 *Ann. Rev. Fluid Mech.* **53**

[24] Kerswell R R 2005 *Nonlinearity* **18** R17–R44

[25] Wedin H, Bottaro A and Nagata M 2009 *Phys. Rev. E* **79**(6) 065305

[26] Okino S, Nagata M, Wedin H and Bottaro A 2010 *J. Fluid Mech.* **657** 413–429

[27] Okino S and Nagata M 2012 *J. Fluid Mech.* **693** 57–68

[28] Uhlmann M, Kawahara G and Pinelli A 2010 *Phys. Fluids* **22** 084102

[29] Jiménez J and Moin P 1991 *J. Fluid Mech.* **225** 213–240

[30] Guckenheimer J and Holmes P J 1983 *Nonlinear Oscillations, Dynamical Systems, and Bifurcations of Vector Fields* 1st ed (*Applied Mathematical Sciences* vol 42) (Springer)

[31] Wiggins S 2003 *Introduction to Applied Nonlinear Dynamical Systems and Chaos* 2nd ed (*Texts in applied mathematics* vol 2) (Springer-Verlag New York)

[32] Toh S and Itano T 2003 *J. Fluid Mech.* **481** 67–76

[33] Biau D and Bottaro A 2009 *Philos. Trans. R. Soc. A* **367** 529–544

[34] Okino S 2014 Spatially localized solutions in rectangular duct flow *Proc. Fluids Eng. Conf.* (The Japan Society of Mechanical Engineers) p 0802

[35] Avila M, Mellibovsky F, Roland N and Hof B 2013 *Phys. Rev. Lett.* **110**(22) 224502

[36] Haldenwang P, Labrosse G, Abboudi S and Deville M 1984 *J. Comput. Phys.* **55** 115–128

[37] Rai M M and Moin P 1991 *J. Comp. Phys.* **96** 15–53

[38] Verzicco R and Orlandi P 1996 *J. Comput. Phys.* **123** 402–414

[39] Saad Y and Schultz M H 1986 *SIAM J. Scien. Statist. Comp.* **7** 856–869

[40] Trefethen L N and Bau D 1997 *Numerical linear algebra* (*Other titles in applied mathematics* vol 50) (Philadelphia: SIAM)

[41] Chandler G J and Kerswell R R 2013 *J. Fluid Mech.* **722** 554–595

[42] Dennis J E and Schnabel R B 1996 *Numerical Methods for Unconstrained Optimization and Nonlinear Equations* (SIAM)

[43] Viswanath D 2009 *Philos. Trans. R. Soc. A* **367** 561–576

[44] Willis A P 2017 *SoftwareX* **6** 124–127

[45] Duguet Y, Willis A P and Kerswell R R 2008 *J. Fluid Mech.* **613** 255–274

[46] Okino S, Nagata M, Wedin H and Bottaro A 2012 Travelling wave solutions in square duct flow *Progress in Turbulence and Wind Energy IV, Proc. iTi Conf. Turb. 2010* ed Oberlack M, Peinke J, Talamelli A Castillo L and Hölling M (Springer Berlin Heidelberg) pp 135–138

[47] Pringle C C T, Duguet Y and Kerswell R R 2009 *Philos. Trans. R. Soc. A* **367** 457–472

[48] Kawahara G and Kida S 2001 *J. Fluid Mech.* **449** 291–300

[49] Kim H T, Kline S J and Reynolds W C 1971 *J. Fluid Mech.* **50** 133–160

[50] Holmes P, Lumley J L and Berkooz G 1996 *Turbulence, Coherent Structures, Dynamical Systems and Symmetry* 1st ed Cambridge monographs on mechanics (Cambridge Univ. Press)

[51] Aubry N, Holmes P, Lumley J L and Stone E 1988 *J. Fluid Mech.* **192** 115–173

[52] Armbruster D, Guckenheimer J and Holmes P 1988 *Physica D: Nonl. Phen.* **29** 257–282

[53] Guckenheimer J and Holmes P 1988 *Math. Proc. Cambridge Philos. Soc.* **103** 189–192

[54] Kreilos T, Veble G, Schneider T M and Eckhardt B 2013 *J. Fluid Mech.* **726** 100–122

[55] Budanur N B, Short K Y, Farazmand M, Willis A P and Cvitanović P 2017 *J. Fluid Mech.* **833** 274–301

[56] Azimi S, Ashtari O and Schneider T M 2022 *Phys. Rev. E* **105**(1) 014217

Numerical analysis of the bistable propagation of lean premixed hydrogen flames

Alba Domínguez-González[1], **Miguel P. Encinar**[2], **Daniel Martínez-Ruiz**[1]

[1] Universidad Politécnica de Madrid, 28040, Madrid, Spain
[2] Universidad Carlos III de Madrid, 28911, Leganés, Spain

E-mail: a.dominguezg@upm.es

Abstract. Premixed hydrogen-air flames that propagate in slender channels have been recently found to be bistable in the limit of fuel-lean mixtures and non-negligible heat losses. Specifically, two stable configurations conformed by either a circular or a double-cell flame front arise for the same combination of controlling parameters (fuel mixture, heat transfer). Nevertheless, this multiplicity of solutions lacks a satisfactory explanation that predicts the formation of either one or another structure given a particular ignition transient. In this work, the analysis of a set of numerical simulations is offered to unveil the underlying physics controlling the bistable behavior. The formulation employed here proves that simplified one-step Arrhenius reaction and quasi-two-dimensional equations suffice to reproduce the complex dynamics, rendering three-dimensional effects and detailed chemical kinetics as second-order contributions that may be important, however, to provide exact quantitative results. Specifically, the initial temperature profiles and subsequent expansion of the flow field prescribe the growth of the flame front leading to different curvatures and sizes of the kernel that control the evolution into one of the canonical structures. Moreover, the achievement of double-cell flames require splitting and reorientation processes prior to pairing of isolated flames, which call for specific initialization conditions that enable these transient dynamics.

1. Introduction

Flame dynamics are characterized by strong non-linear effects, including the transport, production and consumption of reactive species and the local release of heat. Reactive flows are, thus, complex and dependent of fluid dynamics, heat and mass transport, and chemical kinetics.

Therefore, to clarify the nature of these interactions, researchers have relied on simple, canonical configurations such as planar and spherical flames [1, 2, 3]. Although intrinsic flame instabilities of diverse nature are widely recognized in combustion science, some of these canonical stable regimes of premixed hydrogen-air flames have been reproduced in controlled, simplified experiments under microgravity conditions or within thin gaps [4, 5].

In addition, premixed hydrogen-air flames are the simplest combustion process regarding the complexity of detailed chemical kinetics [6], which can be accurately reduced to a one-step global scheme in fuel-lean conditions [7]. However, the large mass diffusivity of hydrogen plays a much more relevant role in reactant transport than for most hydrocarbons, and can modify their characteristic flame front.

It is under these lean hydrogen-air mixture conditions that recent studies have found the coexistence of two stable isolated flame structures that consume the available reactants within

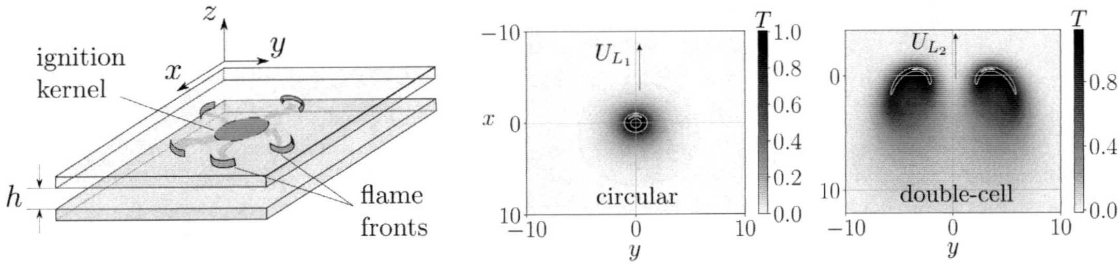

Figure 1. Sketch of the narrow-channel ignition and flame structure generation with non-adiabatic plates separated a small distance h apart (left). Characteristic dimensionless temperature field $T(x, y)$ for circular (center) and double-cell flames (right) with superimposed reaction-rate isolines (white contours).

a narrow channel (Re $\simeq$ 1) with conductive heat losses through the walls [8]. New experimental evidence [9] shows the development of both structures during the same run, which confirms the multiplicity of stable solutions in real-life scenarios propagating at different speeds (U_{L1}, U_{L2}) for identical global parameters (domain, heat losses, fuel properties and mixture). Therefore, local effects must determine the formation of either case via symmetry-breaking details during ignition.

The present work addresses the ignition transient through different initial conditions that try to mimic particular dynamics for different real ignition scenarios. The main objective is to answer fundamental questions on how these structures are originated, while it is unclear which degrees of freedom control the final state of the flame front. Although the flow here is fully laminar, the presence of the reaction non-linearity generates a rich set of chaotic solutions, much like the challenges posed by turbulence.

2. Formulation

The problem at hand considers the reactive flow between two horizontal conductive plates, which mimics experimental Hele-Shaw setups, in the limit of small gaps h compared to the characteristic flame thickness $\delta_T = D_T/S_L$, with S_L the planar propagation speed and D_T the thermal diffusivity coefficient. The formulation is simplified through consideration of quasi-2D flow for the narrow-channel limit [10], which enables the expansion of flow variables $\psi = \psi_0 + \epsilon\psi_1 + O(\epsilon^2)$ over a small coefficient $\epsilon = h/\delta_T \ll 1$, when dimensionless variables are defined for space coordinates $x = x'/\delta_T$, $y = y'/\delta_T$, $z = z'/h$ and time $t = t'/(D_T/S_L^2)$. Figure 1 shows the three-dimensional configuration of confined reactions and the two different quasi-2D isolated front structures propagating at different speeds over the quiescent flow [8].

Integration of leading-order terms of the x and y momentum equations renders Poiseuille profiles for the in-plane x-y dimensionless components of velocity $U_x/S_L = 6z(1 - z)u_x$, $U_y/S_L = 6z(1 - z)u_y$, and null transverse component $U_z = 0$. The first-order solution of the energy and mass conservation equations yield constant profiles of gas temperature $T = T(x, y; t)$ and species concentrations $Y = Y(x, y; t)$. Moreover, the following set of conservation equations with constant transport coefficients are integrated in the transverse direction between plates $z \in (0, 1)$, such that second-order transverse variations of velocity, temperature and mass fraction are accounted for in the quasi-2D problem. This integral method allows to impose off-plane conductive heat losses from the gas at wall-boundary conditions for the in-plane dimensionless conservation of energy,

$$\rho\left[\frac{\partial T}{\partial t} + u_x\frac{\partial T}{\partial x} + u_y\frac{\partial T}{\partial y}\right] = \nabla^2 T + \Omega - bT \qquad (1)$$

Journal of Physics: Conference Series **2753** (2024) 012014 doi:10.1088/1742-6596/2753/1/012014

where $T = (T' - T'_u)/(T'_a - T'_u)$ is the dimensionless temperature variable, being T'_u the unburnt temperature, T'_a the adiabatic flame temperature prescribed by the mixture heat release $q = (T'_a - T'_u)/T'_u$; the heat-loss parameter $b = 2h\lambda_w/(\epsilon^2 h_w \lambda_u)$ arises naturally after transverse integration and involves the wall thickness h_w, conductivity of wall λ_w and that of unburnt gas λ_u. Gas density $\rho = \rho'/\rho'_u$ relates to temperature through the equation of state $\rho(1 + qT) = 1$.

Conservation of the deficient reactive species (H_2) referred to the upstream value $Y = Y'/Y'_u$ with constant transport reads,

$$\rho\left[\frac{\partial Y}{\partial t} + u_x\frac{\partial Y}{\partial x} + u_y\frac{\partial Y}{\partial y}\right] = \frac{1}{Le}\nabla^2 Y - \Omega \tag{2}$$

where the Lewis number for hydrogen mixtures is $Le = D_T/D_i$=0.3 everywhere, based on the quotient of thermal to species diffusivities. A combination of mass, state and energy equations yields the equation

$$\nabla^2 p = -q[\nabla^2 T + \Omega - bT], \tag{3}$$

for the hydrodynamic pressure $p = (p' - P')/(\rho'_u S_L^2/\epsilon^2)$, being the thermodynamic pressure P' a constant value in the low-Mach limit not altering the fluid motion. In-plane velocity components are a function of the pressure gradient through the momentum equation with dominating viscous effects (Re $\lesssim 1$), and unity Prandtl number,

$$u_x = -\frac{\partial p}{\partial x}, \qquad u_y = -\frac{\partial p}{\partial y}. \tag{4}$$

It should be noted that eqs. (3)-(4) accommodate instantaneously the pressure and velocity fields.

The chemical reaction is modelled with a one-step irreversible Arrhenius expression, reducing the computational cost compared to full detailed kinetics and intermediate species consideration [11]. The dimensionless reaction rate

$$\Omega = \rho^2\beta^2(1+q)^2\frac{Y}{2Le}\exp\left(\frac{\beta(T-1)}{1+\frac{q}{1+q}(T-1)}\right) \tag{5}$$

where $\beta = E_a(T'_a - T'_u)/(RT'^2_a)$ is the Zel'dovich number. The chemical properties are defined for lean hydrogen flames [7], be $\beta = 10$ comparable to global activation temperature $E_a/R = T'_{ac} \simeq 20 \times 10^3$ K, and dimensionless heat release $q = (T'_a - T'_u)/T'_u \simeq 2 - 5$, such that the adiabatic temperature range is $T'_a = (1+q)T'_u \simeq 900 - 1800$ K, well bellow that of stoichiometric H_2-air mixtures at ambient conditions. This global reaction rate model has reproduced the intricate hydrogen flame-splitting phenomenon previously [10, 12], and shows adequate agreement with 3D detailed-chemistry simulations in narrow gaps [13]. The simplified chemical kinetics contains the main physical information to qualitatively reproduce the formation of the flame cells under investigation, as shown below.

2.1. Numerical setup

Keeping in mind the objective of performing ensembles of simulations to extract the relevant information that explains the arising of each structure, a fast and efficient numerical code is required. We use an in-house pseudospectral method with Fourier-Fourier discretization of 2048^2 modes in a doubly periodic computational domain of dimensionless size $L \times L = 30\pi \times 30\pi$, which provides a minimum resolution of 15-point flame thickness.

Fourier transformation of the governing equations leads to,

$$\frac{\partial \widehat{T}}{\partial t} + \widehat{\mathbf{u}\cdot\nabla T} = -\left(\alpha_T k^2 + \gamma b\right)\widehat{T} + \widehat{(1-\alpha_T+qT)\,\nabla^2 T} - b\widehat{(1-\gamma+qT)T} + \widehat{\Omega\,(1+qT)}, \tag{6}$$

$$\frac{\partial \widehat{Y}}{\partial t} + \widehat{\mathbf{u}\cdot\nabla Y} = -\frac{1}{Le}\alpha_Y k^2 \widehat{Y} + \frac{1}{Le}\widehat{(1-\alpha_Y+qT)\,\nabla^2 Y} - \widehat{\Omega\,(1+qT)} \tag{7}$$

$$k^2\widehat{p} = q\left(-k^2\widehat{T} - b\widehat{T} + \widehat{\Omega}\right), \tag{8}$$

$$\widehat{\mathbf{u}} = -\mathrm{i}\mathbf{k}\widehat{p}, \tag{9}$$

where the caret denotes Fourier transformation, $\mathbf{k} = [k_x, k_y]$ is the wavevector and $k = |\mathbf{k}|$ its magnitude. Note the introduction of the parameters α_T, α_Y and γ, which separate a somewhat arbitrary portion of the nonlinear Laplacian and proportional terms into a linear part in the energy and species equations.

On the one hand, eqs. (5), (8) and (9) are ligatures for intermediate variables that appear in the RHS of eqs. (6) and (7). In fact, eq. (5) is fully nonlinear, and will be solved in real space as part of the pseudospectral method. On the other hand, the evolution eqs. (6) and (7), require initial conditions in temperature and species at $t = 0$, and can be grouped with the introduction of the vector field $\psi = [T, Y]$,

$$\frac{\partial\widehat{\psi}}{\partial t} + \widehat{\mathbf{u}\cdot\nabla\psi} = -\left(\mathsf{D}_{\boldsymbol{\alpha}}k^2\mathsf{D}_{Le} + \gamma\mathsf{D}_b\right)\widehat{\psi} + \mathsf{D}_{Le}\widehat{(1-\boldsymbol{\alpha}+q\boldsymbol{T})\,\nabla^2\psi} - \mathsf{D}_b\widehat{(1-\gamma+q\boldsymbol{T})\psi} + \widehat{\boldsymbol{\Omega}\,(1+qT)}, \tag{10}$$

where $\boldsymbol{\Omega} = [\Omega, -\Omega]$, $\boldsymbol{T} = [T, T]$, and the diagonal matrices $\mathsf{D}_{\boldsymbol{\alpha}}$, D_{Le} and D_b, have the diagonals $\boldsymbol{\alpha} = [\alpha_T, \alpha_Y]$, $[1, 1/Le]$ and $[b, 0]$ respectively.

This vectorial equation is marched in time from selected initial conditions for temperature and species. The nonlinear Laplacian operator is marched in time semi-implicitly, by separating a linear part ($\boldsymbol{\alpha} > 0$). There are families of integrators for this purpose, most of them benefiting from a separate implementation of the linear and nonlinear terms. We use the third-order Runge-Kutta from [14], although a review with other suitable (potentially better) integrators can be found in [15]. For this purpose, terms in eq. (10) can be regrouped as,

$$\frac{\partial\widehat{\psi}}{\partial t} + \mathcal{L}\widehat{\psi} = \mathcal{N}(\psi)\,, \tag{11}$$

where

$$\mathcal{L} = \mathsf{D}_{\boldsymbol{\alpha}}k^2\mathsf{D}_{Le} + \gamma\mathsf{D}_b \tag{12}$$

$$\mathcal{N}(\psi) = \widehat{-\mathbf{u}\cdot\nabla\psi + \mathsf{D}_{Le}\,(1-\boldsymbol{\alpha}+q\boldsymbol{T})\,\nabla^2\psi - \mathsf{D}_b(1-\gamma+q\boldsymbol{T})\psi} + \widehat{\boldsymbol{\Omega}\,(1+qT)}. \tag{13}$$

Computation of the explicit/implicit versions of $\mathcal{L}$ are trivial, and the speed of the algorithm relies on computing $\mathcal{N}$ efficiently. The time step used in the simulations is $\Delta t = 1e-6$, which is reaction-rate limited in a standard 3/2-dealiased grid.

2.2. Postprocess

The analysis of the dynamics and ignition transients of these flame kernels calls for the definition of the structure of interest. Flame structures are identified by thresholding the reaction rate field Ω/ρ, which is the main forcing term of the temperature and species material derivatives in eqs. (1) and (2). The segmentation of the field enables a temporal tracking of every front, identifying each structure detected at each time step and using an adjacency matrix system that is based on their relative position to structures in the previous time step. Structures are defined

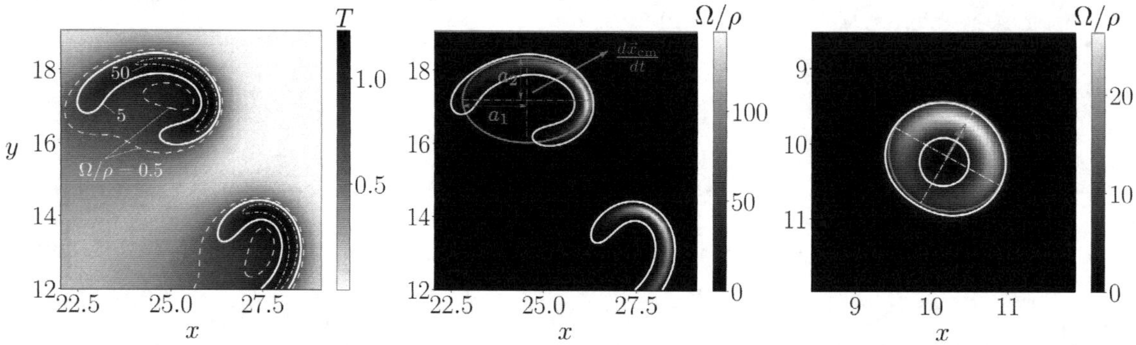

Figure 2. Sample temperature map and reaction-rate iso-contour for $\Omega/\rho = [0.5, 5, 50]$ (left). Reaction-rate map and detection operation for a double structure (mid) and a circular flame (right). White contour: threshold segmentation at $\Omega/\rho = 5$; red: ellipse fitting; red arrow: propagation direction.

using standard 4-connectivity in two dimensions. The entries of the adjacency matrix are set to one if the structures intersect at at least one pixel between consecutive frames, and to zero if they do not. Note that this is equivalent to using 6-connectivity in the three dimensional space comprising time. Additionally, it provides the evolution of structure-averaged variables (mean, maximum and minimum temperature and species), area $\sigma = \int_{\text{surf}} d\sigma$, centroid $\vec{x}_c(t) = \int_\sigma \vec{x} d\sigma/\sigma$ and propagation velocity $\vec{v} = \Delta\vec{x}_c(t)/\Delta t$.
The chosen intensity level for the threshold is set to a fixed value $\Omega/\rho = 5$, as an adequate representation of the main combustion front at intense reaction rate and non-negligible temperature rise regions. The left panel of Fig. 2 shows the threshold for three different values of the reaction rate $\Omega/\rho = [0.5, 5, 50]$ in white dashed, solid and dot-dashed lines respectively. The results are robust respect to changes in the value of the threshold, as long as this one represents the flame front adequately.

Moreover, each flame is fitted with an ellipse in a least-squares sense, which enables the fast calculation of the approximate curvature of the detected front over the facing region of the flame,

$$\kappa = \frac{1}{a_1^2 a_2^2}\left(\frac{\cos^2(\theta_e)}{a_1^2} + \frac{\sin^2(\theta_e)}{a_2^2}\right)^{-\frac{3}{2}}, \tag{14}$$

with a_1 and a_2 the semi-major and semi-minor axis respectively, and θ_e the parametric angle of the ellipse. Therefore, the maximum curvature is aligned with the semi-major axis, with $\theta_e = 0$, $\kappa_{\max} = a_1/a_2^2$, and the minimum curvature, with $\theta_e = \pi/2$, is $\kappa_{\min} = a_2/a_1^2$. The curvature at the intersection between propagation direction vector $\vec{v}/|\vec{v}|$ departing from the centroid and the fitting ellipse, κ_L, will be analyzed as a sign of the front dynamics. Finally, we choose to fit both heads separately for the double-cell cases, so that the information in following graphs of double cells corresponds to each of the semi-fronts.

The middle and right panels of Fig. 2 display two elliptic-fit examples (white lines) of the flame structure (red contour), which provides an approximate representation of the characteristic shape of either the double or circular flames. The curvature at the point aligned with the propagation direction (red arrow) and the fitting ellipse lays in between the minimum and maximum values of curvature for double-cell flames, whereas in circular flames it is aligned with semi-minor axis, where the ellipse practically degenerates into a circle.

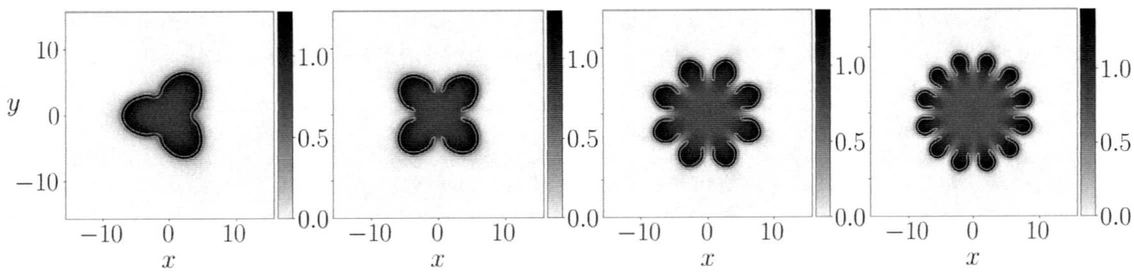

Figure 3. Temperature profiles shortly after initiation at $t = 0.1$, for $q = 5$, $b = 1$, $N = [3, 4, 8, 12]$ and $T_0 = 1.5$. White lines indicate isocontours of the reaction rate $\Omega/\rho = 5$, as used for the detection threshold.

2.3. Initial conditions

In the following, either N-lobe point-like initiation will be used to study the mechanisms that determine each structure. Besides the initial field distribution of the flow variables, the evolution will be characterized by the physical parameters presented above. To begin with, $Le = 0.3$ and $\beta = 10$, such that diffusive transport and chemical kinetics are prescribed, and variable combinations of $[q, b]$ that imply modifications in the equivalence ratio of the mixture and relative heat losses owing to the thermal properties of the chamber. The computations shown below aim to shed some light on the stabilization of the multiple canonical structures by analyzing various ignition kernel shapes, maximum temperatures, flow thermal expansion and random perturbations.

The set of simulations makes use of a concentrated heat source and local fuel depletion as the trigger of the reactive flow. In order to imitate experimental perturbations and accelerate the corrugation and splitting of the flame structure, lobular distributions are proposed with N lobes,

$$T(t = 0) = T_0 \exp\left(-0.1r^2 \left[1 + \frac{\cos(N\theta)}{2}\right]\right), \tag{15}$$

$$Y(t = 0) = 1 - \exp\left(-0.1r^2 \left[1 + \frac{\cos(N\theta)}{2}\right]\right), \tag{16}$$

with T_0 the initial peak temperature, $r = (x^2 + y^2)^{1/2}$ and $\theta = \arctan(y/x)$. Figure 3 represents characteristic temperature profiles for $N = [3, 4, 8, 12]$ and $T_0 = 1.5$, moments after initiation, $t = 0.1$. The reaction rate isolines are drawn in white, depicting the first splitting of flame fronts into several isolated cells of different sizes.

3. Results

The cases considered herein are gathered in Table 1, where the parametric set $[q, b, T_0, N]$ was varied to analyze the effect of different physical properties on the final state after the ignition process. In fact, a very wide range of transient events are identified, unsteady seeding of small structures from the main front, division of the structures to single circular kernels, unsteady flame oscillations, pairing between kernels to produce double structures or extinction of the isolated reaction fronts. The detailed study of their role on the formation of each stable solution, circular or double-cell flames, is presented below. Careful selection of parametric sets serve as adequate comparisons on the intensity of heat losses (b), exothermicity (q) owing to lower or higher equivalence ratios, ignition intensity (T_0) linked to the initial thermal expansion, or geometry

Table 1. Results of the parametric analysis of lobular ignition. Identified events: splittings (s), oscillations (o) and extinctions (e), to produce circular (c) flames, pairing to double-cell (d) kernels or unstable (u) fronts.

q	b	T_0	$N = 3$	$N = 4$	$N = 8$	$N = 12$
2	0.5	1.1	s → **c**	s → **c**	s → **e**	s → **c**
2	0.7	1.1	s → **c**	s → **c**	s → **c**	**c**
2	1.0	1.1	**c**	**c**	**e**	**c**
5	1.0	1.1	s, o → **d**	s, o → **d**	s → **d**	s → **d**
5	1.0	1.5	s → **d**	s, o → **u**	s, o → **d**	**c**
5	1.0	1.8	s → **d**	s, o → **u**	s → **d**	**c**

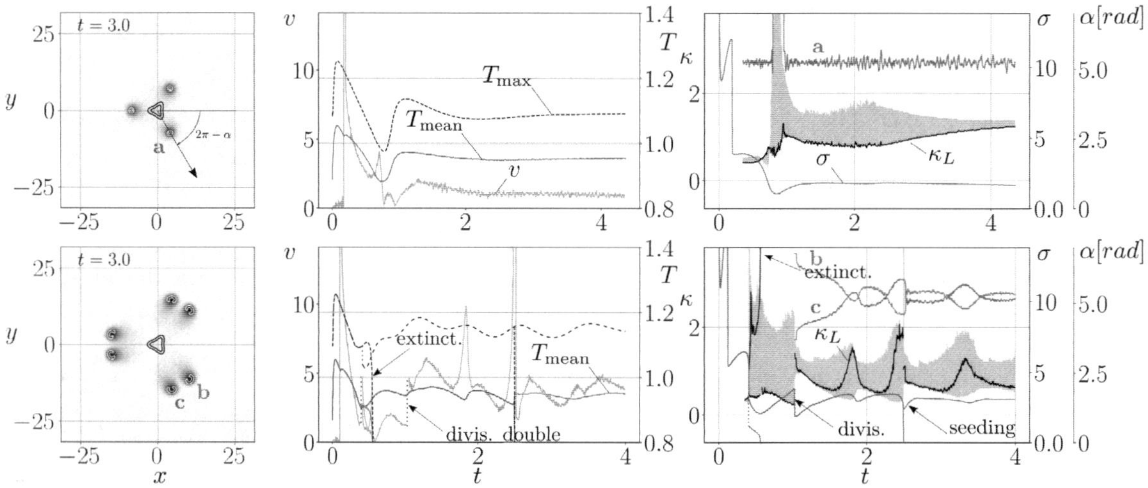

Figure 4. Comparison of the structure evolution between $q = 2$ (top row) and $q = 5$ (bottom row) for $b = 1$, $N = 3$ and $T_0 = 1.1$. Temperature distributions for $t = 3.0$ (left), evolution of maximum and mean temperature of each structure with their propagation velocity v (center), and their area σ, propagation direction α and curvature κ (right).

of the kernel, be size and curvature, through the number of lobes (N). Specially, the dynamics of the transient process under different combinations are addressed to unveil common physical features that determine the type of steady-state propagating flame.

3.1. Heat release effect

First, the exothermicity effect is considered by modifying the heat release parameter $q = (2, 5)$ so that the comparison is made between equivalent leaner and richer mixtures for a heat-loss parameter value of $b = 1$ and given $N = 3$ initial temperature profiles with peak value $T_0 = 1.1$, slightly higher than the adiabatic flame temperature. Figure 4 shows the comparison of the temperature fields at $t = 3.27$ for $q = 2$ (top left) and $q = 5$ (bottom left). A leaner mixture conditions produces weaker circular kernels that evolve steadily, contrarily to the larger heat release case, with structures dividing towards a double cell front. Additionally, the evolution in time of the mean $T_{\text{mean}}(t)$ and max temperatures $T_{\max}(t)$ inside each detected structure and their propagation velocity $v(t) = |\vec{v}(t)|$ is shown in the middle panels. The areas of the detected structures $\sigma(t)$, their propagation direction α, their curvature regarding the ellipse fitting longitudinal propagation $\kappa_L(t)$ (black solid) and $[\kappa_{\min}, \kappa_{\max}]$ (gray envelope) are shown in the right panels of Fig. 4.

The growth of temperatures right after initiation is followed by a pronounced decay owing to heat losses and the increase of extension. The kink in T_{mean} at $t \simeq 0.2$ is related to the first splitting from a continuous tri-lobular front (see Fig. 3) to three smaller structures, identified in the jump of σ in both cases. Moreover, the $q = 5$ case undergoes a second splitting at $t \simeq 0.6$, with smaller flames that extinguish and bigger flames that endure, and a third split at $t \simeq 1$ to produce fronts that propagate as a double structure when their propagation directions α reach an alignment. In this case, a propagation speed v of about three times than the isolated circular flames for $q = 2$ is found at final stages when the steady structures are formed. Additionally, the latter shows a curvature κ_L close to the minimum value of the fitted ellipse, while $\kappa_{\max}$ and $\kappa_{\min}$ converge towards the same value, indicative of the circular shape. In turn, the oscillations observed in the lower panels accommodate progressively towards a stable propagation, where κ_L and σ indicate the deformation and alignment of the front.

Owing to all these differences, effects related to the intensity of heat production are relevant to the splitting process and determine the increase in propagation velocity.

3.2. Heat losses effect

Moreover, the effect of relative heat losses is explored for variations of $b = [0.5, 0.7, 1.0]$ for weak heat release $q = 2$, see Table 1. The specific case for $N = 3$ and $T_0 = 1.1$ is presented in Fig. 4 (top). In general, low heat-loss coefficients enable the arising of splitting events, which are related to stronger reaction rates and lead to the formation of circular configurations under the small heat release. The higher the heat losses the weaker the reaction kernels, where production and cooling are balanced to a steady-state thermal equilibrium without splitting events. This trend is consistent with previous studies in similar narrow-channel configurations [10]. In the absence of conductive heat losses, the initiation would produce a continuous, but wrinkled, flame front. Increasing the losses parameter b allows the fragmentation of the flame front, which can give rise to the two configurations under analysis. The coexistence of solutions was shown to appear within a range of b, which depends on q and the other controlling parameters [8]. Finally, a sufficiently high value of b only allows the formation of the circular structure (as obtained here with $q = 2$), and even higher values of b induce the extinction of the reacting front [16].

As seen in Table 1, the ignition configuration also plays a major role in the extinction case of $N = 8$, with complete extinction dynamics for $b = 0.5$ and $b = 1$, but splitting-free weakly propagating circles for $b = 0.7$. Although these trends do not seem to be significant in the rest of cases explored here, the specific distribution of initial temperature and subsequent thermal expansion induce extinction dynamics that need to be further studied and explained. Additional simulations with variations of the initial conditions and thermochemical parameters are defined for their analysis in future work.

3.3. Initial temperature effect

Regarding the strong effect of convection to produce the splitting in fast-propagating cases, the initial expansion caused by the radial temperature profiles at $t = 0$ should be analyzed. Therefore, a comparative variation of the maximum temperature at ignition $T_0 = (1.1, 1.8)$, under the parametric set $q = 5$, $b = 1$ and $N = 12$, is offered in Fig. 5. Note that an evolution to more intense propagation under the shape of double cells is found in the top-row panels, which is a counter-intuitive result corresponding to the initiation with lower temperature $T_0 = 1.1$. In fact, a faster initial expansion is found for stronger ignition conditions with $T_0 = 1.8$ with a noticeably higher T_{mean} for $t \lesssim 0.5$ in the lower-row panels. Nevertheless, this early dynamics blow rapidly the reaction kernels to colder regions and the mean temperature rapidly decays

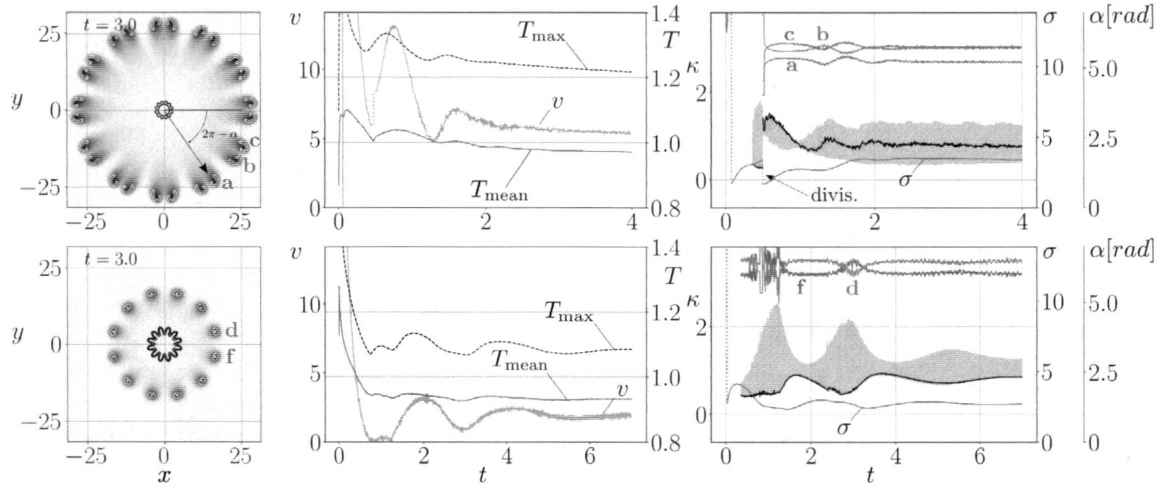

Figure 5. Comparison of the structure evolution between $T_0 = 1.1$ (top row) and $T_0 = 1.8$ (bottom row) for $q = 5$, $b = 1$ and $N = 12$.

below $T_{\text{mean}} = 1$, which also translates into a nearly complete deceleration of the structures and producing a secondary evolution of the weaker circular kind.

The double-cell structures of Fig. 5 (top left) are formed after a splitting event and develop similar characteristics to what was found in the previous section. These flames show faster propagating velocities than the circular counterpart and a longitudinal curvature κ_L somewhere in between the maximum and minimum available values, which indicates the tilted nature of the elliptic fitting. Whereas, the circular fronts show again a strong coupling of the propagation direction with the semi-minor axis. During the evolution of these cases with a large number of lobes, $N = 12$, no extinctions were identified. The spacing of cold regions available at the starting temperature distribution seems to be crucial for the elimination of such events.

In terms of the intensity of initial expansion, two critical T_0 values can be expected. First, a limit below which the ignition fails and the reaction cannot be sustained. Second, an intermediate value of T_0, which in turn depends on the combination of the rest of controlling parameters, that separates the transient behavior towards a double or a single structure.

3.4. Perturbation shape effect

Finally, it is essential to acknowledge the effect of the number of lobes N on the generation of different local environments surrounding the reaction kernels and the dynamics of interaction between them. The selection of parameters $q = 5$, $b = 1$ and $T_0 = 1.5$ shows double structures for $N = 3$ and isolated circular fronts developing for $N = 12$, see Fig. 6.

The proximity of neighboring cells acts as a constraint of the flame growth in the direction transverse to propagation. Top-row panels show the evolution for $N = 3$, with strong oscillations in velocity, temperature and shape, area and curvature of the structures, that can be accommodated in the unoccupied surrounding region. Moreover, the initial fragmentation of the flames from the starting front produces sizes $\sigma(t \simeq 0)$ of the isolated kernels that are one third smaller for $N = 12$ (bottom right) than for $N = 3$ (top right). This fact also defines the subsequent evolution of each case towards a double-cell structure in the former, and circular flames in the latter. Again, the strong differences that characterize these different stable solutions are recovered under the variation of the initiation shape.

It should be noted in Fig. 6 (bottom) that there are two kinds of circular flames here that

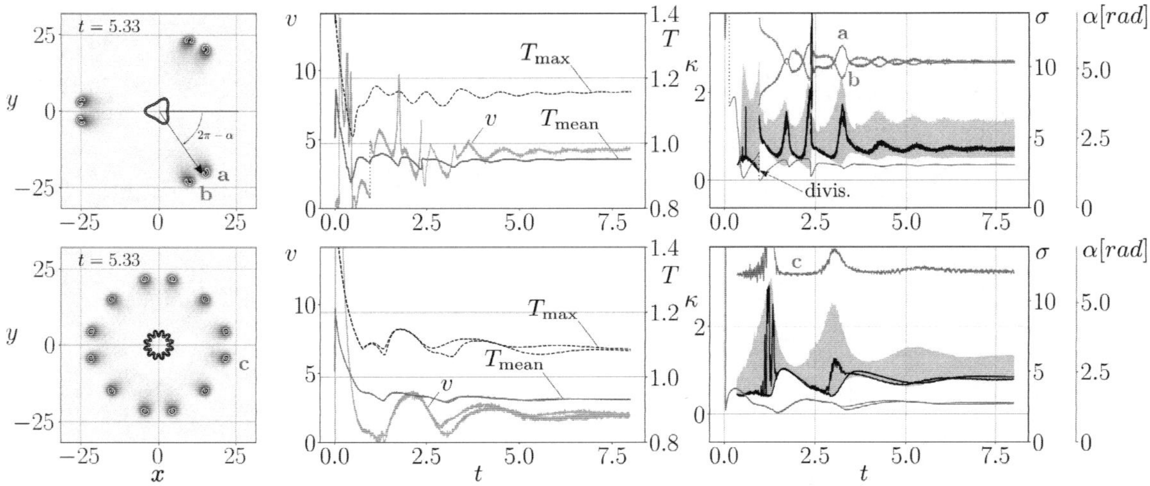

Figure 6. Comparison between $N = 3$ (top row) and $N = 12$ (bottom row) with $q = 5$, $b = 1$ and $T_0 = 1.5$.

evolve with slight differences. Mainly, they are divided in those propagating along the diagonals and the rest. The directionality of the algorithm is ruled out as possible cause, as the simulations with $N = 12$ and different T_0 shown in the previous subsection are perfectly axisymmetric and show an equal evolution for all kernels.

4. Discussion

All of the controlling parameters detailed above produce substantial effects on the formation of different stable structures. Therefore, a deeper analysis is required to identify the physical causes acting during the ignition transient that define the final stage of the solution. In fact, more universal symmetry-breaking events should be addressed in order to understand the bifurcation of lean premixed hydrogen flames.

Non-equidiffusional effects and balances between exothermicity and heat losses explain the multiplicity of steady propagating regimes, where the net equilibrium between heat production and evacuation is achieved through different flame configurations. However, the unsteady dynamics of flame splitting, seeding oscillations and structure pairing into double flames are closely related to local convection effects, which are in turn caused by the temperature field through thermal expansion of the flow. Keeping this in mind, key-variable details regarding the initiation transients at hand are addressed to provide a qualitative analysis on the routes to bi-stable propagation.

4.1. Extinction

Particular extinction scenarios have been identified in the previous section as short processes involving the fast decay of the reaction rate due to conductive and convective mechanisms that cool down local temperature profiles. These events are also characterized by a fast decrease of the detected flame propagation speed, area σ and subsequent increase of curvature κ_L. Nevertheless, different variations of the controlling parametric set $[q,\ b,\ T_0,\ N]$ have shown to produce the complete or partial extinction of the reaction, which indicates that subsidiary mechanisms must be involved through the balance of competing production, consumption and transport of energy and species. Figure 7 shows the evolution of maximum temperature, species concentration, area and propagation speed of reaction-rate structure detected for diverse combinations of parameters

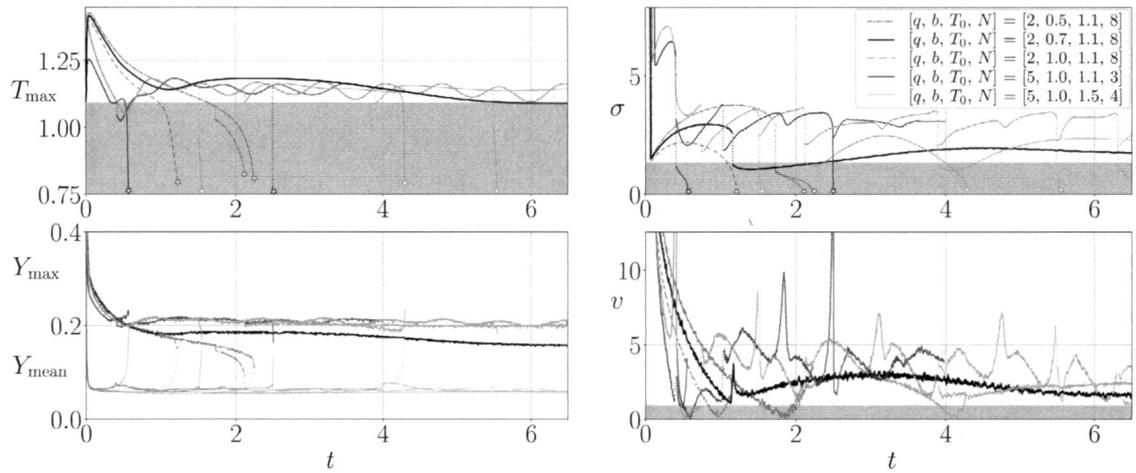

Figure 7. Extinction and splitting events for evolution of T_{max} (top left), flame area σ (top right), species concentration Y_{max} and Y_{mean}, and propagation velocity v for a combination of cases, see legend. Gray shaded areas denote proposed thresholds for extinction.

(see legend) that compare common characteristics in flame extinction.

First example, a non-extinguishing case in the very weak propagation limit, where isolated flame circles are formed, is shown for comparative purposes with $[q, b, T_0, N] = [2, 0.7, 1.1, 8]$ (solid black). It should be noted that decreasing or increasing heat losses with respect to the latter, $b = 0.5$ (dash-dotted red) and $b = 1$ (dashed red) respectively, produce complete extinction with a rapid decrease of T and σ (red stars). Although both cases also involve a progressive decrease in available reactants Y_{max}, the dynamics are somewhat different. For greater heat losses, $b = 1$, there is a progressive decrease in temperature and kernel size as it decelerates, while for $b = 0.5$ a splitting of the front in various pieces takes place at $t \simeq 1.7$ prior to complete extinction.

The one-step Arrhenius model used to describe the reactive problem defines two possible extinction scenarios, which are classified as heat-loss controlled process or lack of reactant species. The first kind is observed when an isolated flame starts to decrease in size while gaining curvature and slowing down, an imbalance between the exothermic reaction term and the heat loss causes a flame wipe out. This event appears as a progressive reduction of temperature (mean and max.) and contour area, as shown for $[q, b, T_0, N] = [2, 1, 1.1, 8]$ (dashed red). The second type of extinction takes place when the available reactant is consumed by neighboring structures. Some small kernels shed from major structures propagate slowly and remain immersed in the wake of products of the originating front, which moves faster, hindering reaction survival owing to the lack of fuel, such as the extinction shown in Fig. 7 for $[q, b, T_0, N] = [5, 1, 1.1, 3]$ (solid blue) after several oscillations at $t \simeq 2.5$ (See bottom row of Fig. 9 for a visualization of the seeding structure). This case also shows a first division for $t \simeq 0.4$, which produces two different kinds of kernels. The first kind is a low-temperature large-area structure that continues to propagate and progressively recovers T_{max} to stable values. The other, a nearly stagnant higher-temperature and lower area type that rapidly cools down and extinguishes. Finally, $[q, b, T_0, N] = [5, 1, 1.5, 4]$ (solid orange) is presented, with multiple divisions followed by sudden extinction events, and one slow extinction process that ranges in time from $t \simeq 2.2$ to $t \simeq 4.2$.

Gray areas are marked as tentative thresholds for T_{max}, σ and v to produce reaction annihilation. It should be noted that some surviving structures may trespass one of the thresholds, but never all of them simultaneously. Extinction is ensured over these cases for

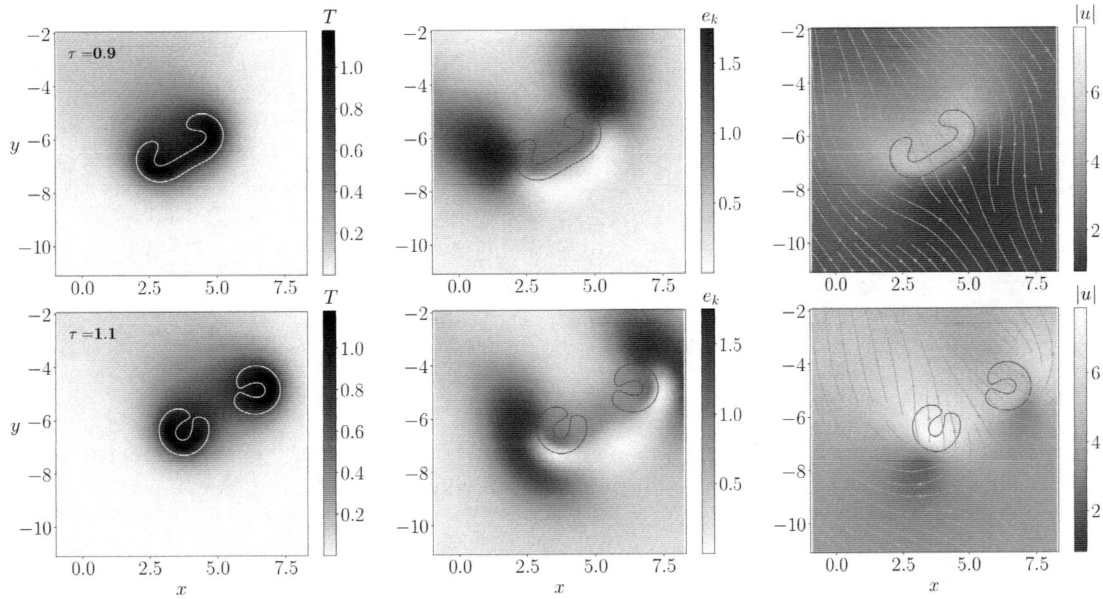

Figure 8. Splitting event at $t = 1$ for $[q,\ b,\ T_0,\ N] = [5,\ 1,\ 1.1,\ 3]$. Instantaneous field of T (left), e_k (center), and velocity in the flame-moving reference frame for the central structure $|u|$ (right). Isocontour $\Omega/\rho = 5$ depicts the reaction kernel detection in all fields.

combined situations of temperature drop, area decrease and propagation slow down. Splitting events into small enough subdivisions conform a natural precursor of extinction that confirm the combined threshold condition. Moreover, a secondary extinction cause arises when negative slopes of all the controlling variables are produced owing to reduced values of $q \simeq 2$ and larger values of $b \simeq 1$, again crossing all the thresholds via thermal extinction.

4.2. Splitting

We shall consider now the process during which a detected structure is divided into two or more substructures. During the propagation process, the structure may grow in size as a result of heat production and loss unbalance, while curvature decreases in one part of the front. This process reduces temperature due to the greater surface subject to conductive losses and lower preferential diffusion effects, which also diminishes the flame propagation velocity due to the weakening of the reaction. The divergence in $\kappa_{\max}$ and $\kappa_{\min}$ is enlarged due to this transverse deformation, until local extinction of that part of the front is achieved and the adjacent subdivided pieces gain curvature intensifying their heat production.

The characteristic size of the front after the initial expansion $\sigma(t \simeq 0)$, together with the convective field induced during that stage, plays an important role in the number, frequency and sequence of divisions. Figure. 7 shows several splitting events for $q = 5$ cases (blue and orange) that are easily identified through vertical dotted lines that join together the evolution stages of a given case. Some divisions produce weak kernels that rapidly extinguish as analyzed previous section during a seeding process, while some others produce two structures of stable temperature and size that may still evolve for a given period of time.

Oscillations of the precursor flame due to reaction unbalance and convective effects seem to act as the common root that originates these events (see temperature, area and curvature oscillations in the transient dynamics shown above). Specially, a splitting event for $[q,\ b,\ T_0,\ N] = [5,\ 1,\ 1.1,\ 3]$ at $t = 1$ can be followed as the discontinuity of area from the blue line in Fig. 7

(top right) and is depicted in Fig. 8. The temperature field (left), shows the local decrease in the divided front, while the kinetic energy distribution $e_k = \frac{1}{2}\rho u^2$ (center), depicts the regions where convective effects are enhanced in their role to stretch and separate the structures. Finally, speed in the flame-moving reference frame together with streamlines (right), are shown for the instants before (top) and after (bottom) the event (notice reorientation detail through streamlines). The velocity of this frame is within numerical precision of the individual segmentations, as both fronts are paired and propagate with the same velocity. This first division is practically symmetric with respect to the propagating direction and no noticeable discontinuity in $T_{\max}$ is thus produced. However, a strong transverse deformation that gives way to the break-up is produced via a local stagnation point. Then, the fragmented substructures are separated and start to move in different directions.

4.3. Pairing

Finally, individual cells formed after the splitting of previous reacting fronts can interact and form together double-cell flames that propagate as a fixed structure with greater speed and intensity. After this structure association, higher curvature areas of the flame are aligned with the propagation direction while the zone between heads remains chemically frozen.

However, the formation of the twin propagation is not always the outcome after a symmetric division, when the fragment kernels travel away from each other. Additional dynamics have been identified to be necessary in the achievement of the double-cell flame, namely a reorientation prior to pairing. This process is granted either by neighbor-flame confinement (large N cases, see Fig. 5 (top)) or by seeding of negligibly small kernels from one side of each substructure, which reorients their propagation direction as shown in Fig. 4 (bottom row, propagation angle oscillations for kernels b and c). There are successive oscillations of the propagation direction, α, of both substructures, which progressively adjust until their convergence. The seeding process is considered a different event than splitting, with a main front that continues to propagate unaffected and with a minor change in area, while a small (seed) separates and is left behind, typically extinguishing shortly. Seeding occurs in the case at hand for $[q,\ b,\ T_0,\ N] = [5,\ 1,\ 1.1,\ 3]$ at $t = 2.48$, where Fig. 9 shows the local fields just before and after the event. The flow field reacts producing an acceleration of the propagation and more intense alignment of the streamlines (right bottom), which enable the parallel propagation of the two fragments. Initial splitting, second division and reorientation towards the final stage propagation can also be identified in Fig. 4 (bottom).

5. Conclusions

Multiplicity of stable solutions in flame fronts is a long-lived topic of discussion mostly attributed to competing intermediate chemical reactions [17], but never confirmed in experiments out of hysteresis loops owing to parametric variation. However, numerical multiplicity has been achieved in the problem at hand by means of a one-step irreversible reaction scheme, which rules out the former cause.

The appearance of these stable configurations has been reported experimentally elsewhere [16]. However, additional understanding is required to unveil the physical process that determine the formation of each structure during the ignition process. Subtle perturbations in the initiation due to inhomogeneities of the flow, spark plug, or construction of the experimental device, may induce transient variations as studied here in a more organized and controlled manner.

A simplified formulation for low-Reynolds, low-Mach, reactive flows in channels with conductive heat losses at the enclosing plates has allowed the computation of a number of sets in a quasi-2D form. In addition, a parametric exploration regarding the chemical heat release, relative conductive heat losses, maximum initiation temperature and shape of the ignition kernel was presented and analyzed.

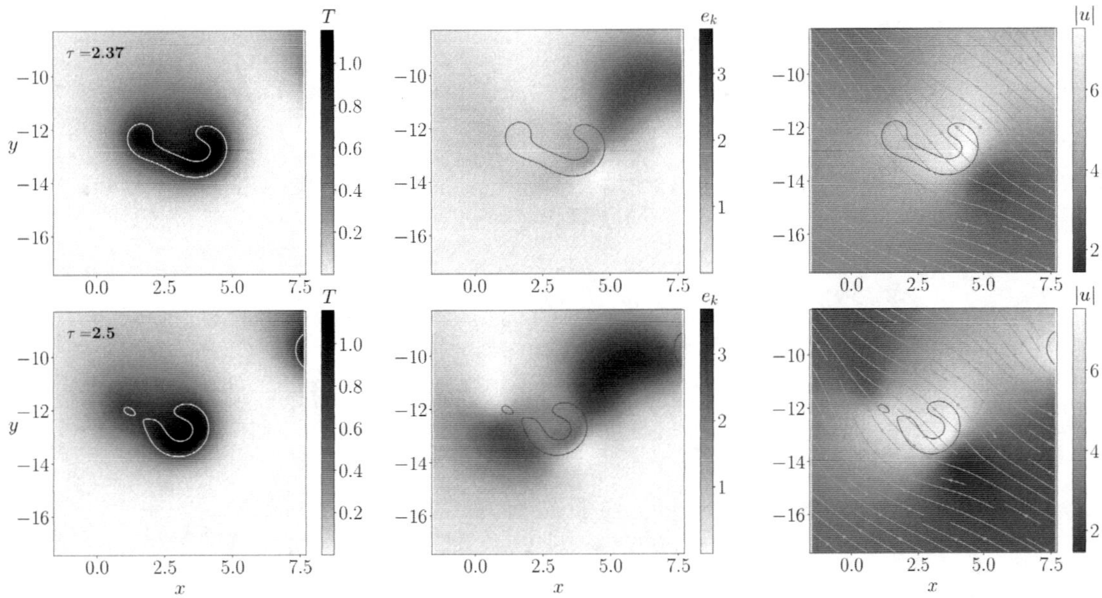

Figure 9. Seeding-reorientation event at $t = 2.48$, prior to double-cell merging, for $[q, b, T_0, N] = [5, 1, 1.1, 3]$. Instantaneous field of T (left), e_k (center), and velocity in the flame-moving reference frame $|u|$ (right). Isocontour $\Omega = 5$ depicts the reaction kernel detection in all fields.

First, it is shown that the one-step Arrhenius reaction rate and simplified off-plane conductive heat losses are enough to qualitatively predict the dynamics and stable configurations found in the experiments. In addition, the adequate combination of heat release q to heat losses b parameters needs to be provided to enable the multiple solutions of stable propagating flames, circular and double. Finally, the subtle changes or inhomogeneities expected in real ignition process have been here addressed in a simplified way through the spatial distribution (N lobes) and maximum temperature of the initial integration conditions. We have shown that very large values of initial peak temperature are inefficient in producing double-cell intense structures, because of the large thermal expansion induced at the transient initial instants.

These preliminary simulations have rendered useful information to understand the possible pathways to double-cell structure formation. Particularly, local extinction, flame front splitting, and kernel seeding processes have been identified to be key in the transient dynamics of flames. Moreover, splitting and reorientation dynamics, related to the interaction between kernels in large N cases or seeding processes, are required to produce an effective pairing of fragments into double-cell structures. These configurations are characterized by a higher burning rate and propagation velocity than the non-merging isolated counterpart of nearly circular shape.

This study serves as a first exploration to anticipate the parametric space and postprocess details that may produce a future extended analysis through ensembles of simulations [18]. This strategy can answer fundamental questions on how these structures are originated and unveil which unknown degrees of freedom control the future state of the flame front. Additional simulations including line-like ignition setups have been performed to produce an ensemble of simulations with initial profiles perturbed with random noise. These new results should be treated statistically in future work as the natural extension of this study.

Acknowledgments
The authors would like to thank M. Sánchez-Sanz and R. Palomeque for fruitful discussions regarding experimental findings. This work was performed in part during the Fifth Madrid Summer Workshop, funded by the European Research Council under the Caust grant ERCAdG-101018287 A.D.G and D.M.R. work was also supported by project TED2021-129446B-C43 funded by MCIN/AEI / 10.13039 / 501100011033 and European Union NextGenerationEU/PRTR.

References
[1] Zeldovich Y B 1944 *Theory of Combustion and Detonation in Gases* (USSR Academy of Science)
[2] Markstein G H 1953 Instability phenomena in combustion waves Tech. rep. Cornell Aeronautical Lab. Buffalo NY
[3] Liñán A and Williams F A 1993 *Fundamental aspects of combustion* (Oxford University Press)
[4] Ronney P D, Wu M S, Pearlman H G and Weiland K J 1998 *AIAA J* **36** 1361–1368
[5] Veiga-López F, Kuznetsov M, Martínez-Ruiz D, Fernández-Tarrazo E, Grune J and Sánchez-Sanz M 2020 *Phys. Rev. Lett.* **124** 174501
[6] Sánchez A L and Williams F A 2014 *Prog. Energy Combust. Sci.* **41** 1–55
[7] Fernández-Galisteo D, Sánchez A, Liñán A and Williams F 2009 *Combust. Flame* **156** 985–996 ISSN 0010-2180
[8] Domínguez-González A, Martínez-Ruiz D and Sánchez-Sanz M 2022 *Proc. Combust. Inst.*
[9] Martínez-Ruiz D and Sánchez-Sanz M 2023 Experimental evidence of multiple solutions in hydrogen-air premixed flames personal communication
[10] Martínez-Ruiz D, Veiga-López F, Fernández-Galisteo D, Kurdyumov V N and Sánchez-Sanz M 2019 *Combustion and Flame* **209** 187–199
[11] Chung J D, Zhang X, Kaplan C R and Oran E S 2020 *Science Advances* **6** eaba0827
[12] Fernández-Galisteo D, Kurdyumov V N and Ronney P D 2018 *Combustion and Flame* **190** 133–145
[13] Melguizo-Gavilanes J, Fernández-Galisteo D and Dejoan A 2021 *International Conf. Hydrogen Safety* **9**
[14] Spalart P R, Moser R D and Rogers M M 1991 *Journal of Computational Physics* **96** 297–324
[15] Kassam A K and Trefethen L N 2005 *SIAM Journal on Scientific Computing* **26** 1214–1233
[16] Palomeque R, Domínguez-González A, Martínez-Ruiz D, Rubio-Rubio M and Sánchez-Sanz M 2024 *JFM Rapids* under consideration
[17] Clavin P, Fife P and Nicolaenko B 1987 *SIAM J. Appl. Math.* **47** 296–331
[18] Jiménez J 2020 *J. Fluid Mech.* **904** A39

Journal of Physics: Conference Series **2753** (2024) 012015 doi:10.1088/1742-6596/2753/1/012015

On the growth of localised perturbations in isotropic turbulence

Miguel P. Encinar
Department of Aerospatial Engineering, Univ. Carlos III de Madrid, 28911 Leganés, Spain
E-mail: miguel.perez.encinar@uc3m.es

Abstract. Assessing the impact of localised perturbations is staple for the problems of data assimilation and control of turbulent flows. In the present work we exploit the public database of [1] that contains the growth and location of about 10^5 perturbations of approximately thirty Kolmogorov units in isotropic turbulence at moderate Reynolds numbers. All the terms in the evolution equation for the perturbation kinetic energy are analysed statistically, disregarding in the process the effect of viscosity. Of the relevant terms, only the stretching of the perturbation by the mean flow can result in production of kinetic energy, whereas every other term must only transport the perturbation across the flow field. However, it is shown that if the problem of interest is the coarse-grained propagation of the perturbation, i.e. its radius, the dynamical terms related to incompressibility are of equal importance for the growth. A simplified model based on the balance between these terms is proposed to explain the quasi-linear growth of the radius with time.

1. Introduction

Turbulence can be considered the canonical example of natural chaotic phenomena. This implies that any uncertainty in the initial conditions of a turbulent flow will result in a divergent prediction of its state, even when the evolution equations are computed correctly. In the dynamical systems framwork, a turbulent flow evolving in time can be seen as a trajectory in a highly-dimensional phase space, where chaos translates into a divergence of the trajectory from its path.

Although perturbations are always present and in principle every trajectory is constantly being deviated from an ideal one, it is useful to think of numerical trajectories as being part of an inertial manifold [2]. Despite the presence of truncation errors, numerical trajectories are reproducible by computation, which allows us to think of them as 'base' trajectories. Small perturbations added to the initial conditions make trajectories diverge from the base one, and this divergence can be measured for various purposes. Perhaps the most well-known example that makes use of perturbed trajectories is the algorithm for the computation of the largest Lyapunov exponent of the system [3]. These have been computed for turbulent flows [4, 5, 6], and relying on measuring the separation between a base trajectory and one where the norm of the perturbation is always kept small, but it is free to align with the most unstable direction of the system. Other examples include the extension of the previous procedure to finite-time [7, 8] or finite-size (nonlinear) [9] Lyapunov exponents.

Alternatively, perturbations can be introduced in the flow affecting only some degrees of freedom of the system, in particular, at some spatial locations. This is typically the case in data assimilation, where the state of the flow is measured at a few locations and then 'nudged' into the assimilated field [10]. Another important scenario that involves localised perturbations is the problem of controlling the flow field. Most often, practical considerations limit control to the wall of wall-bounded flows [11] or even to a few actuators [12, 13], which introduce localised perturbations to the flow in order to drive it to a different state. Localised perturbations can be used also to explore the sensitivity of the flow dynamics in differen regions of the flow. This procedure has been used to analyse the structures of the flow in two dimensional turbulence [14], and more recently in three dimensional turbulence [1]. In these works, the perturbations are classified based on their growth and the properties of the flow in the sensitive/unsensitive regions are used to identify flow structures that have a large impact in the dynamics. In [1], the authors identify that regions where strain dominates over enstrophy are more efficient at driving the flow dynamics than their converse.

Despite their importance, little is known about the dynamics of localised perturbations themselves. Other works have studied the global properties of perturbations [15], but involve global perturbations, sometimes already in steady-state. The present work analises the properties of localised perturbations in isotropic flows, studying its dynamics from a fundamental point of view. We exploit the numerical databases used in [1], but instead of focusing in the properties of the flow, we focus on the properties of the perturbations.

The remainder of this proceeding is organised as follows. Section §2 introduces the numerical database and the characteristic scales of the problem. Section §3 explains the research methodology and the results obtained and §4 closes.

2. Numerical Database

We utilize the databases sourced from [1] and herein describe their principal attributes, directing the reader to the primary publication for supplementary information. The database comprises 1862 perturbation trajectories of decaying homogeneous isotropic turbulence (HIT), each running for approximately two turnover times. It is essential to note that, due to the decaying nature of the flow, the time unit in turnovers is defined as,

$$\tau = \int_0^T \frac{q'\,\mathrm{d}t}{L_\mathcal{E}}, \tag{1}$$

where $(q')^2$ is the time-dependent turbulent kinetic energy and $L_\mathcal{E}$ is the time-dependent turnover time, both defined as in [16].

These trajectories correspond to the top 5% and bottom 5% of growth from ten-fold perturbation experiments that sampled fifty different initial flow conditions, all of which are also included in this database. The nominal size of the perturbations at the initial time is $30\eta_0$, where η_0 is the initial Kolmogorov scale. The perturbations try to zero the velocity vector within a region of that size, softly defined using a Gaussian kernel. Such perturbation would not satisfy the continuity requirement, which ends up preventing them from fully stopping the flow.

The flow employs a grid with $N^3 = 256^3$ collocation points, with a resolution of $k_{\max}\eta_0 = 1$, where k represents the wavenumber magnitude. This combination results in an initial Reynolds number of approximately $Re_\lambda \approx 190$, ensuring a reasonable separation of scales. The evolution of this flow is computed using a standard pseudo-spectral code [17].

The quantities shown in this paper consist mostly on triple products of the perturbation fields, which are computed in a three-fold zero-padded grid to prevent aliasing errors. All statistics are computed using solely the top five percent perturbations, as the computational resources did not allow for the computation of the same quantities in the bottom five percent. Nevertheless, it was shown in [1] that this set is the most important for the flow dynamics, making it more apealing for the subsequent research.

3. Methods and Results

Consider the time evolution of a flow field $\boldsymbol{u}(\boldsymbol{x}, t)$, evolved from $\boldsymbol{u}(\boldsymbol{x}, 0)$. A perturbed trajectory, $\tilde{\boldsymbol{u}}(\boldsymbol{x}, t)$ can be generated by evolving the perturbed initial condition,

$$\tilde{\boldsymbol{u}}_i(\boldsymbol{x}, 0) = \boldsymbol{u}(\boldsymbol{x}, 0) + \boldsymbol{\psi}_i(\boldsymbol{x}, 0), \tag{2}$$

where the subindex 'i' referers to a given perturbation from an ensemble. The perturbation trajectory can be obtained from both base and perturbed trajectories with the identity,

$$\boldsymbol{\psi}_i(\boldsymbol{x}, t) = \tilde{\boldsymbol{u}}_i(\boldsymbol{x}, 0) - \boldsymbol{u}(\boldsymbol{x}, 0). \tag{3}$$

The perturbation kinetic energy (PKE), $\psi = \frac{1}{2}(\boldsymbol{\psi} \cdot \boldsymbol{\psi})$, is a scalar field representing how far apart the perturbed trajectory has diverged locally from the base one in phase space. We study this quantity through its evolution equation,

$$\mathrm{D}_t \psi = -\nabla \cdot [\tilde{p}\boldsymbol{\psi} + \psi\boldsymbol{\psi} - 2\nu \boldsymbol{S}\boldsymbol{\psi}] - 2\nu(\boldsymbol{S}^{\psi} : \boldsymbol{S}^{\psi}) - \boldsymbol{\psi}\boldsymbol{S}\boldsymbol{\psi}, \tag{4}$$

where D_t stands for the material derivative, $\tilde{p}$ is the perturbation pressure, μ is the kinematic viscosity, $\boldsymbol{S}$ is the rate-of-strain tensor of the base trajectory and $\boldsymbol{S}^{\psi}$ is the rate-of-strain tensor of the perturbation. The first three terms in the right hand side of the equation come from the divergence of a vector, and thus they are conservative, i.e. they do not produce net perturbation growth. They, however, can move the perturbation around and thus be responsible for the spatial growth of the perturbation, even if this implies lowering its energy density. We will refer to them as 'transport' terms, to stress their conservative nature, namely the pressure transport, the PKE transport and the viscous transport. The fourth term represents the perturbation dissipation and it can only destroy perturbation magnitude. The last term is the most important for the perturbation dynamics, as it represents the stretching of the perturbation by the base trajectory. It is the only term capable of producing perturbation and thus it has to be positive on average. Due to its dynamical relevance it is the most studied term [15], at it stays relevant even when the perturbation is not localised.

Unless otherwise stated, every quantity is normalised with the time-dependent turbulent kinetic energy $(q')^2$ and the time-dependent turnover time scale $L_{\mathcal{E}}/q'$.

Figures 1 and 2 show snapshots of the different terms in (4), at two different times. First, the PKE is shown in 1(a, b). Notably, two observations emerge. First, the perturbation maintains localization throughout its evolution, a characteristic that may not be readily apparent due to the global spatial influence of the term $\tilde{p}\boldsymbol{\psi}$. Second, the energy density stays approximately constant and of the order of $(q')^2$, with fluctuations comparable to those of the velocity field. The net growth of this quantity is controlled by the other two quantities shown in figure 1, the dissipation (c, d) and the stretching (e, f). The dissipation limits how much the perturbation grows and its of the same order of magnitude that the energy. Note that the dissipation has units od power, and thus it takes times of order of one turnover to change the PKE apreciably. On the other hand, the stretching is apreciably stronger, and the positive mass is larger than the negative one, thus resulting in net production of the perturbation. Visually, the dissipation seems of similar scale and it is collocated with the perturbation energy, which shows in their correlation which is typically -0.6. In contrast with that, the stretching is in scales smaller than the perturbation and not collocated with it, with a very small (≈ 0.25) correlation between the two.

Figure 2 displays the three transport terms, each possessing a zero mean and therefore encompassing equal positive and negative masses. The pressure and kinetic energy transports exhibit magnitudes on par with the stretching term, while the viscous transport is notably weaker, even more so than dissipation. Additionally, the viscous transport operates at a significantly smaller scale, with limited overall impact.

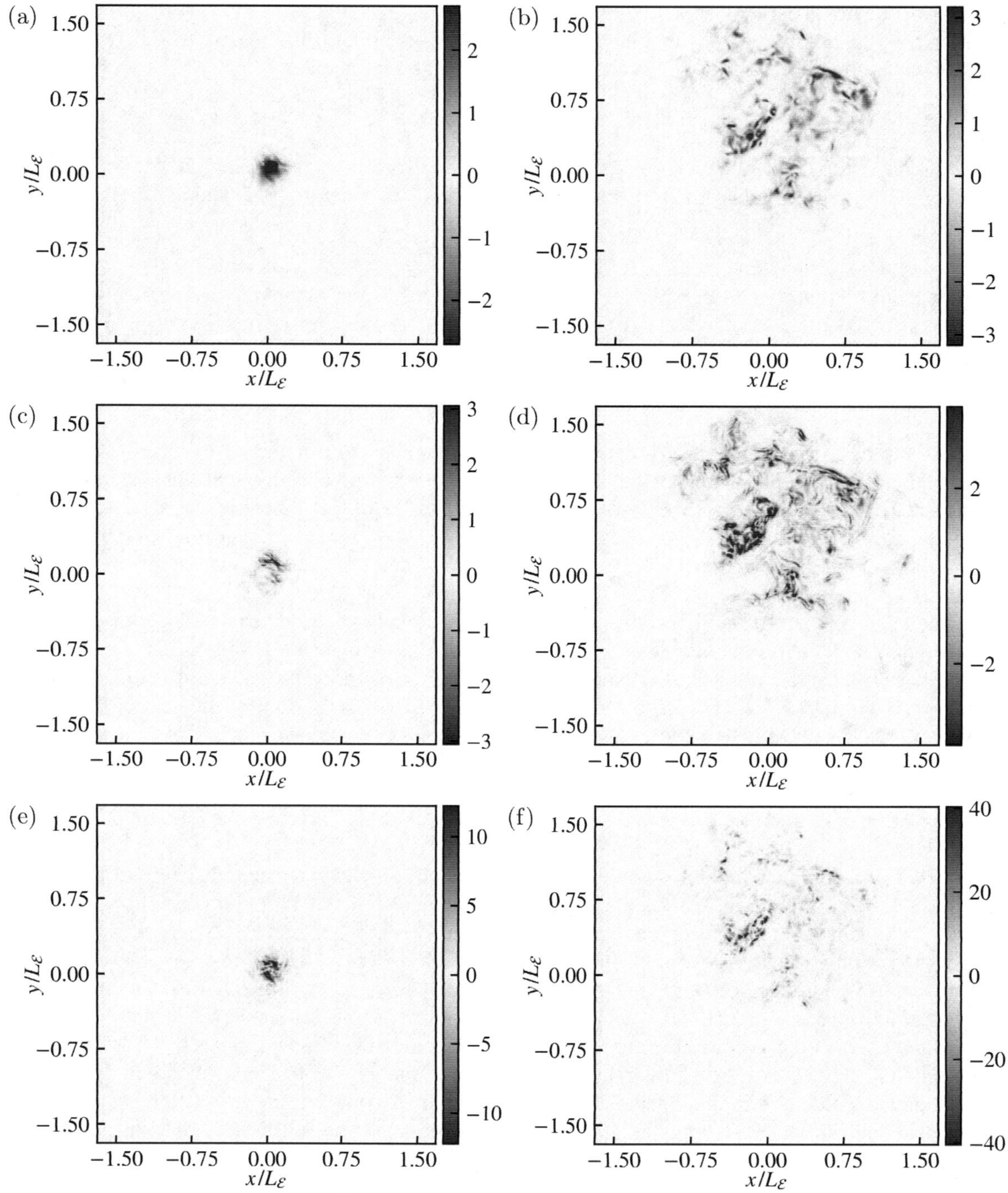

Figure 1. Snapshots of a perturbation for different quantities, (a, b) perturbation kinetic energy, (c, d) perturbation dissipation, (e, f) perturbation stretching. (a, c, d) Initial state. (b, d, f) After $\tau = 1.5$. All quantities are normalised with q' and τ.

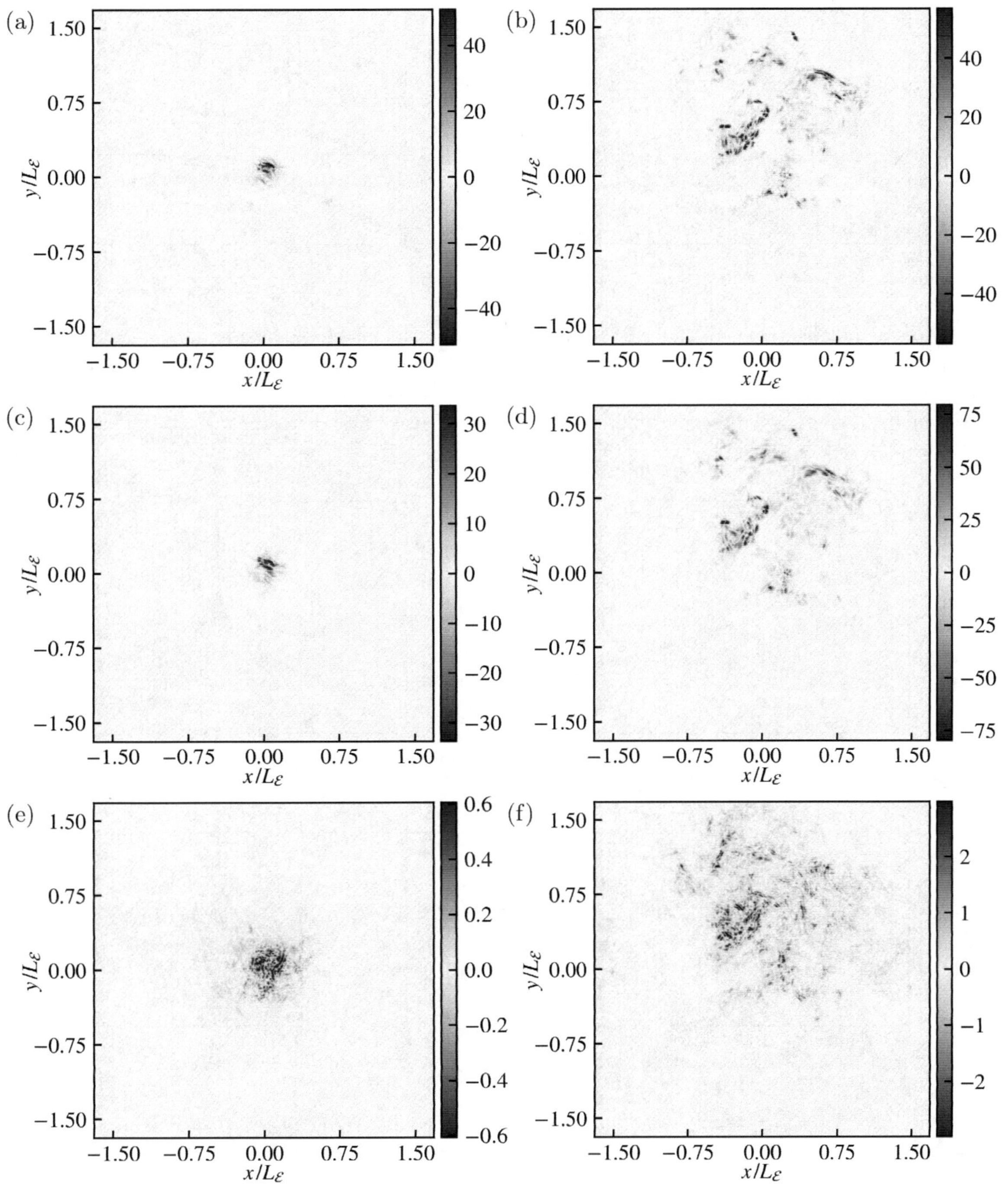

Figure 2. Snapshots of a perturbation for different quantities, (a, b) pressure transport, (c, d) perturbation kinetic energy transport, (e, f) viscous transport. (a, c, d) Initial state. (b, d, f) After $\tau = 1.5$. All quantities are normalised with q' and τ.

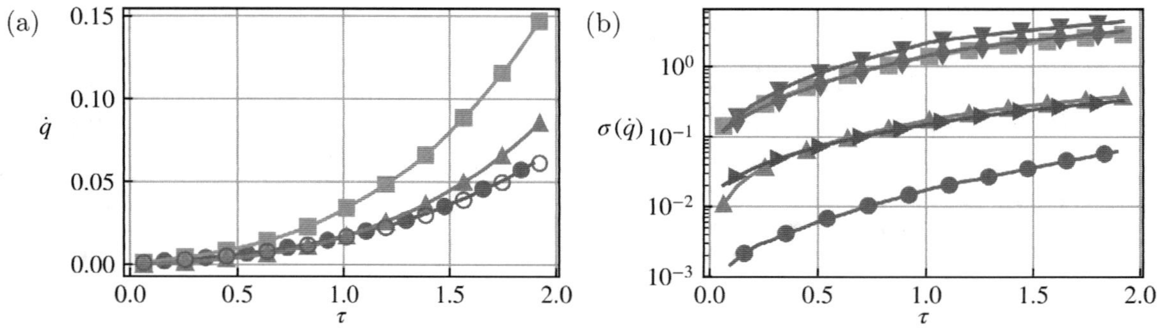

Figure 3. (a) Total values of —■— stretching, —▲— dissipation (opposite sign), —●— perturbation growth rate. Open circles are the sum of stretching and dissipation. (b) Standard deviation of —■— stretching, —▲— dissipation, —▼— pressure transport, —◆— perturbation kinetic energy transport, —▶— viscous transport. —●— Is the perturbation growth rate as in (a), to serve as a reference.

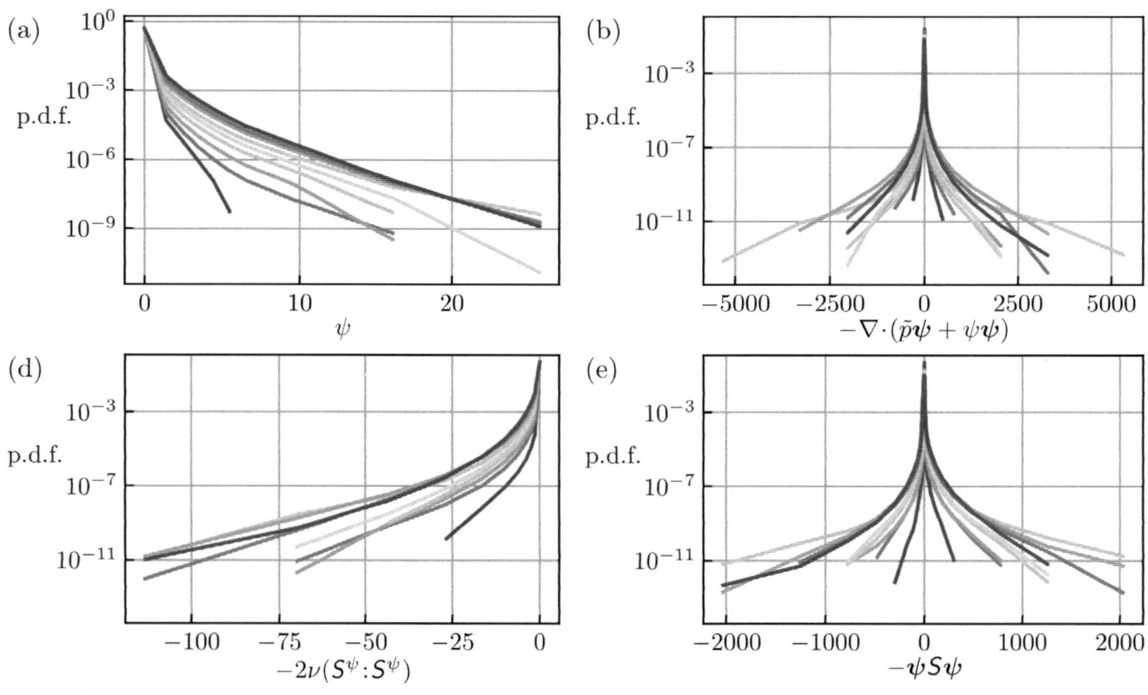

Figure 4. Probability density functions (pdfs) for the different terms in equation (4) as a function of time. (a) Perturbation kinetic energy. (b) Transport term. (c) Dissipation. (d) Stretching. Time goes from blue (initial) to red ($\tau = 1.5$)

The former two transport terms share a similar scale with stretching and are non-collocated with kinetic energy. Consequently, they can transport kinetic energy within times of the order of the perturbation's production via stretching. These two terms exhibit a notable negative correlation, typically around -0.7. This correlation is unsurprising, given that the kinetic energy term serves as a source in the perturbation pressure equation, explaining the observed cancellation between them.

Journal of Physics: Conference Series **2753** (2024) 012015 doi:10.1088/1742-6596/2753/1/012015

In order to quantify properly the properties of these six scalar fields, we need to take statistics over all the perturbations available. Figure 3(a) shows the total production, stretching and dissipation of our ensemble. For comparison, the total production of the field is order unity, but is distributed over a much larger volume (from a hundred- to ten-fold larger during the evolution of the perturbation). The production is computed by using a simple finite difference in the average PKE signal. The total stretching is positive and the largest magnitude of the three which is compensated, by the total dissipation (shown here with the opposite sign). The sum of the two is shown in open symbols, and collapses well with the total production. This agreement validates the temporal resolution of the samples and spatial resolution of the production terms.

The magnitude of the fluctuations of the five production terms is shown in 3(b), by means of their standard deviation. Note that the viscous transport and dissipation terms are ten times weaker than the stretching and the pressure and kinetic energy transports. The latter three are of comparable magnitude, and their fluctuations are a hundred-fold stronger than the average growth of the PKE. This suggests that the three terms should be considered for the dynamics of the perturbation. Based on these results, we neglect the viscous transport and remove it from our analysis, and group pressure and kinetic energy together in a 'transport' term. The latter choice also helps at removing the cancellation between the two from our analysis.

More information is provided by the probability density functions (pdfs) of the different terms, depicted in figure 4. All the pdfs are very heavy tailed, with flatness factors of the order of a thousand. Those values are extreme, but consider that the stretching term, for example, is a cube that contains the velocity gradients times the velocity diffenence twice, and the magnitude of each already has a heavy-tailed distribution. The pdf of the stretching is also skewed, with typical skewness factors of the order of a hundred. Note that the values of these moments decay approximately as an exponential of the time, and the values given are typical for times after $\tau \approx 1$. Because these moments are not universal, but depend on the characteristics of the initial perturbation, we do not pursue their accurate computation. What is important here is that both the transport term and the stretching term attain extremely high values, much higher than the extrema of PKE. Thus, it is reasonable to believe that they are an important component of the apparent diffusion process that makes the perturbation grow in size.

So far we have seen the statistical properties of the production terms, which suggest that the transport terms may be important for the spatial growth of the perturbation. In order for them to expand the perturbation in space, they need to be intense in regions where the perturbation is not. A similar reasoning holds for the stretching, which increases the energy density when is positive in regions of high energy, and the perturbation size when is in regions where the perturbation is weak.

Figure 5(a) shows a snapshot of a perturbation at $\tau = 1.5$. The perturbation is shown in light gray and can be seen stretched to many times its size, similar to what it is shown in figure 1(b), but in three dimensions. The contour contains 90% of the perturbation mass. This implies that the other isovolumes shown are in weak perturbation regions when they are outside this level. The red isovolume is the positive contribution of the transport term, and the blue one of the stretching. Both terms use the same contour level, so they can be compared directly. We only focus on the positive part of the both terms because it may be responsible for the spatial growth of the perturbation. While the negative part reduces the PKE in intense regions, only the positive part may extend it spatially. The transport terms are organised in small sheets that extend outside the perturbation, whereas the stretching tends to from rounded 'blobs', and visually seems to be located inside the perturbation.

We measure the previous qualitative observations by computing the intersection ratio of the transport terms with the kinetic energy, defined as,

$$I_t(t) = \frac{(\psi < \psi|_{90}) \cap (\mathcal{T} > 0)}{(\psi < \psi|_{90}) \cup (\mathcal{T} > 0)}, \tag{5}$$

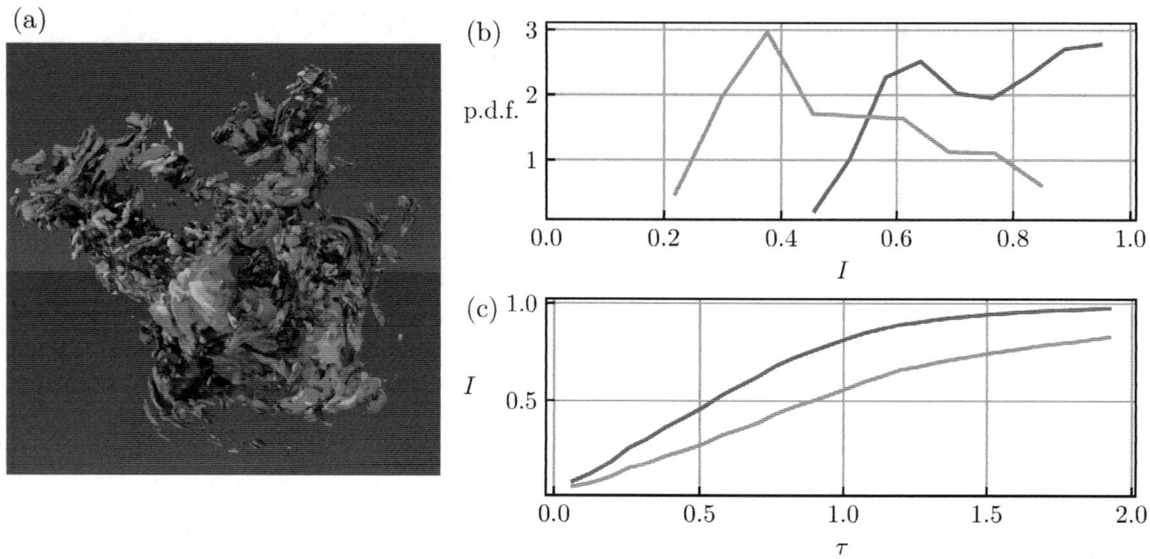

Figure 5. (a) Snapshot of a perturbation at $\tau = 1.2$. The translucent gray isovolume contains 90% of the PKE. The red isovolume contains 99% of the positive mass of the transport term, and the blue one is the same threshold value but for the positive stretching. (b) P.d.f. of the intersection ratios at $\tau = 1.5$. (c) Average intersection ratio as a function of time. (b, c) Blue corresponds to the stretching and orange to the transport

where $\mathcal{T} = -\nabla \cdot [\tilde{p}\boldsymbol{\psi} + \psi\boldsymbol{\psi}]$ stands for the transport term, $\psi|_{90}$ corresponds to the level that contains 90% of the ψ mass, and the subindex 'i' corresponds to each of the structures. Equation (5) is employed analogously to evaluate the stretching term. I is zero if the positive contribution of either term is not collocated with the most intense PKE, and one otherwise.

Figure 5(b) shows the p.d.f. of these intersections for $\tau = 1.0$. Most of the probability mass of the stretching is concentrated towards high intersections, and its modal value is close to one. Conversely, the transport term exhibits a peak at an intersection ratio of $I_{\mathcal{T}} \approx 0.4$, and most of the probability mass is skewed towards the left. This shows that the transport term is statistically more relevant in regions where the perturbation is very weak.

In figure 5(c) it is demonstrated that this behaviour is sustained as a function of time, as the intersection of the stretching term consistently surpasses that of the transport terms. Furthermore, both intersections exhibit linear growth until $\tau < 1.0$ when the growth reaches saturation. This phenomenon could be linked to the perturbation growth reaching a steady state, but it might also relate to the finite simulation domain or the decay of the base flow. Any definitive conclusions drawn from the present simulations remain speculative.

Finally, figure 6 shows the average temporal evolution of the radius of the perturbation. It is defined for various thresholds as the radius of the smallest sphere that contains a given percentage of the PKE. This definition is slightly different than the one that uses the cubic root of the determinant of the covariance tensor, which is often used for passive scalar modelling [18], but results shown in [1] using that metric are similar to the ones shown here. It can be seen that the initial spreading of the perturbation is linear, as is the case for passive scalars immersed in isotropic turbulence [19]. However, the spatial growth of the perturbation here is faster, even if the temporal scaling matches a passive scalar.

We have attempted to obtain an eddy diffusivity model as in [19], but with a different constant that accounts for the transport terms, and a volumetric source term that accounts for the production by the stretching. However, we have not been able to obtain a successful model

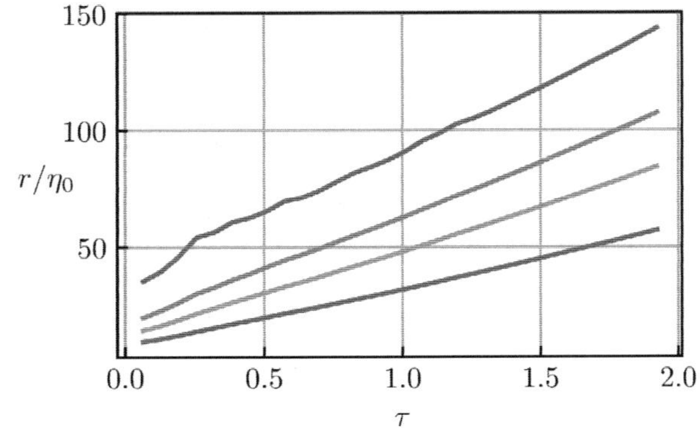

Figure 6. Evolution of the radious that on average contain 25, 50, 75, and 90% of the PKE mass.

without the ad-hoc inclusion of a diffusivity term proportional to the wavenumber modulus, which cannot be justified on physical terms. Part of the difficulty arises from the decaying nature of the flow, coupled with the lack of a steady-state for the perturbation terms, which precludes us from using most of the approximations that can be applied to a passive scalar.

4. Conclusions

In this work we have investigated the magnitude of the different terms that appear in the equation for the perturbation kinetic energy in an isotropic flow. We have focused on localised perturbations which have remainded largely unexplored. We show that at high Reynolds numbers, the relevant terms are the transport due to the pressure of the perturbation, and the stretching term. The former transports the perturbation through the flow field, and the latter destroys and produces the perturbation at different places, with an overall production effect that sustains the perturbation, but also increasing its effective transport. Both quantities are highly intermittent, and distributed in space in very small scales. We have shown that the transport terms are more relevant in regions where the perturbation is weak and the stretching term is more relevant in regions where the perturbation is already intense. Finally, we report that the perturbation diffuses linearly with time, similar to a passive scalar, but with a larger eddy diffusivity. Attempts to provide a reasonable estimation to the eddy diffusivity have failed, in part because the decaying nature of the flow complicates the identification of a steady-state for the perturbation growth.

References

[1] Encinar M P and Jiménez J 2023 *J. Fluid Mech.* **965** A20
[2] Bartuccelli M, Constantin P, Doering C R, Gibbon J D and Gisselfält M 1989 *Physics Letters A* **142** 349–356
[3] Wolf A, Swift J B, Swinney H L and Vastano J A 1985 *Physica D: nonlinear phenomena* **16** 285–317
[4] Yamada M and Ohkitani K 1988 *Prog. Theor. Phys.* **79** 1265–1268
[5] Keefe L, Moin P and Kim J 1992 *J. Fluid Mech.* **242** 1–29
[6] Nikitin N 2018 *J. Fluid Mech.* **849** 942–967
[7] Vastano J A and Moser R D 1991 *J. Fluid Mech.* **233** 83–118
[8] Aurell E, Boffetta G, Crisanti A, Paladin G and Vulpiani A 1996 *Phys. Rev. Lett.* **77** 1262
[9] Aurell E, Boffetta G, Crisanti A, Paladin G and Vulpiani A 1997 *J. Phys. A-Math Gen.* **30** 1
[10] Di Leoni P C, Mazzino A and Biferale L 2020 *Phys. Rev. X* **10** 011023
[11] Choi H, Moin P and Kim J 1994 *J. Fluid Mech.* **262** 75–110
[12] Pastor R, Vela-Martín A and Flores O 2020 *J. Phys. Conf. Ser.* **1522** 012014
[13] Lehmkuhl O, Lozano-Durán A and Rodriguez I 2020 *J. Phys. Conf. Ser.* **1522** 012017
[14] Jiménez J 2020 *J. Turbul.* **21** 544–566

Journal of Physics: Conference Series **2753** (2024) 012015 doi:10.1088/1742-6596/2753/1/012015

[15] Tsinober A and Galanti B 2003 *Phys. Fluids* **15** 3514–3531
[16] Batchelor G K 1953 *The theory of homogeneous turbulence* (Cambridge U. Press)
[17] Vela-Martín A and Jiménez J 2021 *J. Fluid Mech.* **915** A36
[18] Batchelor G 1949 *Australian Journal of Chemistry* **2** 437–450
[19] Batchelor G K 1952 *Mathematical Proceedings of the Cambridge Philosophical Society* **48** 345–362

Causality between streaks and bursts in wall-bounded turbulence

Yuenong Ling (凌越农), Adrian Lozano-Durán

Department of Aeronautics and Astronautics, Massachusetts Institute of Technology, Cambridge, MA 02139, USA

E-mail: `lingyn@mit.edu`

Abstract. We use an information-theoretic method, referred to as information flux, to quantify the causal relationships between streaks and bursts in a non-intrusive manner. Within this framework, causality is quantified as the flux of Shannon information from the present of the quantities of interest to their future. We also use the so-called information leak to measure the information that is not accounted for due to unobserved variables. We investigate data from a direct numerical simulation of turbulent channel flow at the friction Reynolds number $Re_\tau \approx 180$. The spatial distribution of causality is investigated as two time scales based on 50% value of the information leak and maximum of the normalized cross-induced information flux. It is found that the most causal spatial configuration for streaks and bursts is always streamwise-aligned. Furthermore, four dominant causal spatial configurations between streaks and bursts are identified. At the short time scale around 10 viscous units (when information flux is 50%), streaks and bursts have comparable causality to one another. At the longer time scale (corresponding to the maximum cross-induced causality), there is greater causality from streaks to bursts.

1. Introduction

Structural models are important for the theoretical understanding of wall-bounded turbulence. In these models, it is assumed that coherent structures exist and evolve rather independently despite the chaotic nature of turbulence. This view enables the reduced-order description of wall-bounded turbulence, usually too complex and chaotic to be analyzed as a whole. Ever since the discovery of possible candidates for coherent structures, from streaks by Kline et al. [1] to horseshoe vortices by Adrian [2], myriad models have been proposed to describe wall turbulence. It is generally agreed upon that two distinctive but interacting entities exist: streaks and bursts [3]. Streaks are streamwise elongated structures that move slowly or fast relative to the mean flow velocity, referred to as low and high velocity streaks, respectively. On the other hand, bursts are structures carrying wall-normal velocity exemplified by quasi-streamwise vortices, rollers, intense quadrant events, etc. The prevailing understanding is that both structures are involved in a self-sustaining cycle [4, 5]. However, there is no consensus on the exact mechanism of the self-sustaining cycle [6]. To bridge the gap between theoretical models and real-world phenomena and enhance existing models, it is essential to understand the causal relationships between streaks and bursts.

There has long been interest in elucidating the causality in wall-bounded turbulence. Jiménez and Pinelli [7] confirmed the interwoven nature of vortices and streaks by observing the relaminarization after suppressing either structure. Lozano-Durán et al. [6] investigated the cause and effect of linear mechanisms in wall-bounded turbulence by studying the Navier-Stokes equations with modified linear dynamics. Hwang and Cossu [8] corroborated the self-sustainability of large-scale motions by quenching small-scale motions using over-damped large-eddy simulations. Farrell et al. [9] used a restricted nonlinear simplification of NS equations to understand large-scale motions in plane Poiseuille flow. Jiménez [10] showed that turbulence can be sustained without the necessity of long streaks by examining the system with long streaks removed. While these are valuable attempts to understand turbulence, these conventional causal inference methods rely on interventions (i.e., modification of the governing equations) that might result in misleading causal relations by unintentionally altering dynamics of the system. Furthermore, the cost of interventions cannot scale up to higher Reynolds numbers when the dynamics becomes increasingly complex and computationally expensive to resolve. To bypass these limitations, we elucidate the causal relation in a non-intrusive manner using tools from information theory.

Information theory has been a crucial tool in the study of many complex systems [11]. With its roots in communication systems [12], it provides an invaluable tool for quantifying the uncertainty of variables and the amount of information transmitted between them. The seminal work by Schreiber [13] popularized transfer entropy to quantify the causality between variables in a dynamical system. A comprehensive review of existing methods based on transfer entropy can be found in Bossomaier et al. [14]. Recently, Lozano-Durán et al. [15] proposed an information-theoretic method, referred to as information flux, to quantify causality in a multivariate dynamical system. In this paper, we use this tool to quantify the causality between streaks and bursts.

The paper is organized as follows. Section 2 describes the methodology of extracting time-signals from direct numerical simulation (DNS) data and inferring causality between streaks and bursts. Section 3 presents the results on the causality at different time scales as a function of space. Section 4 discusses the most causal configurations, the implications of the results, and the estimation of information-theoretic quantities. Finally, Section 5 concludes the paper.

2. Methodology

2.1. Data

We investigate data from a DNS of turbulent channel flow at the friction Reynolds number $Re_\tau \approx 180$. The computational domain is $L_x \times L_y \times L_z = 8\pi h \times 2h \times 4\pi h$, where h is the half channel-width, and x, y and z correspond to the streamwise, wall-normal and spanwise directions, respectively. The plus units based on viscosity ν and friction velocity u_τ are used throughout the paper. Quantities normalized by plus units will be denoted as $(\cdot)^*$ to differentiate them from quantities in the future denoted by $(\cdot)^+$. The grid spacing is uniform along the streamwise and spanwise directions with $\Delta x^* \approx 9$ and $\Delta z^* \approx 4$. A stretching grid is used in the wall-normal direction with $\Delta y^*_{\text{min}} \approx 0.16$ close to the wall. The time step is set to be $\Delta t^* = 0.011$ but snapshots are only stored every $\Delta t^* = 1$ to achieve a balance between enough resolution and low storage. The total time history measured by flow-through time is $TU_c/L = 14.5$.

The simulation is performed using an incompressible Navier-Stokes solver with a staggered second-order finite difference scheme [16]. Periodic conditions are imposed in the streamwise and spanwise directions, and the no-slip conditions in the wall-normal direction. The code has been validated in canonical turbulent channel flows [17, 18].

2.2. Characterization of bursts and streaks using time signals

We use the streamwise velocity fluctuation u' as the proxy signal of streaks and v' as the proxy signal of bursts. They are averaged over the confining boxes to generate $B(t)$ and $S(t)$ characterizing bursts and streaks, respectively. In particular, the instantaneous value of $S(t)$ is computed as

$$S(t) = \frac{1}{V_S} \int_S u'(x, y, z, t) \mathrm{d}x \mathrm{d}y \mathrm{d}z, \tag{1}$$

where the integration is carried out over the domain covered by the streak box and V_S is its volume. All the boxes are attached to the wall and drift with the local averaged wall-parallel velocity, also defined as in equation (1). Figure 1 shows the relative size of each box and an example in which they are randomly distributed in the flow. The size of the streak boxes and burst boxes is chosen to be $L_x^+ \times L_y^+ \times L_z^+ = 540 \times 54 \times 100$ and $L_x^+ \times L_y^+ \times L_z^+ = 36 \times 36 \times 50$, respectively. The aspect ratio and the relative size follow the results from previous studies [19, 20]. A total amount of 200 streak boxes and 800 burst boxes were used to generate data. The initial positions of the boxes were randomly selected. The number of boxes considered is large enough so that the whole domain has been covered at all times. An example of the time history of $B(t)$ and $S(t)$ within one burst and streak box, respectively, is shown in Figure 2.

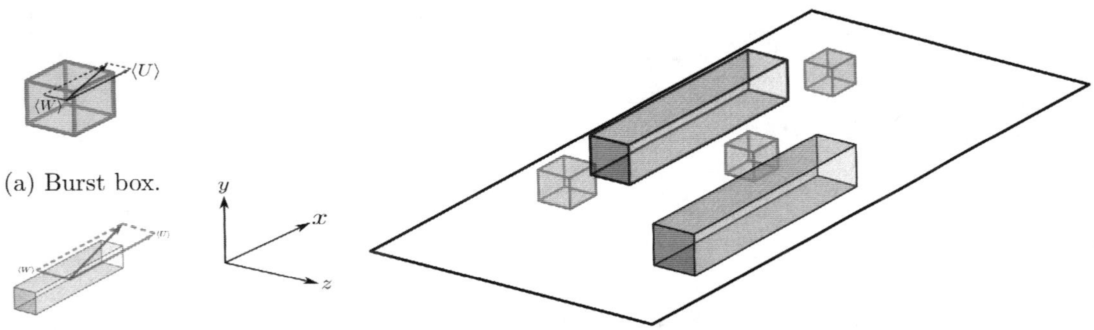

(a) Burst box.

(b) Streak box. (c) Flow domain with boxes randomly distributed. Only a few boxes in the bottom wall are shown for clarity.

Figure 1: (a) Burst boxes and (b) streak boxes are randomly distributed attached to the wall moving with the averaged wall-parallel velocity shown in (c).

While previous streak-identification methods involve analyzing the kinetic energy spectrum [21], extracting positions where the locally average velocity satisfies certain criteria [22, 23] or using wall-normal vorticity as a proxy [7, 24], the current method is preferred due to its simplicity and ability to extract time-resolved signals suitable for causal analysis. Although signals $S(t)$ and $B(t)$ may not capture the complete dynamics of individual streaks and bursts, they are expected to represent the statistical properties of these structures, especially when a large number of boxes are deployed.

2.3. Information flux

2.3.1. Information theory We briefly introduce some key concepts of information theory. For a more comprehensive introduction, the reader is referred to Cover et al. [25]. Proposed by Shannon [12], the information of observing an event q is defined as

$$h(q) = -\log_2[p(q)], \tag{2}$$

where $p(q)$ is the probability mass function (pmf). Intuitively, $h(q)$ is a measure of the 'surprise' of an event q. Its unit depends on the base of the logarithm which are 'bits' for base 2. The Shannon information of a discrete random variable Q is defined as the average amount of information over all its possible outcomes:

$$H(Q) = -\sum_{q \in Q} p(q) \log_2[p(q)]. \tag{3}$$

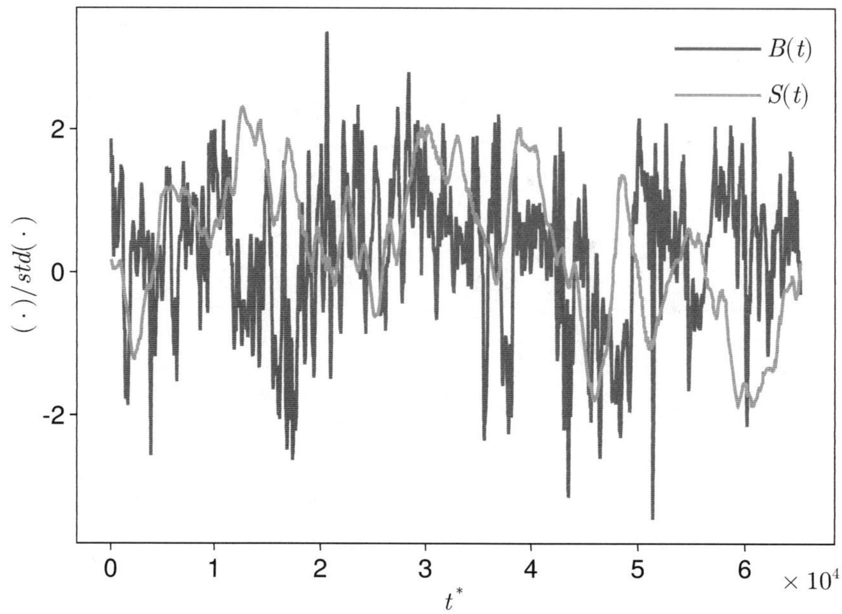

Figure 2: Time history of $S(t)$ for one streak box and $B(t)$ for one burst box, normalized by their respective standard deviations.

Similarly, the conditional entropy of Q_1 given Q_2 is defined as

$$H(Q_1|Q_2) = -\sum_{q_1 \in Q_1, q_2 \in Q_2} p(q_1, q_2) \log_2[p(q_1|q_2)]. \tag{4}$$

The quantities above can be generalized to continuous random variables by replacing the summation with integration and probability mass function with probability density function (pdf).

2.3.2. Formulation The information flux aims to quantify the transfer of information among variables in the system [15]. Here, we only consider the case of two variables $B(t)$ and $S(t)$, as defined in Section 2.2. We introduce the notation $S^+ = S(t + \Delta t)$ to denote the future of the variable S, where $\Delta t > 0$ is an arbitrary time lag. Assuming that all the information of one streak S comes from the past, we can write the Shannon information of S^+ as

$$H(S^+) = T_{S\to S} + T_{B\to S} + T_{B,S\to S} + T_{\text{leak},S}, \tag{5}$$

where $T_{S\to S}$ is the self-induced causality from one variable to itself, $T_{B\to S}$ is the cross-induced causality from one variable to another, $T_{B,S\to S}$ is the synergistic causality, which is the causality emerging by the combined effect of both variables, and $T_{\text{leak},S}$ is the information leak, representing the information not captured by $S(t)$ and $B(t)$. The definitions of the four terms are as follows:

$$T_{S\to S} = H(S^+|S) - H(S^+|S,B), \tag{6}$$

$$T_{B\to S} = H(S^+|B) - H(S^+|S,B), \tag{7}$$

$$T_{B,S\to S} = H(S^+) - H(S^+|B) - H(S^+|S) + H(S^+|S,B), \tag{8}$$

$$T_{\text{leak},S} = H(S^+|S,B). \tag{9}$$

In this work, we focus on the self-induced causality, cross-induced causality and the information leak. The synergistic term represented by equation (8) will be the subject of future exploration.

2.3.3. Spatial decomposition of causality The terms for the information fluxes described in equations (6)-(9) can be interpreted as expectations across all states. However, in certain applications, we might want to quantify causality under specific conditions. Here, we are investigating the spatial distribution of causality between streaks and bursts. Specifically, we aim to determine the spatial configurations where causality is the strongest.

To calculate the spatially conditioned information flux, we compute the probability distributions constrained to bursts (or streaks) when the streaks (or bursts) are confined within a specific distance to the reference object along the streamwise and spanwise directions, $d_x \in D_x = [d_x^{\text{min}}, d_x^{\text{max}}]$ and $d_z \in D_z = [d_z^{\text{min}}, d_z^{\text{max}}]$, where d_x^{min} and d_x^{max} as well as d_z^{min} and d_z^{max} are specified limits. The equation for spatially conditioned cross-induced information flux from bursts to streaks can be expressed as:

$$t_{B\to S}(d_x \in D_x, d_z \in D_z) = \sum_{S^+,S,B} p_d(s^+, s, b)\frac{\log_2[p_d(s^+ \mid s, b)]}{\log_2[p_d(s^+ \mid b)]}, \tag{10}$$

where p_d is the pmf when the variables satisfy the criteria $d_x \in D_x = [d_x^{\text{min}}, d_x^{\text{max}}]$ and $d_z \in D_z = [d_z^{\text{min}}, d_z^{\text{max}}]$. The same procedure can be applied to the self-induced causality and the information leak.

3. Results

3.1. Causality as a function of time lags

The first quantity of interest is the normalized information leak

$$\hat{T}_{\text{leak},S} = \frac{T_{\text{leak},S}}{H(S)}, \tag{11}$$

representing the fraction of information in the future of S not contained in the past of S and B. A similar definition applies to $\hat{T}_{\text{leak},B}$. The second causality of interest is the normalized cross-induced causality, defined as

$$\hat{T}_{B\to S} = \frac{T_{B\to S}}{T_{S\to S}}, \tag{12}$$

which indicates the relative strength of cross-induced causality compared to self-induced causality (similarly for $\hat{T}_{S\to B}$).

As shown in Figure 3, the information leak increases monotonically as a function of the time lag. This is due to the fact that less information can be inferred from the past variables when the time lag increases. Although there is no clear demarcation between small and large information leaks, we choose the time lag where the information leak is 50% of the maximum information leak, that is, $T_{leak}/H = 0.5$, as the representative time scale. This time scales corresponds to the moment when half of the causality can be inferred from the variables considered. The time lags based on 50% information leak for both streaks and bursts are marked by red dashed lines in Figure 3.

The second interesting time scale is identified as the time at which the maximum normalized cross-induced causality occurs, as depicted in Figure 4. The figure shows a clear peak in the normalized cross-induced causality, which is indicated by the red vertical line. This peak represents the maximum amount of information transmitted between streaks and bursts across all time lags. The two aforementioned time scales, the 50% information leak time-scale and the maximum cross-induced causality time-scale, are utilized to evaluate the information flux in the following section.

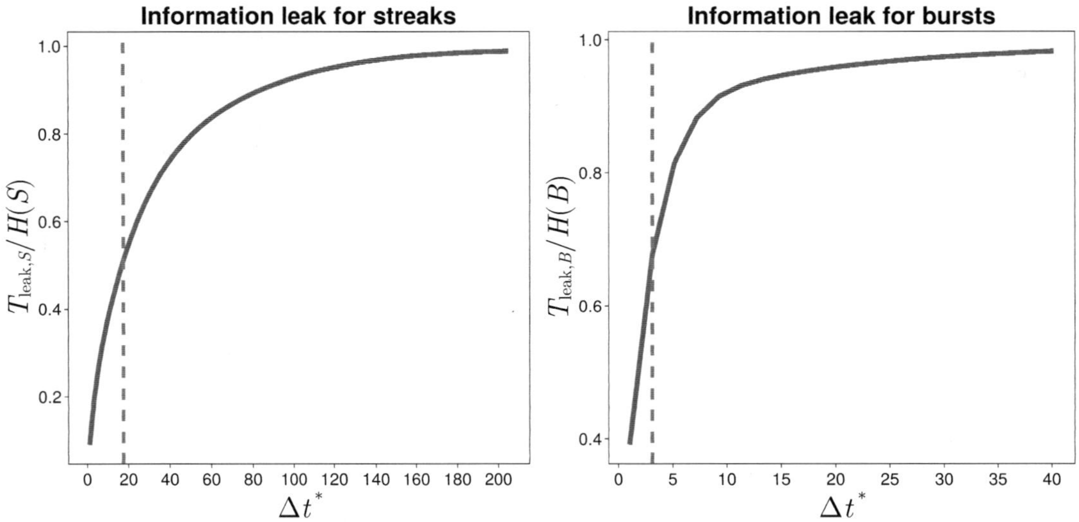

Figure 3: Information leak as a function of time lag in plus units for both streaks (left) and bursts (right). The red dashed line corresponds to the time lag when $T_{leak}/H = 0.5$.

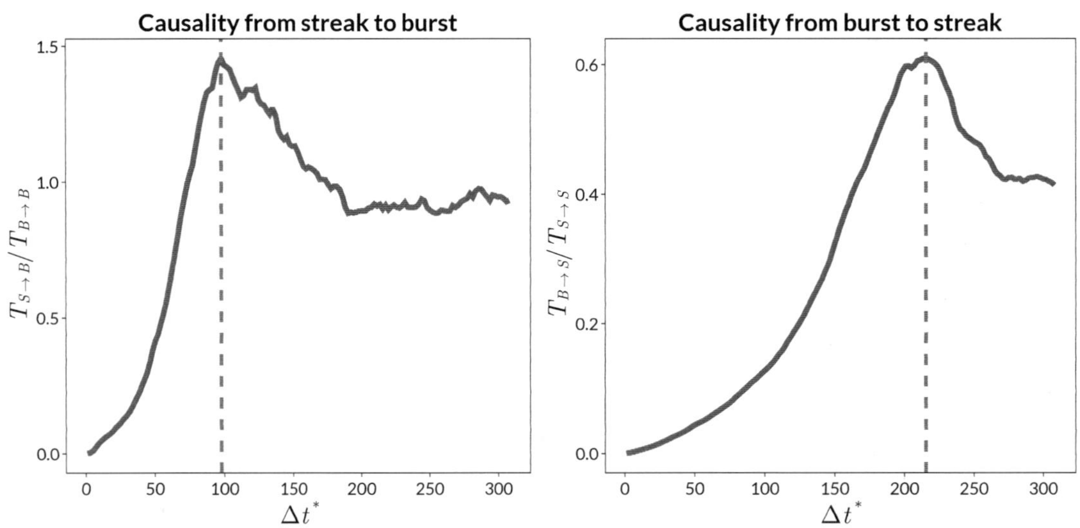

Figure 4: Normalized cross-induced information flux as a function of time lag in plus units. The red vertical dashed line corresponds to the time lag for the maximum cross-induced causality.

3.2. Causality as a function of proximity

We first define the distance between streaks and boxes and its normalization. The streamwise distance d_x and the spanwise distance d_z are defined as the streamwise and spanwise separation between the centroids of the confining streak and burst boxes, as shown in Figure 5a. Note that d_x can be positive or negative, whereas we always set d_z to be positive. We follow the convention that d_x is positive when the causal entity is downstream of the reference object. For example, when we investigate the causal effect from streak to burst, positive d_x represents the case where

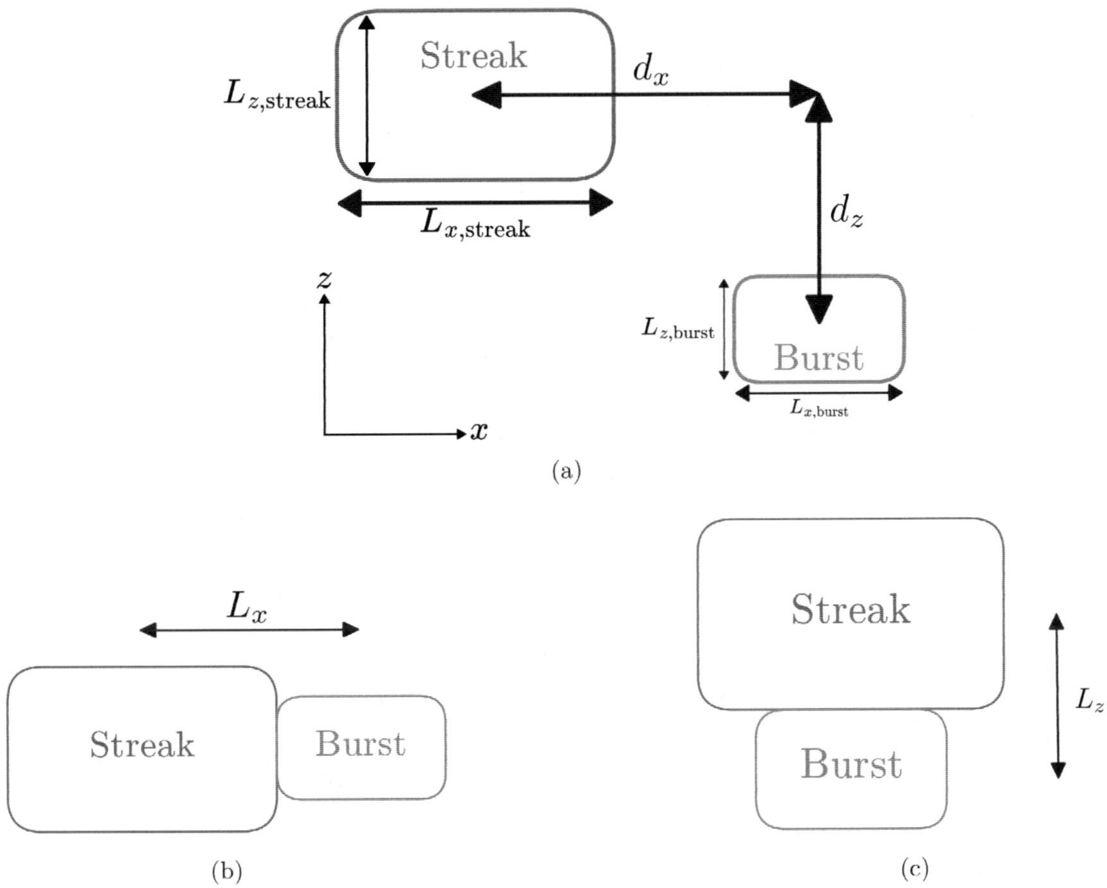

Figure 5: (a) Definition of streamwise distance d_x and spanwise distance d_z between boxes and their dimensions, (b) streamwise normalization coefficient L_x, and (c) spanwise normalization distance L_z.

the streak is downstream. Distances d_x or d_z are normalized by L_x or L_z defined as

$$L_x = \frac{L_{x,\text{streak}} + L_{x,\text{burst}}}{2}, L_z = \frac{L_{z,\text{streak}} + L_{z,\text{burst}}}{2}, \tag{13}$$

as shown in Figure 5b and Figure 5c. This provides an intuitive measure of the relative distance between streaks and bursts. If $d_x/L_x > 1$ and $d_z/L_z > 1$, then the streak box has no overlap with the burst box. Conversely, if $d_x/L_x < 1$ or $d_z/L_z < 1$, the streak and burst are close to each other.

Based on these normalizations and the decomposition of causality, we computed the cross-induced causality normalized by self-induced causality at various levels of proximity between streaks and bursts. The causal strength at different proximity levels for the two time scales of interest is shown in Figure 6 and Figure 7. A common observation for both time scales is that the most causal spatial configuration for streaks and bursts is streamwise aligned, i.e., when d_z/L_z is smaller than one. However, the most causal streamwise distance differs. For the short time scale corresponding to 50% information leak, the most causal relation occurs when the causal object is barely upstream of the effect object, either overlapping or just slightly. On the long time scale with maximum causality, the most causal relation occurs when the causal object

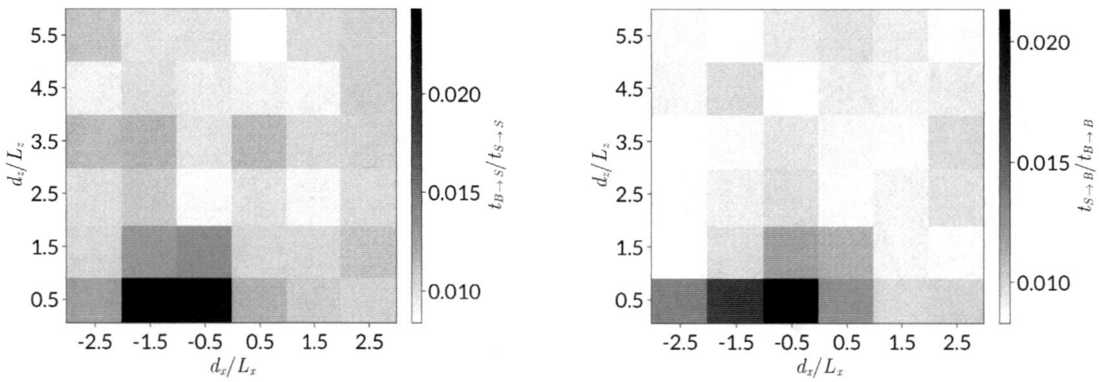

(a) Burst to streak (b) Streak to burst

Figure 6: Causal relation between burst and streak at the time lag representative of fifty-percent-leak. Each block corresponds to the causal strength when the distance satisfies certain distance criteria. For example, the left bottom corner corresponds to $d_x/L_x \in [-3.0, 2.0]$ and $d_z/L_z \in [0.0, 1.0]$. The causal direction is shown in the title of each figure.

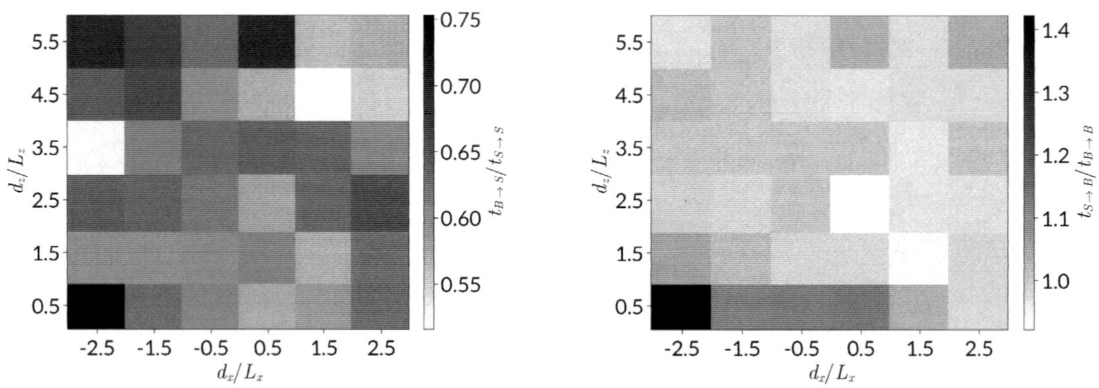

(a) Burst to streak (b) Streak to burst

Figure 7: Causal relation between burst and streak at the maximum cross-induced causality time lag. Follow the same convention as detailed in Figure 6.

is further upstream. It is also noteworthy that while at the 50% information leak time-lag, the magnitude of the information flux is comparable, there is a clear bias towards streak-to-burst causality at the maximum-causality time lag.

Although we only focus on the length scale corresponding to the maximum causality in the previous analysis, we also notice that there are some non-negligible causal relations corresponding to other spatial configurations. Specifically, the causal strengths are significant at the spanwise distance $d_z/L_z \in [5.0, 6.0]$ as shown in Figure 7a. This is currently not well understood and may be attributed to the inaccurate estimation of information flux at this time scale. More studies are needed to determine whether this is a numerical artifact or a physical phenomenon.

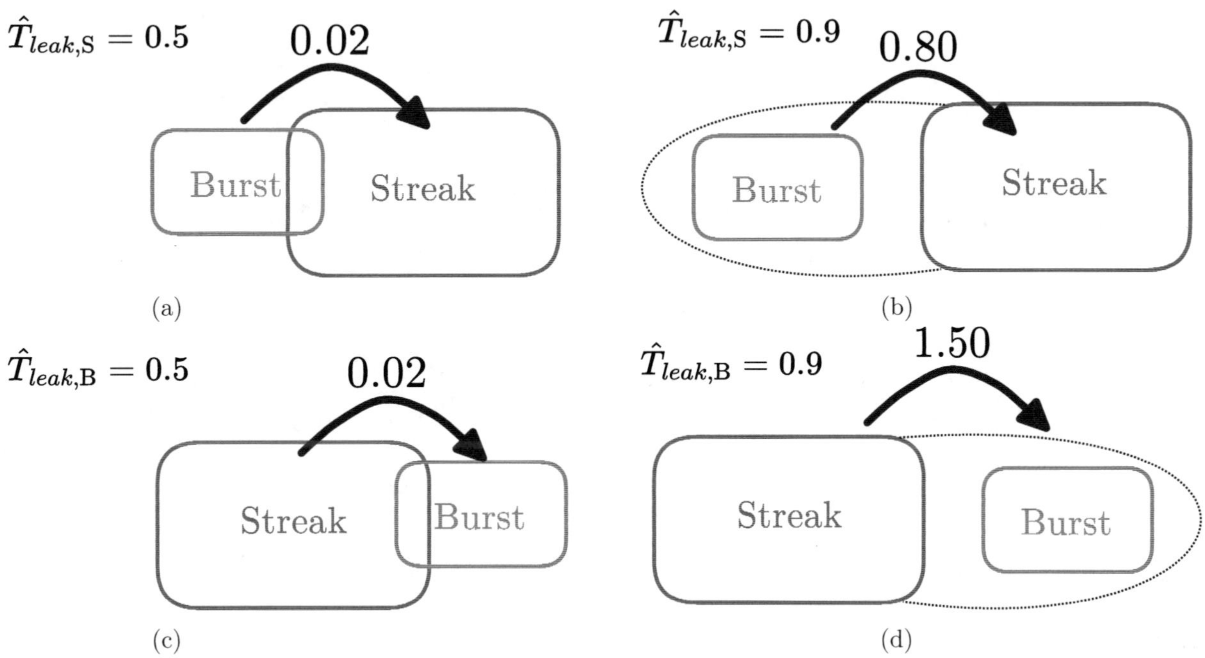

Figure 8: Summary of most causal states: (a) burst to streak at 50% information leak time scale; (b) burst to streak at maximum causality time scale; (c) streak to burst at 50% information leak time scale; (d) streak to burst at maximum causality time scale. The numbers on the left top show the normalized information leak defined in the equation (11) at the corresponding time scale. The values on the arrow represent the normalized cross-induced information flux defined in the equation (12).

4. Discussion

We can identify the most causal relationships between streaks and bursts by synthesizing the causal relations at different time scales and proximity levels. The summary is presented in Figure 8. Four dominant causal patterns are discerned. At the 50% information leak time scale (depicted in Figure 8a and Figure 8b), streaks and bursts exhibit the highest causality when they are streamwise aligned and in close proximity. The causal strength is also similar between them. At the maximum-causality scale, where 90% of the information is missing, they remain most causal to each other when streamwise aligned, but they are positioned further apart. Additionally, there is a noticeable asymmetry in the causal strength, favoring streak-to-burst causality. It is important to note that for the maximum causal time scale, although the most causal relation occurs when the causal object is further upstream based on the current fixed streak box size, it is hypothesized that a much longer streak exists so that the burst and streak still overlap when the causal mechanism is effective.

The analysis above yields three key implications from the current work. Firstly, regardless of the time scale considered, the most causal relationship occurs when streaks and bursts are streamwise aligned. This suggests a localized nature of causal effects. Consequently, in structural models, emphasis should be placed on interactions local in the streamwise direction. Secondly, the magnitude of information leaks in both scenarios indicates the necessity of incorporating more variables. As discussed in [15], a large information leak implies missing significant effects, necessitating the inclusion of additional variables to correctly infer causality. To improve the current method and incorporate more information, one simple modification is to account for the causality from multiple streaks or bursts. For instance, exploring the causality among three closest located streaks and bursts could provide additional insights into the causal interactions. In this scenario, the synergistic term may gain more weight and could become relevant in the analysis. Lastly, the predominant causality from streaks to bursts at the maximum-causality time scale aligns with previous observations in DNS data, indicating that bursts tend to align with long streaks. This observation implies that streamwise streaks act as seeding points for bursts [26]. The information flux from bursts to streaks aligns with the well-known lift-up/Orr linear mechanism through which bursts contribute to the generation of streaks.

Finally, the estimation of information fluxes could be challenging. In this work, about 300,000 samples are collected to estimate the information flux in each block shown in Figures 6 and 7 using a bin-based method. While there is evidence that other estimation methods such as kernel methods, k-nearest-neighbors-based methods or neural network-based methods could be more sample-efficient [27], the bin-based method is chosen in this work due to its flexibility to incorporate the decomposition in proximity. The trade-off is that more samples are needed to obtain a reliable estimation.

5. Conclusion

The causality of dynamically significant structures in wall-bounded turbulence is crucial to understand due to its importance in structural modeling. Previous studies primarily focused on inferring causality by directly intervening in the system. In this work, a non-interventional method based on information theory, known as information flux, was employed to examine the causality between streaks and bursts in channel flows. Using a DNS dataset with $\mathrm{Re}_\tau \approx 180$ in a large channel, time signals representative of streaks and bursts were extracted. The information flux was computed at two distinct time scales: one corresponding to a fifty percent leak and another to the maximum causality. These computations were performed for varying proximity between streaks and bursts.

The results at both time scales revealed that the most causal spatial configuration occurred when streaks and bursts were streamwise aligned. However, the specific streamwise distance varied depending on the time scale: it was further upstream for the maximum-causality time scale and shorter for the 50% information leak counterpart. Although the magnitude of the maximum information flux was comparable at the short time scale, streak-to-burst causality dominated at the longer time scale. Future studies should focus on including more objects to reduce information leak and enhance the estimation of information flux.

Acknowledgments

This work was supported in part by the European Research Council under the Caust grant ERC-AdG-101018287 and the National Science Foundation under Grant No. 2140775 and MISTI Global Seed Funds and UPM. The authors also acknowledge the MIT SuperCloud and Lincoln Laboratory Supercomputing Center for providing HPC resources.

References

[1] Kline S J, Reynolds W C, Schraub F A and Runstadler P W 1967 *J. Fluid Mech.* **30** 741–773
[2] Adrian R J 2007 *Phys. Fluids* **19** 041301
[3] Jiménez J 2018 *J. Fluid Mech.* **842** P1
[4] Hamilton J M, Kim J and Waleffe F 1995 *J. Fluid Mech.* **287** 317–348
[5] Waleffe F 1997 *Phys. Fluids* **9** 883–900
[6] Lozano-Durán A, Bae H J and Encinar M P 2020 *J. Fluid Mech.* **882** A2
[7] Jiménez J and Pinelli A 1999 *J. Fluid Mech.* **389** 335–359
[8] Hwang Y and Cossu C 2010 *Phys. Rev. Lett.* **105** 044505
[9] Farrell B F, Ioannou P J, Jiménez J, Constantinou N C, Lozano-Durán A and Nikolaidis M A 2016 *J. Fluid Mech.* **809** 290–315

[10] Jiménez J 2022 *J. Fluid Mech.* **945** R3
[11] Runge J, Bathiany S, Bollt E, Camps-Valls G, Coumou D, Deyle E, Glymour C, Kretschmer M, Mahecha M D, Muñoz-Marí J, van Nes E H, Peters J, Quax R, Reichstein M, Scheffer M, Schölkopf B, Spirtes P, Sugihara G, Sun J, Zhang K and Zscheischler J 2019 *Nat Commun* **10** 2553
[12] Shannon C E 1948 *Bell Syst. Tech. J.* **27** 379–423
[13] Schreiber T 2000 *Phys. Rev. Lett.* **85** 461–464
[14] Bossomaier T, Barnett L, Harré M and Lizier J T 2016 *Transfer Entropy* (Springer) pp 65–95
[15] Lozano-Durán A and Arranz G 2022 *Phys. Rev. Research* **4** 023195
[16] Orlandi P and Moreau R (eds) 2000 *Fluid Flow Phenomena* (*Fluid Mechanics and Its Applications* vol 55) (Springer Netherlands)
[17] Bae H J, Lozano-Durán A, Bose S T and Moin P 2019 *J. Fluid Mech.* **859** 400–432
[18] Lozano-Durán A and Bae H J 2019 Turbulent channel with slip boundaries as a benchmark for subgrid-scale models in LES *Annual Research Briefs* (Center for Turbulence Research) pp 97–103
[19] Sillero J 2014 *High Reynolds Numbers Turbulent Boundary Layers* Ph.D. thesis U. Politécnica de Madrid
[20] Lozano-Durán A and Jiménez J 2014 *J. Fluid Mech.* **759** 432–471
[21] Hoyas S and Jiménez J 2006 *Phys. Fluids* **18** 011702
[22] Toh S and Itano T 2005 *J. Fluid Mech.* **524** 249–262
[23] Zhou Z, Xu C X and Jiménez J 2022 *J. Fluid Mech.* **940** A23
[24] Jiménez J, Del Álamo J C and Flores O 2004 *J. Fluid Mech.* **505** 179–199
[25] Cover T M and Thomas J A 2005 *Elements of Information Theory* 1st ed (Wiley)
[26] Lozano-Durán A, Flores O and Jiménez J 2012 *J. Fluid Mech.* **694** 100–130
[27] Hlavackovaschindler K, Palus M, Vejmelka M and Bhattacharya J 2007 *Phys. Rep.* **441** 1–46

Linear and nonlinear Granger causality analysis of turbulent duct flows

B Lopez-Doriga[1], M Atzori[2], R Vinuesa[3], H J Bae[4], A Srivastava[1], and S T M Dawson[1]

[1]Mechanical, Materials & Aerospace Engineering Department, Illinois Institute of Technology, Chicago IL 60616, USA
[2]Dipartimento di Scienze e Tecnologie Aerospaziali, Politecnico di Milano (DAER), Milano 20156, Italy
[3]FLOW, Engineering Mechanics, KTH Royal Institute of Technology, Stockholm 10044, Sweden
[4]Graduate Aerospace Laboratories, California Institute of Technology, Pasadena CA 91125, USA

E-mail: `blopezdorigacostales@hawk.iit.edu`

Abstract. This research focuses on the identification and causality analysis of coherent structures that arise in turbulent flows in square and rectangular ducts. Coherent structures are first identified from direct numerical simulation data via proper orthogonal decomposition (POD), both by using all velocity components, and after separating the streamwise and secondary components of the flow. The causal relations between the mode coefficients are analysed using pairwise-conditional Granger causality analysis. We also formulate a nonlinear Granger causality analysis that can account for nonlinear interactions between modes. Focusing on streamwise-constant structures within a duct of short streamwise extent, we show that the causal relationships are highly sensitive to whether the mode coefficients or their squared values are considered, whether nonlinear effects are explicitly accounted for, and whether streamwise and secondary flow structures are separated prior to causality analyses. We leverage these sensitivities to determine that linear mechanisms underpin causal relationships between modes that share the same symmetry or anti-symmetry properties about the corner bisector, while nonlinear effects govern the causal interactions between symmetric and antisymmetric modes. In all cases, we find that the secondary flow fluctuations (manifesting as streamwise vorticial structures) are the primary cause of both the presence and movement of near-wall streaks towards and away from the duct corners.

1. Introduction

There exist a variety of modal analysis methods that can be used to identify coherent structures in fluid flows [1, 2]. However, additional analysis is often required to understand the causal mechanisms that govern the formation and interaction between such structures, particularly true for data-driven decomposition methods such as the proper orthogonal decomposition (POD).

A number of recent studies have introduced quantitative causality analyses for fluid mechanics problems [3–7]. Among numerous frameworks, Granger causality [8] represents a relatively simple and robust scheme, although its original formulation is limited to linear interactions between the time series of different variables/quantities. However, this linear formulation enables direct connections with other linear methods used in fluids, such as dynamic mode decomposition [9], which can be used to model the underlying linear dynamics assumed to describe the system for such analysis. A major drawback of Granger causality is its linearity, which may render it insufficient to expose causal behavior for nonlinear systems. On the other hand, transfer entropy causality analyses [10] have successfully identified relevant dynamics in the field of fluids [11–13]. One of the main strengths of transfer entropy is that it is agnostic to the form of the system dynamics, although a significantly large amount of data and a high computational cost are required to achieve converged results [14–16]. Here, we opt for Granger causality analysis and extend its scope to include second-order quadratic interactions. This makes the framework more amenable for the study of nonlinear systemswhile simultaneously exploiting the advantages and simplicity of the method. Furthermore, the proposed method will enable disambiguation between linear and nonlinear causal mechanisms.

This work focuses on turbulent flows through a square duct. This flow configuration has been the subject of a number of prior investigations [17–26]. One characteristic feature that distinguishes such flows from infinitely-wide channel flow is the presence of secondary mean flow components perpendicular to the direction of the bulk flow [27,28], which typically features a pair of counter-rotating streamwise vortices in each corner, and therefore increases the streamwise velocity near the corner. Moreover, the configuration of this secondary flow does not settle until $Re_\tau \approx 300$ [29], although all the intermediate configurations share a similar pattern that involves counter-rotating streamwise vortices. While it has been postulated for several decades that this secondary flow emerges due to components of the Reynolds stress [18,30], there remains uncertainty as to how the dynamics and causal relationships between near-wall coherent structures contribute to this phenomena. This work will seek to advance our understanding of such mechanisms.

The paper is structured as follows. Section 2 provides a description of the main features of the turbulent duct flow. This is followed by an overview of POD in Section 3. Section 4 presents the linear/standard formulation of Granger causality analysis and introduces a nonlinear formulation of the framework. The results of the present investigation are divided into: Section 5.1, where we discuss the accuracy of linear and nonlinear Granger causality analyses on the low-order turbulence model introduced by Moehlis et al. [31], and compare with the results presented in Martínez-Sánchez et al. [7]; and Section 5.2, in which we examine the coherent structures identified by POD of the turbulent data and discusses the results obtained by both linear and nonlinear Granger causality analyses. Lastly, Section 6 reviews the main findings of this work and discusses future prospects.

2. Turbulent duct flow

We focus on flow through a square duct, governed by the incompressible Navier–Stokes equations

$$\frac{\partial \boldsymbol{u}}{\partial t} = -\boldsymbol{u} \cdot \nabla \boldsymbol{u} - \nabla p + \frac{1}{Re} \Delta \boldsymbol{u}, \tag{1}$$

$$\nabla \cdot \boldsymbol{u} = 0, \tag{2}$$

where $\boldsymbol{u} = [u(\boldsymbol{x},t), v(\boldsymbol{x},t), w(\boldsymbol{x},t)]^T$ represents the instantaneous velocity field, $\boldsymbol{x} = (x,y,z)^T$ are the spatial coordinates and $p = p(\boldsymbol{x},t)$ denotes pressure. These enforce conservation of momentum and continuity, respectively. Here x denotes the streamwise direction, and the other two dimensions $(y,\, z)$ are confined to the rectangular domain $\Omega = \{y \in [-h,h]\} \times \{z \in [-L,L]\}$, where $A = L/h$ is the aspect ratio of the rectangular cross-section. This study will only consider the case $A = 1$. The gradient operator is given by $\nabla = [\partial_x, \partial_y, \partial_z]^T$ and the Laplacian operator is defined as $\Delta = \nabla \cdot \nabla$. No-slip and no-penetration conditions are imposed at all wall-boundaries. In the numerical simulations, periodic boundary conditions are applied in the streamwise direction. Note that velocity is nondimensionalized by the friction velocity $u_\tau = \sqrt{\tau_w/\rho}$, with τ_w denoting the wall-shear stress, and the friction Reynolds number is defined as $Re_\tau = hu_\tau/\nu$, where ν corresponds to the kinematic viscosity.

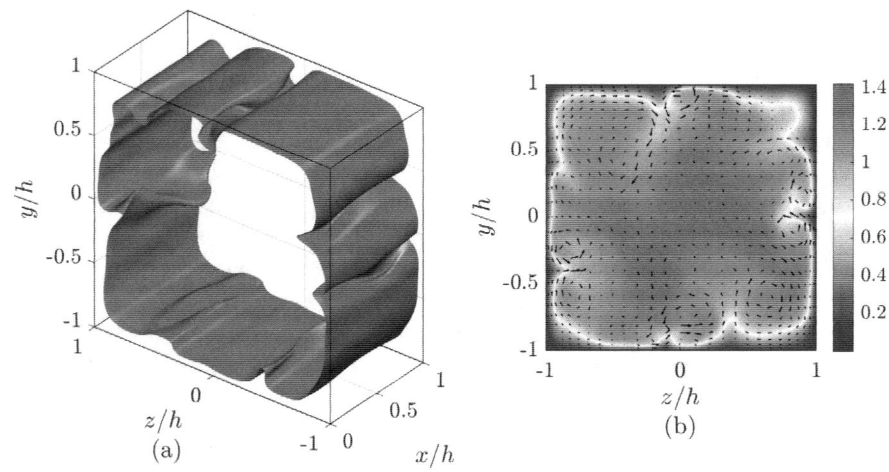

Figure 1: (a) Contour level of a sample snapshot of the instantaneous streamwise velocity component $u(x, y, z, t)$ and (b) streamwise component and vector field of the (y, z) components of the streamwise average of a single snapshot $\boldsymbol{u}(\overline{x}, y, z, t)$ for a minimal turbulent duct, with $Re_\tau = 180$ and $A = 1$.

The majority of the analysis will consider data from a short direct numerical simulation of flow through a square duct of streamwise length h, and cross-section $2h \times 2h$. We use 4,000 snapshots of data, collected with at a friction Reynolds number $Re_\tau = 180$ with a time-step $\Delta t = 1$ (nondimensionalized by the bulk velocity U_b and the lengthscale h). Figure 1 (a) shows a representative snapshot of this system. Further details regarding this configuration, and the numerical methods used for the DNS, can be found in Refs. [32, 33]. Due to the very small streamwise extent of this particular configuration, we refer to this case as the short duct. While the short streamwise extent eliminates some of the phenomena and complexity present in a longer duct, it still retains some of its key features. In particular, the mean secondary flow includes two counter-rotating vortices in each of the corners of the duct, oriented such that mean flow is directed from the duct center towards the corner. As a result, the streamwise mean component features non-convex contour levels near these regions [34]. Note that while some of these vortices can be identified on the individual snapshots (see Figure 1 (b)), the four pairs are rarely identified in one single instance, and in fact only appear altogether when the temporal mean is computed. This secondary mean flow, usually referred to as Prandtl's secondary flow of the second kind [28], is shown in Figure 2. In addition to showing the mean velocity for the short duct configuration, in Figure 2(c) we also show the mean flow for a much longer duct, with streamwise length of $25h$ (the corresponding streamwise mean flow is not shown here since there are no significant differences with respect to Figure 2(a)). While the magnitude of the secondary flow is larger for the longer duct (reaching a maximum amplitude of approximately 2% of the streamwise mean, in comparison to approximately 0.4% for the short duct), we observe that both configurations share secondary flow profiles with the same key features. As such, this short duct configuration represents a simplified system that can be used both to test the causality

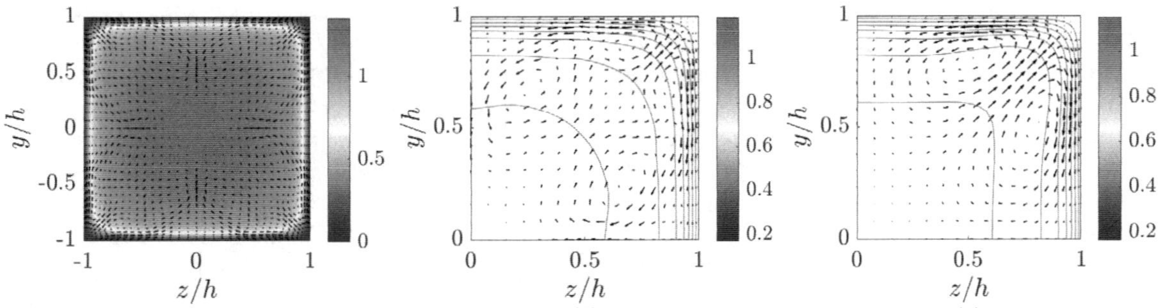

Figure 2: (a) Filled contour plot of the streamwise component and vector field of the secondary components of the turbulent mean $\boldsymbol{U}(y,z)$ of the minimal (short) duct; (b) detail of the mean $\boldsymbol{U}(y,z)$ in the upper-right corner of the minimal duct, and (c) the equivalent plot for the long duct, all with $Re_\tau = 180$ and $A = 1$.

analysis methods that are developed in the present work, and to obtain insight into the physical mechanisms underlying turbulent duct flow.

3. Proper orthogonal decomposition

This section provides a description of the modal decomposition technique used in this paper to extract the dominant coherent structures that arise within turbulent duct flows, which will be the subject of the causality analysis performed in the second part of this study. POD [35–38] identifies modal spatial structures from a vector field in a manner such that each of the coherent modes captures an optimal fraction of the total energy content. In fact, the term *proper* or optimal alludes to the fact that the the kinetic energy captured by the first n spatial modes is maximized, providing a low-order representation of the field of interest. In this work, this technique will naturally apply to turbulent flows, although it has been fruitful in several other applications [39–41]. For the purposes of this research, let us assume that a vector field $\boldsymbol{u}(\boldsymbol{x},t)$ (*e.g.* a time-varying fluid velocity field, such as a turbulent flow) can be decomposed as

$$\boldsymbol{u}(\boldsymbol{x},t) = \boldsymbol{U}(\boldsymbol{x}) + \sum_{i=1}^{\infty} \boldsymbol{\psi}_i(\boldsymbol{x})\boldsymbol{\phi}_i(t), \tag{3}$$

that is, the sum of a reference state $\boldsymbol{U}$ (oftentimes the time-average of the velocity field), and an orthonormal basis of spatial functions $\{\boldsymbol{\psi}_i(x)\}_{i=1}^{\infty}$. Each of these spatial modes maximizes the captured energy content according to the following condition

$$\int_\Omega \mathbb{E}[\boldsymbol{u}'(\boldsymbol{x},t)\boldsymbol{u}'(\boldsymbol{x}',t)]\boldsymbol{\psi}_i(\boldsymbol{x}')d\boldsymbol{x}' = \lambda_i\boldsymbol{\psi}_i(\boldsymbol{x}), \tag{4}$$

where $\mathbb{E}$ represents the expected value, and $\boldsymbol{u}' = \boldsymbol{u} - \boldsymbol{U}$ is the mean-subtracted velocity field. The temporal coefficients $\boldsymbol{\phi}_i(t)$ corresponding to each of the spatial modes can be computed from the integral

$$\boldsymbol{\phi}_i(t) = \int_\Omega \boldsymbol{u}(\boldsymbol{x},t)\boldsymbol{\psi}_i(\boldsymbol{x},t)d\boldsymbol{x}. \tag{5}$$

Notice that while these definitions yield POD coefficients $\boldsymbol{\phi}_i(t)$ that are uncorrelated by design, there can still be causal relationships between these coefficients. In this work, this formulation is adapted in order to manipulate finite-dimensional data. This can be formulated as performing a singular value decomposition (SVD) on a data matrix $\boldsymbol{Y}$, where each column represents one snapshot of the flow field (*e.g.* the snapshot shown in Figure 1).

Here we only consider streamwise-constant structures over the length of the short duct, and therefore we take the average in the x-direction. The square duct has a total of 8 geometric (*i.e.* reflection and rotation) symmetries, which we can enforce to increase the total size of our dataset.

The data matrix $\boldsymbol{Y}$ has dimensions $[3N_yN_z \times 8N_t]$, where the factor of 3 accounts for all velocity components, and the factor of 8 accounts for all transformations that respect the geometric symmetries. The SVD of the discretized operator $\boldsymbol{Y}$ is now written as

$$\boldsymbol{Y} = \boldsymbol{\Psi}\boldsymbol{\Sigma}\boldsymbol{\Phi}^*, \tag{6}$$

where the right-hand side represents the product of discrete operators. Each of the entries σ_i in the diagonal matrix $\boldsymbol{\Sigma}$ is related to the energy content of the spatial mode $\boldsymbol{\psi}_i$ (the i-th column of $\boldsymbol{\psi}$) associated with a temporal coefficient $\boldsymbol{\phi}_i$ (the i-th column of $\boldsymbol{\Phi}$). The POD modes for the duct configuration will be shown and discussed in Section 5.2 (Figures 5–6).

4. Causality analysis

Identifying and characterizing the mechanisms that exist within dynamically-complex fluid flows can be an arduous task. However, the study of causal relationships between some of the identified mechanisms or features can potentially unveil relevant features or processes that would be otherwise difficult to trace. We ultimately seek to identify relationships of causality among the streamwise-constant structures that play a part in sustaining and driving turbulence within a rectangular/square straight duct. These relationships could potentially provide further insight concerning the energy flux between scales and velocity components, and also could inform us about the role of the secondary motions.

In this work, we consider a multivariate formulation of Granger causality analysis (MVGC) [42, 43] to study the pairwise causal relationships between the temporal coefficients ($\boldsymbol{\phi}_i(t)$ in Equation (6)) of the streamwise-constant coherent structures within the minimal duct. This framework provides a quantitative estimation of the amount of information that is transferred from one variable (captured by its time-series) to another, while accounting for the effect of knowledge of all other variables. We first discuss this for standard (linear) Granger causality, before describing a nonlinear extension to this method.

4.1. Linear multivariate Granger causality analysis

Granger causality analysis is a statistics-based test first developed by Granger [8] to measure the causal relationship between econometric time series or variables [44, 45]. Here, the concept of causality is studied through the lens of predictability: the causal relationship between two variables $\boldsymbol{\phi}_i$ and $\boldsymbol{\phi}_j$ measures how much the previous knowledge of $\boldsymbol{\phi}_j$ improves the prediction of $\boldsymbol{\phi}_i$. Let us formalize this concept and consider a set $\boldsymbol{\phi}(t) = [\boldsymbol{\phi}_1(t), \boldsymbol{\phi}_2(t), ..., \boldsymbol{\phi}_n(t)]$ of temporal signals (in this work they will coincide with the temporal coefficients of the POD modes). In this framework, the prediction of the current state of a given signal $\boldsymbol{\phi}_i(t)$ at a time t can be written as

$$\boldsymbol{\phi}_i(t) = \sum_{r=1}^{p} A_{i,i}^{(r)}\boldsymbol{\phi}_i^{(r)} + \sum_{r=1}^{p} A_{i,j}^{(r)}\boldsymbol{\phi}_j^{(r)} + \sum_{\substack{k=1\\ k\notin\{i,j\}}}^{n} \sum_{r=1}^{p} A_{i,k}^{(r)}\boldsymbol{\phi}_k^{(r)} + \epsilon_{i,j}(t), \tag{7}$$

where here we use the shorthand $\boldsymbol{\phi}_i^{(r)} = \boldsymbol{\phi}_i(t - r\Delta t)$ to denote the previous-time measurements, where we use a total of p prior timesteps in the model for the current state. Note that this parameter p will be referred to in the results section as the time lag. The sum over k involves all indices from 1 to n excluding i and j, which makes the causal analysis between i and j conditional on the effect of all other variables. This definition prescribes a linear relationship between a temporal signal ϕ_i, and past values of this and all other variables (across a time horizon $p\Delta t$). In this work, the coefficients A in the linear model are fitted from data. The term $\epsilon_{i,j}(t)$ denotes the error in this fitted linear model. Notice that the contribution of the variable ϕ_j, where $j \neq i$, has been isolated on the second term of the right-hand side, and if we were to neglect it, we could write a modified prediction of $\phi_i(t)$ as

$$\phi_i(t) = \sum_{r=1}^{p} A_{i,i}^{(r)} \phi_i^{(r)} + \sum_{\substack{k=1 \\ k \notin \{i,j\}}}^{n} \sum_{r=1}^{p} A_{i,k}^{(r)} \phi_k^{(r)} + \epsilon_{i,\not{j}}(t). \tag{8}$$

Note that the fitted coefficients A in equation 8 will be different to those in equation 7, owing to having fewer terms on the right hand side. These two expressions allow us to infer the pairwise causal relationship between the temporal signals ϕ_i and ϕ_j by looking at the difference between the size of the residuals $\epsilon_{1,j}(t)$ and $\epsilon_{i,\not{j}}(t)$. Hence, in this framework, causality between two variables ϕ_i and ϕ_j is measured as the difference between the errors in the predictions of ϕ_i including and excluding previous knowledge of ϕ_j.

One of the most relevant implications of this method is the fact that causality can only be found in the history of a time series, and therefore can be sensitive to the observed time window of the temporal signals. Moreover, this formulation restricts the causality analysis to linear mechanisms, potentially missing relevant nonlinear dynamics. It can be additionally shown that Granger causality is equivalent to transfer entropy (which has been used for causality analysis of fluids problems in several previous works [7, 46, 47]) in the case of Gaussian variables [48].

4.2. Nonlinear multivariate Granger causality analysis

One significant drawback of Granger causality analysis is that it is based on a linear modeling assumption. Consequently, the residuals $\epsilon_{i,j}$ and $\epsilon_{i,\not{j}}$ in Equations (7) and (8) likely include the contribution of nonlinear terms involving the variables ϕ_i, ϕ_j, and ϕ_k. To make this approach viable for nonlinear systems, we propose an extension of the original formulation of multivariate Granger causality that allows for the presence of nonlinear terms. In particular, the proposed formulation can be extended to systems with quadratic nonlinear terms while largely maintaining the benefits and simplicity of its linear counterpart. Hence, the causal relationships are still quantified as prediction errors, but after accounting for the presence of selected nonlinear terms. Hence, we can write the general form of the second-order prediction of the current state of a time series $\phi_i(t)$ as a function of the history of the variables ϕ_i, ϕ_j and $\{\phi_k\}_{k \notin \{i,j\}}$

$$\begin{aligned}\phi_i(t) = {} & \sum_{r=1}^{p} A_{i,i}^{(r)} \phi_i^{(r)} + \sum_{r=1}^{p} A_{i,j}^{(r)} \phi_j^{(r)} + \sum_{\substack{k=1 \\ k \notin \{i,j\}}}^{n} \sum_{r=1}^{p} A_{i,k}^{(r)} \phi_k^{(r)} + \sum_{r=1}^{p} Q_{i,i,i}^{(r)} (\phi_i \phi_i)^{(r)} + \\ & \sum_{r=1}^{p} Q_{i,i,j}^{(r)} (\phi_i \phi_j)^{(r)} + \sum_{\substack{k=1 \\ k \notin \{i,j\}}}^{n} \sum_{r=1}^{p} Q_{i,i,k}^{(r)} (\phi_i \phi_k)^{(r)} + \sum_{r=1}^{p} Q_{i,j,j}^{(r)} (\phi_j \phi_j)^{(r)} + \\ & \sum_{\substack{k=1 \\ k \notin \{i,j\}}}^{n} \sum_{r=1}^{p} Q_{i,j,k}^{(r)} (\phi_j \phi_k)^{(r)} + \sum_{\substack{k'=1 \\ k' \notin \{i,j\}}}^{n} \sum_{\substack{k=1 \\ k \notin \{i,j\}}}^{n} \sum_{r=1}^{p} Q_{i,k',k}^{(r)} (\phi_{k'} \phi_k)^{(r)} + \epsilon'_{i,j}(t).\end{aligned} \tag{9}$$

Here, the A-terms capture the linear interactions and the Q-terms model the quadratic interactions. We can similarly write a prediction of the current state $\phi_i(t)$ while excluding all terms that require knowledge of the history of the variable ϕ_j as follows

$$\begin{aligned}\phi_i(t) = \sum_{r=1}^{p} A_{i,i}^{(r)} \phi_i^{(r)} + \sum_{\substack{k=1 \\ k \notin \{i,j\}}}^{n} \sum_{r=1}^{p} A_{i,k}^{(r)} \phi_k^{(r)} + \sum_{r=1}^{p} Q_{i,i,i}^{(r)} (\phi_i \phi_i)^{(r)} + \\ \sum_{\substack{k=1 \\ k \notin \{i,j\}}}^{n} \sum_{r=1}^{p} Q_{i,i,k}^{(r)} (\phi_i \phi_k)^{(r)} + \sum_{\substack{k'=1 \\ k' \notin \{i,j\}}}^{n} \sum_{\substack{k=1 \\ k \notin \{i,j\}}}^{n} \sum_{r=1}^{p} Q_{i,k',k}^{(r)} (\phi_{k'} \phi_k)^{(r)} + \epsilon'_{i,\not{j}}(t),\end{aligned} \tag{10}$$

The causal relationship between ϕ_i and ϕ_j is then again estimated from the difference in magnitude between the residuals of these models, $\epsilon'_{i,j}(t)$ and $\epsilon'_{i,\not{j}}(t)$. While this methodology has been formulated in terms of the POD coefficients, in Section 5 we will also consider the case where the variables in the identified models are instead the squares of the POD coefficients, ϕ_i^2. The dynamics of these variables may indeed need higher order nonlinearities to be fully modeled, though such analysis is beyond the scope of the present investigation. Finding a low-dimensional nonlinear model from data can be performed in a number of ways. Here, this is done by fitting coefficients to data, though it would also be possible to obtain such models by Galerkin projection of the governing equations onto POD modes [49]. This data-driven identification of a nonlinear model also shares similarity with the sparse identification of nonlinear dynamics (SINDy) methodology, which incorporates sparsity-promoting methods to reduce the number of nonzero coefficients [50].

5. Results

Here, we discuss the main findings of this research. Section 5.1 first present preliminary results that validate the formulation of the nonlinear Granger causality analysis on the nine-equation model for turbulent shear flows introduced by Moehlis et al. [31]. This is followed by a discussion of the observed POD modes and the consequent Granger causality analysis using the corresponding temporal coefficients in Section 5.2.

5.1. Linear and nonlinear Granger causality analysis on a low-order turbulence model

In order to assess the validity and capabilities of the proposed nonlinear Granger causality analysis, we first test its performance on the low-dimensional model of wall-bounded turbulent flows introduced by Moehlis et al. [31]. This model is a set of 9 coupled differential equations, featuring quadratic nonlinearities. The variables correspond to features of wall-bounded turbulence, including the mean streamwise velocity, spanwise flow, streamwise streaks, streamwise vortices and so forth. The temporal coefficients $a_j(t)$ associated to each of the entries in the low-dimensional model have the closed solutions in the following form

$$\begin{aligned}
\dot{a}_1(t) &= \frac{\beta^2}{Re}(1-a_1) - \sqrt{\frac{3}{2}}\beta\gamma\left(\frac{a_6 a_8}{\kappa_{\alpha\beta\gamma}} - \frac{a_2 a_3}{\kappa_{\beta\gamma}}\right), \\
\dot{a}_2(t) &= -\frac{a_2}{Re}\left(\frac{4\beta^2}{3} + \gamma^2\right) + \frac{\gamma^2}{\kappa_{\alpha\gamma}}\left(\frac{5\sqrt{2}}{3\sqrt{3}} a_4 a_6 - \frac{1}{\sqrt{6}} a_5 a_7\right) - \frac{\alpha\beta\gamma}{\sqrt{6}\kappa_{\alpha\gamma}\kappa_{\alpha\beta\gamma}} a_5 a_8 - \sqrt{\frac{3}{2}}\frac{\beta\gamma}{\kappa_{\beta\gamma}}(a_1 a_3 - a_3 a_9), \\
\dot{a}_3(t) &= -\frac{\beta^2 + \gamma^2}{Re} a_3 + \frac{2}{\sqrt{6}}\frac{\alpha\beta\gamma}{\kappa_{\alpha\gamma}\kappa_{\beta\gamma}}(a_4 a_7 + a_5 a_6) + \frac{\beta^2(3\alpha^2 + \gamma^2) - 3\gamma^2(\alpha^2 + \gamma^2)}{\sqrt{6}\kappa_{\alpha\gamma}\kappa_{\beta\gamma}\kappa_{\alpha\beta\gamma}} a_4 a_8, \\
\dot{a}_4(t) &= -\frac{3\alpha^2 + 4\beta^2}{3Re} a_4 - \frac{\alpha}{\sqrt{6}} a_1 a_5 - \frac{10}{3\sqrt{6}}\frac{\alpha^2}{\kappa_{\alpha\gamma}} a_2 a_6 - \sqrt{\frac{3}{2}}\frac{\alpha\beta\gamma}{\kappa_{\alpha\gamma}\kappa_{\beta\gamma}} a_3 a_7 - \sqrt{\frac{3}{2}}\frac{\alpha^2\beta^2}{\kappa_{\alpha\gamma}\kappa_{\beta\gamma}\kappa_{\alpha\beta\gamma}} a_3 a_8 - \frac{\alpha}{\sqrt{6}} a_5 a_9,
\end{aligned}$$

$$\dot{a}_5(t) = -\frac{\alpha^2+\beta^2}{Re}a_5 + \frac{\alpha}{\sqrt{6}}a_1a_4 + \frac{\alpha^2}{\sqrt{6}\kappa_{\alpha\gamma}}a_2a_7 - \frac{\alpha\beta\gamma}{\sqrt{6}\kappa_{\alpha\gamma}\kappa_{\alpha\beta\gamma}}a_2a_8 + \frac{\alpha}{\sqrt{6}}a_4a_9 + \frac{2}{\sqrt{6}}\frac{\alpha\beta\gamma}{\kappa_{\alpha\gamma}\kappa_{\beta\gamma}}a_3a_6,$$

$$\dot{a}_6(t) = -\frac{3\alpha^2+4\beta^2+3\gamma^2}{3Re}a_6 + \frac{\alpha}{\sqrt{6}}a_1a_7 + \sqrt{\frac{3}{2}}\frac{\beta\gamma}{\kappa_{\alpha\beta\gamma}}(a_1a_8+a_8a_9) + \frac{10}{3\sqrt{6}}\frac{\alpha^2-\gamma^2}{\kappa_{\alpha\gamma}}a_2a_4 - 2\sqrt{\frac{2}{3}}\frac{\alpha\beta\gamma}{\kappa_{\alpha\gamma}\kappa_{\beta\gamma}}a_3a_5 + \frac{\alpha}{\sqrt{6}}a_7a_9,$$

$$\dot{a}_7(t) = -\frac{\alpha^2+\beta^2+\gamma^2}{Re}a_7 - \frac{\alpha}{\sqrt{6}}(a_1a_6+a_6a_9) + \frac{1}{\sqrt{6}}\frac{\gamma^2-\alpha^2}{\kappa_{\alpha\gamma}}a_2a_5 + \frac{1}{\sqrt{6}}\frac{\alpha\beta\gamma}{\kappa_{\alpha\gamma}\kappa_{\beta\gamma}}a_3a_4,$$

$$\dot{a}_8(t) = -\frac{\alpha^2+\beta^2+\gamma^2}{Re}a_8 + \frac{2}{\sqrt{6}}\frac{\alpha\beta\gamma}{\kappa_{\alpha\gamma}\kappa_{\alpha\beta\gamma}}a_2a_5 + \frac{\gamma^2(3\alpha^2-\beta^2+3\gamma^2)}{\sqrt{6}\kappa_{\alpha\gamma}\kappa_{\beta\gamma}\kappa_{\alpha\beta\gamma}}a_3a_4,$$

$$\dot{a}_9(t) = -\frac{9\beta^2}{Re}a_9 + \sqrt{\frac{3}{2}}\frac{\beta\gamma}{\kappa_{\beta\gamma}}a_2a_3 - \sqrt{\frac{3}{2}}\frac{\beta\gamma}{\kappa_{\alpha\beta\gamma}}a_6a_8,$$

where $\kappa_{\alpha\gamma} = \sqrt{\alpha^2+\gamma^2}$, $\kappa_{\beta\gamma} = \sqrt{\beta^2+\gamma^2}$, $\kappa_{\alpha\beta\gamma} = \sqrt{\alpha^2+\beta^2+\gamma^2}$. Here, α and γ represent the streamwise and spanwise wavenumbers, respectively, and $\beta = \pi/2$. This system of nonlinear equations can be linearized about the turbulent mean, resulting in equations of the form $\dot{\boldsymbol{a}}(t) = \boldsymbol{A}\boldsymbol{a}(t) + \boldsymbol{f}(t)$, where $\boldsymbol{f}(t)$ is a forcing to the linearized dynamics that can include the effect of nonlinear and constant terms. The mean state of this system is nonzero in the first and ninth states. This leads to quadratic terms involving a_1 and a_9 giving nonzero off-diagonal entries of $\boldsymbol{A}$, as shown in Figure 3.

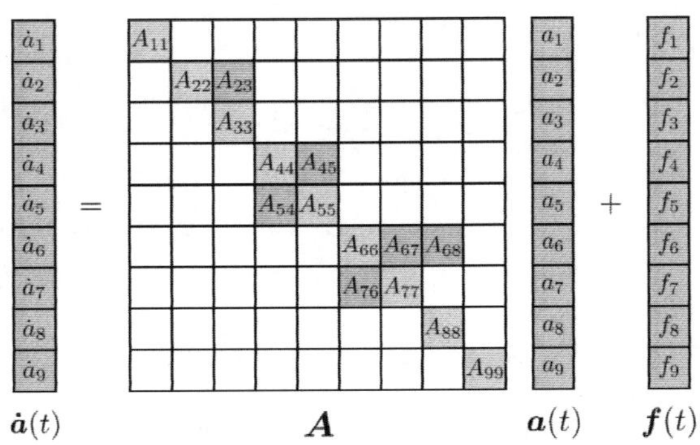

Figure 3: Schematic of the linearized form of the temporal coefficients $a_j(t)$ of the nine equation model [31]. The elements A_{ij} in the linear operator $\boldsymbol{A}$ that are shaded in grey represent the linear viscous terms, those in blue represent the linear elements arising due to the mean flow (*i.e.* analogous to mean shear), and the white elements correspond to zero entries.

Due to the presence of nonlinear terms in these equations, a causality analysis limited to linear interactions would likely not be sufficient for identifying pertinent causal mechanisms. We use linear and nonlinear Granger causality methods to study the interactions between the nine variables of this model, and the results are produced in the form of causality maps. That is, square diagrams in which each of the entries represents one variable, the horizontal axis indicates the origin of the information flux and the vertical axis represents the destination of this flux (*e.g.* a large entry in the element (2,3) of the causality map indicates that there is a relevant causal interaction between variables 3 and 2, in which 3 causes 2). See [42, 43] for more details concerning how such maps are computed from data via identification of the equations described in Section 4.

The causality maps shown in Figure 4 are computed considering 600 different trajectories in a time window $0 \leq t \leq 4,000$ with $\Delta t = 1$, where the first 100 snapshots are discarded. The results described in Martínez-Sánchez et al. [7] are taken as reference, where a transfer entropy

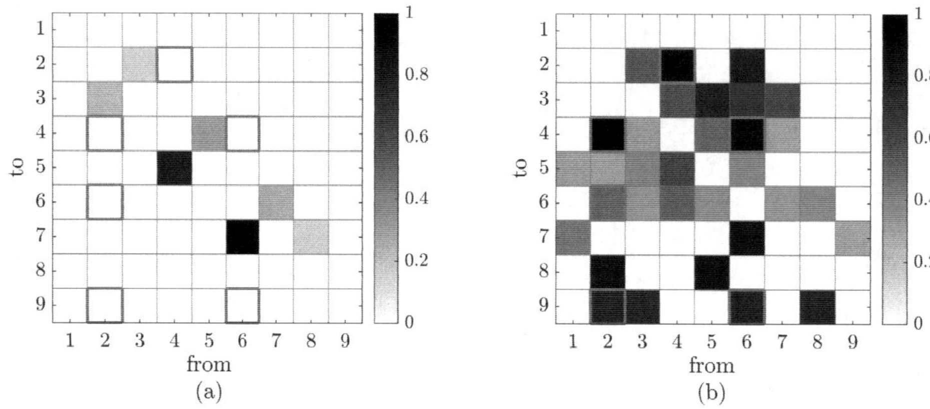

Figure 4: Causality maps produced for the nine-equation model described in Moehlis et al. [31] using (a) linear Granger and (b) nonlinear Granger pairwise multivariate causality analyses. A time lag p corresponding to one time step of size $\Delta t = 1$ was considered. The magnitude of the causality maps has been normalized by the largest entry. Entries highlighted in red are the principal causal relationships identified in Martínez-Sánchez et al. [7].

causality analysis was performed. We observe that the results provided by the linear formulation of Granger causality analysis in Figure 4(a) are not sufficiently accurate, mainly due to the fact that most of the nonlinear interactions are missed. Indeed, the causal interactions identified here correspond to several of the nonzero terms in the linearized form of this model (*e.g.* the terms A_{76} and A_{54} in Figure 3). Note that while the causal mechanisms in the linear Granger analysis (of nonlinear data) do not exactly correlate to all of the nonzero terms in the mean-linearized model, the causality map in figure Figure 4(a) has the same block diagonal structure as the mean-linearized dynamics.

In contrast, the nonlinear formulation of Granger causality analysis is able to identify many additional causal relationships between the variables. In particular, the six major causal interactions identified using transfer entropy in Martínez-Sánchez et al. [7], highlighted in red in Figure 4), are also identified as causal mechanisms here. These entries involve some of the nonlinear term (off-diagonal entries) in Figure 3, and represent mechanisms such as: the effect of streamwise streaks and wall-normal vortices modifying the mean flow in entries (9,2) and (9,6), respectively; the influence of streamwise streaks on the wall-normal vortices in the element (6,2); the excitation of the spanwise flow by the wall-normal vortices in entry (4,6); and the interactions between the streamwise streaks and the spanwise flow by the elements (4,2) and (2,4).

The nonlinear Granger method also identifies a number additional significant causal mechanisms not identified by transfer entropy. The reasons for these differences are the subject of ongoing study. This also highlights an aspect of causality analysis that will repeatedly emerge in the following analysis: the precise causal mechanisms identified can be highly sensitive to the methods used.

5.2. Causal interactions between streamwise-constant coherent structures in turbulent duct flow

Here we first present the spatial POD modes $\boldsymbol{\psi}_j$ (computed according to Equation (6)), extracted from the 4000×8 (accounting for all eight geometric rotations) snapshots of available data $\boldsymbol{Y}$ within the minimal duct. Notice that each of these snapshots has been averaged in the streamwise direction, since the goal of this study is to focus on streamwise-elongated structures. Due to the relations of symmetry within the (y, z) plane, the decomposition is restricted to the lower-left corner of the duct, in order to isolate the relevant coherent structures without accounting for the symmetrical or anti-symmetrical counterparts that may arise in other regions of the cross-section (It would also be possible to instead study only one eighth of the duct using similar symmetry arguments).

The first set of results is produced via POD of the full set of snapshots formed by all three velocity components simultaneously, which yields the full spatial POD modes depicted in Figure 5. Each of these modes contains regions of high and low (negative) streamwise velocity fluctuation, indicative of streaks located in the duct corner (as in modes 2 and 4), or of opposite sign on adjacent walls (modes 1 and 3). Note that modes 1 and 3 are antisymmetric about the bisector of the corner, while modes 2 and 4 are symmetric. Modes 1 and 3 have a secondary (v and w) flow characterized by a single streamwise vortex between the fast and slow streaks, while modes 2 and 4 have a pair of counter-rotating vortices. Modes 2 and 4 therefore resemble the structure of the secondary mean flow (Figure 2 (b)). We emphasize that the mean flow has been subtracted before computing POD, so modes 2 and 4 would correspond to transient fluctuations in the nature and size of the structures giving rise to the secondary mean.

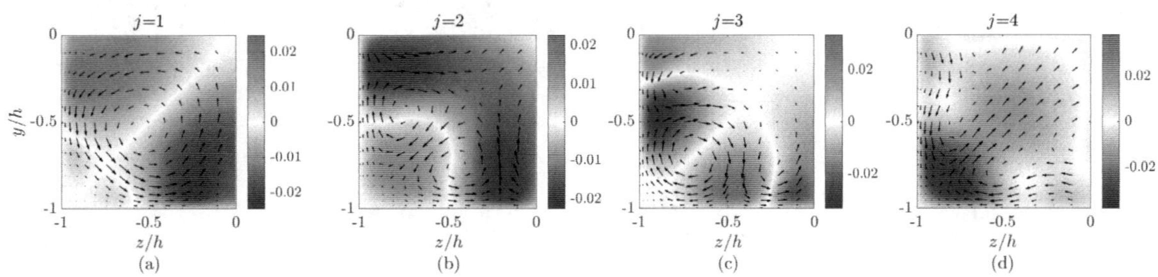

Figure 5: Spatial modes $\boldsymbol{\psi}_j$ with j={1,2,3,4} extracted from the full snapshots in the lower-left corner of the minimal duct, with $Re_\tau = 180$ and $A = 1$. The filled contour plots represent the streamwise component, and the secondary velocity components are depicted as vector fields.

The second set of results is obtained via POD of the decoupled snapshots, that is, the decomposition is performed on the streamwise component u and the (v, w) components of the snapshots separately. This is done to enable causal analysis between the streamwise and secondary velocity components. Spatial modes $\boldsymbol{\psi}_j$ with $j=\{1,2,3,4\}^u$ and $j=\{1,2,3,4\}^{vw}$ from each of the decoupled sets are shown in Figure 6. Notice that the streamwise velocity components of these modes are similar to those shown in each of the four modes in Figure 5. The secondary components are also generally similar, particularly for modes 1 and 3. For mode 2 (Figure 6(f)) the part of streamwise vorticies is larger than in Figure 5(b). For mode 4, the pattern depicted in Figure 6(h) differs from the secondary components of the mode in Figure 5(d). This highlights the fact that when POD modes are computed using all components, the identified structures are driven by the energetically-dominant streamwise velocity component. Also note that since the computations are performed separately, now there can also be correlation between the coefficients of the streamwise and secondary velocity modes.

Figure 7 shows a portion of several of the time series of the temporal signals associated with modes shown in Figures 5–6. Note that while this correlation between the streamwise and secondary velocity modes is perhaps observable in Figure 7(b), this correlation is by no means perfect, indicating some independence between the streamwise and secondary components of the modes shown in Figure 5. These figures also show that coherent features can persist in the

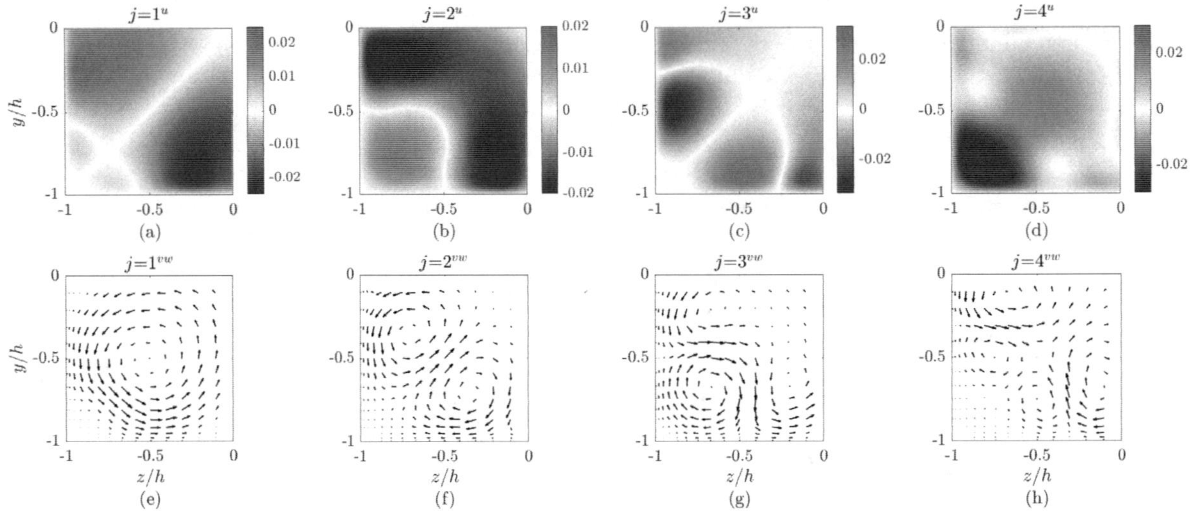

Figure 6: Spatial modes ψ_j with j={1,2,3,4}^{u}(a)-(d) and j={1,2,3,4}^{vw}(e)-(h) extracted from the split snapshots, in the lower-left corner of the minimal duct, with $Re_\tau = 180$ and $A = 1$. The filled contour plots represent the streamwise component, and the secondary velocity components are depicted as vector fields.

flow over long time horizons, for example the negative value of the coefficient of the first mode in Figure 7(a) is maintained over the entire time horizon shown in the figures.

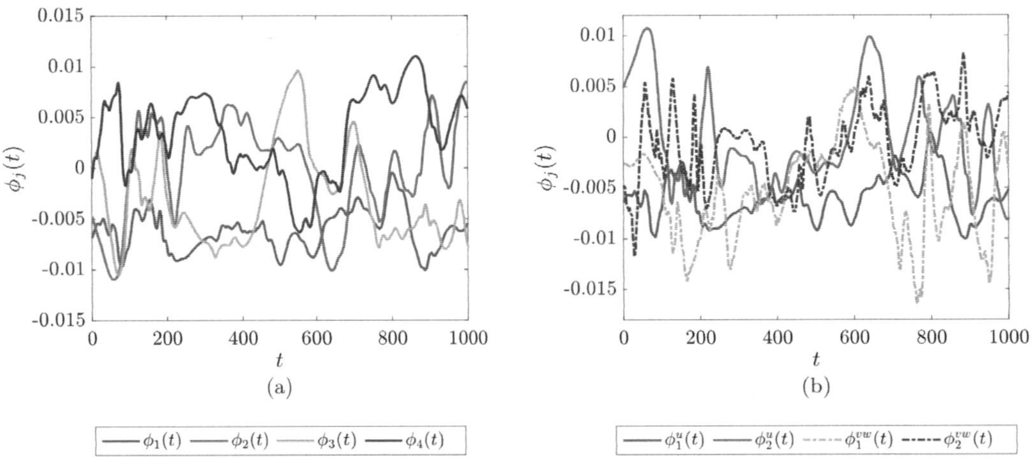

Figure 7: Temporal coefficients $\phi_j(t)$ associated with POD modes ψ_j with j={1,2,3,4} in Figure 5 and POD modes ψ_j with j={1,2}^{u} and j={1,2}^{vw} in Figure 6, over a time window $0 \leq t \leq 1,000$ with $\Delta t = 1$.

The causality maps generated using the temporal coefficients ϕ_j associated to the full modes in Figure 5 are shown in Figure 8. These causality analyses were performed using both the coefficients extracted from the POD $\phi_j(t)$ and the square of these temporal signals $\phi_j^2(t)$. The

Journal of Physics: Conference Series **2753** (2024) 012017 doi:10.1088/1742-6596/2753/1/012017

latter analysis was performed to see if there were causality mechanisms that are independent of the sign of the cause or effect variable. This may be the case for relationships between symmetric and antisymmetric modes (e.g. a symmetric mode may have equal chance of causing a positive or negative change in the coefficient of an antisymmetric mode, which could prevent detection if using the coefficients directly). The square of the temporal coefficients also represents a measure of the kinetic energy of these coherent structures, hence the identified causal relations could also provide insight concerning the energy transfer mechanisms that drive this system. The chosen time lag for the causality analyses presented here is $p = 1$, and different values were not found to have a significant effect on the observed results.

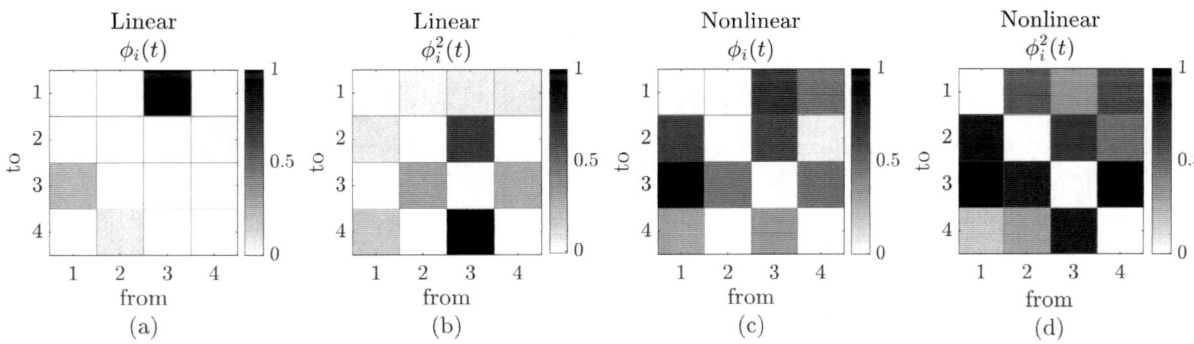

Figure 8: Causality maps produced for the full temporal coefficients $\phi_j(t)$ associated with the spatial modes in Figures 5, as well as the square of the given coefficients, via linear and nonlinear Granger pairwise multivariate causality analyses. The chosen time lag is $p = 1$ with Δt=1 and the entries have been normalized by the largest entry.

The linear formulation of Granger causality analysis in Figure 8(a) highlights one main causal relationship from mode 3 to 1, representing the movement of near-wall streaks and a streamwise vortex along a sidewall away from the corner. There is also a weaker causal analysis in the opposite direction between these two modes. Using the square of the POD coefficients, ϕ_j^2, as inputs to the linear Granger analysis identifies entirely different causal mechanisms, as shown in Figure 8(b). In particular, this analysis identifies causal relationships between symmetric and antisymmetric modes, with the the strongest causality identified as mode 3 influencing modes 2 and 4. This mechanism seemingly corresponds to one of the two streaks on adjacent sides of the corner moving into the corner itself, with the streamwise vortex similarly translating. The other causal mechanisms identified in Figure 8(b) follow similar mechanisms, and the reverse behavior whereby a streak moves out of the corner along one of the adjacent walls. As mentioned previously, we postulate that the reason why this mechanism is entirely missing from Figure 8(a) is because of the interaction between symmetric and antisymmetric modes, where here due to the symmetry of the geometry it is equally likely for a corner streak/vortex to move towards either of the adjacent sides. This would "cancel out" causal relationships when using the coefficients as variables (which can be positive or negative), but is clearly identified when using their energy content, which is agnostic to the sign of the coefficients.

The nonlinear Granger analysis of the full POD mode coefficients shown in Figure 8(c) extracts the same mechanisms identified from the linear Granger analyses of both the POD coefficients and their squared values discussed above (Figures 8(a)-(b)). This is perhaps expected, since the form of the nonlinear model includes both linear and quadratic terms. For completeness, we also show the results of performing nonlinear Granger analysis on the ϕ_j^2 terms in Figure 8(d). This identifies an even broader set of causal relationships, though interpretation of these results must be treated with caution, as the evolution of these energy contents would involve cubic nonlinear terms, which are not included in the nonlinear model.

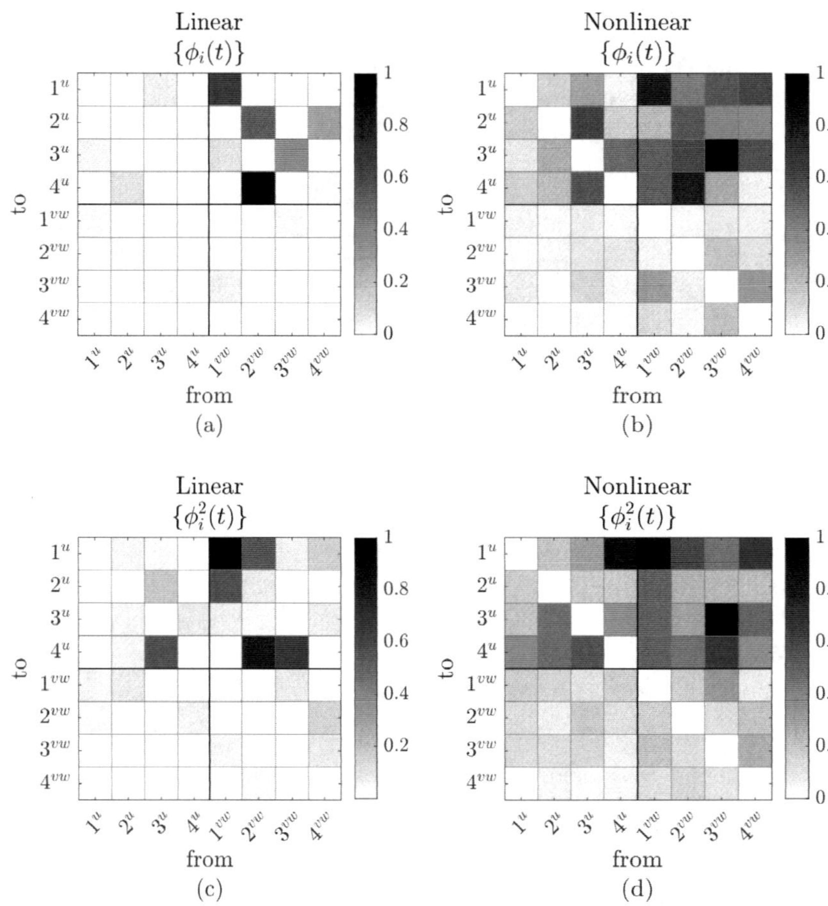

Figure 9: Causality maps produced for the split temporal coefficients $\{\phi_j(t)\}_{split}$ associated with the spatial modes in Figures 6, as well as the square of the given coefficients, via linear and nonlinear Granger pairwise multivariate causality analyses. The chosen time lag is $p = 1$ with Δt=1 and the entries have been normalized by the largest entry.

The causality maps in Figure 9 correspond to the causality analysis performed on the temporal signals ϕ_j and ϕ_j^2 associated with the split spatial modes shown in Figure 6. The temporal coefficients were computed separately in this case to possibly infer relations of causality between velocity components within each mode. In all cases, the upper-right quadrant dominates the causal behavior, emphasizing on the role of the secondary velocity components in driving the formation and dynamics of near-wall and near-corner streamwise streaks. This behavior is likely closely related with the well-known lift-up mechanism [51–53] describing how perturbations in wall-normal velocity can lead to amplified streamwise fluctuations. The causal mechanisms identified by linear Granger causality analysis on the POD coefficients is largely limited to those components corresponding to the same mode for the combined POD, except for the entries corresponding to $2^{vw} \rightarrow 4^u$ and $4^{vw} \rightarrow 2^u$. In particular, the linear analysis once again only identifies causal relationships that preserve the symmetry properties of the associated structures. On the other hand, both the linear analysis of the POD energies (squares of POD coefficients) and the nonlinear Granger analyses additionally identify (presumably nonlinear) causal relationships from the secondary to the streamwise fluctuations that transfer between the symmetric and antisymmetric modes. In particular, this suggests that the similar behavior observed in the combined POD analysis is primarily caused by the presence and motion of the streamwise vortex in all cases.

The causality analyses in Figure 9(b)-(d) also identify several additional mechanisms that are not in the top right quadrant, such as $3^u \rightarrow 2^u$ and/or 4^u, representing a streamwise streak translating along a wall into the corner. The reverse causal relationship, with a corner streak causing sidewall streaks (*e.g.* $4^u \rightarrow 1^u$ and 3^u), is also observed in Figure 9(b) and (d). These mechanisms are analogous to those also identified in the analysis of the combined POD modes (Figure 8), though here we are able to establish the extent to which such mechanisms are driven by the secondary velocity components, as opposed to the streaks themselves. In particular, the results suggest that while the secondary velocity components (streamwise vortices) are primarily responsible for the dynamics of streamwise streaks, there are nonlinear mechanisms whereby streamwise streaks affect their own dynamics.

6. Discussion and conclusions

In this work, we have formulated a nonlinear extension of linear Granger causality analysis that allows for the explicit consideration of nonlinear interactions (of a prescribed form) when quantifying the causal relationships between them. This methodology has been applied to study the dynamics of coherent structures present in turbulent flow through a square duct. It was demonstrated that the causal mechanisms identified are highly sensitive to both the choice of variables and the form of the model used in the analysis.

For standard linear Granger causality applied to POD coefficients, the only causal mechanisms that are identified are between modes that share the same symmetry properties. These mechanisms govern the translation of streamwise vortices and streaks towards and away from the duct corner. Applying linear Granger causality to the POD energy contents reveals a different set of causal mechanisms, which allows for changes in the symmetry properties of the underlying modes. This is attributed to the fact that such mechanisms are agnostic to the sign of the coefficient of the cause or effect variable, so the cumulative causal effect cancels out unless only a quantity that removes dependence on the sign of the coefficient is used. The proposed nonlinear extension is able to capture both of these classes of interactions, however insight is gained when also performing linear analysis, in order to distinguish between linear and nonlinear effects. It is possible that more sophisticated methods to disambiguate between linear and nonlinear relationships could also be incorporated into this analysis workflow [54]. Note that the causal effects of various linear mechanisms can also be studied by intrusively removing or inhibiting certain terms within numerical simulations [5].

By considering the streamwise and secondary mode components separately, we were also able to determine the causal relationships between velocity components within a single mode. It was determined that secondary velocity fluctuations are the primary causal driver of the dynamics in all cases. While the linear mechanisms associated with this are consistent with the lift-up mechanism, the nonlinear mechanisms represent a generalization of this well-known effect.

Note that several of the analysis variants incorporated here are similar to methods used in past causality studies of turbulent flows. For example, Martínez-Sánchez et al. [7] use POD coefficients of modes incorporating all velocity components, while Lozano-Durán et al. [12] use energy content of different streamwise and spanwise wavenumbers (thus not distinguishing the sign/direction of the underlying structures).

There are several possible directions of future work. While the models used for linear and nonlinear Granger causality analysis are here identified from data, their simple form means that they could also potentially be found from physics-based analysis, such as from projecting the linearized or full nonlinear governing equations onto identified POD modes. In the present work, we only presented results studying four modes at a time, which all corresponded to streamwise-invariant structures. Further analysis could also consider additional modes, including those for nonzero streamwise wavenumbers. Here the modes were computed after subtracting both the streamwise and secondary components of the mean. The causal mechanisms leading to the secondary mean flow could possibly be more directly studied by not subtracting this secondary mean component, and instead including it as a mode with time-varying coefficient.

Acknowledgments
This work was performed in part during the Fifth Madrid Summer Workshop, funded by the European Research Council under the Caust grant ERCAdG-101018287. HJB, STMD and BLD acknowledge support from Air Force Office of Scientific Research grant FA9550-22-1-0109.

References

[1] Taira K, Brunton S L, Dawson S T M, Rowley C W, Colonius T, McKeon B J, Schmidt O T, Gordeyev S, Theofilis V and Ukeiley L S 2017 *AIAA J.* **55** 4013–4041
[2] Taira K, Hemati M S, Brunton S L, Sun Y, Duraisamy K, Bagheri S, Dawson S T M and Yeh C 2020 *AIAA J.* **58** 998–1022
[3] Tissot G, Lozano-Durán A, Cordier L, Jiménez J and Noack B R 2014 *J. Phys.: Conf. Ser.* **506** 012006
[4] Lozano-Durán A, Bae H J and Encinar M P 2020 *Journal of Fluid Mechanics* **882** A2
[5] Lozano-Durán A, Constantinou N C, Nikolaidis M A and Karp M 2021 *Journal of Fluid Mechanics* **914**
[6] Lozano-Durán A and Arranz G 2022 *Phys. Rev. Res.* **4**(2) 023195
[7] Martínez-Sánchez Á, López E, Le Clainche S, Lozano-Durán A, Srivastava A and Vinuesa R 2023 *Journal of Fluid Mechanics* **967** A1
[8] Granger C W J 1969 *Econometrica* **37**(3) 424—438
[9] Gunjal R, Nayyer S S, Wagh S R, Stankovic A M and Singh N M 2023 *Electr. Pow. Syst. Res.* **225** 109865
[10] Schreiber T 2000 *Physical Review Letters* **85**(2) 461
[11] Bae H J, Encinar M P and Lozano-Durán A 2018 *Journal of Physics: Conference Series* **1001** 012013
[12] Lozano-Durán A, Bae H J and Encinar M P 2020 *Journal of Fluid Mechanics* **882** A2
[13] Wang W, Chu X, Lozano-Durán A, Helmig R and Weigand B 2021 *Journal of Fluid Mechanics* **920**
[14] Kozachenko L F and Leonenko N 1987 *Problemy Peredachi Informatsii* **23**(2) 9–16
[15] Kraskov A, Stögbauer H and Grassberger P 2004 *Phys. Rev. E* **69**(6) 066138
[16] Duan P, Yang F, Chen T and Shah S L 2013 *IEEE transactions on control systems technology* **21**(6) 2052–2066
[17] Pinelli A, Uhlmann M, Sekimoto A and Kawahara G 2010 *Journal of Fluid Mechanics* **644** 107–122
[18] Vinuesa R, Noorani A, Lozano-Duran A, Khoury G K E, Schlatter P, Fischer P F and Nagib H M 2014 *Journal of Turbulence* **15** 677–706
[19] Vinuesa R, Schlatter P and Nagib H M 2015 *Journal of Turbulence* **16** 588–606
[20] Vinuesa R, Schlatter P and Nagib H M 2018 *Physical Review Fluids* **3** 054606
[21] Matin R, Hellström L H O, Hernández-García A, Mathiesen J and Smits A J 2018 *Int. J. Heat Fluid Flow* **74** 144–153
[22] Pirozzoli S, Modesti D, Orlandi P and Grasso F 2018 *Journal of Fluid Mechanics* **840** 631–655
[23] Orlandi P and Pirozzoli S 2019 *arXiv preprint arXiv:1907.07987*
[24] Modesti D, Pirozzoli S and Grasso F 2019 *Int. J. Heat Fluid Flow* **76** 130–140
[25] Khan H H, Anwer S F, Hasan N and Sanghi S 2020 *Physics of Fluids* **32** 045106
[26] Lopez-Doriga B, Dawson S T M and Vinuesa R 2022 *AIAA Aviation 2022 Forum* 3334
[27] Nikuradse J 1930 *Ing.-Arch.* **1** 306–332
[28] Prandtl L 1931 *Akademische Verlagsgesellschaft*
[29] Gavrilakis S 2019 *Journal of Fluid Mechanics* **859** 731–753
[30] Gessner F B 1973 *Journal of Fluid Mechanics* **58** 1–25
[31] Moehlis J, Faisst H and Eckhardt B 2004 *New Journal of Physics* **6** 56
[32] Vidal A, Vinuesa R, Schlatter P and Nagib H M 2018 *Int. J. Heat Fluid Flow* **72** 317–328
[33] Atzori M, Vinuesa R, Lozano-Durán A and Schlatter P 2021 *Int. J. Heat Fluid Flow* **89** 108802
[34] Nikuradse J 1926 *V.D.I. Forschungsheft* **70** 1229–1230

[35] Lumley J L 1967 *Proc. Int. Colloq. on the Fine Scale Structure of the Atmosphere and its Influence on Radio Wave Propagation* ed Yaglam A and Tatarsky V (Doklady Akademii Nauk SSSR, Moscow, Nauka)
[36] Berkooz G, Holmes P and Lumley J L 1993 *Annual Review of Fluid Mechanics* **25** 539–575
[37] Karhunen K 1946 *Ann. Acad. Sci. Fennicae Ser. A1* **34** 1–7
[38] Loève M 1955 *Probability Theory* (Princeton, N.J.)
[39] Georgiou I T and Schwartz I B 1999 *SIAM Journal on Applied Mathematics* **59** 1178–1207
[40] Kappagantu R and Feeny B F 1999 *Journal of Sound and Vibration* **224** 863–877
[41] Ma X H, Azeez M F A and Vakakis A F 2000 *Mechanical Systems and Signal Processing* **14** 37–48
[42] Barnett L and Seth A K 2014 *Journal of Neuroscience Methods* **223**
[43] Barnett L and Seth A K 2015 *Phys. Rev. E* **91** 040101
[44] Berchtold A and Raftery A E 2002 *Statistical Science* **17**(3) 328–356
[45] Ahmadi S, Nair N N and Weyer E 2020 *IFAC-PapersOnLine* **53**(2) 506–511
[46] Lozano-Durán A and Arranz G 2022 *Physical Review Research* **4.1**(023195)
[47] Srivastava A 2021 *Mechanics of Materials* **154** 103710
[48] Barnett L, Barrett A B and Seth A K 2009 *Physical Review Letters* **103** 238701
[49] Holmes P, Lumley J L and Berkooz G 1996 *Turbulence, Coherent Structures, Dynamical Systems and Symmetry* Cambridge Monographs on Mechanics (Cambridge University Press)
[50] Brunton S L, Proctor J L and Kutz J N 2016 *Proc. Nat. Acad. Sci. USA* **113** 3932–3937
[51] Ellingsen T and Palm E 1975 *The Physics of Fluids* **18** 487–488
[52] Landahl M T 1980 *Journal of Fluid Mechanics* **98** 243–251
[53] Brandt L 2014 *European Journal of Mechanics-B/Fluids* **47** 80–96
[54] Baddoo P J, Herrmann B, McKeon B J and Brunton S L 2022 *Proc. Roy. Soc. A* **478** 20210830

Journal of Physics: Conference Series **2753** (2024) 012018 doi:10.1088/1742-6596/2753/1/012018

Karhunen–Loève decomposition of high Reynolds number turbulent pipe flows: a Voronoi analysis

D. Massaro[1], J. Yao[2,3], S. Rezaeiravesh[4], F. Hussain[3], P. Schlatter[1,5]

[1]SimEx/FLOW, Engineering Mechanics, KTH Royal Institute of Technology, Stockholm 100 44, Sweden

[2]Advanced Research Institute of Multidisciplinary Sciences, Beijing Institute of Technology, Beijing 100081, China

[3]Department of Mechanical Engineering, Texas Tech University, Lubbock 79409, Texas, USA

[4]Department of Fluids and Environment/MACE, The University of Manchester M139PL, Manchester, UK

[5]Institute of Fluid Mechanics (LSTM), Friedrich–Alexander–Universität (FAU) Erlangen–Nürnberg, Erlangen 91058, Germany

E-mail: dmassaro@kth.se

Abstract. We perform the Karhunen–Loève decomposition of the data from direct numerical simulations pertaining to incompressible turbulent pipe flow at various Reynolds numbers, in order to identify large-scale coherent structures. A novel approach based on the Voronoi diagram is introduced to estimate the energy distribution along the radial direction as a function of the geometrical properties of the modes. In contrast to previous classifications, no user-defined criterion, threshold or *ad-hoc* separation of the eddies is required since the two most energetic branches are inherently present as the Reynolds number increases. Details about the Voronoi analysis are provided, together with a comprehensive validation and comparison with previous classifications at low Reynolds numbers. We discuss the results and potential of the presented approach at $Re_\tau = 180$ and 5200, commenting on the two classes of coherent structures with a varying and constant size in the radial direction appearing at $Re_\tau = 5200$.

1. Introduction

Turbulent flows constitute an intricate forest of small- and large-scale structures, *i.e.* organised and coherent patterns, the evolution of which affects the mean and fluctuating fields. While the existence of a "self-sustaining cycle" near the wall is a well-established concept [1, 2], there is a need for a more in-depth comprehension of the large-scale structures in the logarithmic and outer layers. In the literature, several categories of large-scale eddies exist, *e.g.* large-scale motions (LSM) [3], very-large-scale motions (VLSM) [4, 5] superstructures [6], *etc.* These categories describe a nomenclature which lacks consistency, despite being used in different studies. Terms like eddy, motion, and structure are often used interchangeably but have distinct meanings. Generally, the *eddy* is related to the statistics, the *motion* to the kinematics and the *structure* includes both motion and dynamics. Therefore, (coherent) structures can play different roles: they can act as "engines", taking in energy from external factors; act as "sinks", dissipate energy, or act as "repositories", storing energy long enough to affect the overall energy balance of the flow [7]. One approach to identifying the so-called statistical eddies and structures is the Proper Orthogonal Decomposition (POD) which uses the two-point spatial (temporal) correlation tensor. The POD modes include the space eigenfunctions of the correlation tensor and the associated dynamics, represented by the time coefficients. As such, throughout the remainder of this manuscript, we employ the terms "eddy" and "structure" synonymously when referring to the POD modes.

Apart from POD, several different techniques have been developed and applied to classify these structures in the past. The first model goes back to Townsend (1951) [8] who introduced the attached eddy hypothesis (AEH), *i.e.* geometrically self-similar attached eddies (AE) which exist over a range of scales limited by the Reynolds number. The AEH does not provide any information about the shape and organisation of the eddies, making it challenging to accurately classify the attached eddies. In this context, numerous approaches have been explored in the literature. Del Álamo *et al.* (2006) [9] categorised geometrically self-similar vortex clusters, based on a threshold of the discriminant of the velocity gradient tensor. Later, Lozano-Duràn *et al.* (2012) [10] employed quadrant structures to quantify intense Q events that make the most significant contributions to the Reynolds shear stress. Similarly, Sharma & McKeon (2013) [11] proposed an approach to extract coherent structures as the fundamental objects of response modes. Cheng *et al.* (2019) [12] extracted modes resembling attached eddies via the bidimensional empirical mode decomposition (BEMD). More recently, Hu *et al.* (2020) [13] focused on extracting regions of the flow where statistical properties are captured by the attached eddies. Nonetheless, each of these different approaches requires an adequate filtering of the velocity [14, 15, 16, 17] or defining intense events based on a specific threshold [10, 18, 19]. In this perspective, we aim to use the POD, also denoted below as Karhunen–Loève decomposition, to categorise energy-ranked structures. The statistical approach of the POD has already been confirmed to be effective in the detection of different kinds of eddies [20, 21]. The spatial orthogonality condition and the fluctuating energy hierarchical description of the statistical structures are among the key features of such modal decomposition. Hellström *et al.* (2016) [22] provided evidence of a geometric self-similarity of the same POD modes. However, to achieve the separation of eddies based on their energy content, without relying on user-defined thresholds or criteria, it is essential to perform a comprehensive analysis of the spatial organisation of the modes.

The present study aims to introduce an adequate spatial representation of the energy, taking into account the density of POD modes along the radial direction. The simulations and data sets of turbulent pipe flows are detailed in Section 3, followed by the discussion of the results in Section 4. In particular, we discuss the advantages of the Voronoi-based classification, aiming to use this novel energy-based classification to enhance the comprehension of coherent structures appearing in high Reynolds turbulent pipe flows. Eventually, concluding remarks and a future outlook are presented.

2. Pipe flow simulations

The direct numerical simulation of incompressible turbulent pipe flows is performed using the pseudo-spectral code `Openpipeflow` [23]. The solver uses a cylindrical coordinate system, where the radial, axial, and azimuthal directions are denoted by r, z, and θ, respectively, with corresponding velocity components u_r, u_z, and u_θ. The axial length of the computational domain is $L_z = 10\pi R$, where R is the pipe radius. The flow is driven by a pressure gradient which is adjusted at each time step of the simulations to ensure a constant mass flux through the pipe.

For the spatial discretisation, a Fourier decomposition is used in the axial (z) and azimuthal (θ) directions, whereas a central finite difference scheme with a nine-point stencil is used in the radial direction. The number of grid points in the radial direction is denoted by N_r, and the number of Fourier modes in the axial and azimuthal directions are N_z and N_θ, respectively. In physical space, the number of grid points in the axial and azimuthal directions is increased by a factor of 3/2 to account for dealiasing. The grid points are distributed in the radial (r) direction using a hyperbolic tangent function, which allows for accurate resolution of high wall-normal velocity gradients in the viscous sublayer. Additionally, the points near $r = 0$ are clustered together to maintain the high-order accuracy of the finite difference scheme near the centerline of the pipe. Periodic boundary conditions are set in the axial (streamwise) z direction. At the wall, the no-slip and impermeability conditions are imposed. For marching in time, a second-order semi-implicit time-stepping scheme is adopted. Further details are given by Willis (2017) [23].

A set of five different simulations is performed from $Re_\tau \approx 180$ to $Re_\tau \approx 5200$, where $Re_\tau = u_\tau R/\nu$ is the friction Reynolds number based on the friction velocity u_τ, the radius of the pipe R, and the kinematic viscosity ν. In particular, here we mainly consider the lowest and highest Reynolds number, *i.e.* $Re_\tau = 180$ and 5200. See Yao *et al.* (2023) [24] for further details of the simulations.

3. Karhunen–Loève decomposition

The Karhunen–Loève (KL) modal decomposition is a bi-orthogonal stochastic process expansion, also known as principal component analysis (PCA) and proper orthogonal decomposition (POD). The KL decomposition seeks to approximate a generic vector $\boldsymbol{u}(\boldsymbol{x}, t)$, with $\boldsymbol{x} = (r, \theta, z)$, over a domain of interest $\mathcal{D} = \Omega \times [0, T]$ as a finite sum of separated functions of $\boldsymbol{x}$ and t. The KL expansion generates orthogonal basis functions (modes) representing the optimal projection, in energy terms, of the most important features. Our approach follows the snapshot POD by Sirovich (1987) [25] as the number of spatial points is larger than the snapshots. Contrary to this approach, the KL decomposition first exploits the two homogeneities of the flow in the azimuthal and axial directions.

Following the derivation for the minimal channel flow by Webber *et al.* (1997) [26], the Fourier expansion of the velocity field $\boldsymbol{u}(\boldsymbol{x}, t)$ in z and θ is derived as

$$
\begin{aligned}
\boldsymbol{u}(r, \theta, z, t) &= \sum_{\kappa_\theta=-\infty}^{\infty} \sum_{\kappa_z=-\infty}^{\infty} \hat{\boldsymbol{u}}_{(\kappa_\theta, \kappa_z)}(r, t) e^{2\pi i \kappa_\theta \theta / L_\theta} e^{2\pi i \kappa_z z / L_z} \\
&= \sum_{p=1}^{\infty} \hat{\boldsymbol{u}}_p(r, t) e^{2\pi i \kappa_\theta \theta / L_\theta} e^{2\pi i \kappa_z z / L_z} ,
\end{aligned}
\tag{1}
$$

where κ_θ and κ_z are the azimuthal and streamwise (axial) wavenumbers respectively, L_θ and L_z are the domain lengths in the corresponding directions and $\hat{\boldsymbol{u}}_{(\kappa_\theta, \kappa_z)}$ are the Fourier transforms of $\boldsymbol{u}$ in such directions. These latter are functions of the radial distance and time for specific κ_θ and κ_z. In the second row, the expansion is reformulated as a function of the index p, which corresponds to a unique pair of wavenumber numbers $(\kappa_\theta, \kappa_z)$ and spans only one quadrant. In our derivation we consider $(\kappa_\theta \geq 0, \kappa_z \geq 0)$. This is convenient as all the terms can be found from the solutions calculated in the wavenumber space $(\kappa_\theta \geq 0, \kappa_z \geq 0)$ by exploiting the statistical invariance of the pipe flow under axial shift and azimuthal rotation [27]. Thus, the index p varies between one and the number of pairs in the positive wavenumbers space $(\hat{N})$. In the inhomogeneous radial direction, each $\hat{\boldsymbol{u}}_p(r, t)$ is modally decomposed into the spatial and temporal dependencies:

$$
\hat{\boldsymbol{u}}_p(r, t) = \sum_{q=1}^{\infty} \hat{a}_{(q,p)}(t) \hat{\boldsymbol{\Phi}}_{(q,p)}(r) , \tag{2}
$$

where $\hat{a}_{(q,p)}(t)$ and $\hat{\boldsymbol{\Phi}}_{(q,p)}(r)$ are the time coefficients and the space functions, respectively. These latter form an orthonormal basis set, *i.e.* empirical eigenfunctions that minimise, in the least square sense, the residual energy between the non-linear field and its linear projection. The index q is named quantum index [26] and is taken over all realisations and symmetries $(\tilde{N})$.

Note that $\tilde{N}$ is $2N$ and not $4N$ because the azimuthal homogeneity is imposed by considering only the positive quadrant. For a wavenumber pair p, the order of the modes goes from the most ($q = 1$) to the least ($q = \tilde{N}$) energetic one. Therefore, a specific eigenfunction associated with a wavenumber index pair is fully defined by the triplet $\mathbf{k} = (q, \kappa_\theta, \kappa_z) = (q, p)$.

To compute the time coefficients $\hat{a}$ and space functions $\hat{\mathbf{\Phi}}$, we move from the current continuous formulation to a discrete one, where the solution is represented at a finite number of grid points for a number of snapshots, N. The original DNS resolution [24] is unnecessary for large-scale evaluation and impractical due to memory constraints. Therefore, a data dimensionality reduction is necessary. In the radial direction, a downsampling was performed for the highest Reynolds numbers, carefully assessing its effect on the final POD results. In the axial and azimuthal directions, the first 32 modes are kept in each direction. After performing fast Fourier transforms in the axial and azimuthal directions for each snapshot, the snapshot matrix is assembled. Given the two homogeneities, the snapshot matrix $\hat{\mathbf{U}}_p$, is constructed only for each wavenumber pair $p = (\kappa_\theta \geq 0, \kappa_z \geq 0)$, by stacking the Fourier coefficients of each snapshot into a single column vector. For a generic pair p the snapshot matrix reads

$$\hat{\mathbf{U}}_p = [\hat{\boldsymbol{u}}_0, \hat{\boldsymbol{u}}_1, ..., \hat{\boldsymbol{u}}_{m-1}] \in \mathbb{C}^{nxm} , \tag{3}$$

where $n = 3n_r$ is the number of radial points times the number of velocity components and $m = \tilde{N} = 2N$ is two times the number of snapshots. The snapshots were sampled with an interval $\Delta T\ U_b/R = 2$ (U_b is the bulk velocity) and collected for an interval that embeds the slowest frequency. It is analytically proved that the space eigenfunctions are obtained from the singular value decomposition (SVD) of the matrix (3), see Tropea *et al.* (2007) [28]. The orthogonal basis function $\hat{\mathbf{X}}_p \in \mathbb{C}^{nxm}$ and its corresponding time coefficients $\hat{\mathbf{T}}_p$ (note that each column of $\hat{\mathbf{X}}_p$ and $\hat{\mathbf{T}}_p$ corresponds to a specific $\hat{\mathbf{\Phi}}_{(q,p)}(r)$ and $\hat{a}_{(q,p)}(t)$ in the continuous formulation) are computed via SVD of the snapshot matrix as

$$\hat{\mathbf{U}}_p = \hat{\mathbf{X}}_p \mathbf{\Sigma}_p \hat{\mathbf{W}}^*_p, \tag{4}$$

where the time coefficients are retrieved as $\hat{\mathbf{T}}_p = \mathbf{\Sigma}_p \hat{\mathbf{W}}^*_p$, the matrix $\hat{\mathbf{W}}^*_p \in \mathbb{C}^{mxm}$ is the right singular vectors matrix (* indicates the conjugate transpose) and $\mathbf{\Sigma}_p \in \mathbb{R}^{mxm}$ is the diagonal matrix of singular values of $\hat{\mathbf{U}}_p$ (*i.e.* the energies). To guarantee unit energy of each mode, proper normalisation has to be taken into account: $||\hat{\mathbf{X}}_p||_\mathbf{M} = \mathbf{I}$ and $||\hat{\mathbf{W}}_p||_\mathbf{N} = \mathbf{I}$, where $\mathbf{M} \in \mathbb{R}^{nxn}$ and $\mathbf{N} \in \mathbb{R}^{mxm}$ are the mass and the temporal weight matrix, respectively, and $\mathbf{I}$ is the identity matrix. When the snapshots are collected with equidistant time intervals, $\mathbf{N}$ is the identity matrix $\mathbf{N} = \mathbf{I}$. The mass matrix $\mathbf{M}$ contains the volume quadrature weights $\mathrm{d}V = r\mathrm{d}r\mathrm{d}\theta$. The unit energy normalisation is obtained by considering $\mathbf{M}^{1/2}\hat{\mathbf{U}}_p\mathbf{N}^{1/2}$ into equation (4) and decomposing as

$$\mathbf{M}^{1/2}\hat{\mathbf{U}}_p\mathbf{N}^{1/2} = \widehat{\mathbf{X}}_p \mathbf{\Sigma}_p \widehat{\mathbf{W}}^*_p, \tag{5}$$

where $\widehat{\mathbf{X}}^*_p\widehat{\mathbf{X}}_p = \mathbf{I}$ and $\widehat{\mathbf{W}}^*_p\widehat{\mathbf{W}}_p = \mathbf{I}$. Note that $\widehat{\mathbf{X}}_p$ and $\widehat{\mathbf{W}}^*_p$ are different from $\hat{\mathbf{X}}_p$ and $\hat{\mathbf{W}}^*_p$ as the mass and temporal weight matrices are considered. Eventually, the modes are reconstructed as $\hat{\mathbf{X}}_p = \mathbf{M}^{-1/2}\widehat{\mathbf{X}}_p$, with unit energy and orthogonal to each other. While the time coefficients are $\hat{\mathbf{T}}_p = \mathbf{\Sigma}_p \left(\mathbf{N}^{-1/2}\widehat{\mathbf{W}}_p\right)^* = \mathbf{\Sigma}_p \hat{\mathbf{W}}^*_p$, where the energies are given by the diagonal matrix $\mathbf{\Sigma}_p$.

Note that for each wavenumber pair p, the matrix $\mathbf{\Sigma}_p$ has $\tilde{N}$ eigenvalues (energies) $\lambda_{(q,p)}$, that are ordered according to the quantum index q (where $q = 1$ is the most energetic POD mode for a pair $p = (\kappa_\theta, \kappa_z)$). Furthermore, for the overall energy ranking, the proper degeneracy $d^\mathbf{k}$ has to be taken into account since we perform the SVD only for the first quadrant of the wavenumbers space. It is proved that an eigenvalue $\lambda_{(q,(\kappa_\theta,\kappa_z))}$ is equal to eigenvalues in the other quadrants $\lambda_{(q,(\kappa_\theta,-\kappa_z))}$, $\lambda_{(q,(-\kappa_\theta,\kappa_z))}$ and $\lambda_{(q,(-\kappa_\theta,-\kappa_z))}$ [26]. Thus, the contribution to the total kinetic energy (k) of a specific triplet $\mathbf{k} = (q, p)$ is given by $e^\mathbf{k} = d^\mathbf{k}\lambda(q, p)$:

- $d^{\mathbf{k}} = 1$ for $(\kappa_\theta = 0, \kappa_z = 0)$
- $d^{\mathbf{k}} = 2$ for $(\kappa_\theta \neq 0, \kappa_z = 0)$ and $(\kappa_\theta = 0, \kappa_z \neq 0)$
- $d^{\mathbf{k}} = 4$ for $(\kappa_\theta \neq 0, \kappa_z \neq 0)$.

The degeneracy is crucial when reconstructing the flow field. In the interest of clarity, the following derivation is reverted to the continuous formulation. Initially, the spatial eigenfunctions are obtained by the inverse Fourier transform

$$\Phi_{(q,p)}(r,\theta,z) = \hat{\Phi}_{(q,p)}(r) e^{2\pi i \kappa_\theta \theta / L_\theta} e^{2\pi i \kappa_z z / L_z}. \tag{6}$$

Then, for a generic $p = (\kappa_\theta, \kappa_z)$ and a quantum index q, the reconstruction of the original field has to take into account the contributions from each wavenumber pair

$$\begin{aligned} \boldsymbol{u}_{(q,p)}(r,\theta,z,t) &= \hat{a}_{(q,p_1)}(t)\Phi_{(q,p_1)}(r,\theta,z) \\ &+ \hat{a}_{(q,p_2)}(t)\Phi_{(q,p_2)}(r,\theta,z) \\ &+ \hat{a}_{(q,p_3)}(t)\Phi_{(q,p_3)}(r,\theta,z) \\ &+ \hat{a}_{(q,p_4)}(t)\Phi_{(q,p_4)}(r,\theta,z) \end{aligned} \tag{7}$$

where $p_1 = (\kappa_z, \kappa_\theta)$, $p_2 = (\kappa_z, -\kappa_\theta)$, $p_3 = (-\kappa_z, \kappa_\theta)$ and $p_4 = (-\kappa_z, -\kappa_\theta)$. Note that equation (7) is valid when $(\kappa_\theta \neq 0, \kappa_z \neq 0)$. With a degeneracy equal to $d^{\mathbf{k}} = 1$ (or $d^{\mathbf{k}} = 2$), single (or double) contributions are considered. Eventually, the sums over the quantum index q and all the possible pairs $\hat{N}$ in the positive quadrant $(\kappa_\theta \geq 0, \kappa_z \geq 0)$ are performed

$$\boldsymbol{u}(r,\theta,z,t) = \sum_{p=1}^{\hat{N}} \sum_{q=1}^{\tilde{N}} \boldsymbol{u}_{(q,p)}(r,\theta,z,t). \tag{8}$$

4. Results

The space characteristics of the large-scale spatially coherent structures are presented here. Initially, we validate the results of the KL decomposition through a comparison with the prior studies by Duggleby *et al.* (2007, 2009) [29, 30]. Then, a qualitative description of the most energetic modes at the lowest and highest Reynolds number is provided. Eventually, the novel spatial characterisation that makes use of the Voronoi diagram is introduced.

4.1. Validation and comparative assessment

The linear representation of the flow obtained via KL decomposition can potentially depend on the spatial and temporal resolutions. Thus, we begin by assessing the chosen discretisations. Then, a comparison with previous works by Duggleby *et al.* (2007, 2009) [29, 30] is provided.

In the DNS of the pipe flow [24], the spatial resolution was crucial to capture the smallest scales appearing in the flow. However, such a grid spacing is excessive for the extraction of the large-scale structures, *i.e.* the most energetic motions. Furthermore, especially at the highest Reynolds numbers, affording such a resolution is impractical due to memory constraints. Therefore, a data dimensionality reduction was necessary. In particular, the effect of the downsampling in the radial direction has been carefully assessed. At the moderately high Reynolds number $Re_\tau = 1000$, the original DNS resolution ($N_r = 384$) can still be handled when assembling the snapshot matrix. In this manner, the effect of the reduction of radial resolution can be measured. We observe that the coarsening of resolution by a factor 2 or 3 does not significantly alter the eigenvalues, with a tolerance on the estimated energy fraction $f = \lambda_{(q,p)}/k$ below 10^{-6} for 96% of the first 1000 POD modes. In the axial and azimuthal directions, there was no need for downsampling. In fact, only the first 32 modes were retained

Table 1. First 10 most energetic POD modes at $Re_\tau = 550$ ranked in the descending order of energy. The convergence of the POD calculations is assessed by progressively increasing the total number of snapshots. For each $\tilde{N}$: the azimuthal and streamwise wavenumber (κ_θ, κ_z), the quantum index (q) and the fraction of the total fluctuating kinetic energy ($f = \lambda_{(q,p)}/k$), with $p =(\kappa_\theta, \kappa_z)$).

	$\tilde{N} = 76$				$\tilde{N} = 96$				$\tilde{N} = 136$				$\tilde{N} = 202$			
	κ_θ	κ_z	q	$f(\%)$	κ_θ	κ_z	q	$f(\%)$	κ_θ	κ_z	q	$f(\%)$	κ_θ	κ_z	q	$f(\%)$
1	3	1	1	0.847	3	1	1	0.811	3	1	1	0.727	3	1	1	0.721
2	4	2	1	0.630	4	2	1	0.607	2	1	1	0.640	2	1	1	0.600
3	2	2	1	0.462	2	1	1	0.559	4	2	1	0.566	4	2	1	0.524
4	2	1	1	0.457	2	2	1	0.551	2	2	1	0.539	2	2	1	0.479
5	4	3	1	0.445	4	1	1	0.474	4	1	1	0.485	3	0	1	0.453
6	4	1	1	0.439	4	3	1	0.399	3	2	1	0.404	4	1	1	0.442
7	3	3	1	0.409	3	3	1	0.358	1	1	1	0.369	1	1	1	0.389
8	1	1	1	0.371	3	2	1	0.349	3	3	1	0.366	3	2	1	0.362
9	3	2	1	0.335	1	1	1	0.341	3	0	1	0.364	3	3	1	0.340
10	5	1	1	0.328	5	1	1	0.318	4	3	1	0.323	2	3	1	0.310

Table 2. Comparison between the fist 10 POD modes at $Re_\tau = 150$ [30] and $Re_\tau = 180$. For each Reynolds number: the azimuthal and streamwise wavenumber (κ_θ, κ_z), the quantum index (q) and the fraction of the total fluctuating kinetic energy ($f = \lambda_{(q,p)}/\tilde{k}$, with $p =(\kappa_\theta, \kappa_z)$). In the current table, the total kinetic energy is estimated as the sum of the POD energies ($\tilde{k}$), differently from the rest of the manuscript where the total kinetic energy (k) from the statistics is used. In the data by Duggleby *et al.* (2007) [30], we observe the appearance of POD modes with $\kappa_z = 0$, *i.e.* modes with a wavelength longer than their pipe itself ($L_z = 20R$). Note that the pipe length is different in our case, *i.e.* the axial wavenumber is defined by different L_z.

	$Re_\tau = 150$ [30]				$Re_\tau = 180$			
	κ_θ	κ_z	q	$\tilde{f}(\%)$	κ_θ	κ_z	q	$\tilde{f}(\%)$
1	5	1	1	1.62	6	1	1	1.35
2	6	1	1	1.58	4	1	1	1.25
3	3	1	1	1.56	2	1	1	1.22
4	4	1	1	1.42	5	1	1	1.20
5	6	0	1	1.39	3	1	1	0.98
6	7	1	1	1.32	4	2	1	0.90
7	5	0	1	1.28	5	2	1	0.81
8	3	0	1	1.25	3	2	1	0.77
9	2	1	1	1.18	8	2	1	0.73
10	4	0	1	1.11	7	1	1	0.71

for each direction. This approach effectively balances the memory limitations and the assessment of the most energetic structures, which consistently exhibit long wavelengths. In this regard, additional computations at low Reynolds number where the wavenumbers space was expanded, considering $N_z = 64$ and $N_\theta = 64$, corroborate our outcomes. With respect to the temporal resolution, one must take into account the number of samples, which represents the overall time span. At the moderately high $Re_\tau = 550$, the POD has been computed for a progressively larger number of snapshots. Table 1 shows a clear convergence in the last two columns where the order of the first 10 POD modes is almost unchanged and the energy contribution of the most energetic structure is computed with a tolerance of 0.01%. We collect the same number of snapshots at all Reynolds numbers considered and, in all cases, a conservative sampling interval is used ($\Delta T U_b/R = 2$). In addition, the results were not affected by the time interval used for the POD calculation.

Eventually, a comparison with the first 10 modes extracted by Duggleby *et al.* (2009) [30] at $Re_\tau = 150$ is performed. Note that Duggleby *et al.* (2009) [30] computes the energy contribution of each mode by considering the sum of all eigenvalues ($\lambda_{q,p}/\tilde{k}$). Differently, in the rest of the manuscript we use the total kinetic energy (k) from the statistics ($\lambda_{q,p}/k$). However, to have an adequate comparison in table 2, we also divide the energies by $\tilde{k}$. The two estimated total kinetic energies k and $\tilde{k}$ converge as $N \longrightarrow \infty$. The results are in good agreement, *e.g.* the most energetic mode at $Re_\tau = 180$ gives an energy contribution (1.35%) very close to the contribution (1.58%) of the corresponding mode at $Re_\tau = 150$ (Tab. 2). Furthermore, the trend of the energy distribution in the ranked modes is consistent with the increase of the Reynolds number. As Re_τ increases, the most energetic modes give a progressively smaller energy contribution since the energy is distributed among a broader spectrum of scales, see also Table 1 in Duggleby *et al.* (2009) [30]. Minor differences appear in the order and, very likely, they are related to different numerical settings, *e.g.* the number of snapshots, the length of the pipe, *etc.*

4.2. Propagating and non-propagating modes

Classifying the POD modes poses a challenge due to the necessity of considering a multidimensional parameter space, which may include: the azimuthal and streamwise wavenumbers (κ_θ and κ_z), the size of the structure (l_y), radial location of the axial velocity peak ($|u_z|(y_p)$), symmetry/asymmetry of the mode, the energy contribution ($f = \lambda_{(q,p)}/k$), *etc.* Previously, Duggleby *et al.* (2007, 2009) [29, 30] suggested a classification based on two main classes.

The first class consists of propagating modes that have a streamwise wavenumber $\kappa_z \neq 0$, *i.e.* travelling waves with a bursting nature and a nearly constant phase velocity. They are further organised into four different subclasses:

- Wall modes: a travelling wave near the wall with $\kappa_\theta > \kappa_z$ and $\kappa_\theta > 1$. For these structures, the Reynolds stress is generated near the wall and the amplitude of the time history shows its bursting nature.
- Lift modes: a travelling wave with $\kappa_z > \kappa_\theta$ and $\kappa_\theta > 1$, with the extension of the Reynolds stress from the wall to near the centerline.
- Asymmetric modes: $\kappa_\theta = 1$. They show nonzero radial and azimuthal velocities at the pipe centerline, *i.e.* asymmetric. This is physically allowed only for $\kappa_\theta = 1$, since a positive radial velocity at the origin leads to a negative radial velocity when subjected to a rotation of π around the axis (the same for the azimuthal).
- Ring modes: $\kappa_\theta = 0$. Travelling modes with a ring of coherent vorticity in the outer region.

The second class is formed by non-propagating modes ($\kappa_z = 0$) with a non-constant phase speed mode:

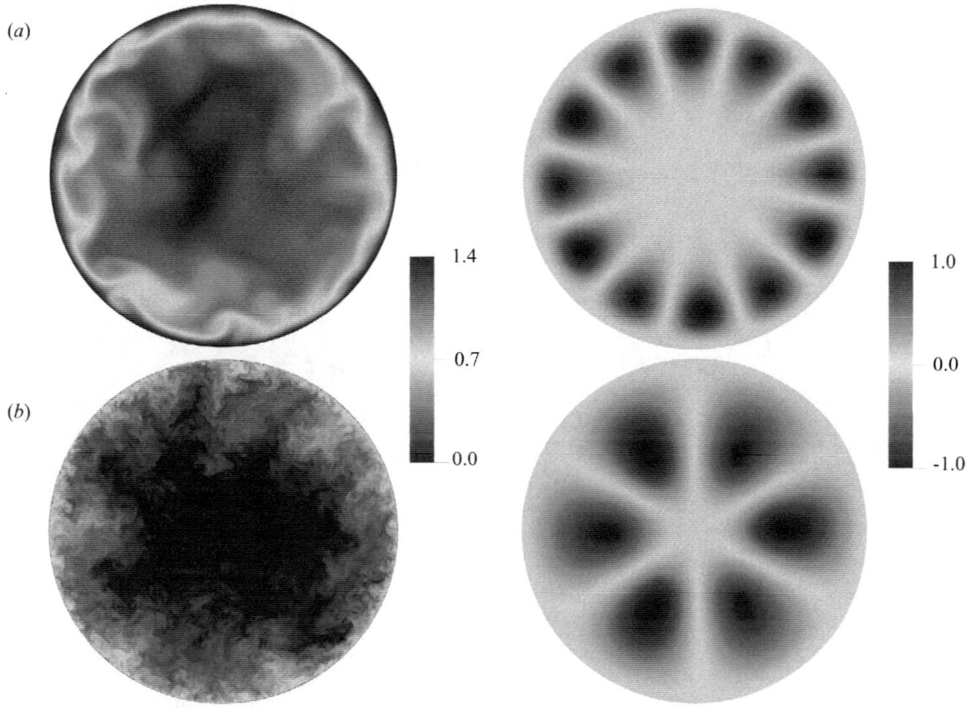

Figure 1. From left to right, cross-stream planes of the streamwise velocity of the instantaneous flow field and of the most energetic POD modes $\hat{\Phi}_{(q=1,(\kappa_\theta,\kappa_z))}$ (normalised by their maximum). The two different Reynolds numbers Re_τ = 180 and 5200 with their corresponding most energetic mode $\hat{\Phi}_{(q=1,(\kappa_\theta=6,\kappa_z=1))}$ and $\hat{\Phi}_{(q=1,(\kappa_\theta=3,\kappa_z=0))}$ are shown in (*a*) and (*b*), respectively.

- Roll modes: $\kappa_z = 0$ and $\kappa_\theta \neq 0$. The near-wall streamwise vortices with an energy decay rate lower than that of the propagating modes.
- Shear modes: $\kappa_z = 0$ and $\kappa_\theta = 0$. The centerline streamwise structures correspond to the fluctuations of the mean flow rate. Note that the mean is not subtracted when the matrix (3) is assembled.

For the current set of simulations, we observed similar results at Re_τ = 180 in terms of the percentage of the contribution of each subclass and energy spectra [29]. In the first instance, we adopt the same classification at the highest Reynolds number (Re_τ = 5200), for which figure 1 shows a comparison between the instantaneous fields and the most energetic structures. The flow is always characterised by a limited number of bulges distributed along the azimuthal direction that represent patterns (with a low azimuthal wavenumber) where high-speed fluid enters from the pipe core and low-speed fluid is ejected from the wall. At Re_τ = 180, the most energetic POD mode corresponds to a propagating mode with $\hat{\Phi}_{(q=1,(\kappa_\theta=6,\kappa_z=1))}$, specifically a wall mode. In fact, the POD mode 1 corresponds to 12 streak pairs: 6 pairs of streamwise velocity streaks, which qualitatively resemble the ten evenly distributed instantaneous low-speed structures in Fig. 1-(*a*), *left*. Especially, the peak of the modes is located at approximately $y^+ \approx 24$, with an azimuthal separation $\Delta R\theta^+ \approx 85$ and axial wavelength $\Delta x^+ = 10\pi/\kappa_z \approx 1000$. The contribution of $\hat{\Phi}_{(q=1,(\kappa_\theta=6,\kappa_z=1))}$ to the total kinetic energy is significant, around 1.1 %. The corresponding three-dimensional contour plots of the radial, azimuthal and streamwise velocity components are shown in figure 2-(*a*). They resemble spatially correlated near-wall streaks, in agreement with the propagating wall mode described by Duggleby *et al.* (2007) [29].

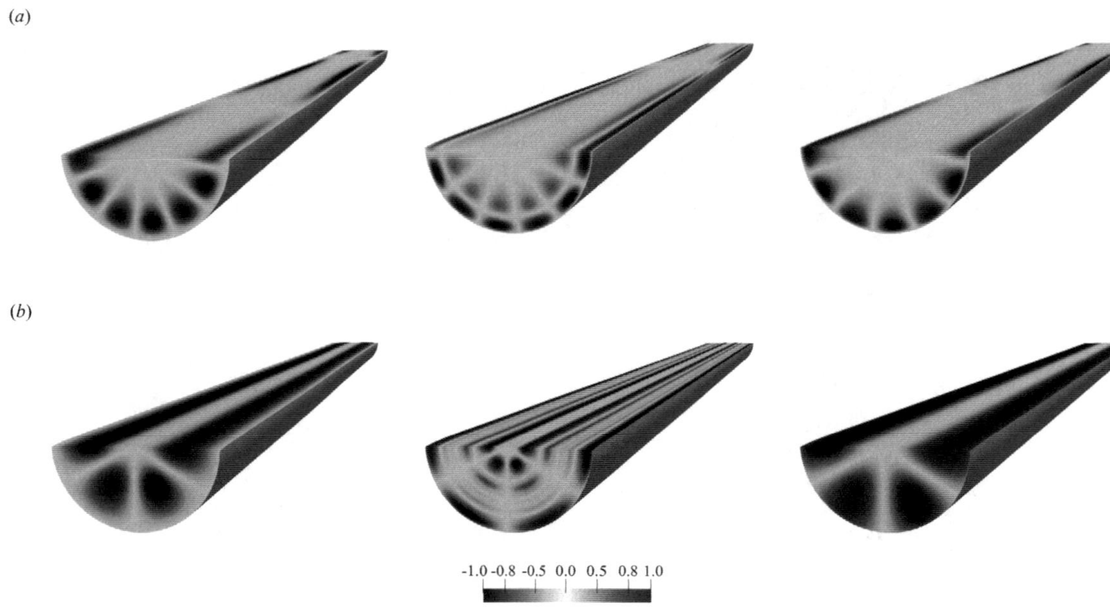

Figure 2. From left to right, the radial, azimuthal and streamwise velocity components (normalised by their maximum) at (*a*) $Re_\tau = 180$ ($\hat{\Phi}_{(q=1,(\kappa_\theta=6,\kappa_z=1))}$) and (*b*) $Re_\tau = 5200$ ($\hat{\Phi}_{(q=1,(\kappa_\theta=3,\kappa_z=0))}$).

At $Re_\tau = 5200$, the instantaneous large scales dominate in the central region of the pipe (Fig. 1-(*b*)). However, the most energetic mode is a non-propagating mode with a rolling nature: ($\kappa_\theta = 3$, $\kappa_z = 0$). The mode is mainly located in the central part of the pipe. In addition, the energy transfer from these roll modes to the wall modes, and then eventually to the lift modes, needs to be carefully assessed. Indeed, the one-dimensional energy spectra representation (see Fig. 16 in Duggleby *et al.* (2007) [29])) does not provide a unified and clear description of the energy distribution in the multidimensional parameters space (κ_θ, κ_z, r, f, l_y).

In conclusion, as the Reynolds number increases, the classification introduced by Duggleby *et al.* (2007) [29] for the low-*Re* data needs to be revised. Henceforth, we seek an adequate spatial characterisation of the POD modes that highlight the primary properties and properly classify the energy distribution across the inhomogeneous radial direction.

4.3. Spatial characteristics

The POD has proved to be valuable to gain insights into energy variations across the inhomogeneous wall-normal direction [22]. In the formulation described in Section 3, the KL decomposition exploits the two homogeneous directions, defining, for a specific triplet $\mathbf{k} = (q, \kappa_\theta, \kappa_z) = (q, p)$, a one-dimensional mode as a function of only the radial direction. As shown in figure 3-(*a*), their shapes can vary in a significant way. Especially at high Reynolds numbers, the classification suggested by Duggleby *et al.* (2007) [29] needs to be revised, and proper characterisation of the radial shape of the modes is also required. In the current classification, we consider just the dominant peak for each mode and only $q = 1$ is used for a triplet, *i.e.* the most energetic structure for a wavenumber pair p. The conclusions below hold independent of including all the pairs ($q = 1, ..., \tilde{N}$) for a wavenumber pair p.

Initially, a `MATLAB` algorithm is used to extract the spatial characteristics of the modes, see

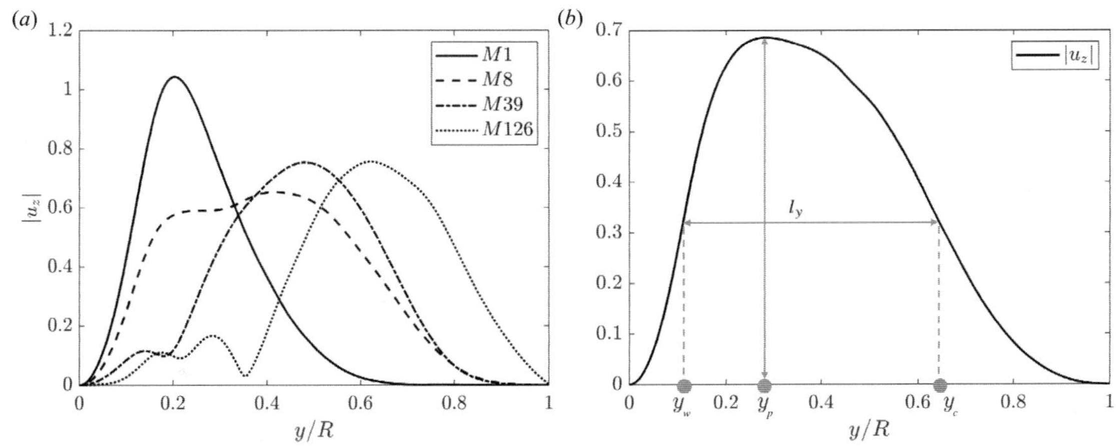

Figure 3. (a) Modulus of the (complex) streamwise velocity for different POD modes at $Re_\tau = 180$. (b) Spatial characteristics of a generic mode: height of the largest peak h (*blue*), the wall-normal distance of the peak y_p, the thickness of the mode l_y (*orange*) and the wall distance y_w and y_c, indicating the beginning and the end of the structure, respectively. All the lengths are normalised by R.

figure 3-(b). Each structure is distinguished by the height of the peak h, the location of the peak y_p and the thickness of the mode l_y (here and hereafter normalised by R). The size $l_y = y_c - y_w$ is crucial since it measures the radial extension, *i.e.* the thickness of the structure. The radial locations y_w and y_c are the coordinates where the peak halves towards the wall ($y = R - r = 0$) and the centre ($y = 1$), respectively. These points indicate the start and end of the structure and also serve as a measure of its degree of asymmetry, as the peak may not be perfectly symmetric. The classification is carried out for the streamwise velocity component u_z in spectral space, *i.e.* each line alike figure 3-(b) corresponds to a specific $\mathbf{k} = (q = 1, p)$. Similar results are obtained by performing the classification in physical space and then averaging in the streamwise and azimuthal directions.

4.4. Voronoi-based tessellation

Decomposing a complex dynamical system, such as turbulent flow, using modal analysis entails a separation between the time coefficients and spatial modes. These latter represent individual spatial patterns of the original system. In particular, the KL expansion generates modes that are orthogonal in space and whose energetic contribution is summed up to reflect their contribution to the total fluctuating kinetic energy. Thus, when observing the radial distribution of energy, the density of the POD modes at a specific location is crucial since their contribution is additive.

As an example, we consider a two-dimensional scatter plot of the energetic content. In figure 4-(a), each point indicates the energy contribution of a mode whose peak is located at the radial location y_p and the height of the peak is $|u_z|(y_p)$. The higher intensity of the colour indicates higher energy levels. For a specific point, the radial shape of the corresponding mode is also reported, see black curve in Fig. 4-(a). The contour plot of the scattered points in figure 4-(b) shows two regions with intense energy at $0.15 < y/R < 0.32$ and $y/R \approx 0.6$ for $|u_z|(y_p) \approx 0.4$. The current representation is already a step forward compared to the previous energy spectra where the energy was shown as a function of a single parameter, like κ_z (see Fig. 16 in [29]). The contour plot is the result of the average energetic intensity, without taking into account the density of the modes. Neglecting this piece of the puzzle leads to misleading results because of

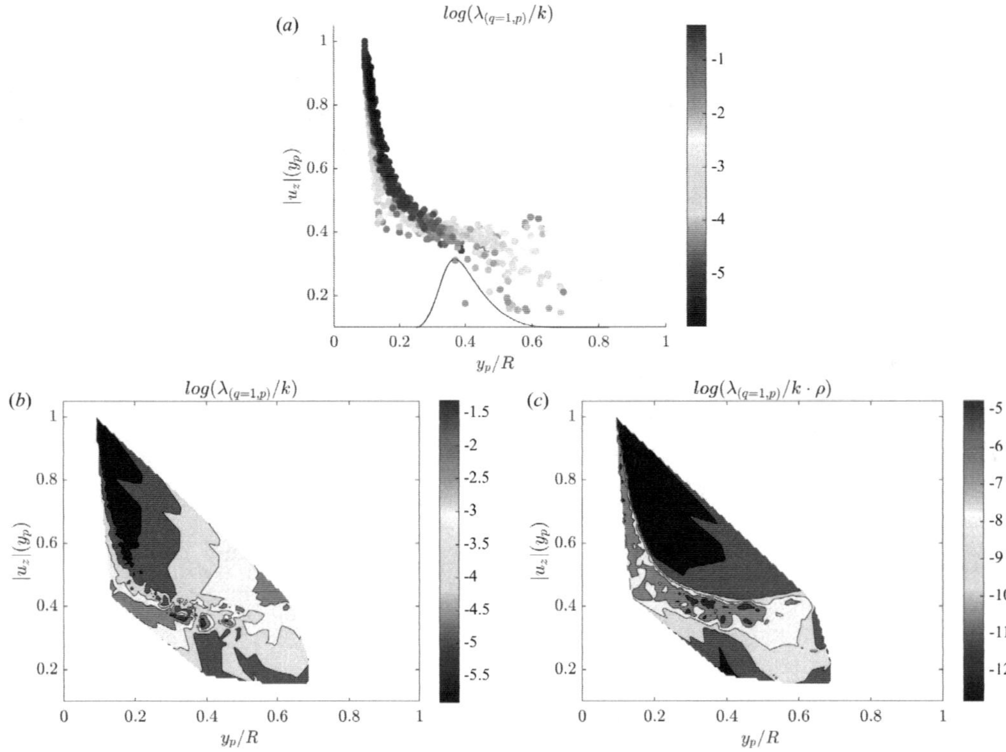

Figure 4. (a) Scatter plot of points corresponding to the maximum peak of the mode at the radial location y_p for $Re_\tau = 550$. The colour of the point indicates the energy content: more intense colour (in the blue-red scale), more energetic content. The black solid line shows the radial variation for a generic mode. (b) Energy contour plot of the scatter plot in panel (a). (c) Energy contour plot of the scatter plot in panel (a) weighted by the density estimated through the Voronoi tesselation.

the cumulative energy property of the POD. Indeed, a cluster of points at a specific height y can give an overall contribution larger than a single point with high energy at the same location. At this stage, this aspect has not been taken into account yet. By weighting the energy contribution with the density of the points ρ, the shape of the contours substantially changes, see Fig. 4-(c). The energy shows a more uniform radial distribution with the most energetic structures having a peak that varies from $y/R \approx 0.15$ to $y/R \approx 0.5$. Specifically, to properly estimate the density of each point, we introduce, for the first time in the POD analysis, the usage of the Voronoi diagram.

Named after the mathematician Georgy Voronoi, the Voronoi diagram is a general n-dimensional geometric construct, that is applied here specifically in a two-dimensional space. Given a discrete set of points $\{x_j, y_j\}$ for $j = 1, ..., N_{TOT}$, the Voronoi diagram creates a region of interest Ω_j around each point (x_j, y_j), guaranteeing that any arbitrary point within the region Ω_j is closer to the point j than any point in the set. The algorithm first computes the Delaunay triangulation of the input points, which provides the connectivity information of the input points, forming polygons, as triangles. The Voronoi edges are defined as the centres of the circles that pass through all the vertices of Delaunay polygons. The intersection of the edges defines the vertices of the Voronoi cells. The region of influence Ω_j, with an area A_j, is called the Voronoi region (or cell) and the collection of all the Voronoi regions is the

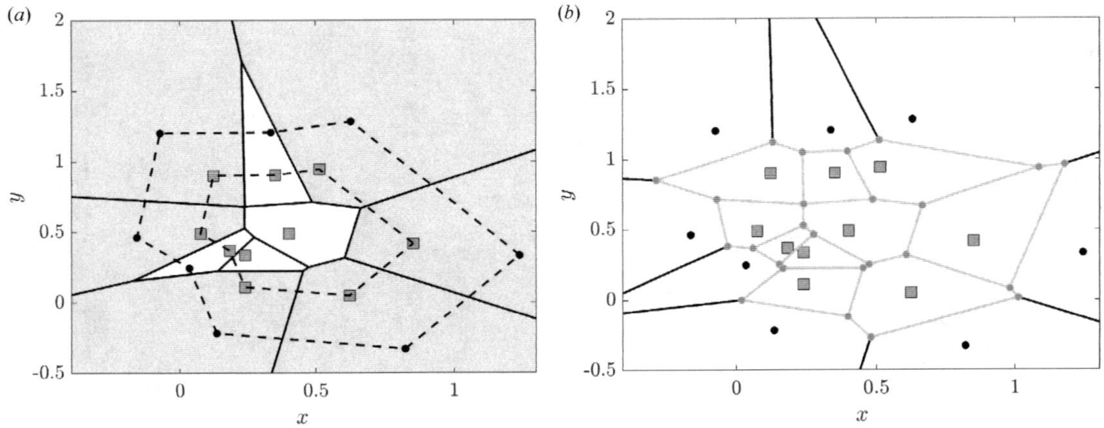

Figure 5. (a) Discrete set of points $\{x_j, y_j\}$ (red squares) with the boundaries connected by dashed black lines. The Voronoi diagram generates open areas A_j (light blue) for the boundary points. These latter are expanded by a safety factor (black dots with dashed black lines). (b) The Voronoi diagram is recomputed for a new set of points where the expanded boundary is included. Closed areas (with the corresponding perimeter in green) are generated for the original set of points (red squares). The set of green points is used from the Delaunay triangulation method to form an area-wide network of triangles, *i.e.* the areas A_j. The Voronoi diagram is also known as the dual graph of the Delaunay triangulation.

Voronoi diagram. A few, but crucial, modifications were necessary in the `MATLAB` function to properly estimate the density. Examining the set of points depicted in Fig. 5-(a), the Voronoi algorithm creates regions of influence. Nevertheless, when it comes to the points situated along the boundary, the algorithm produces open regions with infinite area A_j, *i.e.* zero density. These boundary regions require particular attention to prevent the algorithm from generating null-density weights. Thus, the boundary points are expanded by a safety factor, forming an auxiliary set of points. The rate of expansion is chosen conservatively since we aim to assign a large, but still finite, area. At this stage, the Voronoi diagram is computed again, guaranteeing closed regions of interest for any point, see Fig. 5-(b). The inverse of the area A_j, *i.e.* the density ρ_j, measures the two-dimensional concentration of points. The Voronoi diagram was previously used by the fluid mechanics' community mainly for describing the spatial structure of the dispersed phase, analysing the particle acceleration statistics, and averaging particle-conditioned averaging [31, 32]. Notably, to the best of the authors' knowledge, here we use the Voronoi diagram for the first time to represent the energy spatial distribution of the POD modes. The differences between the unweighted and weighted results are shown in figures 4-(b) and (c): the region with high energy content results in an elongated branch from $y/R \approx 0.15$ to $y/R \approx 0.5$, differently from the two isolated spots in the unweighted case, panel (b).

4.5. Energetic structures with a constant and varying size

The density-weighted contour plot effectively quantifies the energy, or more precisely, the logarithm of the energy contribution weighted by the associated density, across the radius of the pipe. A first application is shown in Fig. 6 where the contour plot is obtained by considering the energy distribution across the radial section of the large-scale structures with size l_y at $Re_\tau = 180$ and $Re_\tau = 5200$. Figure 6-(a) can be interpreted as the following: a point at (0.2, 0.2) (*red region*), *i.e.* a structure with maximum located at $y_p/R = 0.2$ and a size $l_y = 0.2$, has a higher energy content than an eddy with the same size ($l_y = 0.2$) but the peak located at

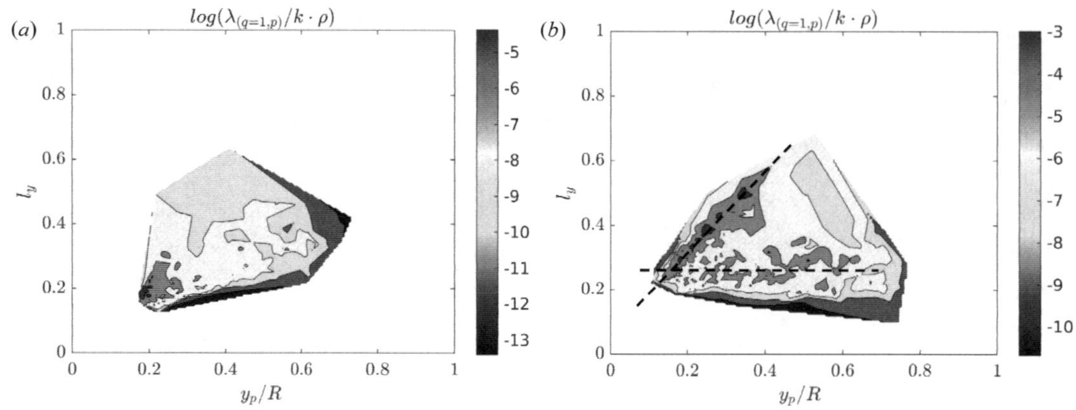

Figure 6. Density-weighted energy contour of the scattered points (y_p, l_y): y_p is the radial location of the largest peak and l_y is a measure of the structures' size. The panels (a) and (b) refer to $Re_\tau = 180$ and $Re_\tau = 5200$, respectively.

$y_p/R = 0.4$ (*yellow region*).

As described above, the energy contribution, *i.e.* the ratio between the eigenvalue obtained by the KL decomposition $\lambda_{q,p}$ to the total kinetic energy k, is weighted by the Voronoi density ρ to consider the density of the points in a region of the (l_y, y_p/R) space. The logarithmic colour scheme is used for $\lambda_{q,p}/k \cdot \rho$ as the data spans several orders of magnitude with small fluctuations. Note that a specific y/R point corresponds to the radial location of the largest peak of a structure that extends radially (not symmetrically) with a thickness denoted as l_y. At the lowest Reynolds number considered ($Re_\tau = 180$), the energy is highly concentrated in structures relatively small ($l_y \approx 0.2R$) and close to the wall: the location of the largest peak y_p varies between $0.15R$ and $0.25R$. At the Reynolds number $Re_\tau = 5200$, the energy is distributed across structures of varying sizes, and its peak is located within a broad range of radial positions, specifically within $0.1 < y_p < 0.6$. Remarkably, in the contour plot, two distinct high-energy branches appear: the first is inclined extending up to $y/R \approx 0.4$, and the second reaches up to $y/R \approx 0.6$ with a slope close to zero. The inclined line refers to a class of POD modes, highly energetic, whose thickness (l_y) scales linearly with the wall-normal distance of their peak y_p. In particular, this class I has a slope $l_y \approx 1.201y/R$ at $Re_\tau = 5200$. Differently, the horizontal line identifies a class of structures (class II) with an almost constant size as the peak is moving towards the centre of the pipe, $l_y \approx 0.26R$ at $Re_\tau = 5200$. Structures with a linearly scaling or constant size along the wall-normal direction are usually named attached (class I) and detached eddies (class II), respectively.

5. Conclusions

This work presents the Karhunen–Loève (KL) modal decomposition of turbulent pipe flows at various Reynolds numbers. In contrast to previous studies where a classification based on the streamwise and azimuthal wavenumbers was used [29], we introduce a novel methodology that exploits the properties of the Voronoi diagram to characterise the energy distribution among the modes.

First, a comprehensive description of the KL decomposition, also known as proper orthogonal decomposition (POD), is provided. A detailed assessment of the results w.r.t. to the time and space resolutions is performed, followed by a comparison with the classification by Duggleby *et al.* (2007) [29] between propagating and non-propagating modes. At a low Reynolds number considered here (Re_τ = 180), the results agree with Duggleby *et al.* (2007) [29]. However, as Re_τ increases, the classification needs to be revised. Indeed, the complexity arising at higher Reynolds numbers requires a new spatial characterisation of the structures. For each POD mode, we extract the height of the peak h, the location of the peak y_p and the thickness of the mode l_y. In addition, the radial locations y_w and y_c are used to measure the beginning and the end of the structure, providing a measure of the degree of asymmetry. Eventually, a density-weighted energy contour plot is introduced for different quantities of interest, such as $|u_z|(y_p)$ and l_y. The novel application of the Voronoi diagram consists of defining a density to each point of the diagram such that the cumulative property of the kinetic energy of the POD modes is properly taken into account. Two distinct energetic branches appear: structures with a size that scales linearly with the wall-normal distance (class I) and those which have an almost constant size (class II). This finding suggests the appearance of attached and detached eddies among the most energetic structures at Re_τ = 5200. In future works, we aim to characterise these structures and describe their evolution with Re_τ, especially at high Reynolds numbers where an adequate separation of scales is guaranteed. Nevertheless, a crucial finding is that the POD, with a novel Voronoi-weighted energy contour plot, allows to automatically identify classes of structures. It is worth noting that no threshold or *ad-hoc* criterion was necessary to separate the two branches.

Acknowledgments
This work was performed in part during the Fifth Madrid Summer Workshop, funded by the European Research Council under the Caust grant ERCAdG-101018287. Funding by the Knut and Alice Wallenberg (KAW) Foundation is gratefully acknowledged. Computational and visualisation resources provided by Texas Tech University HPCC, TACC Frontera and the Swedish National Infrastructure for Computing (SNIC) are acknowledged. We are grateful to Dr Taygun R. Güngör for his careful review of the manuscript.

References

[1] Jiménez J and Pinelli A 1999 *J. Fluid Mech.* **389** 335–359
[2] Schoppa W and Hussain F 2002 *J. Fluid Mech.* **435** 57–108
[3] Hellström L H O, Ganapathisubramani B and Smits A J 2015 *J. Fluid Mech.* **779** 701–715
[4] Del Álamo J C, Jiménez J, Zandonade P and Moser R 2004 *J. Fluid Mech.* **500** 135–144
[5] Monty J P, Stewart J A, Williams R C and Chong M S 2007 *J. Fluid Mech.* **589** 147–156
[6] Hutchins N and Marusic I 2007 *J. Fluid Mech.* **579** 1–28
[7] Jiménez J 2018 *J. Fluid Mech.* **842** P1
[8] Townsend A A 1951 *Math. Proc. Camb. Philos. Soc.* **47** 375–395
[9] Del Álamo J C, Jiménez J, Zandonade P and Moser R D 2006 *J. Fluid Mech.* **561** 329–358
[10] Lozano-Duràn A, Flores O and Jiménez J 2012 *J. Fluid Mech.* **694** 100–130
[11] Sharma A S and McKeon B J 2013 *J. Fluid Mech.* **728** 196–238
[12] Cheng C, Weipeng L, Lozano-Duràn A and Hong L 2019 *J. Fluid Mech.* **870** 1037–1071
[13] Hu R, Yang X I A and Zheng X 2020 *J. Fluid Mech.* **885** 1037–1071
[14] Perry A E and Marusic I 1995 *J. Fluid Mech.* **298** 361–388
[15] Baars W J and Marusic I 2020 *J. Fluid Mech.* **882** A25
[16] Baars W J and Marusic I 2020 *J. Fluid Mech.* **882** A26
[17] Lee M and Moser R 2019 *J. Fluid Mech.* **860** 886–938
[18] Moisy F and Jiménez J 2004 *J. Fluid Mech.* **513** 111–133
[19] Atzori M, Vinuesa R, Lozano-Durán A and Schlatter P 2018 *J. Phys.: Conf. Ser.* **1001** 012008
[20] Wang L, Pan C, Wang J and Gao Q 2022 *J. Fluid Mech.* **944** A26
[21] Wang L, Pan C and Wang J 2022 *Phys. Fluids* **34** 095124
[22] Hellström L H O, Marusic I and Smits A J 2016 *J. Fluid Mech.* **792** R1
[23] Willis A P 2017 *SoftwareX* **6** 124–127 URL `https://www.sciencedirect.com/science/article/pii/S2352711017300158`
[24] Yao J, Rezaeiravesh S, Schlatter P and Hussain F 2023 *J. Fluid Mech.* **956** A18

[25] Sirovich L 1987 *Q. Appl. Math.* **45**
[26] Webber G A, Handler R A and Sirovich L 1997 *Phys. Fluids* **9** 1054–1066
[27] Sirovich L 1987 *Quart. Appl. Math.* **45** 573–582
[28] Tropea C, Yarin A L and Foss J 2007 *Springer Handbook of Experimental Fluid Mechanics* (Springer Berlin, Heidelberg) 1st ed
[29] Duggleby A, Ball K S, Paul M R and Fischer P F 2007 *J. Turb.* **8** N43
[30] Duggleby A, Ball K S and Schwaenen M 2009 *Phil. Trans. R. Soc.* **367** 473–488
[31] Monchaux R, Bourgoin M and Cartellier A 2010 *Phys. Fluids* **22** 103304
[32] García-Villalba M, Kidanemariam A G and Uhlmann M 2012 *Int. J. Multiphase Flow* **46** 54–74

Inter-scale Causality Relations in Wall Turbulence

J de Salis Young, Z Hao, R Garcia-Mayoral

Department of Engineering, University of Cambridge, Trumpington St, Cambridge, CB2 1PZ, UK

E-mail: r.gmayoral@eng.cam.ac.uk

Abstract. We investigate the inter-scale energy transfer in the buffer layer between different streamwise and spanwise lengthscales. In a periodic channel, inter-scale transfer of momentum in the periodic directions is only possible through wavenumber interactions in the non-linear advection term. The aim of this study is to map the inter-scale interactions which transfer energy through advection to different lengthscales of interest. The non-linear advection term has nine contributions, which correspond to three different advecting and advected velocity components. Because the flow is anisotropic, each of these contributions can be viewed as portraying a different physical mechanism. Maps of inter-scale transfer are therefore constructed for each of the nine contributions to the advection term in order to understand the relative importance of advecting and advected velocity components in wall turbulence. As a preliminary exploration, two maps of the inter-scale energy transfer in the buffer layer are presented for $Re_\tau = 180$, one for transfer to a 'target' lengthscale dynamically relevant in the near-wall cycle, $\lambda_x^+ = 188, \lambda_z^+ = 94$, and one for transfer to a 'target' lengthscale in the dissipative range, $\lambda_x^+ = 42, \lambda_z^+ = 21$. In the first case, we observe that the transfer of energy is dominated by transfer into streamwise-velocity (u) by spanwise-elongated structures of spanwise-velocity (w) that advect streamwise-elongated structures of streamwise-velocity (u). This is nothing but the process of streak meandering induced by spanwise-elongated (w) structures. In the second case, the inter-scale transfer can be split into three dominant mechanisms. The first and second mechanisms are observed across many of the nine contributions to the advection term. Velocity structures with size similar or larger than streamwise streaks either advect (mechanism 1) or are advected by (mechanism 2) velocity structures with lengthscales close to the target one . The third mechanism is only significant for streamwise (u) velocity structures advecting streamwise (u) and spanwise (w) velocity, and involves structures with an intermediate scale which is in-between the dissipation scale and the near-wall cycle dynamic lengthscale, but still in the dissipation range. In this mechanism, intermediate-scale streamwise-velocity structures advect intermediate-scale streamwise- and spanwise-velocity structures.

1. Introduction

The turbulent energy cascade is central to our understanding of turbulence. It describes how in turbulent flow, energy is typically introduced to the flow at large lengthscales and transferred through inter-scale interactions to smaller and smaller lengthscales until it is dissipated at the smallest lengthscales in the flow. The core mechanism behind the turbulent energy cascade is the transfer of energy between different lengthscales. Inter-scale transfer is fundamentally a non-linear effect: it can only occur through non-linearity as linear operators preserve wavenumber. In the Navier-Stokes equations, the non-linear term which facilitates inter-scale transfer is the advection term.

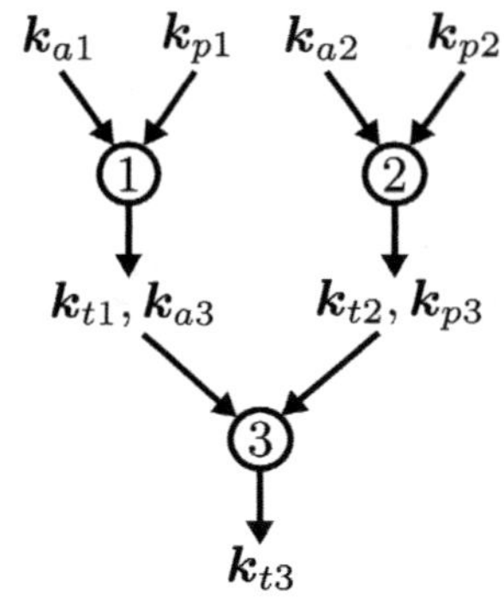

Figure 1. Hierarchy of causal advection interactions.

The non-linear advection term in the Navier-Stokes equations has nine contributions, which arise from the products of three different advecting and advected velocity components. Each of the nine contributions describe how one velocity component advects another velocity component. Wall-bounded turbulence is anisotropic, so each of these nine contributions represents a different physical process; for example, $\partial uv/\partial y$ represents the wall-normal velocity v advecting the streamwise velocity u and contributes to the streamwise (x) momentum equation. This is a completely different process to, for example, $\partial uw/\partial x$, which represents the streamwise velocity u advecting the spanwise velocity w and contributes to the spanwise (z) momentum equation. For inter-scale interactions, both the size of the eddies and the advecting/advected velocity components involved give physical insight.

Inter-scale transfer of momentum through the advection term $\partial u_i u_j/\partial x_j$ occurs through dyadic wavenumber interactions: advecting waves of wavenumber $\boldsymbol{k}_a$ advect advected waves of wavenumber $\boldsymbol{k}_p$ to transfer momentum to target waves with wavenumber $\boldsymbol{k}_t = \boldsymbol{k}_a + \boldsymbol{k}_p$. Here, in an abuse of notation, we use as subindicies a for 'active', p for 'passive' and t for 'target', in an analogy with passive scalar transport. These dyadic interactions are causal in the sense that a velocity structure of wavenumber $\boldsymbol{k}_a$ advects a velocity structure with wavenumber $\boldsymbol{k}_p$ which causes an inter-scale transfer of momentum to a velocity structure with wavenumber $\boldsymbol{k}_t$.

These causal advection interactions are what must drive the turbulent energy cascade as they are the only mechanism through which momentum can be transferred between different lengthscales. What this means is that there must be a hierarchy of causal advection interactions which are responsible for the turbulent energy cascade. A sketch of this hierarchy is shown in Figure 1. A velocity structure at a given target lengthscale $\boldsymbol{k}_{t1}$ will receive significant inter-scale momentum transfer caused by a wavenumber dyad pair of $\boldsymbol{k}_{a1}, \boldsymbol{k}_{p1}$ through advection interaction 1. While this lengthscale was the target lengthscale in advection interaction 1, it may also act as the advecting or advected velocity structure in subsequent advection interactions. In Figure 1 the target lengthscale in advection interaction 1, $\boldsymbol{k}_{t1}$, is shown to subsequently be the advecting lengthscale in advection interaction 3. Figure 1 only shows a small part of full hierarchy: in reality each target lengthscale will have many advection interactions which contribute momentum to it, and each target lengthscale would also act as the advecting or advected lengthscale in many other advection interactions. If a full hierarchy was constructed for a particular flow, then this would give insight into the routes through which momentum is transferred between scales. This could potentially be used for flow control applications, as if there was a particular 'bottleneck' advection interaction in the hierarchy which was important to the turbulent energy cascade then this could be targeted in order influence the turbulent energy cascade.

In this paper, we establish a framework for mapping the streamwise-spanwise advection interactions which contribute momentum to a particular target lengthscale. As a proof of concept, we demonstrate how this framework can be used to identify the dominant advection processes which transfer momentum to two target lengthscales in the buffer layer of an $Re_\tau = 180$ minimal flow unit channel flow.

As discussed later in Section 2.1, in order to measure the relative importance of inter-scale advection interactions, we propose a metric which measures the average inter-scale energy transfer from a particular advection interaction. This metric is analogous to the advection term in the turbulent kinetic energy transport equation, which is a triadic wavenumber interaction. However, the focus of this paper is on dyadic wavenumber interactions, and this triadic interaction is just a consequence of how we choose to measure the inter-scale transfer from a particular advection interaction. We use the term 'inter-scale transfer' to refer to any transfer of momentum between velocity structures of different streamwise and spanwise lengthscales. When using the inter-scale energy transfer metric, which is analogous to the turbulent kinetic energy transport equation, this means that we consider the terms which are typically referred to as 'wall normal transport' in literature as being inter-scale transfer, as the advection process they represent facilitates the inter-scale transfer of momentum.

The evolution of turbulent kinetic energy, including inter-scale transfer through wavenumber triads, has been the subject of much research [1, 2, 3]. Previous studies have mapped dyadic and triadic interactions in wall-bounded turbulence in various different forms. These include: maps of the bispectrum and biphase of streamwise wave triads [4]; spanwise triadic wave interactions involved in turbulent energy transfer [5]; spanwise and streamwise triadic wave interactions involved in turbulent energy transfer [6]; spanwise and streamwise wave dyad interactions into the principal resolvent mode [7]; and causal maps of inter-scale flux of information between energy-transfer signals [8]. Inter-scale transfer has also been analysed in the spatial domain through the use of structure functions and the Kármán–Howarth–Monin–Hill equation [9, 10]. One common aspect of the listed previous works is that no distinction is made between each of the nine contributions to the advection term in the Navier-Stokes equations. As discussed previously, the anisotropy of wall-bounded turbulence means that physical insight can be gained by identifying which advecting and advected velocity components are most important for inter-scale transfer. In this paper, maps of inter-scale transfer are constructed for each of the nine contributions to the advection term in order to identify which advecting and advected velocity components are most significant.

The paper is organised as follows. Section 2 describes the methods used to map the inter-scale transfer in a Fourier $x - z$ channel. Section 3 examines maps of the inter-scale energy transfer to two different scales: one scale that is relevant for the dynamics of the near wall cycle, and one scale that is in the dissipative range. Section 4 summarises and concludes the paper.

2. Methods

The Navier-Stokes equations in the spatial domain are

$$\frac{\partial \boldsymbol{u}}{\partial t} + \nabla \cdot (\boldsymbol{u}\boldsymbol{u}) = -\nabla p + \nu \nabla^2 \boldsymbol{u}, \qquad \nabla \cdot \boldsymbol{u} = 0, \tag{1}$$

where $\boldsymbol{u} = [u, v, w]^T$ is the velocity vector, ν is the kinematic viscosity and p pressure.

The Navier-Stokes equations in Fourier space describe the evolution in time of the velocity modes. The streamwise-spanwise Fourier transform of Equation 1 is

$$\frac{\partial \hat{\boldsymbol{u}}}{\partial t}(\boldsymbol{k}_t, y, t) + \sum_{\boldsymbol{k}_a + \boldsymbol{k}_p = \boldsymbol{k}_t} \hat{\nabla} \cdot (\hat{\boldsymbol{u}}(\boldsymbol{k}_a, y, t)\hat{\boldsymbol{u}}(\boldsymbol{k}_p, y, t)) = -\hat{\nabla}\hat{p}(\boldsymbol{k}_t, y, t) + \nu \hat{\nabla}^2 \hat{\boldsymbol{u}}(\boldsymbol{k}_t, y, t), \qquad \hat{\nabla} \cdot \hat{\boldsymbol{u}}(\boldsymbol{k}_t, y, t) = 0, \tag{2}$$

where the Fourier transform of the del operator is given by $\hat{\nabla} = [\mathrm{i}k_x \; \partial/\partial y \; \mathrm{i}k_z]^T$. We note that while the linear operators $\partial/\partial t, \nabla, \nabla^2$ are trivially Fourier transformed, the non-linear product in the physical domain in $\nabla \cdot (\boldsymbol{uu})$ becomes a convolution in the wavenumber domain.

The different terms that contribute to the evolution of the Fourier modes in Equation 2 can be split into three categories: linear spatial operators on $\hat{\boldsymbol{u}}$, $\boldsymbol{L}$, the pressure gradient, $\boldsymbol{P}$, and non-linear terms, $\boldsymbol{N}$,

$$\frac{\partial \hat{\boldsymbol{u}}(\boldsymbol{k}_t)}{\partial t} = \hat{\boldsymbol{L}}(\boldsymbol{k}_t) + \hat{\boldsymbol{P}}(\boldsymbol{k}_t) + \hat{\boldsymbol{N}}(\boldsymbol{k}_t), \tag{3}$$

where

$$\begin{aligned} \hat{\boldsymbol{L}}(\boldsymbol{k}_t) &= -\hat{\nabla}(\boldsymbol{U}\hat{\boldsymbol{u}}(\boldsymbol{k}_t)) - \hat{\nabla}(\hat{\boldsymbol{u}}(\boldsymbol{k}_t)\boldsymbol{U}) + \nu\hat{\nabla}^2\hat{\boldsymbol{u}}(\boldsymbol{k}_t), \\ \hat{\boldsymbol{P}}(\boldsymbol{k}_t) &= -\hat{\nabla}\hat{p}(\boldsymbol{k}_t), \\ \hat{\boldsymbol{N}}(\boldsymbol{k}_t) &= \sum_{\substack{\boldsymbol{k}_a+\boldsymbol{k}_p=\boldsymbol{k}_t \\ \boldsymbol{k}_a,\boldsymbol{k}_p\neq 0}} -\hat{\nabla}(\hat{\boldsymbol{u}}(\boldsymbol{k}_a)\hat{\boldsymbol{u}}(\boldsymbol{k}_p)). \end{aligned} \tag{4}$$

The advection terms involving the mean velocity $\boldsymbol{U} = \hat{\boldsymbol{u}}(\boldsymbol{0})$ are linear in $\boldsymbol{u}(\boldsymbol{k}_t)$ and therefore are included in $\boldsymbol{L}$. Note that the dependence on y and t is from here on not explicitly shown. In this form, the cause of in-plane inter-scale transfer can be appreciated. Firstly, the only influence on the evolution of the velocity wave at wavenumber $\boldsymbol{k}_t$ from the linear terms is also from waves with wavenumber $\boldsymbol{k}_t$. Linear operators do not transfer momentum between scales. The cause of the inter-scale transfer lies then on the inter-scale interactions in the non-linear term.

We note that in the present analysis we use the conservative form of the advection, $\nabla \cdot (\boldsymbol{uu})$. The reason is that we are interested in the nature of inter-scale interactions through this term, and the latter is its form as derived from first principles, i.e. from the momentum balance on an infinitesimal fluid parcel. For incompressible flows, the advection in this original form is often combined with continuity to yield its non-conservative form, $\boldsymbol{u} \cdot \nabla \boldsymbol{u}$. While both forms yield the same net contribution to the momentum equations, their scale-by-scale contributions need not be equal. Take for instance the contribution into v of $\partial(uv)/\partial x$, or $u\partial v/\partial x$, for a given set of k_t, k_a and k_p. While in conservative form we would have $\mathrm{i}k_x(\boldsymbol{k}_t)\, u(\boldsymbol{k}_a)\, v(\boldsymbol{k}_p)$, in non-conservative form we would have $u(\boldsymbol{k}_a)\, \mathrm{i}k_x(\boldsymbol{k}_p)\, v(\boldsymbol{k}_p)$. Of the two, only the conservative form reflects the true contribution to $v(\boldsymbol{k}_t)$ from $u(\boldsymbol{k}_a)$ advecting $v(\boldsymbol{k}_p)$.

In the frame of Equation 3, the pressure term has contributions from wavenumber $\boldsymbol{k}_t$ into wavenumber $\boldsymbol{k}_t$ only. In the Navier-Stokes equations, the role of the pressure is to redistribute energy/momentum between different velocity components in order to enforce continuity [6]. In Equation 3, only pressure waves of wavenumber $\boldsymbol{k}_t$ appear in the evolution equation for $\boldsymbol{u}(\boldsymbol{k}_t)$, so in this sense the pressure acts as a linear term. It is common to derive an equation for the pressure by combining Equation 3 with the continuity equation, yielding

$$\hat{\nabla}^2\hat{p}(\boldsymbol{k}_t) = \hat{\nabla} \cdot \hat{\boldsymbol{N}}(\boldsymbol{k}_t), \tag{5}$$

which can be interpreted as the pressure having a non-linear source. In essence, $\hat{p}$ ensures that the solenoidal part of $\hat{\boldsymbol{N}}$ is balanced out of Equation 3 so that the resulting velocity is solenoidal. We note, however, that the non-linear/inter-scale source in Equation 5 is the same as in Equation 3, where it appears in it's first-principles form, in the conservation laws that govern the flow. In the present work we analyse it in that framework, leaving out of our scope it's analysis in Equation 3 with the understanding that it would be analogous.

2.1. Inter-scale transport through dyadic interactions

As discussed above, the inter-scale momentum transport is facilitated by the non-linear advection of velocity; specifically, two velocity waves combine to form a wave at a new wavenumber. In

Equation 4, the advecting wavenumber $\boldsymbol{k}_a$ and the advected wavenumber $\boldsymbol{k}_p$ form a wavenumber dyad $\boldsymbol{k}_a + \boldsymbol{k}_p = \boldsymbol{k}_t$. The non-linear term is made up of the sum of contributions $\hat{\boldsymbol{n}}$,

$$\hat{\boldsymbol{n}}(\boldsymbol{k}_a, \boldsymbol{k}_p) = -\hat{\nabla} \cdot (\hat{\boldsymbol{u}}(\boldsymbol{k}_a)\hat{\boldsymbol{u}}(\boldsymbol{k}_p))/(\Delta k_x \Delta k_z), \tag{6}$$

for all possible wavenumber dyads $\boldsymbol{k}_a + \boldsymbol{k}_p = \boldsymbol{k}_t$,

$$\hat{\boldsymbol{N}}(\boldsymbol{k}_t) = \sum_{\substack{\boldsymbol{k}_a+\boldsymbol{k}_p=\boldsymbol{k}_t \\ \boldsymbol{k}_a,\boldsymbol{k}_p \neq 0}} \hat{\boldsymbol{n}}(\boldsymbol{k}_a, \boldsymbol{k}_p)\Delta k_x \Delta k_z. \tag{7}$$

The factor $\Delta k_x \Delta k_z$ is required so that $\hat{\boldsymbol{n}}(\boldsymbol{k}_a, \boldsymbol{k}_p)$ is a spectral density which integrates to $\hat{\boldsymbol{N}}(\boldsymbol{k}_t)$ in Equation 7. This ensures that $\hat{\boldsymbol{n}}(\boldsymbol{k}_a, \boldsymbol{k}_p)$ is independent of simulation domain size. The non-linear advection term has nine contributions, which correspond to three different advecting and advected velocity components. The different contributions from u, v, w advecting u, v, w appear in the momentum equation in the following form,

$$\begin{aligned}
\hat{\boldsymbol{n}}(\boldsymbol{k}_a, \boldsymbol{k}_p) &= \begin{bmatrix} \hat{n}_1(\boldsymbol{k}_a, \boldsymbol{k}_p) \\ \hat{n}_2(\boldsymbol{k}_a, \boldsymbol{k}_p) \\ \hat{n}_3(\boldsymbol{k}_a, \boldsymbol{k}_p) \end{bmatrix} \\
&= \begin{bmatrix} \hat{n}_{uu}(\boldsymbol{k}_a, \boldsymbol{k}_p) + \hat{n}_{vu}(\boldsymbol{k}_a, \boldsymbol{k}_p) + \hat{n}_{wu}(\boldsymbol{k}_a, \boldsymbol{k}_p) \\ \hat{n}_{uv}(\boldsymbol{k}_a, \boldsymbol{k}_p) + \hat{n}_{vv}(\boldsymbol{k}_a, \boldsymbol{k}_p) + \hat{n}_{wv}(\boldsymbol{k}_a, \boldsymbol{k}_p) \\ \hat{n}_{uw}(\boldsymbol{k}_a, \boldsymbol{k}_p) + \hat{n}_{vw}(\boldsymbol{k}_a, \boldsymbol{k}_p) + \hat{n}_{ww}(\boldsymbol{k}_a, \boldsymbol{k}_p) \end{bmatrix} \\
&= \begin{bmatrix} -\mathrm{i}k_{t,x}\hat{u}(\boldsymbol{k}_a)\hat{u}(\boldsymbol{k}_p) - \frac{\partial}{\partial y}(\hat{v}(\boldsymbol{k}_a)\hat{u}(\boldsymbol{k}_p)) - \mathrm{i}k_{t,z}\hat{w}(\boldsymbol{k}_a)\hat{u}(\boldsymbol{k}_p) \\ -\mathrm{i}k_{t,x}\hat{u}(\boldsymbol{k}_a)\hat{v}(\boldsymbol{k}_p) - \frac{\partial}{\partial y}(\hat{v}(\boldsymbol{k}_a)\hat{v}(\boldsymbol{k}_p)) - \mathrm{i}k_{t,z}\hat{w}(\boldsymbol{k}_a)\hat{v}(\boldsymbol{k}_p) \\ -\mathrm{i}k_{t,x}\hat{u}(\boldsymbol{k}_a)\hat{w}(\boldsymbol{k}_p) - \frac{\partial}{\partial y}(\hat{v}(\boldsymbol{k}_a)\hat{w}(\boldsymbol{k}_p)) - \mathrm{i}k_{t,z}\hat{w}(\boldsymbol{k}_a)\hat{w}(\boldsymbol{k}_p) \end{bmatrix} /(\Delta k_x \Delta k_z).
\end{aligned} \tag{8}$$

The vector $\hat{\boldsymbol{n}}(\boldsymbol{k}_a, \boldsymbol{k}_p)$ has three components, $\hat{n}_1(\boldsymbol{k}_a, \boldsymbol{k}_p)$, $\hat{n}_2(\boldsymbol{k}_a, \boldsymbol{k}_p)$ and $\hat{n}_3(\boldsymbol{k}_a, \boldsymbol{k}_p)$, each appearing in one of the three components of the momentum equation that determine u, v and w. Each one is the sum of three advection contributions, e.g. $\hat{n}_1(\boldsymbol{k}_a, \boldsymbol{k}_p)$ is the sum of the three advection contributions arising from u advecting u, v advecting u and w advecting u. In total there are nine such contributions, $\hat{n}_{ij}(\boldsymbol{k}_a, \boldsymbol{k}_p)$, of u_i advecting u_j, as shown in Equation 8. We also define $\hat{N}_{ij}(\boldsymbol{k}_t)$ as the total momentum transfer to wavenumber $\hat{u}_i(\boldsymbol{k}_t)$ caused by u_i advecting u_j across all possible wavelength diads,

$$\hat{N}_{ij}(\boldsymbol{k}_t) = \sum_{\substack{\boldsymbol{k}_a+\boldsymbol{k}_p=\boldsymbol{k}_t \\ \boldsymbol{k}_a,\boldsymbol{k}_p \neq 0}} \hat{n}_{ij}(\boldsymbol{k}_a, \boldsymbol{k}_p)\Delta k_x \Delta k_z. \tag{9}$$

The total non-linear term $\hat{\boldsymbol{N}}(\boldsymbol{k}_t, y, t)$ is then a sum of wave dyads $\hat{\boldsymbol{n}}(\boldsymbol{k}_a, \boldsymbol{k}_p, y, t)$. The wave dyad $\hat{\boldsymbol{n}}$ is a function of $\boldsymbol{k}_a, \boldsymbol{k}_p, y, t$, which complicates its graphical portrayal. To tackle this, we reduce the scope of the analysis by considering in-plane interactions at a fixed wall-normal distance, y. We also focus on a specific target wavenumber, $\boldsymbol{k}_t$, implicitly eliminating the separate dependence on $\boldsymbol{k}_a$ and $\boldsymbol{k}_p$, as $\boldsymbol{k}_a + \boldsymbol{k}_p = \boldsymbol{k}_t$.

The final degree of freedom that can be reduced is time t. This analysis is aimed at statistically steady flows, so a suitable statistical measure of $\hat{\boldsymbol{n}}(\boldsymbol{k}_a, \boldsymbol{k}_p, y, t)$ over time can be used. A simple way to quantify the inter-scale effects of the non-linear term is to just consider the typical magnitude of the terms in Equation 8 by looking at their mean squared value. For example for w advecting u, the mean squared value is given by $\langle|\hat{n}_{wu}|^2\rangle$ where $\langle\rangle$ denotes a time average. Since the mean squared value of a quantity is always positive, this metric will always be positive.

Journal of Physics: Conference Series **2753** (2024) 012019 doi:10.1088/1742-6596/2753/1/012019

One shortcoming of this approach is that the value of $\hat{n}_{wu}$ at a given advecting and advected wavenumber is a phasor, which means that there can be misalignment and cancellation between the different wave dyads in Equation 7. For this reason, we also consider a metric based on the average transfer of energy, which can account for such effects. This metric makes use of the time-averaged projection $\mathcal{M}$ of b onto a,

$$\mathcal{M}(a,b) = \Re\langle a^* b\rangle, \tag{10}$$

where $\Re$ denotes real part. The time-averaged projection is linear for fixed a, e.g $\mathcal{M}(a, b+c) = \mathcal{M}(a,b) + \mathcal{M}(a,c)$. The time-averaged change in energy of the target wave can be expressed using a time-averaged projection,

$$\begin{aligned}\underbrace{\left\langle \frac{\partial}{\partial t}\left(\frac{1}{2}|\hat{u}(\boldsymbol{k}_t)|^2\right)\right\rangle}_{\text{Average change in energy}} &= \mathcal{M}\left(\hat{u}(\boldsymbol{k}_t), \frac{\partial \hat{u}(\boldsymbol{k}_t)}{\partial t}\right) \\ &= \underbrace{\mathcal{M}(\hat{u}(\boldsymbol{k}_t), \hat{L}_1(\boldsymbol{k}_t)) + \mathcal{M}(\hat{u}(\boldsymbol{k}_t), \hat{P}_1(\boldsymbol{k}_t))}_{\text{Linear and pressure terms}} + \underbrace{\mathcal{M}(\hat{u}(\boldsymbol{k}_t), \hat{N}_1(\boldsymbol{k}_t))}_{\text{Non-linear advection term}}.\end{aligned} \tag{11}$$

The total contribution of the non-linear advection to the time-averaged change in energy can then be expressed as a sum of inter-scale interactions,

$$\mathcal{M}(\hat{u}(\boldsymbol{k}_t), \hat{N}_1(\boldsymbol{k}_t)) = \sum_{\substack{\boldsymbol{k}_a+\boldsymbol{k}_p=\boldsymbol{k}_t \\ \boldsymbol{k}_a,\boldsymbol{k}_p\neq 0}} \mathcal{M}(\hat{u}(\boldsymbol{k}_t), \hat{n}_1(\boldsymbol{k}_a, \boldsymbol{k}_p))\Delta k_x \Delta k_z. \tag{12}$$

This is analogous to taking the time average of the equation for the evolution of turbulent kinetic energy. The distinction is that the terms which contribute to the energy of u, v, w are treated separately, rather than being combined into one equation for the total turbulent kinetic energy. This metric can be applied to each of the nine components of the non-linear advection term which correspond to different advecting and advected velocity components. For example, the inter-scale transfer of energy into the streamwise velocity which occurs through w waves advecting u waves is given by

$$\mathcal{M}(\hat{u}(\boldsymbol{k}_t), \hat{N}_{uw}(\boldsymbol{k}_t)) = \sum_{\substack{\boldsymbol{k}_a+\boldsymbol{k}_p=\boldsymbol{k}_t \\ \boldsymbol{k}_a,\boldsymbol{k}_p\neq 0}} \mathcal{M}(\hat{u}(\boldsymbol{k}_t), \hat{n}_{wu}(\boldsymbol{k}_a, \boldsymbol{k}_p))\Delta k_x \Delta k_z. \tag{13}$$

For a given target wavenumber of interest, 2D maps of inter-scale transfer can be portrayed in terms of advected wavenumber $\boldsymbol{k}_a = [k_{a,x}, k_{a,z}]^T$, with the advecting wavenumber given implicitly by $\boldsymbol{k}_p = \boldsymbol{k}_t - \boldsymbol{k}_a$. When using the above metric, the integral of the map for an advection component gives the total energy transfer to the target wave by that advection component, as demonstrated in Equation 13. The metric based on the mean-squared value, in contrast, satisfies no such property.

Maps can just as well be constructed in terms of advected wavenumber $\boldsymbol{k}_p = [k_{p,x}, k_{p,z}]^T$, but in wavenumber space there is no practical difference between expressing the maps in terms of $\boldsymbol{k}_a$ or $\boldsymbol{k}_p$. The transformation from $\boldsymbol{k}_a$ to $\boldsymbol{k}_p$ corresponds merely to a reflection in the k_x-k_z axes plus a translation by $\boldsymbol{k}_t$, following $\boldsymbol{k}_p = -\boldsymbol{k}_a + \boldsymbol{k}_t$. This means that the two maps show the same information, just flipped and shifted. If the maps were to be represented in wavelength space, however, whether the map was expressed in terms of advecting or advected wavelength would be significant, as the pre-multiplication is different. In this preliminary work only wavenumber maps are considered, and all maps are shown in terms of advecting wavenumber for consistency.

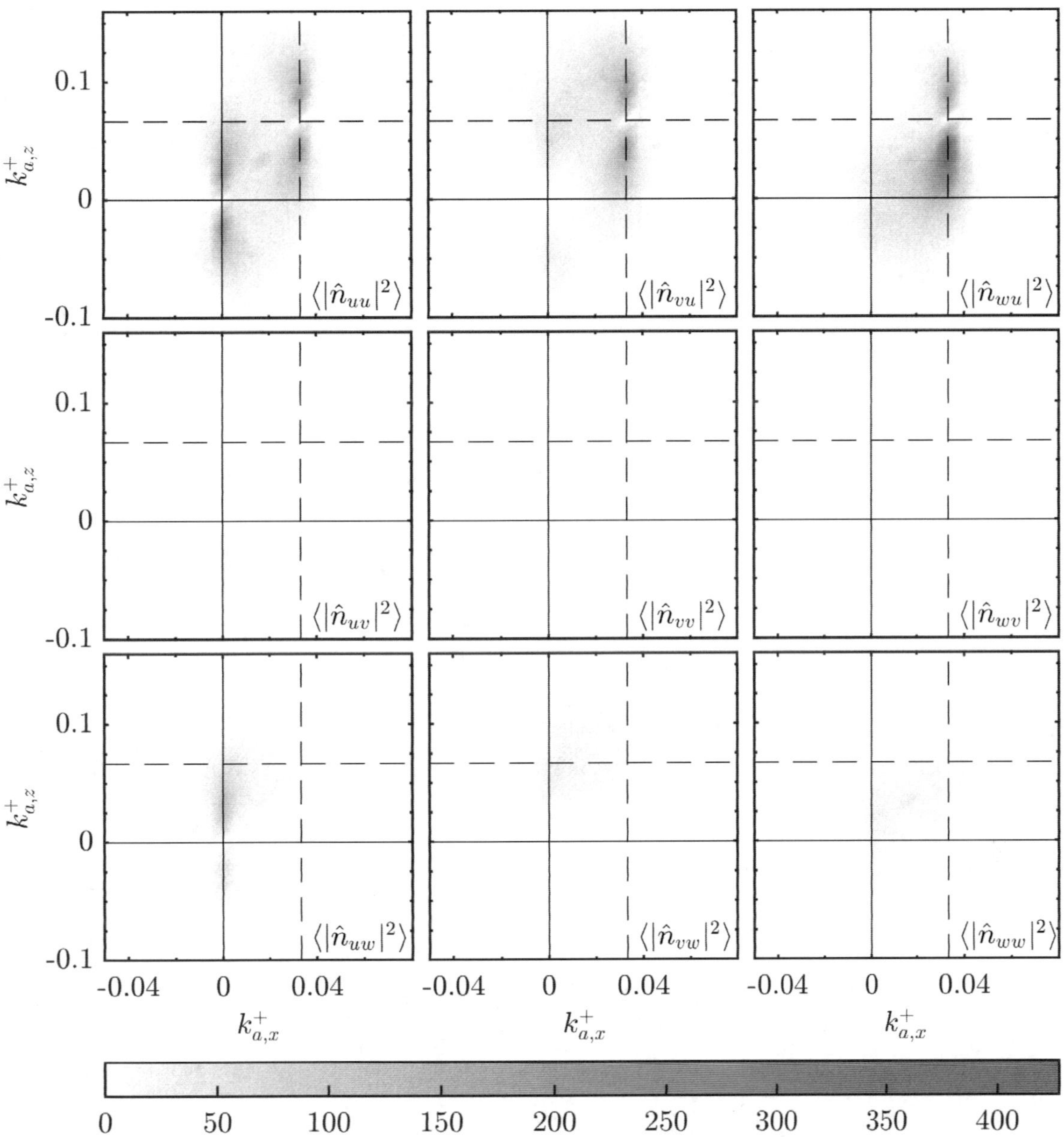

Figure 2. Maps of mean-squared value of the inter-scale transfer to the dynamic lengthscale, $\lambda^+_{t,x} = 188, \lambda^+_{t,z} = 94$ at $y^+ = 15$ from the nine contributions to the advection term for an $Re_\tau = 180$ channel flow. The colour scale in wall units is from 0 (White) to 428 (red). The black dashed lines indicate the target wavenumber $\boldsymbol{k}_t$.

Journal of Physics: Conference Series **2753** (2024) 012019 doi:10.1088/1742-6596/2753/1/012019

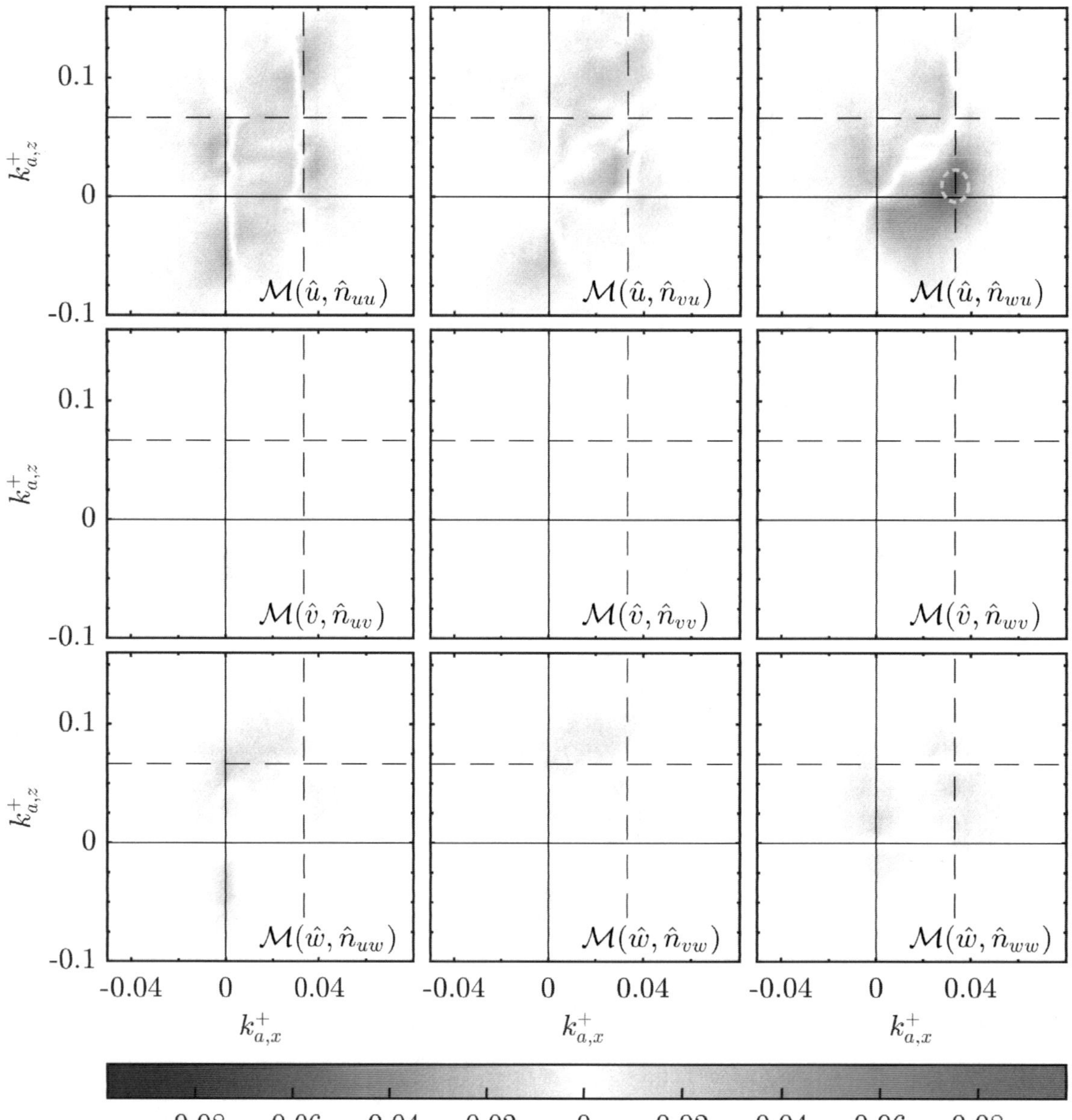

Figure 3. Maps of inter-scale energy transfer to the dynamic lengthscale, $\lambda^+_{t,x} = 188, \lambda^+_{t,z} = 94$ at $y^+ = 15$ from the nine contributions to the advection term for an $Re_\tau = 180$ channel flow. The colour scale in wall units is from -0.1 (blue) to 0.1 (red). The dashed green circle shows the wavenumbers where the strongest inter-scale transfer of energy occurs in the w advecting u term. The black dashed lines indicate the target wavenumber $\boldsymbol{k}_t$.

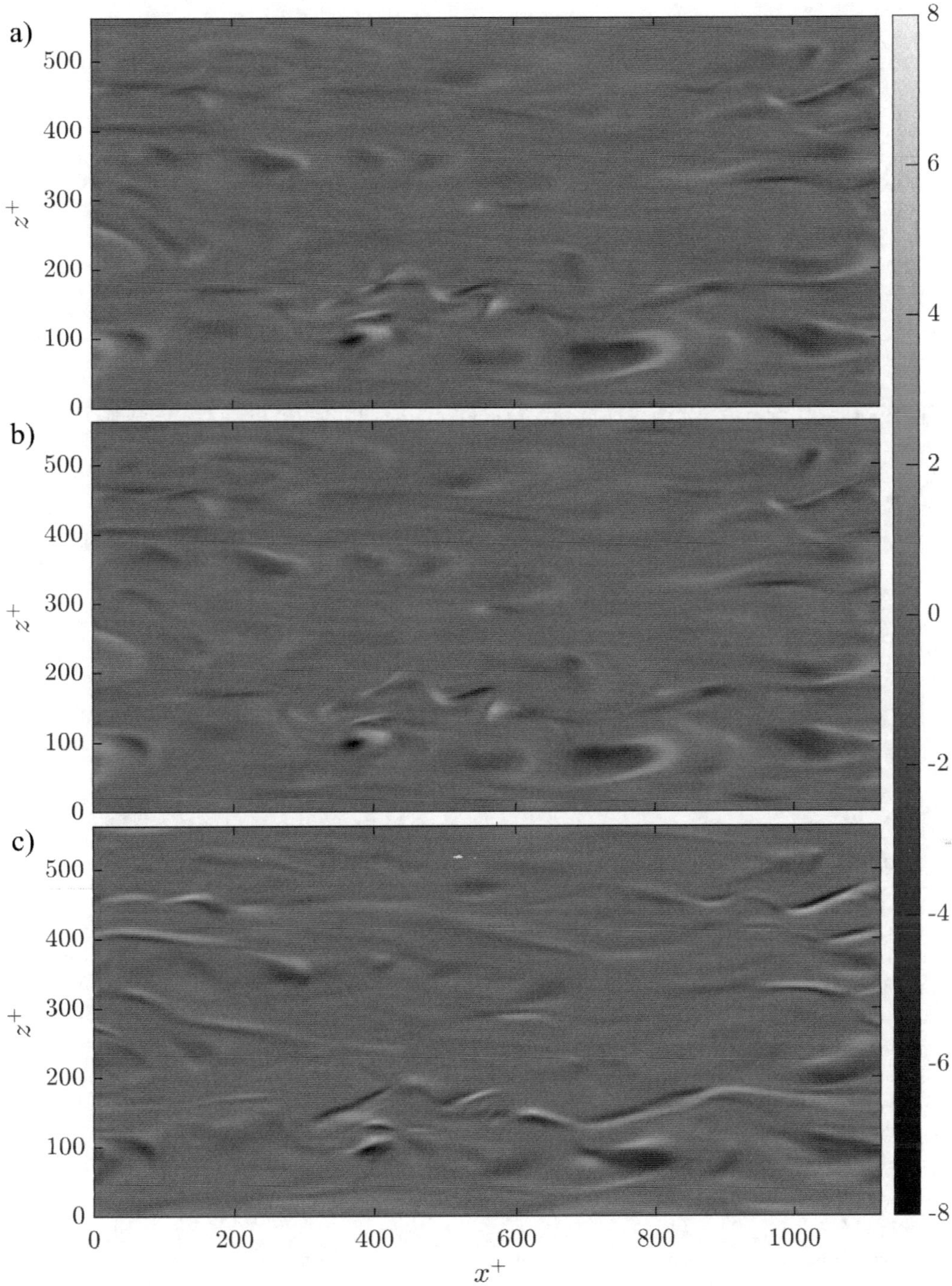

Figure 4. Instantaneous realisation of different transport terms for an $Re_\tau = 180$ channel flow in the buffer layer at $y^+ = 15$. a) Total advection term in the TKE equation, $u_i \partial u_i u_j / \partial x_j$. b) u, v, w advecting u in the TKE equation $u_1 \partial u_1 u_j / \partial x_j$. c) u advecting w in the TKE equation $u \partial wu / \partial z$. From dark to clear, from -8 to 8 in viscous units.

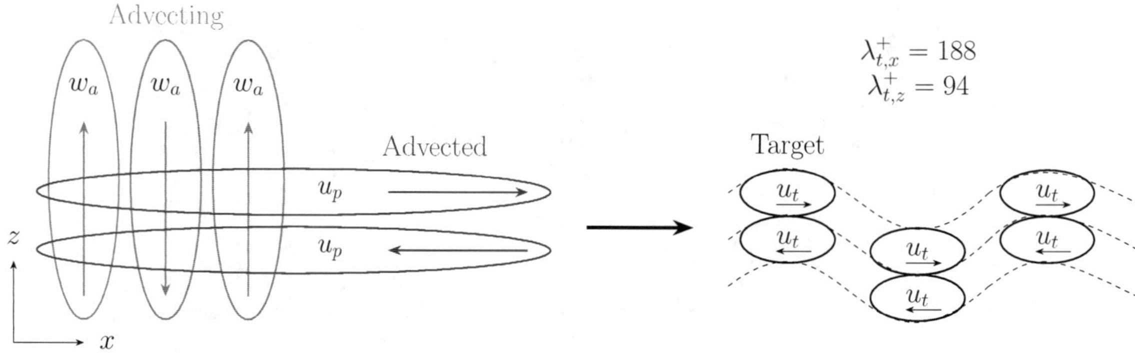

Figure 5. Sketch showing the advection process which dominates energy transfer to the dynamic lengthscale, $\lambda_x^+ = 188, \lambda_z^+ = 94$. Spanwise-velocity ($w$) structures which have a streamwise wavelength typical of quasi-streamwise vortices but are much longer in the spanwise direction advect streamwise-velocity (u) structures with a size typical of streamwise streaks. The dashed black lines show how the target dynamic wavelength is representative of streamwise streak meandering.

3. Results

The inter-scale transfer in the buffer layer was investigated for two target lengthscales for $Re_\tau = 180$. The first lengthscale, $\lambda_x^+ = 188, \lambda_z^+ = 94$, where the + superscript denotes the use of viscous units, was chosen as it is relevant in the near wall cycle. It is a lengthscale representative of quasi-streamwise vortices, but then also of their interactions with streamwise streaks and streak meandering [11, 12, 13]. This will be referred to as the 'dynamic' lengthscale in the remaining discussion. The second lengthscale, $\lambda_x^+ = 42, \lambda_z^+ = 21$, is a lengthscale in the dissipative range [14]. This will be referred to as the 'dissipation' lengthscale.

3.1. Inter-scale transfer to the dynamic lengthscale

Maps of inter-scale transfer to the 'dynamic' lengthscale in the buffer layer are shown for the nine contributions to the advection term $\partial u_i u_j/\partial x_i$ in Figures 2 and 3. In Figure 2, the mean squared value metric is used to quantify the relative importance of $\hat{n}_{ij}(\boldsymbol{k}_a, \boldsymbol{k}_p)$, while in Figure 3, the energy transfer metric is used. The two metrics are relatively consistent in identifying which wavenumber dyads have a strong inter-scale effect. The energy-transfer metric, however, has a positive value for dyads which statistically contribute to the strengthening of the target velocity wave, and a negative value for dyads which decrease the energy of the target wave. This is information that the mean-squared metric cannot provide, as it is always positive and thus only measures magnitude. For this reason, the remaining discussion will be based on the energy transfer metric.

Let us now focus on the information on inter-scale transfer that can be drawn from these maps. As discussed in Section 2.1, the integral of each of the nine maps in Figure 3 is the total energy transfer to the target wave from the mapped advection contribution. In Figure 3, the advection contribution $\hat{n}_{wu}$ has a much larger integral than any of the other advection components. This implies that inter-scale energy transfer to the dynamic lengthscale is dominated by the advection of streamwise-velocity (u) structures by spanwise-velocity (w) structures.

In physical space, the equation for the evolution of turbulent kinetic energy $k = u_i'u_i'/2$ (TKE) is given by

$$\frac{\partial k}{\partial t} = -u_i'\frac{\partial p'}{\partial x_i} - \underbrace{u_i'\frac{\partial u_i'u_j'}{\partial x_j}}_{\text{Advection}} + \nu u_i'\frac{\partial^2 u_i'}{\partial x_j \partial x_j}, \tag{14}$$

where $u_i' = u_i - \langle u_i \rangle$ are velocity fluctuations around the mean. As discussed in Section 2, only the non-linear advection term facilitates inter-scale transfer of energy. If w advecting u is the dominant part of the advection term in Fourier space, then we would also expect to observe this in physical space in the w-advecting-u term $-u\partial uw/\partial z$. This term is compared to the full TKE advection term $-u_i'\partial u_i'u_j'/\partial x_j$, and the sum of terms originating from advection in the streamwise momentum equation $-u_1'\partial u_1'u_j'/\partial x_j$, in Figure 4. This figure shows that the sum of the streamwise momentum equation terms is almost identical to the full advection term and that even the w-advecting-u term alone captures many of the features of the full advection term. This supports the conclusion drawn from the map of inter-scale energy transfer in Figure 3 that the dominant mechanism of inter-scale transfer to the dynamic lengthscale is spanwise (w) velocity advecting streamwise (u) velocity.

The map also gives insight on the dominant lengthscales in this process. Inter-scale transfer is largest for advecting streamwise waves with wavenumbers $0.025 < k_{a,x}^+ < 0.04$, $-0.01 < k_{a,z}^+ < 0.025$, which correspond to wavelengths of $160 < \lambda_{a,x}^+ < 250$, $160 < \lambda_{a,z}^+ < \infty$. These are structures with lengths comparable to quasi-streamwise vortices in the streamwise direction but are much wider in the spanwise direction. In order to satisfy the wavenumber triad $\boldsymbol{k}_t = \boldsymbol{k}_a + \boldsymbol{k}_p$, the wavelength of the corresponding advected velocity structures are in the range $750 < \lambda_{p,x}^+ < \infty, 94 < \lambda_{p,z}^+ < 150$. These correspond to structures with a size comparable to streamwise streaks in the streamwise and spanwise directions.

The physical picture of the dominant mechanism through which energy is transferred to the dynamic lengthscale in the buffer layer is sketched in Figure 5. Spanwise-velocity (w) structures which have a streamwise wavelength typical of quasi-streamwise vortices but are much longer in the spanwise direction advect streamwise-velocity (u) structures with a size typical of streamwise streaks.

What we see is that streamwise streaks, which are a streamwise-velocity (u) structure, are being advected by spanwise-velocity (w) structures in order to transfer energy to dynamic lengthscale streamwise-velocity (u) structures. The dynamic lengthscale is representative of quasi-streamwise vortices which are a spanwise and wall-normal (w and v) structure; however, the dynamic lengthscale also represents the streamwise streak meandering which is a streamwise-velocity (u) structure. Therefore, since the inter-scale transfer of energy to the dynamic lengthscale is almost entirely to streamwise-velocity (u) structures, this suggests that the advection process sketched in Figure 5 is contributing to the meandering of streamwise streaks in the self sustaining process [15, 16]. Specifically, this suggests that streak meandering, which is a key part of the self-sustaining process, occurs when streamwise streaks are advected by spanwise-velocity (w) structures which have a streamwise wavelength typical of quasi-streamwise vortices but are much longer in the spanwise direction.

3.2. Inter-scale transfer to the dissipation lengthscale

Maps of inter-scale transfer of energy in the buffer layer to the dissipation scale are shown for the nine contributions to the advection term $\partial u_i u_j/\partial x_j$ in Figure 6. The inter-scale transfer of energy to this scale is distributed more evenly across the nine contributions to the advection term in comparison with the dynamic lengthscale discussed above.

The regions of significant inter-scale transfer in Figure 6 have been split into three proposed mechanisms which are sketched in Figure 7. In mechanism 1 and 2, velocity structures which are the size of quasi-streamwise vortices or are larger interact with velocity structures in the

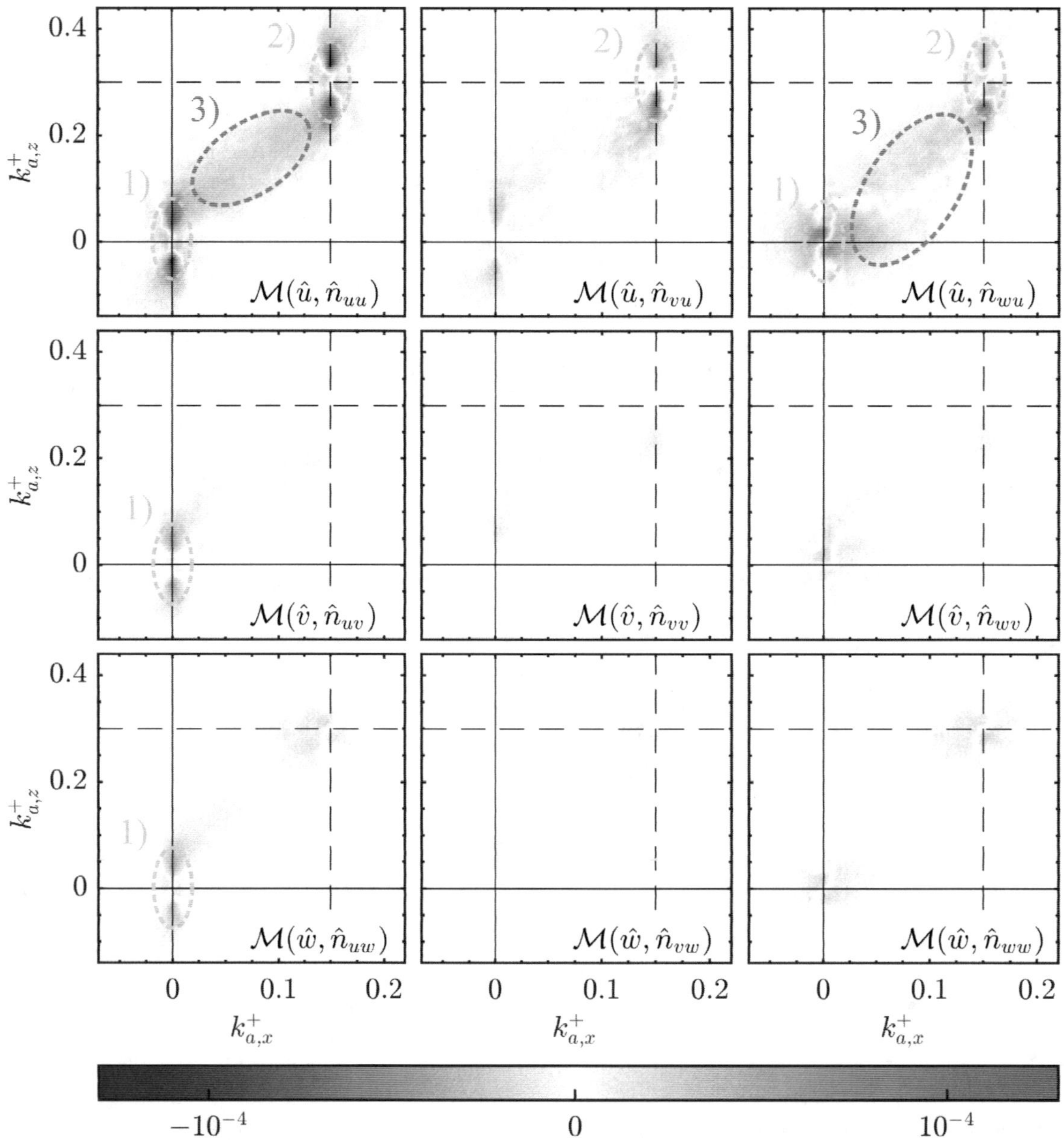

Figure 6. Maps of inter-scale transfer to the dissipation scale, $\lambda^+_{t,x} = 41, \lambda^+_{t,z} = 21$, at $y^+ = 15$ for the nine contributions to the advection term for an $Re_\tau = 180$ channel flow. The colour scale in wall units is from -1.3e-4 (blue) to 1.3e-4 (red). The dashed circles show the advecting wavenumbers of the advection mechanisms 1), 2) and 3) in Figure 7. The black dashed lines show the target wavenumber $\boldsymbol{k}_t$.

Journal of Physics: Conference Series **2753** (2024) 012019 doi:10.1088/1742-6596/2753/1/012019

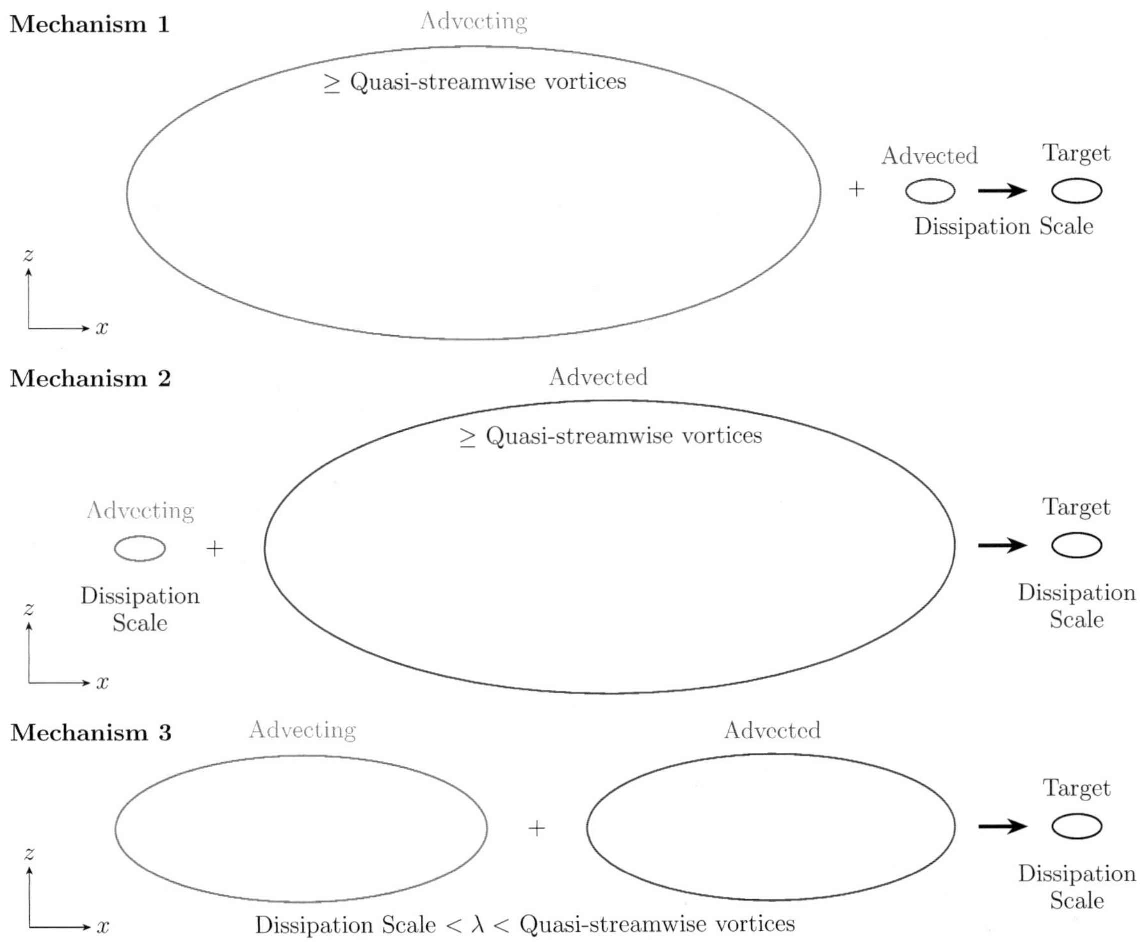

Figure 7. Sketches of three mechanisms through which energy is transferred to the dissipation scale. Mechanism 1: Velocity structures the size of quasi-streamwise vortices or larger advect velocity structures the size of the dissipation scale. Mechanism 2: Velocity structures the size of the dissipation scale advect velocity structures the size of quasi-streamwise vortices or larger. Mechanism 3: Velocity structures the size of an intermediate scale advect velocity structures the size of an intermediate scale. The intermediate scale is in-between the dissipation scale and the dynamic lengthscale, but still in the dissipation range.

dissipation scales. The larger velocity structures either advect (mechanism 1) or are advected by (mechanism 2) the smaller velocity structures. Mechanism 1 is significant in four of the nine contributions to the advection term: u advecting u, u advecting w, v advecting u, and w advecting u. Mechanism 2 is significant in three of the nine contributions to the advection term: u advecting u, u advecting v, and u advecting w.

In mechanism 3, the advecting and advected velocity structures have an intermediate scale in the dissipation range which is larger than the dissipation scale but smaller than the scale of quasi-streamwise vortices. This mechanism is significant in two of the nine advection terms: u advecting u and u advecting w.

Journal of Physics: Conference Series **2753** (2024) 012019 doi:10.1088/1742-6596/2753/1/012019

4. Conclusions

We have proposed a framework for the analysis of wall turbulence, looking at how information is transferred between scales in the Navier-Stokes momentum equations. Inter-scale transfer is facilitated by the non-linear advection term, in which a Fourier lengthscale $\boldsymbol{k}_a$ advects a Fourier lengthscale $\boldsymbol{k}_p$ to cause a transfer of momentum to Fourier lengthscale $\boldsymbol{k}_t$. There are three advecting and advected velocity components, which correspond to nine contributions to the advection term. Because wall turbulence is anisotropic, the proposed framework discriminates between these nine contributions in order to identify the relative importance of different advecting and advected velocity components. This framework can be used to identify the dominant advection mechanisms through which energy is transferred to a chosen target lengthscale.

Using the proposed framework, we have investigated the inter-scale transfer of energy in the buffer layer of an $Re_\tau = 180$ channel flow. Two statistical metrics for evaluating the time-averaged inter-scale transfer through the non-linear term were evaluated: one which looked at the mean-squared value of the non-linear advection term, and one which evaluated the average change in the energy of the target wave from the non-linear advection term. The two metrics are relatively consistent in identifying which wavenumber dyads have a strong inter-scale effect; however, the mean-squared metric does not give any indication of whether a particular wavenumber dyad increases or decreases the energy of the target wave. For this reason, the energy transfer metric is preferred as it is able to identify which wave dyads increase or decrease the energy of the target wave on average.

Two lengthscales were investigated. The first lengthscale, the dynamic lengthscale, was chosen as it is relevant in the near wall cycle. It is a lengthscale representative of quasi-streamwise vortices, but then also of their interactions with streamwise streaks and streak meandering. The second lengthscale, the dissipation scale, is a lengthscale in the dissipative range. The inter-scale transfer of energy was characterised by mapping each of the nine contributions to the non-linear advection term $\partial u_i u_j / \partial x_j$ in Fourier space.

For the dynamic lengthscale, the inter-scale transfer of energy is dominated by spanwise-velocity (w) structures which have a streamwise wavelength typical of quasi-streamwise vortices but are much longer in the spanwise direction advecting streamwise-velocity (u) structures with a size typical of streamwise streaks. In this process, streamwise streaks, which are a streamwise-velocity (u) structure, are being advected by spanwise-velocity (w) structures in order to transfer energy to dynamic lengthscale streamwise-velocity (u) structures. The dynamic lengthscale is representative of streamwise streak meandering; therefore, since the inter-scale transfer of energy to the dynamic lengthscale is almost entirely to streamwise-velocity (u) structures, this suggests that this process is contributing to the meandering of streamwise streaks in the self-sustaining process.

For the scale in the dissipation range, the inter-scale transfer of energy was distributed more evenly over the different contributions to the advection terms, i.e. significant inter-scale transfer was present for different advecting and advected velocity components, not just one mechanism. Three mechanisms are proposed for inter-scale transfer of energy to the dissipation scale. In the first two mechanisms, velocity structures which are the size of streamwise streaks or larger interact with velocity structures the size of the dissipation scale. The larger velocity structures either advect (mechanism 1) or are advected by (mechanism 2) the smaller velocity structures. In mechanism 3, the advecting and advected velocity structures have an intermediate scale in the dissipation range which is larger than the dissipation scale but smaller than the scale of quasi-streamwise vortices.

In conclusion, this framework shows promise for identifying the dominant lengthscales involved in the inter-scale transfer of information to a chosen target lengthscale in wall turbulence. The framework was shown to give results consistent with an analysis of the turbulent kinetic energy equation in physical space, and identify the dominant lengthscales of the velocity structures in the velocity structures.

Acknowledgments

This work was performed in part during the Fifth Madrid Turbulence Summer Workshop, funded by the European Research Council under the Caust grant ERCAdG-101018287. ZH was supported by UK's EPSRC under grant EP/S013083/1. Computational resources were provided by the University of Cambridge Research Computing Service under EPSRC grant EP/P020259/1, and by the UK 'ARCHER2' system under EPSRC, project e776.

References

[1] Domaradzki J A and Rogallo R S 1990 *Physics of Fluids A: Fluid Dynamics* **2** 413–426
[2] Domaradzki J A 1992 *Physics of Fluids A: Fluid Dynamics* **4** 2037–2045
[3] Waleffe F 1992 *Physics of Fluids A: Fluid Dynamics* **4** 350–363
[4] Cui G and Jacobi I 2021 *Physical Review Fluids* **6** 014604
[5] Cho M, Hwang Y and Choi H 2018 *Journal of Fluid Mechanics* **854** 474–504
[6] Lee M and Moser R D 2019 *Journal of Fluid Mechanics* **860** 886–938
[7] Bae H J, Lozano-Durán A and McKeon B J 2021 *Journal of Fluid Mechanics* **914** A3
[8] Lozano-Durán A and Arranz G 2022 *Physical Review Research* **4** 023195
[9] Chiarini A, Mauriello M, Gatti D and Quadrio M 2022 *Journal of Fluid Mechanics* **930** A9
[10] Yao H, Mollicone J P and Papadakis G 2022 *Journal of Fluid Mechanics* **941** A14
[11] Kline S J, Reynolds W C, Schraub F A and Runstadler P W 1967 *J. Fluid Mech.* **30** 741–773
[12] Robinson S K 1991 *The Kinematics of Turbulent Boundary Layer Structure* Ph.D. thesis Stanford University
[13] Jiménez J and Moin P 1991 *Journal of Fluid Mechanics* **225** 213–240
[14] Jiménez J 2012 *Annual Review of Fluid Mechanics* **44** 27–45
[15] Hamilton J M, Kim J and Waleffe F 1995 *Journal of Fluid Mechanics* **287** 317–348
[16] Waleffe F 1997 *Physics of Fluids* **9** 883–900

Journal of Physics: Conference Series **2753** (2024) 012020 doi:10.1088/1742-6596/2753/1/012020

Coherent structures in elastic turbulent planar jets

Christian Amor[1], Adrián Corrochano[2], Giulio Foggi Rota[1], Marco Edoardo Rosti[1†], Soledad Le Clainche[2‡]

[1] Complex Fluids and Flows Unit, Okinawa Institute of Science and Technology Graduate University, 1919 – 1 Tancha, Onna-son, 904 – 0412 Okinawa, Japan

[2] School of Aerospace Engineering, Universidad Politécnica de Madrid, Plaza del Cardenal Cisneros 3, 28040 Madrid, Spain

E-mail: †`marco.rosti@oist.jp`, ‡`soledad.leclainche@upm.es`

Abstract. At low Reynolds numbers, the flow of a Newtonian planar jet remains laminar, thus easy to characterize. In contrast, the presence of elasticity (*e.g.,* attained dissolving polymers in a Newtonian solvent) enables a highly-complex turbulent-like behavior termed *elastic turbulence.* In this work, we run data-driven modal decomposition algorithms on high-fidelity data collected from the simulation of an elastic turbulent planar jet. The large-scale motions are expressed as a finite expansion of modes that condense the dominant dynamics. The modes associated with lower frequencies weight the most on the reconstruction of the original data, thus they are further decomposed in space to investigate their implications on the sustainment of the elastic turbulent state. Our findings suggest that slower dynamics are crucial for the sustainment of elastic turbulence, which is connected to the interaction of spanwise-coherent structures, steady in space, with spanwise-periodic traveling waves, causing the breakdown of the structures close to the inlet.

1. Introduction

The addition of a small amount of long-chain polymers to a Newtonian solvent confers richer dynamics to the flow. The competition among the inertial, elastic and viscous contributions induces the rise of complex behaviors, which leads to a non-trivial flowing pattern. In this work, it is considered turbulence in planar jets when the elasticity of the fluid dominates and the inertial effects are negligible. In this regime, commonly named *elastic turbulence* [1], the elasticity of the fluid causes the turbulent motion in the flow, while the same Newtonian case would be characterized by a steady, laminar flow.

Early direct numerical simulations on laminar planar jets showed that the flow statistics are strongly influenced by the elasticity of the polymer [2]. Additionally, turbulent jet flows exhibited a slower spreading and decaying rates [3] than Newtonian jets, though these effects disappear in the far-field if elasticity remains low [4]. Increasing the elasticity strengthens the non-Newtonian behavior and the flow transitions to a chaotic state at markedly lower values of the Reynolds number Re. Therefore, elasticity promotes an earlier onset of the instability [5], though this behavior is reverted to a more stable state if inertia is enhanced [6]. Experiments in round jets showed the emergence of thin and elongated filaments at the shear-layer that stabilized the flow [7]. Planar jets exhibited a similar behavior, where the transition to turbulence is delayed to greater distances from the nozzle [8]. Nevertheless, this stable regime disappears for sufficiently large polymer elasticity, and the flow transitions to a turbulent-like state sustained equally by the elastic and inertial contributions, termed *elasto-inertial turbulence* [7, 9]. The chaotic state persists at low Re if elasticity remains large. In this case, inertia is no longer relevant and it is negligible compared to elasticity. As a result, the new turbulent state is sustained exclusively by the elastic fluctuations of the polymer. This self-sustained regime was observed in direct numerical simulations on planar jets [10].

The main goal of this work is the identification of the most relevant flow structures related to purely elasticity-driven turbulence in planar jets of a viscoelastic fluid. The complex dynamics are simplified by identifying coherent structures, *i.e.,* structures that are organized in space and persistent in time. These coherent structures are successfully identified by means of the higher order dynamic mode decomposition (HODMD) [11], a purely data-driven method that breaks down the flow into its most dynamically relevant structures. In elastic planar jets, slow dynamics are responsible for the sustainment of a turbulent-like state purely driven by the elasticity of the fluid.

The paper is organized as follows. First, sections §2 and §3 introduce the methodology. Then, section §4 provides the specifics on the numerical simulation. Lastly, results are presented in sections §5 and §6, while section §7 summarizes the main conclusions.

2. Higher order dynamic mode decomposition

The complex dynamics underlying elastic turbulent planar jets are described through modal decompositions. These algorithms have become popular for understanding turbulent flows, where modal decompositions express the data as a superposition of modes. Each mode represents a specific flow structure and it comprises spatio-temporal information connected to their dynamics. The modes are obtained using HODMD [11], which builds on the dynamic mode decomposition (DMD) [13] and introduces a delay embedding procedure [14] that enables the application of the classic DMD algorithm to data described by a broadband temporal spectrum. The algorithm splits the original data into sequences of observables or windows that overlapped constitute the original signal, while keeping the underlying dynamics. This procedure allows for an effective computation of the most dynamically relevant structures in flows whose spectral complexity overcomes the spatial complexity [15, 16].

The modal decomposition algorithm is fed with three-dimensional data regularly sampled in time with a rate Δt. Data are organized in a fifth-order tensor $\mathbf{V}$ that contains the flow variables (streamwise u, jet-normal v and spanwise w fluctuating velocity fields) for a specific set of N_t observables. HODMD (and DMD) reinterprets the data as a superposition of M modes and it fits the reconstruction in a Fourier-like expansion (for a single velocity component):

$$\mathbf{V}\left(x,y,z,t_k\right) \simeq \sum_{m=1}^{M} a_m \boldsymbol{u}_m\left(x,y,z\right) e^{(\delta_m + i\omega_m)t_k}, \quad k=1,\ldots,N_t. \tag{1}$$

The algorithm provides a finite set of modes $\boldsymbol{u}_m$, the so-called *DMD modes*, each related to a single frequency ω_m, and they grow, decay or remain neutral in time with growth rate δ_m. The coefficients a_m quantify the dominance of the DMD modes in the observed data. The inspection of these coefficients is useful for recognizing essential processes in the flow. Therefore, frequencies associated with relevant dynamics have a larger value of a_m, contributing more significantly to the flow, and vice-versa.

To calculate eq. (1), HODMD first performs dimensionality reduction through singular value decomposition (SVD) [17]. This step cuts down the spatial complexity of the data represented by the number of retained SVD modes (note that this is different and smaller to the spatial dimension of the data, given by the grid points and the number of variables analysed using this algorithm). An user-defined threshold ε_1 sets the extent of the compression in space that exploits the filtering ability of SVD to remove noise and filter small uncorrelated flow scales. As a result,

Journal of Physics: Conference Series **2753** (2024) 012020 doi:10.1088/1742-6596/2753/1/012020

the spatial complexity is described by few scales that contain most of the flow energy (thus connected to the main flow dynamics). Then, HODMD combines the DMD routine with the time delayed observables given by the number of windows. At this point, the algorithm chooses the modes that reconstruct most faithfully the original data, *i.e.,* those with the largest amplitude a_m, based on a second threshold ε_2. This last step fixes the spectral complexity, defined by the number of modes M retained in eq. (1).

3. Spatio-temporal Koopman decomposition

Equation (1) suggests an expansion of the data matrix $\mathbf{V}$ as a sum of trigonometric functions. In this way, complex data can be understood as the superposition of spatial structures that evolve in time following a simple function. At this point, HODMD can be carried out in space on every temporal structure, simplifying the description for each DMD mode $\boldsymbol{u}_m$ in the same fashion of eq. (1). This idea conceptualizes the STKD [12]. This algorithm performs sequentially the HODMD routine in time and space and it describes the flow by superposing a finite set of standing and/or traveling waves (*e.g.,* in streamwise x):

$$\mathbf{V}\left(x_i, y, z, t_k\right) \simeq \sum_{m=1}^{M} \sum_{n=1}^{N} a_{mn} \boldsymbol{u}_{mn}\left(y, z\right) e^{(\delta_m + i\omega_m)t_k + (\nu_{mn} + i\alpha_{mn})x_i}, \; i = 1, \ldots, N_x; \; k = 1, \ldots, N_t. \tag{2}$$

STKD expresses the spatial contribution from each DMD mode $\boldsymbol{u}_m$ as a superposition of N modes oscillating in space with wavenumber α_{mn}. Therefore, the spatio-temporal expansion (2) extends the compatibility of HODMD to traveling wave problems and it enables the decomposition of fully nonlinear dynamics whose behavior shows both space and time dependencies in a purely data-driven fashion. Each of the spatio-temporal structures $\boldsymbol{u}_{mn}$ is weighted with a spatio-temporal coefficient a_{mn}. Hence, the larger the coefficient, the more relevant is the spatio-temporal pattern in the reconstruction of the data, then implying a greater contribution to the flow dynamics.

4. Numerical simulation

The flow of a viscoelastic fluid injected through a planar slit of height $2h$ with velocity U is solved using direct numerical simulations (DNS). The incompressible Navier-Stokes equations are solved, accounting for an additional term in the momentum equation to represent the polymer: $\nabla \cdot \tau$, with τ the non-Newtonian stress tensor. τ is modeled with the so-called Oldroyd-B model [18]:

$$\lambda\left(\frac{\partial \tau}{\partial t} + \boldsymbol{v} \cdot \nabla \tau - \left(\tau \cdot \nabla \boldsymbol{v} + \nabla \boldsymbol{v}^\top \cdot \tau\right)\right) = \mu_p\left(\nabla \boldsymbol{v} + \nabla \boldsymbol{v}^\top\right) - \tau \tag{3}$$

with $\boldsymbol{v}$ the flow velocity field, λ the polymer relaxation time and μ_p its dynamic viscosity.

The equations are solved resorting to the extensively validated in-house solver *Fujin* (`https://groups.oist.jp/cffu/code`) [19, 20, 21, 22]. The equations are discretized on a staggered, uniform, Cartesian grid, and all the spatial derivatives are approximated using a second-order central finite differences scheme. However, the transport equation for the extra-stress tensor, eq. (3), requires a more cautious approach since its numerical solution in the presence of high elasticity is prone to numerical instabilities [25]. To tackle this issue, a logarithmic formulation [23, 24] is adopted in conjunction with a high-order weighted essentially non-oscillatory (WENO) scheme [26] to treat the left-hand side of eq. 3. The system is advanced in time using a second-order Adams-Bashforth scheme coupled with a fractional step method [27] to solve the pressure field at every time step.

The dynamical behavior of viscoelastic flows is modelled using two parameters: the Reynolds number Re that represents the relative importance of inertia to viscosity; and the Weissenberg number Wi that compares the elastic property of the fluid to inertia. Comparing the Wi to

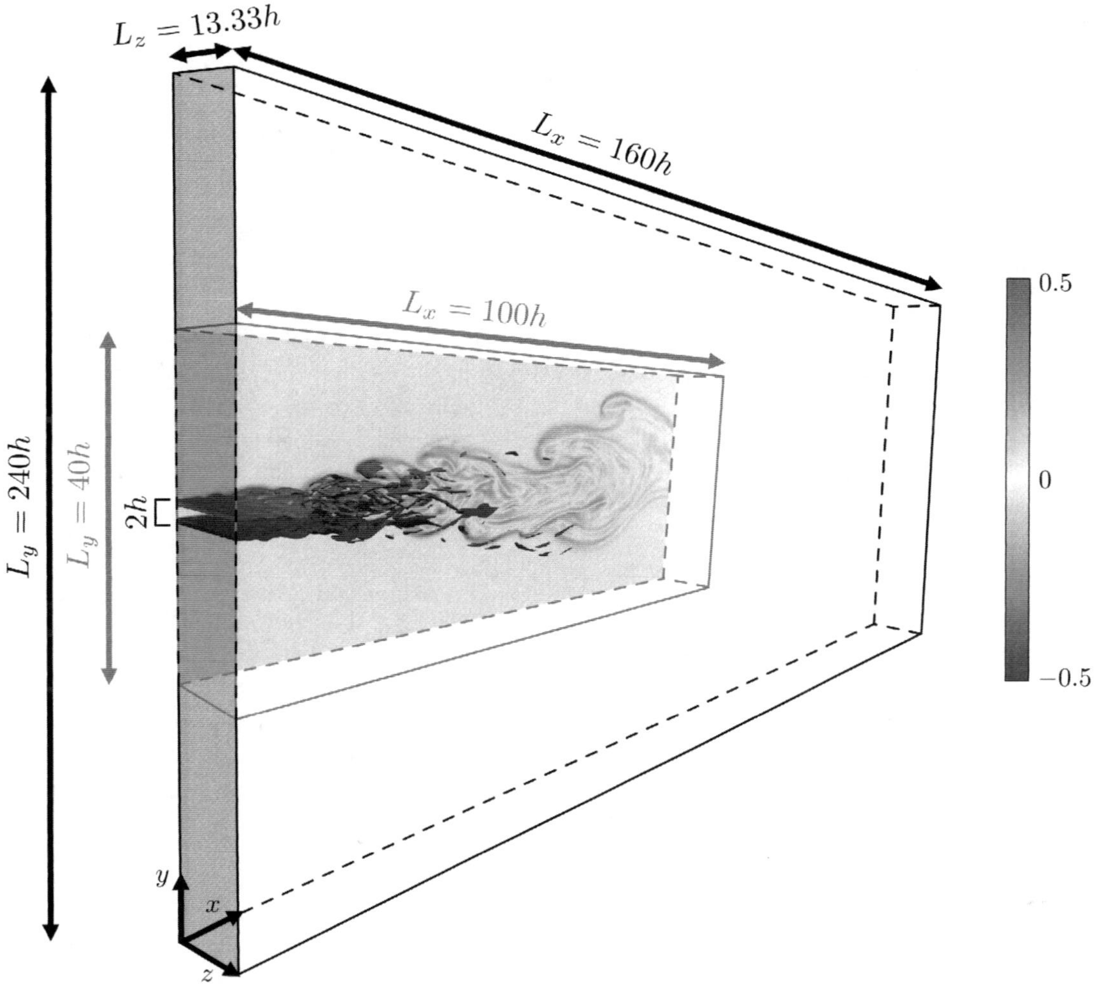

Figure 1. Sketch of the computational setup. The data-analysis is performed on observables extracted in a box within the domain used for simulations (black) that confines the flow near injection (red). The flow is plotted at a specific observable and it is overlapped, for the same field, with the instantaneous spanwise vorticity represented at a xy plane located at $z = 0$, and three-dimensional iso-surfaces of magnitude $-0.5U/h$ and $0.5U/h$.

Re allows to identify distinct regimes of motion. The elastic turbulent regime is characterize by the predominance of elasticity over inertia, so $Wi/Re > 1$. Therefore, $Re = hU/\nu_0 = 20$ is considered, with U the streamwise velocity at the inlet and ν_0 the sum of the kinematic viscosity of the solvent and of the polymer; and $Wi = \lambda h/U = 100$. This value represents a high-elasticity polymer solution.

The computational domain consists of a box with dimensions $L_x \times L_y \times L_z = 160h \times 240h \times 13.3h$ in the streamwise (x), jet-normal (y) and spanwise (z) directions, and it is discretized using $N_x \times N_y \times N_z = 1536 \times 2304 \times 128$ grid points. The domain is periodic in span and no-slip and no-penetration are imposed at the left wall ($x = 0$), though free-slip is allowed at the top and bottom boundaries ($y = 0$ and $y = L_y$). Lastly, the right boundary ($x = L_x$) enforces a non-reflective condition that prevents the flow to re-enter the computational domain. The

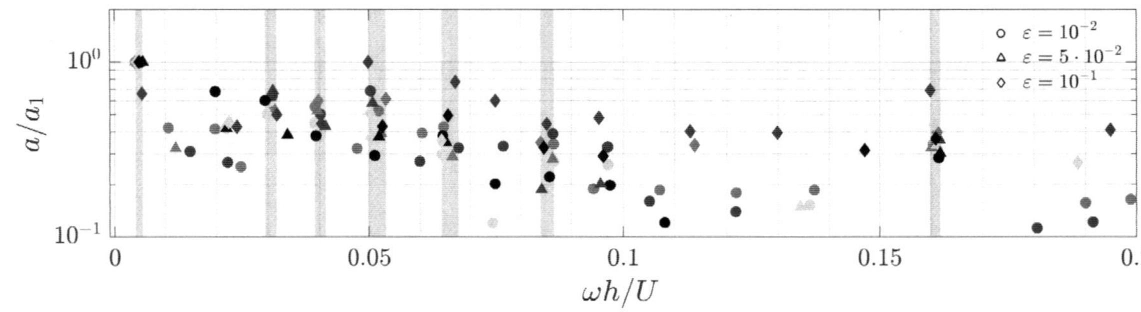

Figure 2. Calibration of the HODMD.The frequencies, ω, and their weighting coefficients, a, (normalized with the largest one for each combination of parameters) are shown. The calibration is carried out considering $d = 30$ (red), 40 (green), 50 (blue), 60 (black), and several thresholds ε on the reconstruction. There are seven groups of modes that satisfy the robustness criteria. The grey bars represent the dispersion on the frequency, measured as the difference between the smallest and the largest values for each group.

three-dimensional velocity fluctuations are collected in a smaller box that encapsulates the flow near injection (see Fig. 1) with size $100h \times 40h \times 13.3h$, thus reducing the grid down to $321 \times 130 \times 43$. The data tensor stores up to $N_t = 120$ observables, collected every $\Delta t = 10$ time units.

Clearly, coarser data lose resolution in space and time, so they do not represent the smallest scales and the fastest dynamics in the flow. Nevertheless, the resolution is still enough to identify the main dynamics of the flow while decreasing the computational burden during the data-analysis. This is true if the most coarse grid does not corrupt the spectral information compared to a more refined one. The Fourier transform is a simple method for checking the spectrum of frequencies describing the temporal dynamics (or wavenumbers, if performed in space). This approach is repeated at a few locations along the jet centerline, where the temporal (or spatial) signal is extracted, decomposed into Fourier space and undersampled (either removing a few observables or grid points). The procedure ends if the spectrum is corrupted because of the undersampling (*i.e.*, the dominant frequencies change). This calibration ensures the most coarse database that contains the main dynamics.

5. Temporal coherent structures

The complexity of elastic turbulent planar jets claims not merely a tool more sophisticated than the standard algorithms for identifying the main structures in the flow (herein the choice for HODMD), but also a more cautious use. The large spectral complexity presents an extra difficulty when recognizing the most relevant patterns in the flow. In this case, a preliminary calibration of the modal decomposition algorithm is compulsory. HODMD is performed recursively for different choices of parameters: (i) the number of windows in the delay embedding, called d, and (ii) the thresholds ε_1 and ε_2 (for simplicity, both are set to the same value ε). The number of modes kept in eq. (1) is intimately related to the choice of these parameters.

Figure 2 summarizes the calibration process. The calibration allows to identify modes that are physically meaningful, so they are not spuriously generated by the fitting algorithm. These modes are calculated for different combinations of parameters. Therefore, it is possible to isolate dynamical processes that are dominant in the flow by recognizing these frequencies. To guarantee the robustness of these modes, their frequency should be dispersed within a set tolerance (up to 20% deviation) across several calibrations (75% of the total). These criteria allow to distinguish the leading modes, that present similar frequencies though their amplitudes could deviate with

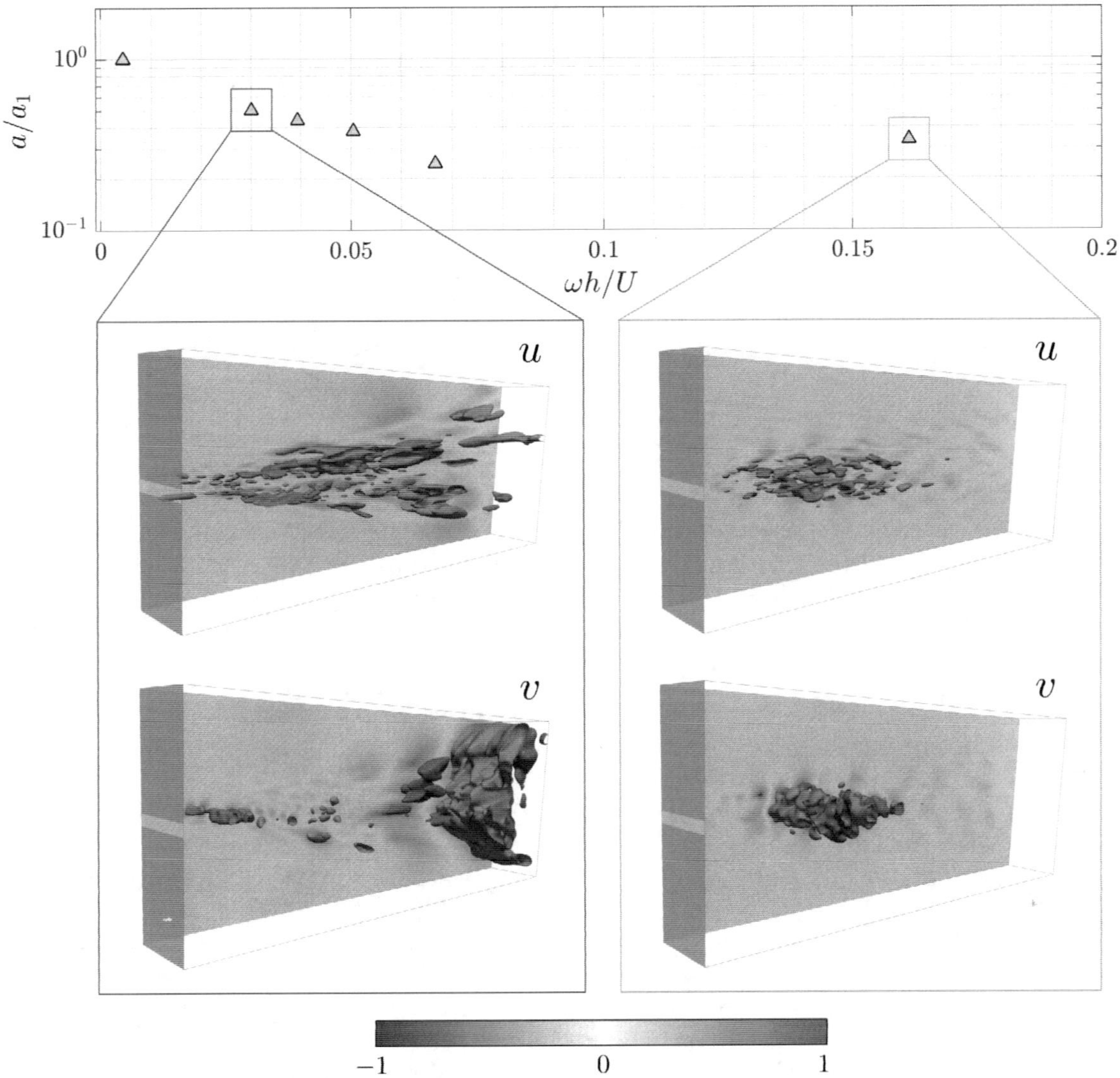

Figure 3. DMD modes. This particular calibration depicts six robust modes (The same color coding than Fig. 2 is adopted). The colored boxes report the shape of two dominant modes, represented by the iso-surfaces of the streamwise and jet-normal velocity components, normalised with their respective maxima, with $u = v = -0.5$ (blue), 0.5 (red), overlapped with a xy plane of the same fields at $z = 0$.

the calibration, from the spurious ones, whose frequencies and amplitudes always change. In fact, the same dominant frequencies are found for several choices of parameters, highlighting the robustness of the method (see [16, 28] for more information regarding the calibration process).

Figure 3 shows these modes for a particular set of parameters. The complex dynamics of the problem can be simplified by considering six robust modes that contain the smallest amount of information needed to properly reproduce the flow dynamics. These modes can be separated into two groups depending on whether their frequencies are large or small; moreover, the very-low non-zero frequency mode gathers the slowest dynamics that are observable, given the extent of the dataset (notice that the frequency of this mode is equivalent to the time span $T = \Delta t \times N_t$;

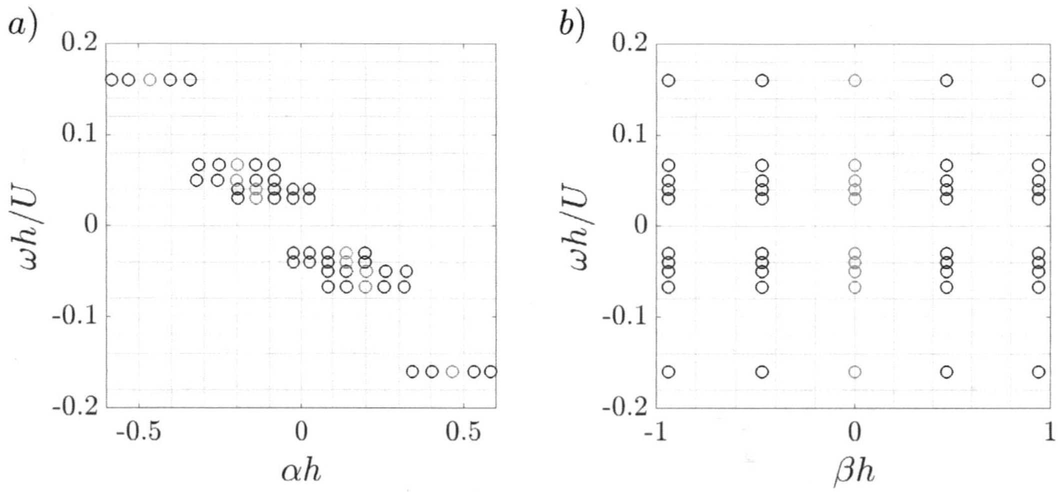

Figure 4. Spatio-temporal decomposition. The five largest-amplitude wavenumbers obtained in the streamwise (**a**) and the spanwise (**b**) spatial decompositions for the robust modes in Fig. 3 are shown. The dominant wavenumbers (*i.e.,* with the greatest coefficient a_{mn} in eq. (3)) are denoted in red for each temporal frequency.

$\omega = 2\pi/T$).

Starting with the slow modes, their collective weight is significant on the reconstruction of the data, *i.e.,* their amplitudes are large, so their dynamics have a major impact in the flow. Furthermore, the shape of these modes is similar. Here, this observation is sampled for the lowest frequency mode (see purple box in Fig. 3). The structures of the low frequency modes extend throughout the domain in the streamwise direction. Nevertheless, there are small size structures coherent in the spanwise direction close to the region of injection, while larger ones can be found downstream the jet. The description of the temporal dynamics is concluded with the high frequency mode. Despite the low Reynolds, wavepackets (see yellow box in Fig. 3) denote the presence of faster dynamics in the jet, represented by . The shape of the mode is significantly different to the low frequency one, and it is dynamically relevant mainly in the enclosed area that matches with the location of the mixing region in the jet. Wavepackets are structures typically found in classical Newtonian turbulent jets [29].

6. Spatio-temporal characterization of the flow

The results from HODMD provide a simple and clear description of the complex dynamics in the elastic turbulent jet, thus achieving a characterization of the flow given a finite set of two groups of modes detailing slow and fast dynamics in the flow. Two dominant modes are identified, each on a different group, where the rest are the result of nonlinear combinations of them. These modes give a description of the near and far fields of the jet, and the low frequencies have a particular relevance on the flow. Now, a more complete but still simple description of these particular flow dynamics is provided, following STKD. The space decomposition provides a characterization of the flow according to streamwise-traveling or spanwise-periodic modes.

The results of the spatio-temporal analysis are summarized in Fig. 4. Each temporal DMD mode $\boldsymbol{u}_m$ in eq. (1) is decomposed in space using HODMD. Therefore, a description based on a superposition of spatial structures oscillating with wavenumber ν_{mn} at each temporal frequency ω_m is fulfilled. As a result, the original signal is described based on spatio-temporal modes, each identified by a pair $\omega - \alpha$ (or β), with α their streamwise-wavenumber (β in

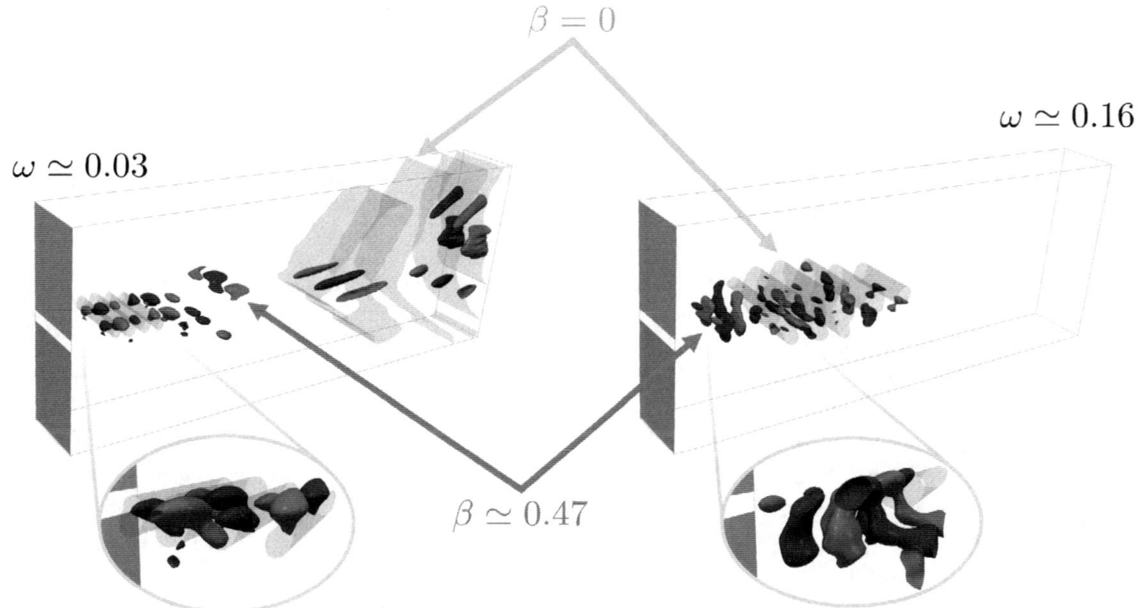

Figure 5. Spanwise-traveling wave structures. The dominant temporal patterns depicted in Fig 3 are decomposed in space. These spatio-temporal structures are either oscillating ($\beta \simeq 0.47$) or remaining steady ($\beta = 0$) in the spanwise direction. The iso-surfaces of the jet-normal velocity field are plotted, normalised with their maxima, with magnitude $|v| = 0.4$ (grey) at $\beta = 0$, and $v = -0.5$ (blue), 0.5 (red) at $\beta \simeq 0.47$.

spanwise). This representation provides a characterization based on periodic traveling waves, each displacing at their own phase velocity ($c_x = -\omega/\alpha$ in streamwise or $c_z = -\omega/\beta$ in spanwise). The spatio-temporal data is then recovered by superposing these groups of traveling waves for every DMD mode, where each temporal pattern is represented as a main wave and some modulations, identified with the red and black circles in Fig. 4, respectively.

The decomposition in streamwise highlights the traveling nature of the flow in this direction. Every spatio-temporal structure is moving with non-zero phase velocity, which somewhat increases with the frequency. Furthermore, there are pairs of frequencies whose dominant wavenumber is the same. These structures are similarly large in space, though their frequency is slightly modulated. On the other hand, the analysis in spanwise gives a different picture. In this case, the traveling waves are coexisting with spanwise-steady structures ($\beta = 0$). The spanwise-steady waves are dominant at every frequency and they are found along with spanwise-periodic waves oscillating with $\beta \simeq 0.47$ (fixed by the width of the computational domain) and their harmonics.

Figure 5 gathers the representation of the steady and unsteady modes in the spanwise decomposition of the flow for the dominant frequencies showed in Fig. 3. The steady mode ($\beta = 0$) show structures that are coherent in the spanwise direction. This coherence is enhanced in the wake at low frequency, whereas it is observed in the mixing region if the frequency is high. Their location is in good agreement with the large-size structures that can be observed in the representation of the temporal patterns. The steady structures are overlapped, and thus interact in space with those traveling in the spanwise direction with dominant wavenumber $\beta \simeq 0.47$. The low frequency mode shows an interaction that is mainly located at the region of injection, where the presence of spanwise-coherent structures contained in the steady mode is found. On the other hand, this interaction is observed mainly at the mixing region for the highest frequency mode.

The traveling wave structures are also located near the inlet, similarly to the low frequency mode, though they are not interacting with the steady mode.

7. Conclusions

This work provides the first characterization of the main flow structures involved in the sustainment of elastic turbulence in viscoelastic planar jets at low Reynolds. A state-of-the-art direct numerical simulation is combined with purely data-driven analysis tools to infer on the most relevant dynamics that characterize the complex flow dynamics.

The proposed methodology allows to simplify the description of the planar jet, thus condensing the most relevant dynamics into two sets of modes. Within these groups, two modes are dominant and thus contribute the most to the reconstruction of the data. In particular, slow dynamics play an important role in preserving the turbulent state, while higher frequencies are mainly related to wavepackets.

The description of these modes is completed by decomposing the temporal structures in traveling waves, that either move along the streamwise or the spanwise direction. Structures moving in the streamwise direction have a phase velocity that scales with their temporal frequency. On the other hand, the spanwise analysis highlights the breakdown of steady waves due to their interaction with periodic traveling waves. This behavior is present at low frequencies and appears disconnected from higher ones.

Acknowledments

The research was supported by the Okinawa Institute of Science and Technology Graduate University (OIST) with subsidy funding from the Cabinet Office, Government of Japan. This work has been partially done during the 2023 Madrid Turbulence Workshop, organized by Prof. J. Jiménez and made possible by the European Research Council under the Caust grant ERC-AdG-101018287. C.A., G.F.R. and M.E.R. acknowledge the computational resources provided by the Scientific Computing section of the Research Support Division at OIST, on Oakbridge-CX provided by the Information Technology Center (The University of Tokyo) through the HPCI System Research Project (Project ID: hp210246) and on SQUID provided by the Cybermedia Center (Osaka University) through the HPCI System Research Project (Project ID: hp220099). The authors also acknowledge Giovanni Soligo (OIST) for the DNS data. A.C. and S.L.C. acknowledge the grant PID2020-114173RB-I00, TED2021-129774B-C21 and PLEC2022-009235 funded by MCIN/AEI/ 10.13039/501100011033 and by the European Union "NextGenerationEU"/PRTR, and S.L.C. acknolwedges the support of Comunidad de Madrid through the call Research Grants for Young Investigators from Universidad Politécnica de Madrid.

References

[1] Groisman A and Steinberg V 2000 Elastic turbulence in a polymer solution flow *Nature* **405** pp 53–55
[2] Parvar S, da Silva C B and Pinho F T 2020 Local similarity solution for steady laminar planar jet flow of viscoelastic FENE-P fluids *J. Non-Newton. Fluid Mech.* **279** 109265
[3] Guimarães M C, Pimentel N, Pinho F T and da Silva C B 2020 Direct numerical simulations of turbulent viscoelastic jets *J. Fluid Mech.* **899** A11
[4] Guimarães M C, Pinho N and da Silva C B 2022 Turbulent planar wakes of viscoelastic fluids analysed by direct numerical simulations *J. Fluid Mech.* **946** A26
[5] Ray P K and Zaki T A 2015 Absolute/convective instability of planar viscoelastic jets *Phys. Fluids* **27** 014110
[6] Rallison J M and Hinch E J 1995 Instability of a high-speed submerged elastic jet *J. Fluid Mech.* **288** pp 311–324
[7] Yamani S, Keshavarz B, Raj Y, Zaki T A, McKinley G H and Bischofberger I 2021 Spectral universality of elastoinertial turbulence *Phys. Rev. Lett.* **127** 074501
[8] Yamani S, Raj Y, Zaki T A, McKinley G H and Bischofberger I 2023 Spatiotemporal signatures of elastoinertial turbulence in viscoelastic planar jets *Phys. Rev. Fluids* **8** 064610

[9] Samanta D, Dubief Y, Holzner M and Hof B 2013 Elasto-inertial turbulence *Proc. Nat. Aca. Sci.* **110** pp 10557–10562
[10] Soligo G and Rosti M E 2023 Non-Newtonian turbulent jets at low-Reynolds number *Int. J. Multiphase Flow* **167** 104546
[11] Le Clainche S and Vega J M 2017 Higher order dynamic mode decomposition *SIAM J. Appl. Dyn. Sys.* **16** pp 882–925
[12] Le Clainche S and Vega J M 2018 Spatio-temporal Koopman decomposition *J. Nonlin. Sci.*, **28** pp 1793–1842
[13] Schmid P 2010 Dynamic mode decomposition of numerical and experimental data *J. Fluid Mech.* **656** pp 5–28
[14] Takens F 1981 Detecting strange attractors in turbulence *Lecture Notes in Math* **898** D.A. Rand and L.S. Young, eds. Springer pp 366–381
[15] Le Clainche S, Vega J M and Soria J 2017 Higher order dynamic mode decomposition of noisy experimental data: the flow structure of a zero-net-mass-flux jet *Exp. Thermal and Fluid Sci.* **88** pp 336–353
[16] Le Clainche S, Izbassarov D, Rosti M E, Brandt L and Tammisola O 2020 Coherent structures in the turbulent channel flow of an elastoviscoplastic fluid *J. Fluid Mech.* **888** A5
[17] Sirovich L 1987 Turbulence and the dynamics of coherent structures, parts I–III *Quarterly of Applied Mathematics* **45** pp 561–571
[18] Oldroyd J G 1950 On the formulation of rheological equations of state *Proc. of the Royal Society A Math., Phys. and Eng. Sci.* **200** pp 523–541
[19] Abdelgawad M S, Cannon I and Rosti M E 2023 Scaling and intermittency in turbulent flows of elastoviscoplastic fluids *Nat. Phys.* **19** pp 1059—1063
[20] Rosti M E, Perlekar P and Mitra D 2023 Large is different: Nonmonotonic behavior of elastic range scaling in polymeric turbulence at large Reynolds and Deborah numbers *Sci. Adv.* **9** eadd3831
[21] Olivieri S, Brandt L, Rosti M E and Mazzino A 2020 Dispersed fibers change the classical energy budget of turbulence via nonlocal transfer *Phys. Rev. Lett.* **125** 114501
[22] Monti A, Olivieri S and Rosti M E 2023 Collective dynamics of dense hairy surfaces in turbulent flow *Sci. Rep.* **13** 5184
[23] Fattal R and Kupferman R 2005 Time-dependent simulation of viscoelastic flows at high Weissenberg number using the log-conformation representation *J. Nonnewton. Fluid Mech.* **126** 23–37
[24] Hulsen M A, Fattal R and Kupferman R 2005 Flow of viscoelastic fluids past a cylinder at high Weissenberg number: Stabilized simulations using matrix logarithms *J. Nonnewton. Fluid Mech.* **127** 27–39
[25] Keunings R 1986 On the high Weissenberg number problem *J. Nonnewton. Fluid Mech.* **20** 209–226
[26] Sugiyama K, Li S, Takeuchi S, Takagi S and Matsumoto Y 2011 A full Eulerian finite difference approach for solving fluid-structure coupling problems *J. Comp. Phys.* **230** pp 596–627
[27] Kim J and Moin P 1985 Application of a fractional-step method to incompressible Navier-Stokes equations *J. Comp. Phys.* **59** pp 308–323
[28] Le Clainche S, Rosti M E and Brandt L 2022 A data-driven model based on modal decomposition: application to the turbulent channel flow over an anisotropic porous wall *J. Fluid Mech.* **939** A5
[29] Towne A, Schmidt O T and Colonius T 2018 Spectral proper orthogonal decomposition and its relationship to dynamic mode decomposition and resolvent analysis *J. Fluid Mech.* **847** pp 821–867

Prediction of turbulent energy based on low-rank resolvent modes and machine learning

Yitong Fan[1,2], **Bo Chen**[1], **and Weipeng Li**[1]

[1] School of Aeronautics and Astronautics, Shanghai Jiao Tong University, Shanghai 200240, China

[2] Department of Mechanical Engineering, University of Melbourne, Parkville, VIC 3010, Australia

E-mail: `liweipeng@sjtu.edu.cn`

Abstract. A modelling framework based on the resolvent analysis and machine learning is proposed to predict the turbulent energy in incompressible channel flows. In the framework, the optimal resolvent response modes are selected as the basis functions modelling the low-rank behaviour of high-dimensional nonlinear turbulent flow-fields, and the corresponding weight functions are determined by data-driven neural networks. Turbulent-energy distribution in space and scales, at the friction Reynolds number 1000, is predicted and compared to the data of direct numerical simulation. Close agreement is observed, suggesting the feasibility and reliability of the proposed framework for turbulence prediction.

1. Introduction

Understanding the wall-bounded turbulence is of great significance for the aerodynamic design and flow control in practical engineering applications. To acquire precise spatial-temporal representations of turbulent motions, high-fidelity numerical simulations, such as direct numerical simulation (DNS) and wall-resolved large eddy simulation (WRLES), and three-dimensional particle image velocimetry (3D-PIV) have been widely performed in recent decades. However, these methods are cost prohibitive. On the other hand, techniques such as Reynolds-averaged Navier-Stokes equations (RANS), hybrid RANS/LES, and low-cost experiments provide access to accurate mean flow fields, but cannot predict perturbations well. Hence, relevant methods to efficiently predict the statistics and features of turbulent structures are highly desired.

To reduce the complexity of the high-dimensional problem of turbulent flows, reduced-order models are often adopted as basis functions to estimate the full flow field, e.g. the widely used proper orthogonal decomposition (POD) [1, 2], spectral proper orthogonal decomposition (SPOD) [3, 4] and the dynamic mode decomposition (DMD) [5]. These methods are technically data-driven and usually require extensive post-processing [6]. To enable the flow prediction in a physics-enhanced way, an equation-based method proposed from the perspective of linear dynamical system theory, named resolvent analysis [7], is adopted in the present research. In this framework, an input-output system is obtained by linearizing Navier-Stokes equations around the steady base flow, with the input (forcing) and output (response) related by a linear resolvent operator. During the recent years, the resolvent analysis has been widely acknowledged in that it successfully provides a promising tool to identify the prominent linear mechanisms and predict the turbulent fluctuations and structures.

Singular value decomposition is conducted upon the resolvent operator, to yield responses to the endogenous stimulus ranked by the amplification rate. The low-rank response modes are able to identify the the features in energy distribution and coherent structures, thus providing a promising tool to gain insight into the origins of the turbulent fluctuations and structures [8] and subsequently to enable the turbulence modelling and flow predictions. This is proved to be especially efficient in the incompressible wall-bounded turbulence. Sharma et al. [9] identified the coherent structures in the turbulent pipe flow, including the velocity streaks and packets of hairpin vortices, based on optimal response modes at representative wavenumber combinations. They confirmed that a simple linear combination of optimal modes is able to constitute the "skeleton" of turbulence. Abreu et al. [10] used resolvent analysis and SPOD to identify the dominant near-wall coherent structures in channel flows. Good agreement was observed between the low-rank representations of SPOD and resolvent analysis, especially at the frequency and wavenumbers where the lift-up mechanism is present. Moarref et al. [11] exploited the optimal resolvent modes, subject to the non-broadband forcing in wave speed, to predict the streamwise velocity fluctuations in high-Reynolds-number channel flows. With a positive weight function amplifying or attenuating the streamwise energy density of the optimal mode, they solved a convex optimization problem so as to match the model-based and the direct-numerical-simulation(DNS)-resolved streamwise velocity fluctuations. Similar methods were applied to estimate the distribution of spanwise and wall-normal velocity fluctuations and the Reynolds shear stress across the wall layer [12, 13]. Further investigations on the influence of the selected numbers of low-rank modes on the prediction capability were conducted by Moarref et al. [13]. They found that optimally forced (rank-1) response mode is able to well capture the streamwise energy. However, to estimate the spanwise and wall-normal energy spectrum and the streamwise-wall-normal co-spectrum, some more higher-order modes might be in necessity. Symon et al. [14] predicted the evolution of turbulence kinetic energy, in plane Poiseuille flows, based on the optimal response modes of velocity fluctuations from resolvent analysis, with the addition of eddy viscosity. The successful usage of the resolvent operator in the fully developed turbulence helps to identify the equation-based models for prediction.

In addition, more other strategies for flow estimation/reconstruction on the basis of resolvent framework have emerged in the recent years, e.g. constructing a suitable transfer function from the measurements to the driving force [15, 16, 17, 18, 6] and appropriately shaping the nonlinear forcing terms in the input-output framework [19, 20, 21]. However, these methods are usually used case by case. To enable the resolvent-based prediction in a computationally feasible and generalizable way, we intend to employ data-driven methods in combination with the resolvent analysis. Recent advances in machine learning have shown great potential for a fast alternative method in fluid mechanics [22], due to its inherent strengths in managing high-dimensional data compression, feature extraction, and generative inversion [23]. Notwithstanding, the interpretability of purely data-driven methods is highly restricted and the training of these models is reliant on a sufficient amount of data. Studies have been seeking to address this issue by integrating existing knowledge of physics with machine learning methods, yielding satisfying outcomes [24, 25, 26, 27].

In the present study, we intend to exploit the resolvent analysis to obtain a low-rank approximation of the nonlinear system as basis functions, which is characteristic of the turbulence cycle from a physics-enhanced perspective. Network models are then reconstructed to predict high-dimensional coefficients of the low-rank forcing modes in the wavenumber-frequency domain, with a set of mean velocity profiles being the input. Linear combination of resolvent response modes and the coefficient matrix, i.e. a weighted sum of the resolvent modes, consequently leads the reconstruction of three-dimensional flow fields. Using the limited database of turbulent channel flows, this paper targets at predicting the wall-normal distribution of turbulent energy and the premultiplied spectra of turbulent fluctuations. The remainder of this paper is outlined as follows. Section 2 introduces the resolvent formulation and the machine-learning framework. Databases that are used for training and testing are also described. In Sec. 3, validation and verification of the resolvent-based reduced-order model is investigated and the the predicted results of the turbulent energy are shown, in comparison with the true results. Finally, concluding remarks are given in Sec. 4.

2. Methodology

The non-dimensionalized Navier-Stokes equations for the incompressible turbulent channel flows are

$$\frac{\partial \boldsymbol{u}}{\partial t} + \boldsymbol{u} \cdot \nabla \boldsymbol{u} = -\nabla p + \frac{1}{Re} \nabla^2 \boldsymbol{u}, \tag{1}$$

$$\nabla \cdot \boldsymbol{u} = 0, \tag{2}$$

where t is time, p is pressure, the vector $\boldsymbol{u} = [u, v, w]^\top$ contains the velocity components in the streamwise (x), wall-normal (y) and spanwise (z) directions, and $\nabla = [\partial/\partial x, \partial/\partial y, \partial/\partial z]^\top$. The bulk Reynolds number Re is defined by $u_b^* h^*/\nu^*$, with u_b being the bulk velocity, h the channel half height, and ν the kinematic viscosity. The superscript $*$ denotes the dimensional values, otherwise the quantities are non-dimensionalized.

2.1. Resolvent formulation

Consider a state variable $\boldsymbol{q} = [u, v, w, p]^\top$ in the statistically steady turbulent channel flows. It can be Fourier transformed in the temporal and homogenous spatial (streamwise and spanwise) directions,

$$\boldsymbol{q}(x, y, z, t) = \iiint_{-\infty}^{\infty} \hat{\boldsymbol{q}}(y; \kappa_x, \kappa_z, \omega)\, \mathrm{e}^{\mathrm{i}(\kappa_x x + \kappa_z z - \omega t)} \mathrm{d}\kappa_x \mathrm{d}\kappa_z \mathrm{d}\omega. \tag{3}$$

The $\hat{(\cdot)}$ denotes the Fourier-transformed variables, κ_x and κ_z are the streamwise and spanwise wavenumbers, ω is the temporal frequency, and $\mathrm{i} = \sqrt{-1}$. The relevant wavelengths in the streamwise and spanwise directions are defined by $\lambda_x = 2\pi/\kappa_x$ and $\lambda_z = 2\pi/\kappa_z$, respectively. The mean state of $\boldsymbol{q}$ varies only in the wall-normal direction, i.e. $\bar{\boldsymbol{q}}(y) = [\bar{u}(y), 0, 0, \bar{p}(y)]^\top$, identical to the $\hat{\boldsymbol{q}}$ at $(\kappa_x, \kappa_z, \omega) = (0, 0, 0)$.

Substituting (3) into the Navier-Stokes equations (1) and (2), they are recast in the Fourier form for each $(\kappa_x, \kappa_z, \omega) \neq (0, 0, 0)$ as:

$$-\mathrm{i}\omega \hat{\boldsymbol{q}}(y; \kappa_x, \kappa_z, \omega) = \boldsymbol{A}(\kappa_x, \kappa_z, \omega)\, \hat{\boldsymbol{q}}(y; \kappa_x, \kappa_z, \omega) + \boldsymbol{B}(\kappa_x, \kappa_z, \omega)\, \hat{\boldsymbol{f}}(y; \kappa_x, \kappa_z, \omega). \tag{4}$$

The $\hat{\boldsymbol{f}}$ is the input forcing containing all the nonlinear contributions. It is comprised of streamwise, wall-normal and spanwise components, i.e. $\hat{\boldsymbol{f}} = [f_x, f_y, f_z]^\top$. $\boldsymbol{A}$ is a linear operator and $\boldsymbol{B}$ is the input matrix restricting the forcing terms to exist only in the momentum equations.

It consequently leads to a linearized form of the governing equations:

$$\begin{aligned} \hat{\boldsymbol{q}}(y; \kappa_x, \kappa_z, \omega) &= (-\mathrm{i}\omega \boldsymbol{I} - \boldsymbol{A}(\kappa_x, \kappa_z, \omega))^{-1} \boldsymbol{B}(\kappa_x, \kappa_z, \omega)\, \hat{\boldsymbol{f}}(y; \kappa_x, \kappa_z, \omega) \\ &= \boldsymbol{H}(\kappa_x, \kappa_z, \omega)\, \hat{\boldsymbol{f}}(y; \kappa_x, \kappa_z, \omega). \end{aligned} \tag{5}$$

$\boldsymbol{I}$ is the identity matrix, and $\boldsymbol{H} = (-\mathrm{i}\omega \boldsymbol{I} - \boldsymbol{A})^{-1} \boldsymbol{B} = \boldsymbol{L}^{-1} \boldsymbol{B}$ is the linear operator relating the input forcing to the output state. To simplify the notations, the variable dependency on

$(y; \kappa_x, \kappa_z, \omega)$ will be dropped hereafter. The expressions of $\boldsymbol{L}$ and $\boldsymbol{B}$ are given by

$$\boldsymbol{L} = \begin{bmatrix} \mathrm{i}k_x\bar{\boldsymbol{u}} - \mathrm{i}\omega\boldsymbol{I} - \frac{1}{Re}\nabla^2 & \frac{d\bar{\boldsymbol{u}}}{d\boldsymbol{y}} & \boldsymbol{0} & \mathrm{i}k_x\boldsymbol{I} \\ \boldsymbol{0} & \mathrm{i}k_x\bar{\boldsymbol{u}} - \mathrm{i}\omega\boldsymbol{I} - \frac{1}{Re}\nabla^2 & \boldsymbol{0} & \frac{d}{d\boldsymbol{y}} \\ \boldsymbol{0} & \boldsymbol{0} & \mathrm{i}k_x\bar{\boldsymbol{u}} - \mathrm{i}\omega\boldsymbol{I} - \frac{1}{Re}\nabla^2 & \mathrm{i}k_z\boldsymbol{I} \\ \mathrm{i}k_x\boldsymbol{I} & \frac{d}{d\boldsymbol{y}} & \mathrm{i}k_z\boldsymbol{I} & \boldsymbol{0} \end{bmatrix}, \tag{6}$$

$$\boldsymbol{B} = \begin{bmatrix} \boldsymbol{I} & \boldsymbol{0} & \boldsymbol{0} \\ \boldsymbol{0} & \boldsymbol{I} & \boldsymbol{0} \\ \boldsymbol{0} & \boldsymbol{0} & \boldsymbol{I} \\ \boldsymbol{0} & \boldsymbol{0} & \boldsymbol{0} \end{bmatrix}, \tag{7}$$

where $\nabla^2 = d^2/d\boldsymbol{y}^2 - (k_x^2 + k_z^2)\boldsymbol{I}$.

In the resolvent formulation (5), even though the nonlinear forcing is unknown, the dominant characteristics of turbulent fluctuations can be captured by the low-rank modes of $\boldsymbol{H}$, as is confirmed by various previous studies [7, 10, 11]. Before we proceed to apply singular value decomposition (SVD) to the resolvent operator, we introduce the definition of the inner product utilized herein, which is able to characterize the turbulent motions from the perspective of turbulent kinetic energy of the whole dynamical system. Let

$$\langle \hat{\boldsymbol{q}}, \hat{\boldsymbol{q}} \rangle = \hat{\boldsymbol{q}}^\dagger \boldsymbol{W_q} \hat{\boldsymbol{q}}, \ \left\langle \hat{\boldsymbol{f}}, \hat{\boldsymbol{f}} \right\rangle = \hat{\boldsymbol{f}}^\dagger \boldsymbol{W_f} \hat{\boldsymbol{f}}, \tag{8}$$

where the superscript † denotes the complex conjugate transpose. The weight matrices are given by

$$\boldsymbol{W_q} = \begin{bmatrix} \boldsymbol{K} & \boldsymbol{0} & \boldsymbol{0} & \boldsymbol{0} \\ \boldsymbol{0} & \boldsymbol{K} & \boldsymbol{0} & \boldsymbol{0} \\ \boldsymbol{0} & \boldsymbol{0} & \boldsymbol{K} & \boldsymbol{0} \\ \boldsymbol{0} & \boldsymbol{0} & \boldsymbol{0} & \boldsymbol{0} \end{bmatrix} \quad \text{and} \quad \boldsymbol{W_f} = \begin{bmatrix} \boldsymbol{K} & \boldsymbol{0} & \boldsymbol{0} \\ \boldsymbol{0} & \boldsymbol{K} & \boldsymbol{0} \\ \boldsymbol{0} & \boldsymbol{0} & \boldsymbol{K} \end{bmatrix}. \tag{9}$$

The $\boldsymbol{K}$ is the Clenshaw-Curtis quadrature weights [28] in the wall-normal direction. Taking the $\boldsymbol{W_q}$ and $\boldsymbol{W_f}$ into consideration, we perform SVD on the corrected resolvent operator to extract the low-rank approximations, that is,

$$\boldsymbol{W_q}^{1/2} \boldsymbol{H} \boldsymbol{W_f}^{-1/2} = \sum_{p=1}^{N} \tilde{\boldsymbol{\psi}}_p \sigma_p \tilde{\boldsymbol{\phi}}_p^\dagger, \tag{10}$$

$$\boldsymbol{\phi}_p = \boldsymbol{W_f}^{-1/2} \tilde{\boldsymbol{\phi}}_p, \tag{11}$$

$$\boldsymbol{\psi}_p = \boldsymbol{W_q}^{-1/2} \tilde{\boldsymbol{\psi}}_p, \tag{12}$$

where N is the number of resolvent modes, $\boldsymbol{\psi}_p$ and $\boldsymbol{\phi}_p$ are pth-rank orthogonal basis functions of the response and forcing modes. The singular value σ_p denotes the energy amplification rate, ranked by $\sigma_p > \sigma_{p+1}$. Based on this modal decomposition, the nonlinear terms and the state variables can be reconstructed from a weighted sum of all the forcing and response modes:

$$\hat{\boldsymbol{f}} = \sum_{p=1}^{N} a_p \boldsymbol{\phi}_p, \tag{13}$$

$$\hat{\boldsymbol{q}} = \sum_{p=1}^{N} \boldsymbol{\psi}_p \sigma_p a_p. \tag{14}$$

The a_p is the projection of the forcing mode onto the nonlinear term.

2.2. Prediction of turbulent energy based on low-rank resolvent modes and machine learning

In wall turbulence, the low-rank nature of the resolvent operator is widely observed, indicating that the understanding of the origins of the turbulent fluctuations and structures can be realized using the rank-1 (optimal) model. Therefore, in the present study, we majorly consider this simple response mode as a basis function of the turbulent fields, which can be recovered with a linear superposition of optimal modes at all wavenumber combinations. The premultiplied spectra of turbulent fluctuations carried by the optimal mode can be obtained by

$$E_{qq}\left(y;\kappa_x,\kappa_z,c\right)=\kappa_x^2\kappa_z\left(\left|\boldsymbol{\psi}_1\right|\left(y;\kappa_x,\kappa_z,c\right)\sigma_1\left(\kappa_x,\kappa_z,c\right)a_1\left(\kappa_x,\kappa_z,c\right)\right)^2. \tag{15}$$

The $\omega-$dependence is replaced by the wave speed c, expressed as $c=\omega/\kappa_x$, according to the grid settings of the resolvent operator which will be accounted for in the following. Integrating (15) over the frequency and spatial wavenumber domain yields the one-dimensional premultiplied spectra and the turbulent energy density in the wall-normal direction, viz.

$$E_{qq}(y,\kappa_x)=\iint E_{qq}\left(y;\kappa_x,\kappa_z,c\right)\mathrm{d}\log\left(\kappa_z\right)\mathrm{d}c, \tag{16}$$

$$E_{qq}(y,\kappa_z)=\iint E_{qq}\left(y;\kappa_x,\kappa_z,c\right)\mathrm{d}\log\left(\kappa_x\right)\mathrm{d}c, \tag{17}$$

$$E_{qq}(y)=\iiint E_{qq}\left(y;\kappa_x,\kappa_z,c\right)\mathrm{d}\log\left(\kappa_x\right)\mathrm{d}\log\left(\kappa_z\right)\mathrm{d}c. \tag{18}$$

In this framework, to predict the $E_{qq}(y,\kappa_x)$, $E_{qq}(y,\kappa_z)$ or $E_{qq}(y)$, the projection coefficients of the resolvent modes onto the real forcing contribution, i.e. the a_1 in (15), are of great significance and to be determined. To quantify a_1 in a full field of spatial wavenumber and frequency pairs in a computationally feasible and generalizable way, we intend to employ data-driven methods herein.

The training network models have been developed, working as a projection function of the input, shown in figure 1. The cases of incompressible turbulent channel flows at various Reynolds numbers are considered. Since only the mean streamwise velocity is required in the resolvent operator (see equation (6)), the mean velocity profiles across the wall layer is fed to the network. In figure 1, we use the architecture of convolutional neural network (CNN), which is especially designed for large-scale structured data. The CNN is mainly composed of convolutional layers, pooling layers, and fully connected layers. The convolutional layer captures the global and local information of the input data, and the pooling layer aims to reduce the dimension of the data to avoid overfitting. We reshape the mean profile in the turbulent boundary layers to satisfy the requirement of input, and make the full-field projection coefficients to be the output. In this way, turbulent energy E_{qq} can be consequently obtained following the low-rank approximation of the resolvent analysis. Beyond that, with the inverse transform, this framework is a quite promising tool to quantify instantaneous three-dimensional flow field as shown in figure 1, as long as sufficient databases are exploited for training and testing. This is out of the scope of this paper's target.

2.3. Databases and computational methods

Incompressible turbulent channel flows, at the friction Reynolds numbers $Re_\tau=180,390,550,1000,2000,4200$ and 5200, are considered in the present study. The mean base flow for input and the energy density for prediction is obtained from the direct numerical simulations in Refs. [29, 30, 31].

To obtain the resolvent modes at all required frequency and spatial wavenumber pairs, a sufficient domain with $N_x\times N_z\times N_c=87\times87\times24$ points is constructed. N_x and N_z are the numbers of grid points for the streamwise and spanwise wavelength, which are logarithmically

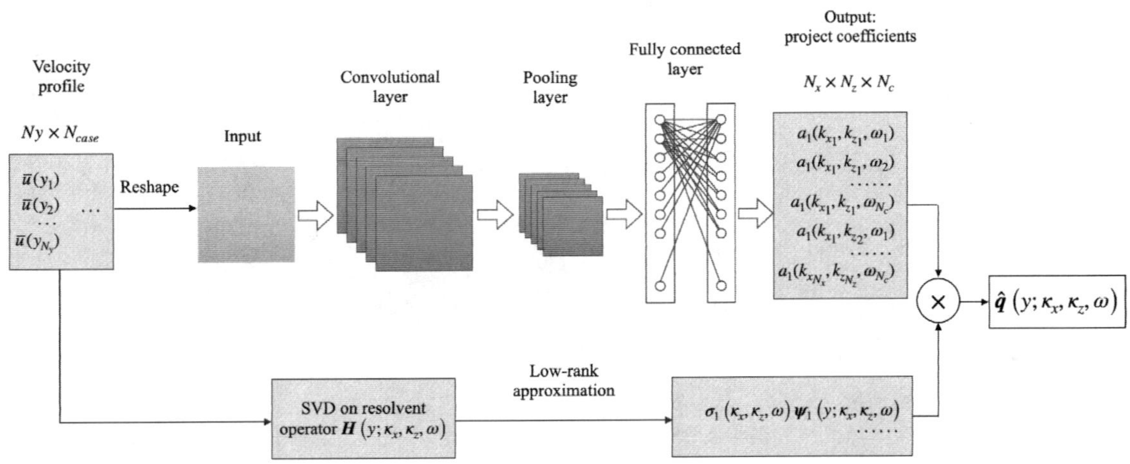

Figure 1. The schematic of the network for prediction of the projection coefficients.

spaced with the grid resolutions being $\Delta \log(\lambda_x) = 0.1$ and $\Delta \log(\lambda_z) = 0.1$, for all the cases. In the frequency domain, N_c denotes the number of grid points for the streamwise wave traveling at the speed of c. Since it has been found that the wave speeds are energetically important within $2 \leq c/u_\tau \leq \bar{u}_c/u_\tau$ [11], where the $\bar{u}_c$ is the mean velocity at the channel centerline and u_τ is the friction velocity, the wave speeds can be linearly chosen in this range to efficiently reduce the computational cost, and the excluded wave parameters will not influence the following results. In the wall-normal direction, N_y grid points are spaced via the Chebyshev collocation method, defined by $y = \cos(\pi j/(N_y - 1))$, where $j = 0, 1, ..., N_y - 1$ and $-1 \leq y \leq 1$. In the present study, we make $N_y = 200$ for all the cases and the second-layer inner-scaled height y^+_{min} restricted within 0.65. The non-slip impermeable boundary condition is applied at the wall.

3. Results and discussion

In this section, we will first extract the optimal response modes in channel flows, evaluating their capability in identifying the coherent structures and recovering the energy distribution. The projection coefficients of the low-rank forcing modes are then obtained through machine learning, adjusting the magnitude of the response, to predict the spectra and wall-normal distributions of turbulent energy.

3.1. Low-rank modelling of the coherent structures

Figure 2 shows the shapes of the velocity and temperature mode at a representative wavenumber combination $(\lambda^+_x, \lambda^+_z, c^+) = (1000, 100, 16.8)$. The superscript + denotes the normalization with inner units. Only the results in the channel flow at $Re_\tau = 1000$ are shown here for brevity, and those in the other cases do not affect the following conclusion. In panels (*a*) and (*b*), positive and negative velocity fluctuations appear alternatively in the streamwise direction, with the u-mode attached to the wall with a narrow angle while the v-mode without an inclination angle to the wall. This is associated with the mild phase variation of v-response in the wall-normal direction. The signs of the u- and v-mode are negatively correlated above the bottom surface, meaning that the low- or high-speed streaks (denoted by the negative or positive u-fluctuations) are carried away from or towards the wall. This phenomenon is more straightforwardly depicted in the cross-stream plane, shown in panel (*c*). The high- and low-speed streaks accompanied by streamwise vortices expose the sweep and ejection events, consistent with the lift up mechanism, capturing

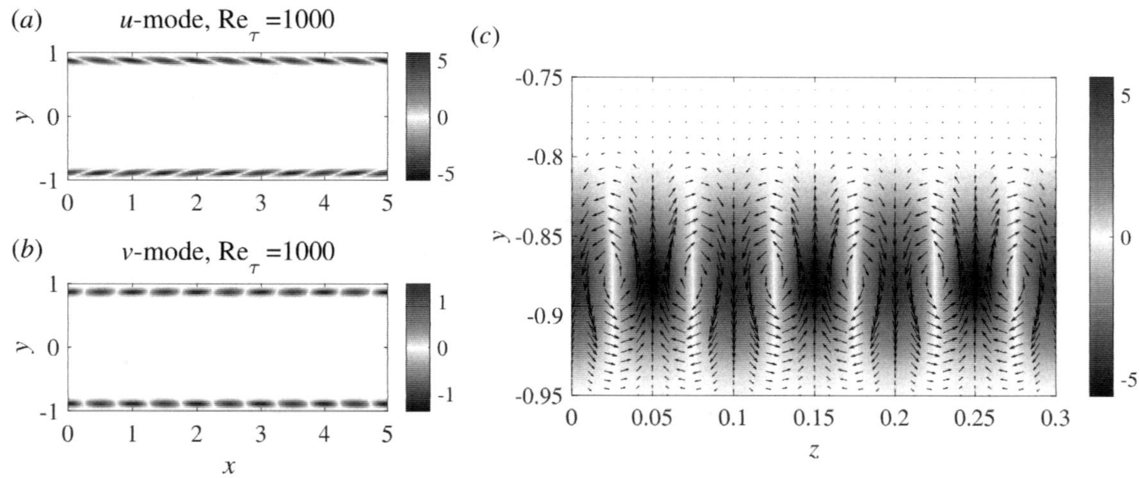

Figure 2. The response mode shapes of the (a,b) streamwise and wall-normal velocity in the streamwise-wall-normal plane, and (c) streamwise velocity in the cross-stream plane at $(\lambda_x^+, \lambda_z^+, c^+) = (1000, 100, 16.8)$, in the channel flow at $Re_\tau = 1000$. Red and blue denote positive and negative disturbances, respectively. The arrows denote the wall-normal and spanwise velocity components of the vortices.

the empirically observed characteristics of the near-wall coherent structures in the wall-bounded turbulence.

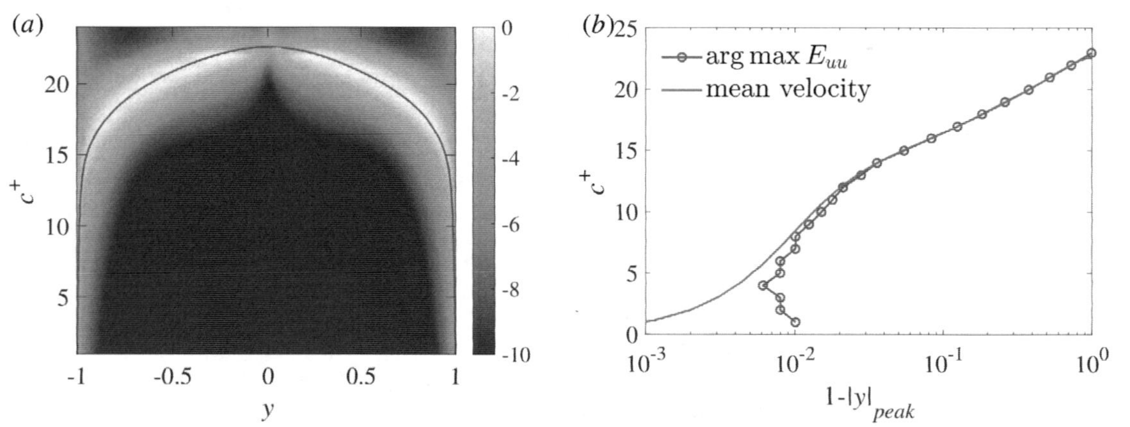

Figure 3. (a) Premultiplied one-dimensional energy spectra of the streamwise viscosity, and (b) the wall-normal distance of the peak streamwise velocity amplitude as a function of the wave speed, in the channel flow at $Re_\tau = 1000$. The solid lines in panel (a) and (b) represent the mean streamwise velocity profile issued from DNS.

Figure 3 shows the premultiplied one-dimensional energy spectra in the optimal streamwise response, given by

$$E_{uu} = \frac{\iint_{-\infty}^{\infty} k_x^2 k_z (\sigma_1 |\psi_{1,u}|)^2 \mathrm{d}\log(\kappa_x)\,\mathrm{d}\log(\kappa_z)}{\max \iint_{-\infty}^{\infty} k_x^2 k_z \sigma_1^2 \mathrm{d}\log(\kappa_x)\,\mathrm{d}\log(\kappa_z)}, \tag{19}$$

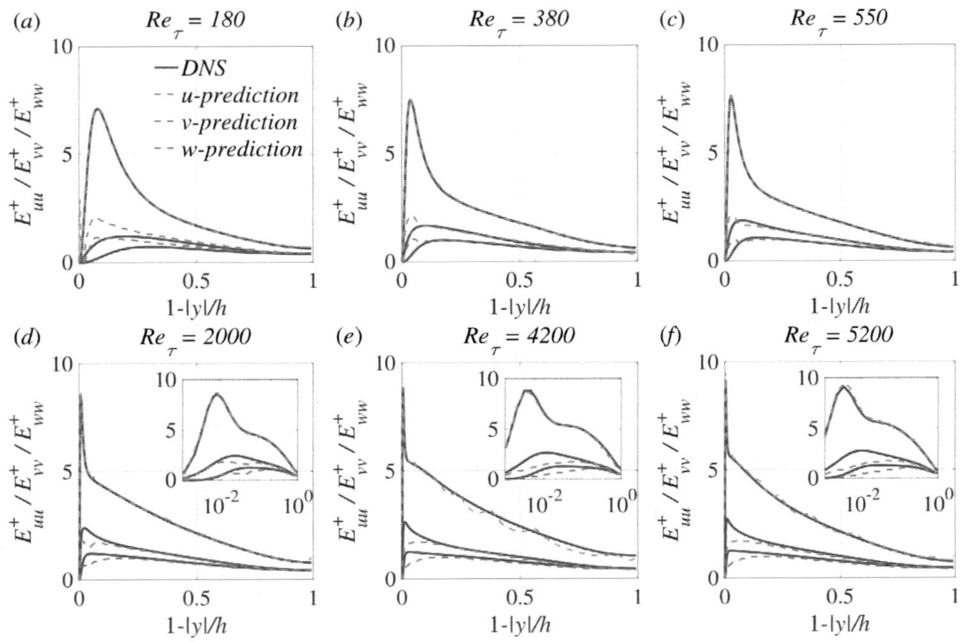

Figure 4. The prediction of E^+_{uu}, E^+_{vv} and E^+_{ww} as a function of wall-normal distance in the training process, when $(\gamma_{uu}, \gamma_{vv}, \gamma_{ww}, \gamma_{pp}) = (1, 0, 0, 0)$, in channel flows at various Reynolds numbers. The DNS results are also shown for comparison. Profiles are also plotted in logarithmic scale in insets of panels (*d-f*), to clarify the inner-layer behaviour at high Reynolds numbers. Legends in panels (*b-f*) refer to (*a*).

where $\psi_{1,u}$ is the streamwise-velocity component of the optimal response mode, and the forcing is simply assumed to be a white-noise signal. In figure 3(a), the premultiplied spectra are plotted in the logarithmic scale. It is observed that the energy density is clustered around the mean velocity profile, suggesting that the response mode follows the Taylor's frozen-turbulence hypothesis, in the sense of critical-layer mechanism [7].

The wall-normal distance of peak streamwise velocity amplitude is quantified as a function of the wave speed c^+ in figure 3(b). The most energetic wave-propagating speed profile is well collapsed onto that of the mean streamwise velocity, which is consistent with the critical-layer observation in figure 3(a). Hence the corresponding mode, which travels at the speed c^+ and is located at the wall-normal distance y_{peak}, is termed the "critical response mode". Exceptions are observed in the very-near-wall region, e.g. $y^+ < 10$ ($y^+ = (1-|y|)/\delta_\nu$ with δ_ν being the viscous length scale), where the $y_{peak} - c^+$ distribution deviates from the mean velocity profile. At this state, the streamwise turbulent energy is finite reaching down to the wall, suggesting that the mode has a footprint down to the wall. In this sense, it is termed the "wall-attached response mode" [32]. Overall, the phenomenon observed here in the optimal response mode is confirmed to be consistent with the classical turbulent structures and energy distribution in wall-bounded flows, hence it is rational to make it as a basis function for predicting the turbulent statistics.

3.2. Prediction of the turbulent energy

As introduced in Sec. 2.2, this section aims to predict the turbulent energy in space and scales, by obtaining the a_1 as a function of (k_x, k_z, c). In the practical process of training, this would

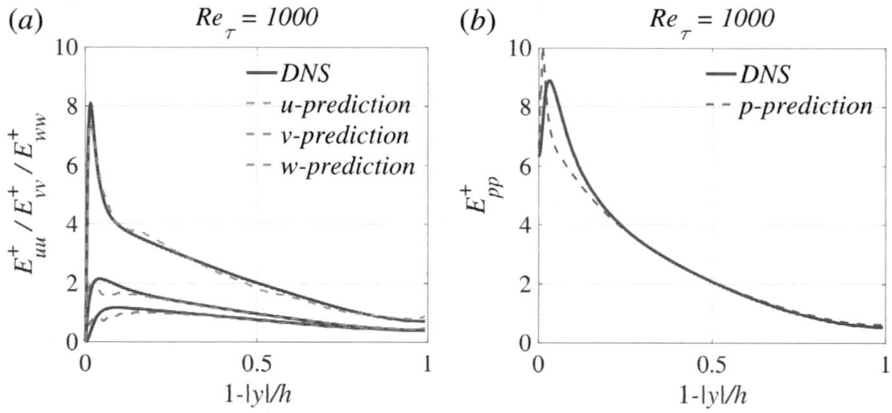

Figure 5. The prediction of (*a*) E^+_{uu}, E^+_{vv} and E^+_{ww} and (*b*) E^+_{pp} as a function of wall-normal distance in the testing process, when $(\gamma_{uu}, \gamma_{vv}, \gamma_{ww}, \gamma_{pp}) = (1, 0, 0, 0)$, in the channel flow at $Re_\tau = 1000$.

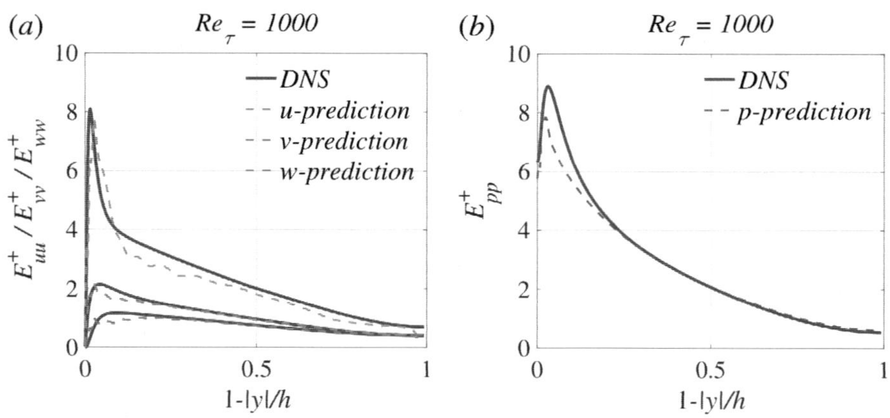

Figure 6. The prediction of (*a*) E^+_{uu}, E^+_{vv} and E^+_{ww} and (*b*) E^+_{pp} as a function of wall-normal distance in the testing process, when $(\gamma_{uu}, \gamma_{vv}, \gamma_{ww}, \gamma_{pp}) = (1, 1, 1, 1)$, in the channel flow at $Re_\tau = 1000$.

be rather difficult with the limited training databases. Hence the spatial broadband-forcing assumption is exploited, which means the dependency dimension of a_1 is reduced, to enable the prediction and generalization of the full-field projection coefficients at a lower computational cost and memory requirement. For example, the spanwise premultiplied spectral density and the wall-normal distribution of the turbulent energy are rewritten, respectively, as

$$E_{qq}(y, \kappa_z) = \int a_1^2(\kappa_z, c) \left[\int \kappa_x^2 \kappa_z |\boldsymbol{\psi}_1|^2 (y; \kappa_x, \kappa_z, c)\, \sigma_1^2(\kappa_x, \kappa_z, c)\, \mathrm{d}\log(\kappa_x) \right] \mathrm{d}c, \qquad (20)$$

$$E_{qq}(y) = \int a_1^2(c) \left[\iint \kappa_x^2 \kappa_z |\boldsymbol{\psi}_1|^2 (y; \kappa_x, \kappa_z, c)\, \sigma_1^2(\kappa_x, \kappa_z, c)\, \mathrm{d}\log(\kappa_x)\, \mathrm{d}\log(\kappa_z) \right] \mathrm{d}c. \qquad (21)$$

It has been confirmed that this reduced-order a_1 is still able to provide sufficient degrees of freedom for prediction [11].

In the present study, a CNN architecture is employed, as shown in figure 1. To optimize the weight and bias parameters in neurons per hidden layer during the training process, the mean squared error (MSE) is defined by a loss function:

$$\begin{aligned} MSE = & \gamma_{uu} \frac{||E_{uu,model} - E_{uu,true}||^2}{||E_{uu,true}||^2} + \gamma_{vv} \frac{||E_{vv,model} - E_{vv,true}||^2}{||E_{vv,true}||^2} \\ & + \gamma_{ww} \frac{||E_{ww,model} - E_{ww,true}||^2}{||E_{ww,true}||^2} + \gamma_{pp} \frac{||E_{pp,model} - E_{pp,true}||^2}{||E_{pp,true}||^2}, \end{aligned} \tag{22}$$

where the $||\cdot||$ represents the L_2 norm, and the subscript "model" and "true" denote the result from the resolvent analysis and DNS, respectively. The γ_{uu}, γ_{vv}, γ_{ww} and γ_{pp} controls different forms of the matching errors. Two kinds of loss functions are considered in the present study to train the network, that is (i) only the matching error of the streamwise energy is penalized, i.e. $(\gamma_{uu}, \gamma_{vv}, \gamma_{ww}, \gamma_{pp}) = (1, 0, 0, 0)$; and (ii) the matching error of all the four state variables is penalized, i.e. $(\gamma_{uu}, \gamma_{vv}, \gamma_{ww}, \gamma_{pp}) = (1, 1, 1, 1)$.

First, we present the training and testing performance in the prediction of the wall-normal distribution of energy, with six cases at $Re_\tau = 180, 380, 550, 2000, 4200, 5200$ used for training, and one case at $Re_\tau = 1000$ for testing. Figure 4 shows the output of streamwise (E^+_{uu}), wall-normal (E^+_{vv}) and spanwise energy (E^+_{ww}) in inner units, in comparison to the DNS results, in the training process with $(\gamma_{uu}, \gamma_{vv}, \gamma_{ww}, \gamma_{pp}) = (1, 0, 0, 0)$. It is observed that the predicted streamwise energy is almost collapsed onto the DNS profiles, especially at $Re_\tau \lesssim 2000$, confirming that the optimization process has been well converged. At higher Reynolds numbers, wiggles in E^+_{uu} are observed in the outer region, which is probably associated with the limited wall-normal grid points. Note that we make $N_y = 200$ for all the cases under scrutiny, which is both computationally feasible and generally sufficient to discretize the differential operators in the wall-normal direction.

Whereas for the prediction of wall-normal and spanwise component, obvious deviations are observed in contrast to the DNS results, which is partly due to that only the matching error of streamwise energy is considered in the training process. For channel flows, the streamwise velocity dominates the other velocity components, leading to significant implications on projecting the resolvent modes onto the real distributions issued from DNS in the streamwise direction. Since the inherent ratio of the streamwise response to the wall-normal and spanwise response in the rank-1 model does not match that in the DNS results, the same projection coefficients that optimally recover E^+_{uu} cannot perfectly yield results of E^+_{vv} and E^+_{ww}. The corresponding energy prediction in the testing process, where $Re_\tau = 1000$, is depicted in figure 5. Similar phenomenon is observed herein, where the streamwise energy is recovered reasonably well whereas the predictive performance of wall-normal and spanwise energy and the pressure perturbation is degraded especially near the wall, when only the error in streamwise energy is minimized.

On the other hand, when $(\gamma_{uu}, \gamma_{vv}, \gamma_{ww}, \gamma_{pp}) = (1, 1, 1, 1)$ is used in the loss function, the prediction results are shown in figure 6. Only the case in the testing process when $Re_\tau = 1000$ is presented herein, for brevity. It is found that, in contrast to the results in figure 5, the predictive performance of E^+_{vv}, E^+_{ww} and E^+_{pp} is improved, to some extent, at the expense of the reducing the predictive accuracy of E^+_{uu}, similar to the performance in ref. [12]. A slight mismatch of the inner peak of E^+_{uu} appears in panel (a). As we mentioned earlier, this might be attributed to the ratio of the streamwise response to the wall-normal, spanwise and pressure response inherent in the rank-1 model. In this case, the matching error of the velocities and pressure cannot be simultaneously small enough, which means that the rank-1 approximation by optimal response modes might lead to insufficient basis functions for modelling. Nonetheless, the predictive performance presented here is still believed to be reasonably acceptable.

To be more quantitative, Table 1 lists the results of MSE of prediction, in terms of each component. Consistent with the qualitative description above, both the loss function used can

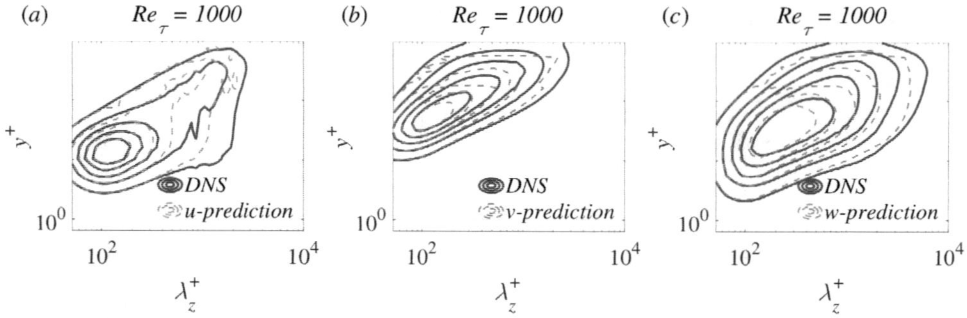

Figure 7. The prediction of spanwise spectra (a) E^+_{uu}, (b) E^+_{vv} and (c) E^+_{ww} in the testing process, when $(\gamma_{uu}, \gamma_{vv}, \gamma_{ww}, \gamma_{pp}) = (1, 0, 0, 0)$, in the channel flow at $Re_\tau = 1000$. The contour lines represent 1/6, 1/3, 1/2, 2/3 and 5/6 of their respective maximum, moving inward.

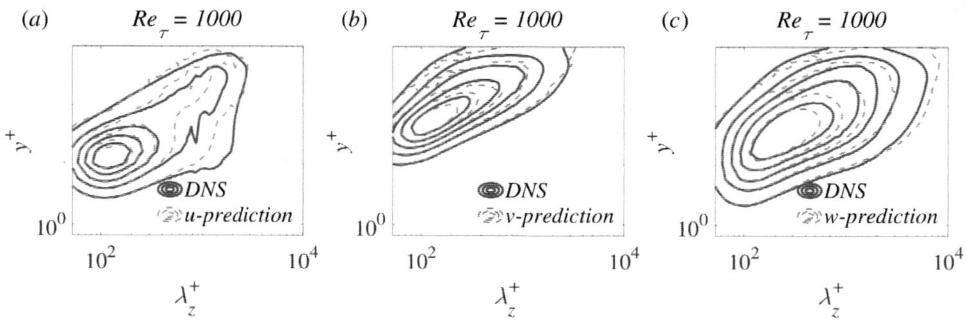

Figure 8. The prediction of spanwise spectra (a) E^+_{uu}, (b) E^+_{vv} and (c) E^+_{ww} in the testing process, when $(\gamma_{uu}, \gamma_{vv}, \gamma_{ww}, \gamma_{pp}) = (1, 1, 1, 1)$, in the channel flow at $Re_\tau = 1000$.

yield acceptable performance in prediction, e.g. $MSE \leq 10\%$, except for the cases at very-low Reynolds numbers where the inner-outer characteristics are not remarkable and can be a bit different from those in the higher-Reynolds-number cases. Specifically, the choice of the loss function can be determined by the prediction target.

Table 1. MSE of each component of wall-normal energy in the training and testing process.

Re_τ		$(\gamma_{uu}, \gamma_{vv}, \gamma_{ww}, \gamma_{pp}) = (1, 0, 0, 0)$				$(\gamma_{uu}, \gamma_{vv}, \gamma_{ww}, \gamma_{pp}) = (1, 1, 1, 1)$			
		$MSE_{uu}(\%)$	$MSE_{vv}(\%)$	$MSE_{ww}(\%)$	$MSE_{pp}(\%)$	$MSE_{uu}(\%)$	$MSE_{vv}(\%)$	$MSE_{ww}(\%)$	$MSE_{pp}(\%)$
Training	180	2.32	66.78	39.45	56.67	5.29	53.78	29.95	10.17
	380	0.62	11.61	7.56	7.97	1.92	8.54	4.66	0.85
	550	0.29	6.37	4.28	7.33	1.13	5.32	2.90	0.90
	2000	0.05	4.38	3.08	2.71	4.83	3.71	2.49	2.07
	4200	0.19	6.97	6.66	4.90	0.10	5.89	2.76	2.62
	5200	0.16	8.03	8.19	4.85	1.09	6.59	2.92	2.30
Testing	1000	0.44	3.27	2.08	2.64	4.72	2.69	1.40	1.46

We also use the resolvent-based modelling (20) in combination with machine learning to predict the spanwise spectra of velocity. The prediction results in the testing process are shown in figures 7 and 8, for the cases when $(\gamma_{uu}, \gamma_{vv}, \gamma_{ww}, \gamma_{pp}) = (1, 0, 0, 0)$ and $(\gamma_{uu}, \gamma_{vv}, \gamma_{ww}, \gamma_{pp}) = (1, 1, 1, 1)$, respectively. In both cases, the inner-layer peaks in the premultiplied spanwise spectra, E^+_{uu}, E^+_{vv} and E^+_{ww}, which are representative of the near-wall streaks and streamwise vortices, can be accurately captured. Whereas for larger scales in the outer-layer cycle, the prediction error increases, which is possibly due to the limitation of the databases used for training as the outer-layer characteristics exist evidently only at high Reynolds numbers, that is $Re_\tau = 2000$ and $Re_\tau = 5200$ in the present study. Similar to the discussion above, the loss function with $(\gamma_{uu}, \gamma_{vv}, \gamma_{ww}, \gamma_{pp}) = (1, 1, 1, 1)$ is able to increase the prediction ability of E^+_{vv} and E^+_{ww}, whereas decrease that of E^+_{uu}, which is straightforwardly exposed by the MSE of each component listed in Table 2. In general, the streamwise energy spectra can be predicted well, as $MSE_{uu} < 6\%$ for all the cases, while the $MSEs$ of the wall-normal and spanwise energy spectra are much larger. Hence the full-field a_1 or higher-order models might be necessary and more databases for training and testing will definitely improve the performance of turbulence prediction.

Table 2. MSE of each component of energy spectra in the training and testing process.

Re_τ		$(\gamma_{uu}, \gamma_{vv}, \gamma_{ww}, \gamma_{pp}) = (1, 0, 0, 0)$			$(\gamma_{uu}, \gamma_{vv}, \gamma_{ww}, \gamma_{pp}) = (1, 1, 1, 1)$		
		$MSE_{uu}(\%)$	$MSE_{vv}(\%)$	$MSE_{ww}(\%)$	$MSE_{uu}(\%)$	$MSE_{vv}(\%)$	$MSE_{ww}(\%)$
Training	180	1.4945	10.5316	11.3235	4.2361	13.9983	20.0548
	550	1.4359	2.3791	9.9905	3.0671	2.1305	4.0066
	2000	0.72158	25.1527	20.2006	1.1954	13.4671	9.146
	5200	4.1169	51.7848	45.8747	1.5947	33.8016	26.9096
Testing	1000	5.941	10.5823	7.2193	5.0774	4.7236	3.2054

4. Conclusion

This paper proposes a modelling framework based on the resolvent analysis and machine learning, and uses it to predict the turbulent energy in incompressible channel flows. In this method, the mean streamwise velocity profiles are required as input to the network. On one hand, the input works to provide efficient optimal response modes as basis functions, since the low-rank behaviour of the resolvent operator is verified. On the other hand, the projection coefficients of the low-rank forcing modes onto the true nonlinear terms are obtained through a CNN architecture, adjusting the magnitude of the linearly superposed response at each wavenumber combination and consequently leading to the reconstruction of the flow perturbations.

To optimize the weight and bias parameters in neurons per hidden layer during the training process, two types of loss functions are considered, with matching error of different components included. Results show that both the methods can yield a reasonably good prediction of the wall-normal profile of turbulent fluctuations, except that better performance is observed in the prediction of the streamwise component when only the matching error of the streamwise energy is penalized. As for the prediction of spanwise spectra of velocity, the inner-layer peaks can be accurately captured whereas the prediction error increases for larger scales in the outer-layer cycle. In general, the proposed method provides a promising tool to predict the turbulent motions based on the mean field obtained from low-cost RANS or experimental measurements. Further work will be undertaken to investigate the effect of increasing the degrees of freedom of the projection coefficient and the numbers of resolvent-based modes on the predictive accuracy.

Acknowledgments
The funding support of the National Natural Science Foundation of China (under the Grant Nos. 91952302 and 92052101) is acknowledged. This work was supported in part by the European Research Council under the Caust grant ERC-AdG-101018287. The authors also thank Prof. Ricardo Vinuesa for his suggestions and comments on this manuscript.

References
[1] Lumley J L 1967 The structure of inhomogeneous turbulent flows *Atmospheric Turbulence and Radio Wave Propagation* ed Yaglom A M and Tartarsky V I pp 166–177
[2] Berkooz G, Holmes P and Lumley J L 1993 *Annu. Rev. Fluid Mech.* **25** 539–575
[3] Taira K, Brunton S L, Dawson S T M, Rowley C W, Colonius T, McKeon B J, Schmidt O T, Gordeyev S, Theofilis V and Ukeiley L S 2017 *AIAA J.* **55** 4013–4041
[4] Towne A, Schmidt O T and Colonius T 2018 *J. Fluid Mech.* **847** 821–867
[5] Schmid P J 2010 *J. Fluid Mech.* **656** 5–28
[6] Arun R, Bae H J and McKeon B J 2023 *Phys. Rev. Fluids* **8**(6) 064612
[7] McKeon B J and Sharma A S 2010 *J. Fluid Mech.* **658** 336–382
[8] McKeon B J 2017 *J. Fluid Mech.* **817** P1
[9] Sharma A S and McKeon B J 2013 *J. Fluid Mech.* **728** 196–238
[10] Abreu L I, Cavalieri A V G, Schlatter P, Vinuesa R and Henningson D S 2020 *Int. J. Heat Fluid Flow* **85** 108662
[11] Moarref R, Sharma A S, Tropp J A and McKeon B J 2013 *J. Fluid Mech.* **734** 275–316
[12] Moarref R, Sharma A S, Tropp J A and McKeon B J 2013 On effectiveness of a rank-1 model of turbulent channels for representing the velocity spectra *43rd Fluid Dynamics Conference*
[13] Moarref R, Jovanović M R, Tropp J A, Sharma A S and McKeon B J 2014 *Phys. Fluids* **26** 051701
[14] Symon S, Illingworth S J and Marusic I 2021 *J. Fluid Mech.* **911**
[15] Towne A, Lozano-Durán A and Yang X 2020 *J. Fluid Mech.* **883**
[16] Amaral F R, Cavalieri A V, Martini E, Jordan P and Towne A 2021 *J. Fluid Mech.* **927** A17
[17] Illingworth S J, Monty J P and Marusic I 2018 *J. Fluid Mech.* **842** 146–162
[18] Martini E, Cavalieri A V G, Jordan P, Towne A and Lesshafft L 2020 *J. Fluid Mech.* **900**
[19] Morra P, Semeraro O, Henningson D S and Cossu C 2019 *J. Fluid Mech.* **867** 969–984
[20] Morra P, Nogueira P A S, Cavalieri A V G and Henningson D S 2021 *J. Fluid Mech.* **907** A24
[21] Holford J J, Lee M and Hwang Y 2023 *J. Fluid Mech.* **961** A32
[22] Brunton S L, Noack B R and Koumoutsakos P 2020 *Annu. Rev. Fluid Mech.* **52** 477–508
[23] Kou J and Zhang W 2021 *Prog. Aerosp.* **125** 100725
[24] Tushar and Chakraborty S 2023 *J. Comput. Phys.* **479** 112004
[25] Ouala S, Brunton S L, Chapron B, Pascual A, Collard F, Gaultier L and Fablet R 2023 *Physica D: Nonlinear Phenomena* **446** 133630
[26] Raissi M, Perdikaris P and Karniadakis G 2019 *J. Comput. Phys.* **378** 686–707
[27] Cai S, Mao Z, Wang Z, Yin M and Karniadakis G E 2021 *Acta Mech. Sin.* **37** 1727–1738
[28] Trefethen L N 2000 *Spectral Methods in MATLAB* (Society for Industrial and Applied Mathematics)
[29] Moser R D, Kim J and Mansour N N 1999 *Phys. Fluids* **11** 943–945
[30] Lozano-Durán A and Jiménez J 2014 *Phys. Fluids* **26** 011702
[31] Lee M and Moser R D 2015 *J. Fluid Mech.* **774** 395–415
[32] McKeon B J, Sharma A S and Jacobi I 2013 *Phys. Fluids* **25** 031301

Active flow control of a turbulent separation bubble through deep reinforcement learning

Bernat Font[1,2,*], Francisco Alcántara-Ávila[3], Jean Rabault[4], Ricardo Vinuesa[3] and Oriol Lehmkuhl[1]

[1] Barcelona Supercomputing Center, 08034 Barcelona, Spain
[2] Faculty of Mechanical Engineering, TU Delft, Delft, The Netherlands
[3] FLOW, Engineering Mechanics, KTH Royal Institute of Technology, Stockholm, Sweden
[4] Independent researcher, Oslo, Norway

E-mail: `b.font@tudelft.nl`

Abstract. The control efficacy of classical periodic forcing and deep reinforcement learning (DRL) is assessed for a turbulent separation bubble (TSB) at $Re_\tau = 180$ on the upstream region before separation occurs. The TSB can resemble a separation phenomenon naturally arising in wings, and a successful reduction of the TSB can have practical implications in the reduction of the aviation carbon footprint. We find that the classical zero-net-mas-flux (ZNMF) periodic control is able to reduce the TSB by 15.7%. On the other hand, the DRL-based control achieves 25.3% reduction and provides a smoother control strategy while also being ZNMF. To the best of our knowledge, the current test case is the highest Reynolds-number flow that has been successfully controlled using DRL to this date. In future work, these results will be scaled to well-resolved large-eddy simulation grids. Furthermore, we provide details of our open-source CFD–DRL framework suited for the next generation of exascale computing machines.

1. Introduction

Wall turbulence, a phenomenon naturally arising in the field of fluid dynamics, has long captivated the curiosity of scientists and engineers alike. Although our knowledge about turbulence has markedly advanced over the past decades, there are still numerous open questions that continue to challenge researchers in the field. In the area of flow control (FC), a better understanding of turbulence is the key to unlock novel control strategies that can help us optimize different processes or systems. FC plays a crucial role in the transportation industry, potentially allowing the reduction of the carbon footprint by improving the efficiency of transport vehicles. By gaining insights into the underlying mechanisms of wall turbulence and its impact on aerodynamic drag and skin friction, FC strategies can be derived and applied in a passive or active manner. The efficacy of this approach lies in the fact that even a minor reduction in drag can result in a substantial decrease in the overall percentage of global emissions. A practical situation in which FC can help reduce CO_2 emissions arises in the aeronautical sector. Consider the upper part of an aircraft wing, also known as the suction side, which exhibits an adverse pressure gradient (APG) that can induce flow separation at high angles of attack. A turbulent separation bubble (TSB) is formed when the flow is separated and then reattached. This situation may occur mostly during take-off and landing on the high-lift devices (flaps, slats), or during cruise when the incoming free-stream flow is highly perturbed. In order to alleviate constraints on the aircraft design and facilitate a more precise optimization of the cruise phase, flow control can potentially help to reduce the TSB hence improving flow attachment. There are countless applications of FC as referenced in the literature, and the reader is referred to Refs. [1, 2] for comprehensive reviews.

A turbulent boundary layer (TBL) subjected to an APG (APG TLB) that generates a TSB resembles the suction side of a wing in a simplified manner, and it is the most suitable first approach to learn the key features to control separation. APG TBLs are characterized by a wider parametric space compared to zero-pressure gradient (ZPG) TBLs. The most influential parameters are [3]: a) the Rotta–Clauser pressure-gradient parameter $\beta = (\delta^*/\tau_w)\mathrm{d}_x P$, where τ_w is the wall-shear stress, P is the static pressure, x is the streamwise direction coordinate, and δ^* is the displacement thickness – this parameter represents the pressure-to-viscous force ratio. b) the friction Reynolds number $Re_\tau = \delta^* u_\tau/\nu$, where $u_\tau = \sqrt{\tau_w/\rho}$ is the friction velocity, ρ is the fluid density, and ν is the kinematic viscosity – this parameter represents the large-scale to small-scale ratio of wall turbulence. c) the acceleration parameter $K = (\nu/u_\infty^2)\mathrm{d}_x u_\infty$, where u_∞ is the local free-stream velocity – this parameter represents the equilibrium state of the boundary layer. A number of examples in the literature have investigated APG TBLs under different parametric regimes, both numerically and experimentally. Some examples include [4, 5, 6, 7, 8] using direct numerical simulation (DNS), [9, 10, 11] using well-resolved large-eddy simulation (LES), and [3, 12, 13] using experiments.

The application of active flow control (AFC) for a TSB is typically performed through surface actuators located upstream of the separation bubble. AFC has been the focus of numerous TSB studies for both canonical and practical flows. The latter refer to wing-like configurations, where the surface curvature and/or angle of attack induce flow separation. You and Moin [14] applied oscillatory synthetic jets, also known as zero-net-mass-flux periodic control (or periodic forcing), which connected the pressure and suction sides of a NACA 0015 airfoil. By harmonic blowing and suction, they found that this control effectively delayed the separation of the boundary layer at high angles of attack. Atzori *et al.* [15] focused on uniform blowing and uniform suction applied to the suction side of a NACA 4412 airfoil, finding that AFC had a larger impact on the APG TBL compared to ZPG TBL. Lehmkuhl *et al.* [16] used LES to investigate periodic control applied on the suction side of a SD7003 airfoil wing and the JAXA standard model. They found that the periodic forcing successfully eliminated the recirculation bubble in both cases. With respect to canonical flows, *i.e.* those simulated in a flat-plate-like configuration in absence of surface curvature, the APG that induces the TSB is commonly generated using suction-blowing (SB) or suction-only (SO) boundary conditions at the top of the domain. The SB approach has the advantage of conserving mass flux across the domain boundaries, and yields a more stable recirculation bubble caused by the favourable pressure gradient (FPG) that follows the APG. For the SO case, the flow is naturally reattached by the turbulent diffusion of momentum, so it can better resemble the separation type found in wings [17]. Cho *et al.* [18] focused on a TSB generated by SB on the top boundary of the domain. Using periodic forcing and sweeping across a range of characteristic frequencies ($St = fL_b/U_\infty$, where L_b is the uncontrolled bubble length and U_∞ is the free-stream velocity), they found that low frequencies $St < 0.5$ tend to reduce the separation bubble and high frequencies $St > 1.56$ induce an intermittent separation/reattachment pattern. Wu *et al.* [17] investigated a TSB generated by SO and the effect of periodic control for a range of St. It was found that the frequency $St = 0.45$ as well as a higher frequency $St = 1.125$ frequency yield a 50% reduction of the TSB. On the other hand, a much higher frequency, $St = 4.5$, had no effect on the TSB.

The proliferation of computational resources has facilitated the broader application of machine learning (ML) to a wider range of problems. Notably, deep reinforcement learning (DRL), a subset of ML, proves particularly fitting to address AFC challenges and formulate effective control strategies as an alternative to the classical fluid-dynamics theory approach. In this work, we perform classical control and DRL control of an APG TBL hence comparing both control strategies. The main motivation behind the use of DRL is to allow the controlling agent certain freedom in selecting the most appropriate action for a given flow state, instead of using a single frequency as in classical periodic control. Therefore, DRL allows for an unconstrained closed loop control strategy that can adapt to the system dynamics based on previously learnt experiences. In a nutshell, DRL consists of two main entities: an agent, which is composed by a neural network (NN), and an environment, which is represented by the flow simulation in the present context. These two entities interact in a trial-an-error process in the form of episodes. This interaction occurs through three communication channels: a) The state, $s(t)$, which is sent from the environment to the agent. The state can be a total or partial observation of the environment at a given time t. In the case of a numerical simulation, s is usually a set of probes intentionally distributed in the domain regions of interest which sample, *e.g.* velocity or pressure. b) The action, $a(t)$, which is sent from the agent to the environment. The action is a modification to the environment, *e.g.* a new boundary condition, which in computational fluid dynamics (CFD) is associated with the control strategy. For example, the action can be the mass flow rate of synthetic jet actuators, a modulation of the temperature profile in a wall, or a nudge in a turbulence model constant, among others. c) The reward, $r(t)$, which is sent from the environment to the agent. The reward is the parameter that represents the environment fitness with respect to an optimization goal, *e.g.* the drag coefficient if the goal is to reduce drag. Therefore, the agent will try to optimize a reward r through the use of an action a which is selected based on a state s, *i.e.* the agent will use a policy $\pi(a|s)$ to optimize the reward. The policy defines a probabilistic mapping from the current state s to the action a. During training, the agent works in the so-called exploration mode by adding noise on the action sampling process. In this way, new dynamics can be explored and learnt by the algorithm. During a training episode, a collection of (s, r, a) triplets is generated, also known as a trajectory. After an episode finalizes, the resulting trajectory is used to update the NN weights so that the accumulated reward expectation is maximized. Once the training is finalized, the agent is used in the so-called deterministic mode, where the most probable actions are selected (the mean of the distribution).

The applications of DRL for AFC have grown exponentially in the recent years. The reader is referred to Ref. [19] for a wide overview of DRL for AFC, where the most representative works of different problems are discussed. Some of the most typical cases studied are drag reduction on a cylinder, both in 2D [20, 21, 22, 23, 24] and 3D [25, 26, 27]; convective heat reduction in Rayleigh-Bénard problems [28, 29]; reduction of the skin-friction coefficient in turbulent channels [30, 31]; and turbulence modelling [32, 33, 34]. The increasing tendency of tackling more computationally expensive cases has motivated the development of novel DRL techniques that can accelerate the training process of a control strategy. A first approach is to simulate multiple environments in parallel, also known as multi-environment DRL. With this approach, the agent trains faster by generating multiple experiences in parallel. This method has an almost perfect scaling as shown by Rabault and Kuhnle [35]. A second approach, independent to the first one, is to use a multi-agent reinforcement learning (MARL) method. First introduced by Belus *et al.* [36] and later named by Vignon *et al.* [29], the MARL method exploits the domain spatial invariants so that the dimensionality of the actuation space is reduced. For example, consider a set of multiple synthetic jets placed along the span of a circular cylinder. Since the flow is statistically invariant along the spanwise direction, every individual jet can be treated separately within its own span subdomain (or pseudo-environment). In this approach, rather than predicting multiple actions simultaneously, individual agents are dedicated to each pseudo-environment, each responsible for predicting a single action. The key is that all these agents share the same policy. The advantage of this approach is to reduce the number of combinations that yield an overall positive action, hence avoiding the curse of dimensionality. As reported in Refs. [25] and [29], using MARL finally allowed the agent to learn the system dynamics and provide positive actions. Defining the number of parallel environments as N_{e} and the number of parallel pseudo-environments within each environment N_{pe}, a total of $N_{\mathrm{e}}N_{\mathrm{pe}}$ trajectories can be sampled in parallel, and the training time of an agent can be greatly reduced.

With the aim of investigating classical periodic control and DRL control on a TSB, the paper has been structured as follows: §2 introduces the mathematical framework for the simulation and the different control strategies. §3 presents and discusses results for classical control in §3.1, and DRL control in §3.2. Finally, conclusions are presented in §4.

2. Methodology

2.1. CFD setup

Following the LES method, the spatially filtered incompressible non-dimensional Navier–Stokes (NS) equations

$$\nabla \cdot \overline{\boldsymbol{u}} = 0 \tag{1}$$

$$\partial_t \overline{\boldsymbol{u}} + (\overline{\boldsymbol{u}} \cdot \nabla)\,\overline{\boldsymbol{u}} = -\nabla \overline{p} + Re^{-1}\nabla^2 \overline{\boldsymbol{u}} - \nabla \cdot \boldsymbol{\tau}, \tag{2}$$

are numerically solved on a discrete domain, where $\overline{\boldsymbol{u}} = (\overline{u}, \overline{v}, \overline{w})$ is the filtered velocity vector field, $\overline{p} = \overline{P}/\rho$ is the density-scaled filtered pressure, and $\boldsymbol{\tau} = \overline{\boldsymbol{u} \otimes \boldsymbol{u}} - \overline{\boldsymbol{u}} \otimes \overline{\boldsymbol{u}}$ is the sub-grid scale (SGS) stress tensor. The deviatoric part of the SGS tensor is modelled using the Boussinesq hypothesis

$$\boldsymbol{\tau}^d = \boldsymbol{\tau} - \frac{2}{3}k\boldsymbol{\delta} = -2\nu_{\mathrm{sgs}}\overline{\boldsymbol{S}}, \tag{3}$$

where $k = \mathrm{tr}(\boldsymbol{\tau})/2$ is the turbulent kinetic energy, $\boldsymbol{\delta}$ is the Kronecker delta, and $\overline{\boldsymbol{S}} = (\nabla \otimes \overline{\boldsymbol{u}} + \overline{\boldsymbol{u}} \otimes \nabla)/2$ is the rate-of-strain tensor. The SGS viscosity is finally closed with the Vreman model [37]. The overbar notation of the LES filtering operation for flow-field quantities is thereafter implied.

The high-order spectral-element-method solver SOD2D [38] is used to simulate an incompressible APG TBL in a computational domain with dimensions $L_x \times L_y \times L_z = 1100 \times 120 \times 125$, corresponding to the streamwise x, wall-normal y, and spanwise z directions. A coarse grid and a well-resolved (fine) grid composed of 5th-order hexahedral elements are considered, and details are given in table 1. The idea behind considering 2 different resolutions is to train the DRL model on the coarse grid, thus significantly reducing the computational cost of training. Afterwards, the model can be trained further on the fine grid via transfer learning, *i.e.* using the optimized NN weights already available as the training starting point on the fine grid. Alternatively, given that similar dynamics are captured in both coarse and fine grids, the controlling agent can be directly applied into the fine-grid simulation without further training.

With respect to the numerical methods implemented in the flow solver, an implicit-explicit Adams–Bransford Crank–Nicolson scheme is used together with the fractional-step method to solve the governing equations. In addition, a matrix free conjugate gradient preconditioned by the diagonal is used to solve the linear part of the NS equations and the pressure equation. The computational domain sets a buffer zone of 50 units in the x direction at the inlet and at the outlet, yielding an effective streamwise domain of 1000 units ($x \in [-50, 950]$). Dimensions are normalized by the displacement thickness at the inlet δ_0^*, and the velocity field is normalized by the streamwise velocity imposed at the top of the domain U_∞. The flow is initialized using a Blasius boundary layer at $Re_{\delta_0^*} = 450$, where $Re_{\delta_0^*} = U_\infty \delta_0^*/\nu$ is the Reynolds number based on the displacement thickness. The inlet boundary condition retains the Blasius profile used as initial condition. At $x = -40$, the laminar boundary layer is tripped using the method explained in Refs. [39, 40] to accelerate the transition to turbulence. On top the domain, the APG is defined using SB similar to Refs. [5, 8]

$$v_{\rm top} = v_{\rm max}\sqrt{2}\left(\frac{x_c - x}{\sigma}\right)\exp\left[\psi - \left(\frac{x_c - x}{\sigma}\right)^2\right], \tag{4}$$

where $v_{\rm max} = 0.4$, $x_c = 306.64$, $\sigma = 110.49$, and $\psi = 0.95$. We note that $v_{\rm max}$ is 20% larger than the value selected by Wu *et al.* [8] so that the TSB is still formed when using the coarse LES grid. Moreover, at the top of the domain, a zero spanwise vorticity condition ($\omega_z = 0$) and a homogeneous Neumann condition for the spanwise velocity ($\partial_y w = 0$) are applied. A periodic boundary condition is imposed in the spanwise direction, and a convective outlet is set on the streamwise end of the domain. On the wall, the lower part of the domain, a classical non-slip boundary condition is imposed. Finally, in the outlet, a pressure-based boundary condition is applied together with the sponge zone. In the fine grid, the a-posteriori computation of the wall-shear stress yields an approximate friction Reynolds number of $Re_\tau = 180$ and $Re_\tau = 750$ at $x = 150$ and $x = 900$, hence right upstream of the TSB and right before the convective outlet of the domain, respectively. A schematic representation of the computational domain which summarizes the setup is shown in figure 1, where vortical structures represented with the Q criterion [41] are depicted too.

	Δx^+	Δz^+	$\Delta y^+\rvert_{\rm min}$	$\Delta y^+\rvert_{\rm max}$	DOFs
Coarse	38.20	38.20	1.27	25.08	4,063,913
Fine	15.28	15.28	0.93	14.51	40,543,349

Table 1: Grid details. Δ refers to grid spacing, the $\cdot^+$ notation refers to viscous units (scaling with u_τ and ν), and DOFs stands for degrees of freedom, *i.e.* the total number of nodes in the spectral-element domain discretization.

Once the flow is initialized, the simulation is run until the TSB is formed and the flow has fully developed. Afterwards, the control surfaces start acting on the flow. There are $N_{\rm ac} = 6$ rectangular actuators located at $x \in [150, 195]$, right upstream of the separation bubble, and each actuator has a spanwise width of $d_{\rm ac} = 10.42$ which is also the length between adjacent actuators. A classical control strategy similar to those reported in Refs. [17, 18] is considered. Classical control is denoted as the harmonic time-forcing of the wall-normal velocity, defined as

$$v_{\rm ac} = A_{\rm ac}\sin(2\pi f_{\rm ac}t). \tag{5}$$

By definition, the free parameters in classical control are the actuation amplitude $A_{\rm ac}$ and the actuation frequency $f_{\rm ac}$. Depending on the APG top-boundary condition type, the actuation amplitude needs to be adjusted. We use $A_{\rm ac} = 0.3U_\infty$, which corresponds to a momentum-flux coefficient of $C_\mu = (v_{\rm ac,rms}/U_\infty)^2 N_{\rm ac} d_{\rm ac}/L_z = 2.25\%$, where 'rms' stands for root mean square, and it is within the effective range $0.01\% < C_\mu < 3\%$ [18, 42]. We note that Wu *et al.* [17] used $C_\mu = 0.0625\%$ in a SO APG setup, and Cho *et al.* [18] used $C_\mu = 1.4\%$ in a SB APG setup, even though the top wall-normal velocity profile spanned the entire streamwise length of the domain in the latter case. The SB APG setup requires a larger actuation amplitude because of the forced reattachement arising from the FPG that follows the APG. On the other hand, sizing of the control surfaces is based on the objective of effectively modulating the low-frequency motion inherent in the uncontrolled flow. These motions are closely associated with the Görtler

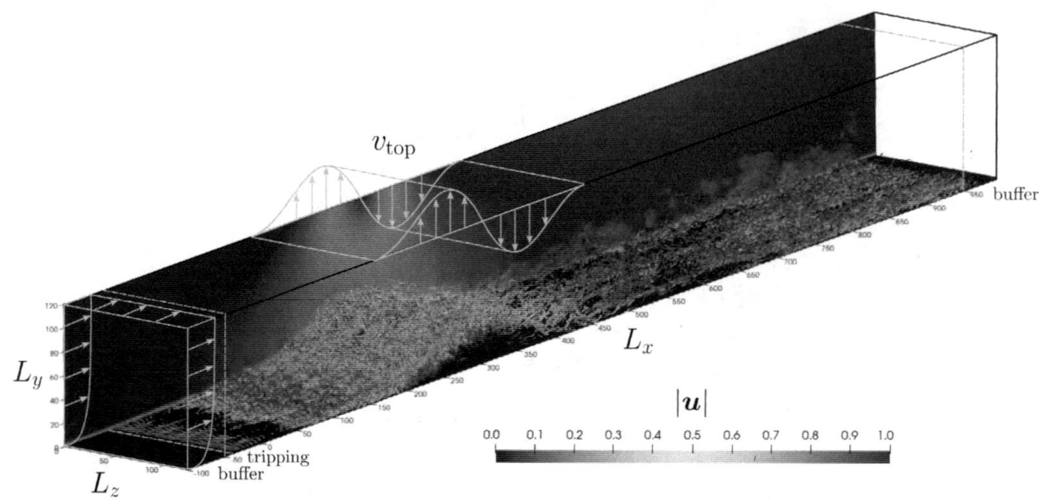

Figure 1: Schematic representation of the computational domain together with an instantaneous snapshot of vortical structures captured by the Q criterion and colored by velocity magnitude.

instability, which exhibits a spanwise wavelength of the same order of magnitude than the boundary-layer thickness and extends by more than 10 times the boundary-layer thickness in the streamwise direction [8]. More details and comparisons with other orientations of the control surfaces or rounded actuators can be read in Refs. [43, 44, 45] and [46, 47, 48], respectively.

In the current setup, an actuation frequency of $f_{\rm ac} = 0.0025$ is selected (normalized by δ_0^* and U_∞). This is the same frequency as identified by Wu *et al.* [8] for both SO-TSB and SB-TSB configurations (named the high frequency). Wu *et al.* [17] also selects this frequency for periodic control in a SO configuration. In this direction, we perform a spectral analysis of the streamwise velocity component at different locations. In figure 2, the spectra of a probe inside the TSB and another probe near the TSB downstream shear layer are displayed. While the downstream probe reaches a peak near $f = 0.0015$, a clear dominant frequency cannot be detected. This process is repeated for all the probes displayed in figure 5 and the v, w, p flow fields (not shown here for the sake of brevity), yielding a similar conclusion. On the other hand, Cho *et al.* [18] observes that the most effective frequency for the reduction of the TSB in a SB setup is $St = 0.5$. In this work, the non-actuated separation bubble has an approximate length of $L_b = 143$ (more results on the next section), which translates our selected actuation frequency to $St = 0.3575$, still within the range of effective actuation frequencies proposed by Cho *et al.* [18].

2.2. DRL setup

Beyond periodic forcing, we also consider DRL control to reduce the TSB length. As we have briefly summarized in the introduction, in a DRL problem we define two main entities: the environment, *i.e.* the system where we will apply the control actions, and the agent, *i.e.* the model in charge of predicting actions. In the current framework, the environment is the simulation performed by the CFD solver, and the agent is a NN that predicts a probability distribution of possible actions. This is depicted schematically in figure 3. The software used to simulate the environment is again the SOD2D CFD solver, and the TF-agents [49] is used for the DRL model part. A challenge that commonly arises when linking high-performance physics solvers (typically written in Fortran/C/C++) and high-level libraries that implement

Journal of Physics: Conference Series **2753** (2024) 012022 doi:10.1088/1742-6596/2753/1/012022

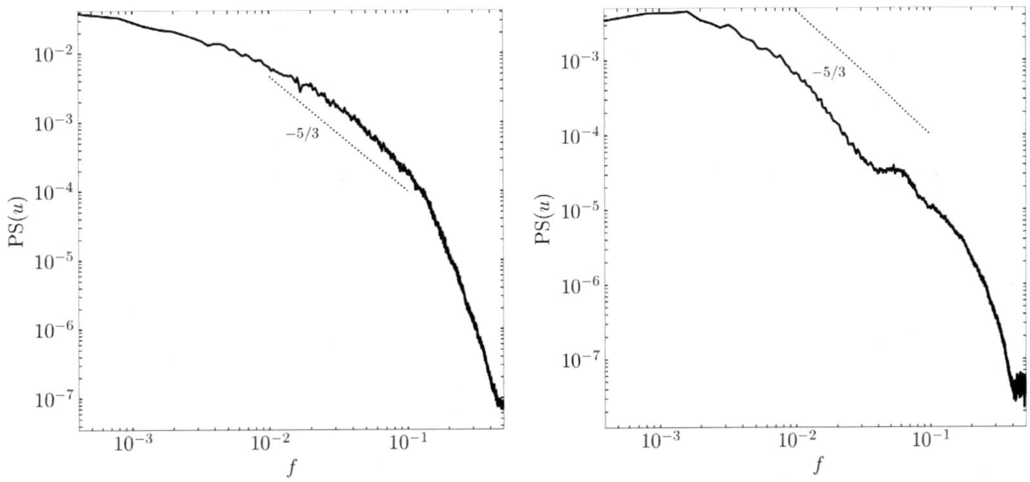

Figure 2: Power spectrum (PS) of the streamwise velocity component u sampled at 2 different locations. Left: $(x, y) = (324, 15)$, corresponds to a point inside the TSB. Right: $(x, y) = (372, 55)$, corresponds to a point near the downstream shear layer of the TSB. The PS is computed using the Welch method for a temporal signal of 15000 time units (TU) in total, split into 6 segments with 50% overlap. Additionally, 6 PS are computed along z for each (x, y) which are then averaged and result in the displayed spectra.

ML models (typically Python) is the communication between the different executable instances, also known as the two-language problem. While this can be accomplished through Unix sockets [50] or message-passing interface (MPI) [30], in the current setup the SmartSim [51] library is employed. SmartSim allows communication between processes through an in-memory Redis database with minimal overhead and, compared to the previously mentioned approaches, it lowers the software complexity of the framework. Other projects such as RELEXI [34, 52] have already successfully used SmartSim to handle communication across CFD solver and DRL model. In the current framework, the main computation workload arises from the CFD temporal integration, so the SOD2D solver is run on graphics-processing units (GPUs) in parallel. On the other hand, the DRL model is based on a small multi-layer perceptron NN and this can be easily handled by the central-processing unit (CPU) part of a cluster node which would remain idle otherwise. A schematic diagram of this setup is depicted in figure 4, noting the multi-environment approach as introduced earlier.

The same actuators as defined for the periodic control are employed in the DRL framework. However, to impose mass conservation both in space and time, we group the actuators in pairs and the DRL model sets a mass flux value on one actuator, and the opposite value on the other actuator. As explained before, the MARL approach can be used when the flow is invariant in a certain direction. In this case, the flow is invariant in the spanwise direction, and a subdomain composed by one pair of actuators is defined as a pseudo-environment. Therefore, a total of $N_{\mathrm{pe}} = 3$ pseudo-environments are defined for each CFD environment hence generating 3 different trajectories that the agent uses during the optimization step. Also, the output dimensionality is reduced from 3 to 1, hence reducing the number of combinations needed to explore all possible control strategies. With the aim of equating the classical and DRL controls, we allow the agent to explore actions with a maximum absolute value equal to the amplitude set in the classic control, $|v_{\mathrm{ac}}|_{\max} = 0.3$. Furthermore, an exponential smoothing function is applied to the discrete actions predicted by the agent, hence forcing a smooth continuous control signal applied at every time

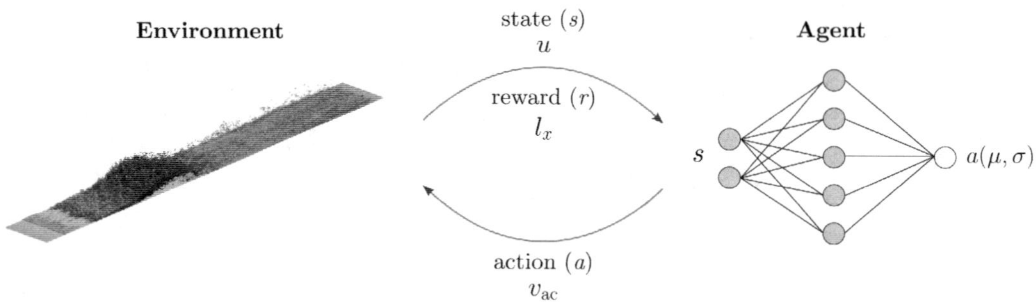

Figure 3: CFD–DRL setup. Trajectories consisting of state-reward-action triplets $\{(s^0, r^0, a^0), (s^1, r^1, a^1), ..., (s^n, r^n, a^n)\}$ are sampled during an episode. The NN weights are then optimized based on the trajectories to find the best action distribution that maximizes the accumulated reward in time.

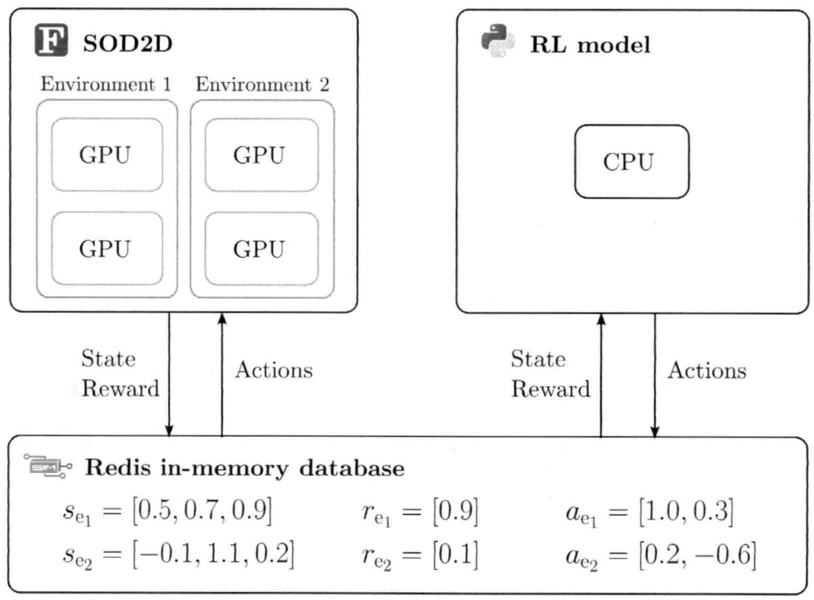

Figure 4: Communication scheme between the multiple CFD environments run in parallel using GPUs and the DRL model run on a CPU. The in-memory Redis database handles the communication of state, reward, and actions across the solver and the model. For the sake of simplicity, only 2 parallel CFD environments are represented, but we typically run up to 8 parallel CFD environments on the cluster.

step in the environment.

The environment state s is defined as a cloud of witness points (probes) that measure the streamwise velocity u. The witness points cloud is composed by an equidistant $6 \times 6 \times 6$ points grid spanning a $240 \times 50 \times 104$ subdomain that captures the non-actuated recirculation bubble region and its vicinity, as depicted in figure 5. Since the MARL approach is used and the computational domain is divided into 3 pseudo-environments, it is important to use a representative state for each subdomain. Therefore, the state of each pseudo-environment is composed of 2 planes of witness points aligned with the actuators in the streamwise direction totaling 72 witness points.

The environment reward r is based on the recirculation length of the turbulent bubble, and the optimization process aims to maximize r so that the recirculation area is reduced. To

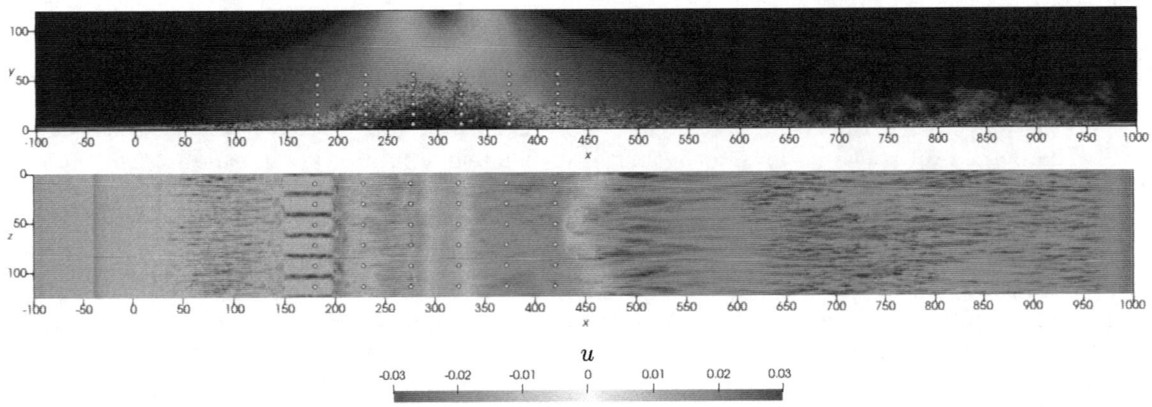

Figure 5: Distribution of witness points. Top: non-actuated xy-plane at $z = L_z/2$ displaying the same flow-field information (and colormap) as in figure 1. Bottom: periodic-forcing xz-plane of the first off-wall node displaying the streamwise u velocity component.

calculate a characteristic length for the recirculation bubble (l_x), the wall-shear stress value is computed at each element face of the wall surface. The area of those elements with a negative wall-shear stress ($A_i|_{\tau_w<0}$), *i.e.* with local recirculation, is integrated and then divided by the span of the pseudo-environment

$$l_x = \frac{N_{\text{pe}}}{L_z} \sum_i A_i|_{\tau_w<0}. \tag{6}$$

The reward is normalized by the time-averaged non-actuated characteristic recirculation length $\overline{l_x^*}$, yielding $r = -l_x/\overline{l_x^*}$ (noting that the overline notation is now related to temporal averages). Additionally, instead of using an instantaneous value, the reward is averaged during an actuation period so that the overall response of the system to a given action is more representative. Last, each pseudo-environment computes a local reward r_l and the global reward r_g is an average of the local rewards. These two quantities are finally merged in a weighted sum, $r = \alpha r_l + (1-\alpha)r_g$, where α is a tunable parameter, and the $\alpha = 0.5$ value is selected because it represents an unbiased trade-off between both rewards.

With respect to the DRL model, the proximal policy optimization (PPO) [53] algorithm is employed as the controlling agent. The PPO method is a policy-gradient method which tries to find an optimal policy, *i.e.* distribution of actions, which maximizes the accumulated reward expectation given the state of the environment. The main advantages of PPO with respect to other DRL algorithms are its small number of tunable parameters and its suitability for continuous control problems [20, 54]. In the current framework, we employ a NN consisting of 2 layers with 128 neurons per layer. An episode duration of $T_e = 4/f_{\text{ac}} = 1600$ is set (4 periods of the periodic-forcing frequency) so that the low frequencies of the system can develop. The actuation frequency of the DRL model is $f_{\text{ac,DRL}} = 10 f_{\text{ac}}$ yielding 40 actuations per episode, which we considered a trade-off between actuating too often (the flow has not enough time to develop after a new actuation) and actuating too seldom (the flow is not acted on when required). Actuating every 10% of the system's lowest frequency is within the order of magnitude that can be found in the literature [20]. We also note that 8 environments are typically run in parallel depending on the cluster availability and architecture. Given that 3 pseudo-environments are defined within each CFD simulation, this results in 24 trajectories sampled in parallel.

3. Results and discussion

3.1. Periodic control

Results of the classical time-periodic forcing control are first compared with the non-actuated TSB. The open-loop control starts actuating after 5000 time units (TU, normalized by δ_0^* and U_∞) and statistics are recorded during 15000TU for both the coarse and fine grids. Figure 6 displays the time-averaged streamwise velocity $\overline{u}$ for the fine-grid case. It is observed that the non-actuated case exhibits a recirculation bubble generated by the SB top-boundary condition that qualitatively spans from $x = 225$ to $x = 350$. On the other hand, the actuated bubble qualitatively spans a smaller region of the domain and new separation bubbles are spotted near the actuators. The reduction of the TSB on its downstream region is a phenomenon also observed by Cho *et al.* [18] in a similar SB APG TBL setup. The bubble length can be better quantified by the time-averaged characteristic recirculation length $\overline{l_x}$. As presented in table 2, a reduction of 6.8% in $\overline{l_x}$ is obtained for the fine grid when using periodic control. It can be noted that the normalized standard deviation of $\overline{l_x}$ is similar to the rms value of the forcing signal, *i.e.* $\sigma(\overline{l_x})/\overline{l_x} \sim A_{\mathrm{ac}}/\sqrt{2}$, and this means that the periodic-forcing control dominates the TSB. This phenomenon can be visualized in figure 7, where the temporal signals of the non-actuated and actuated l_x are shown. Because of the FPG present in the SB setup, the reattachment point of the TSB has less freedom compared to a SO setup. In the latter case, a reduction of the TSB up to 50% can be observed when using periodic control under the correct actuation frequency, as reported by Wu *et al.* [17].

	$\overline{l_x^*}$	$\overline{l_x}$	$1 - \overline{l_x}/\overline{l_x^*}$
Fine	143.0 ± 8.6	133.3 ± 27.2	6.8%
Coarse	153.6 ± 8.3	129.5 ± 33.4	15.7%

Table 2: Time-averaged characteristic length $\overline{l_x}$ and standard deviation of the non-actuated and the periodic-forcing cases for both fine and coarse grids. The reduction of the characteristic recirculation length is also shown.

The instantaneous flow field of the actuated TSB is depicted in figure 8. A large-scale spanwise vortical structure can be observed as a result of the harmonic actuation. The interaction of these structures with the TSB effectively reduces the size of the bubble. Additionally, the spectra of the witness points previously shown for the non-actuated case (figure 2) is now displayed for the periodic forcing case in figure 9. The effect of the actuators is clearly appreciated on the peak arising precisely at f_{ac}, while an harmonic peak is also present at a higher frequency.

The coarse-grid results are discussed next. The main objective of the coarse–fine grid comparison is to assess whether the main flow features are preserved so that the coarse grid can be used for training the DRL model. Coarse-grid results depicting the time-averaged flow field and its recirculation region are displayed in figure 10. In general, a very similar flow structure compared with the fine-grid results can be observed. It can also be noted that the TSB is slightly larger in this case, and this is quantified in table 2. The characteristic recirculation length is reduced by 15.7% when applying periodic actuation on the coarse grid. Similarly to the fine grid, the standard deviation of this metric is significantly increased as a result of the periodic forcing. Spectra of the coarse-grid simulation are also shown in figure 9. While the general trend is in good agreement with the fine-grid results, the spectrum of the probe located downstream of the bubble and farther away from the wall (bottom right) presents some discrepancies. This can be expected since the stretching of the coarse grid significantly reduces the resolution far

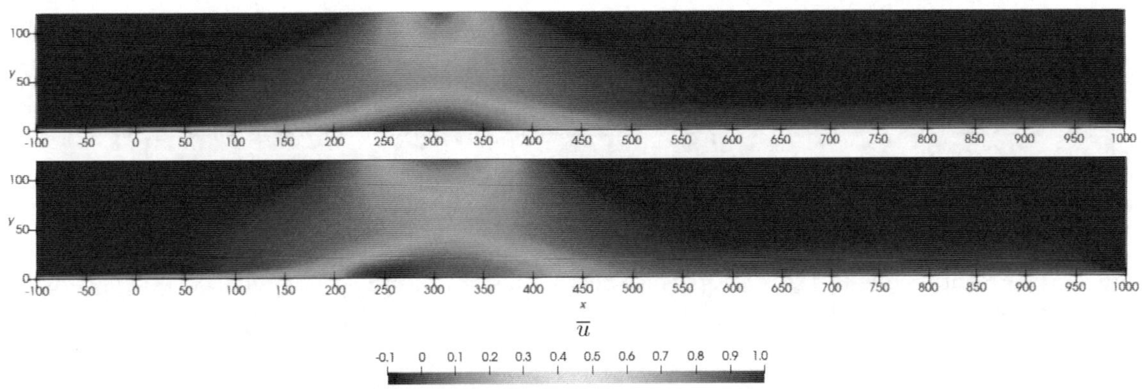

(a) Time-averaged streamwise velocity component $\overline{u}$ at $z = L_z/2$.

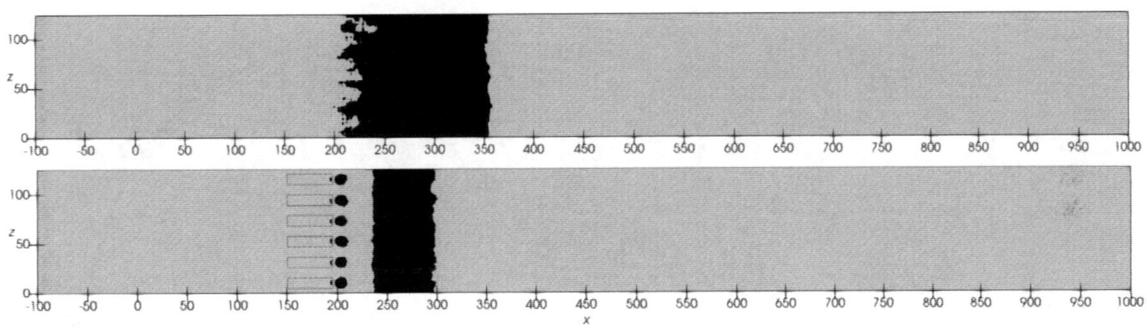

(b) Time-averaged recirculation region ($\overline{u} < 0$) at the xz-plane of the first off-wall node.

Figure 6: Fine-grid time-averaged flow field. Both panels (a) and (b) show non-actuated (top) and periodic control (bottom) results.

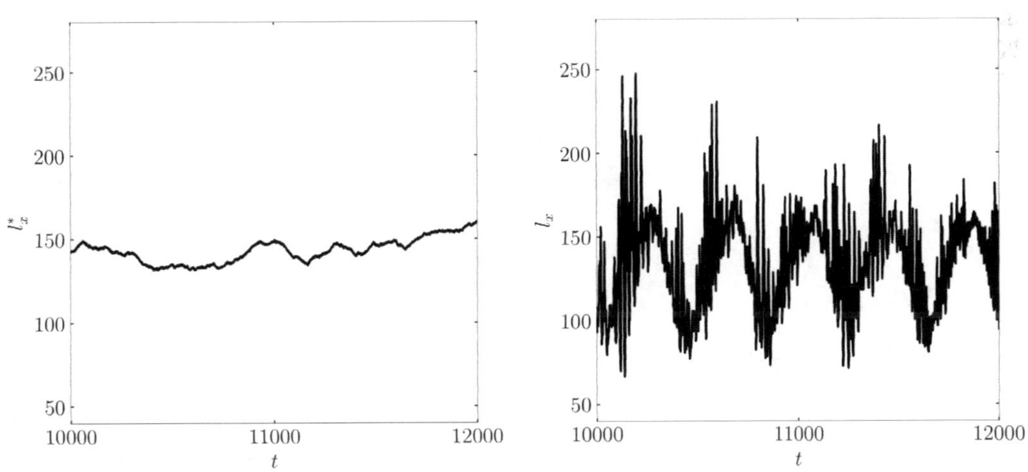

Figure 7: Temporal signal of the characteristic recirculation length for the non-actuated (left) and periodic control (right) fine-grid cases.

from the wall. Nevertheless, the main flow features of the fine-grid simulation are preserved, and we conclude that the coarse grid is adequate for training the DRL model.

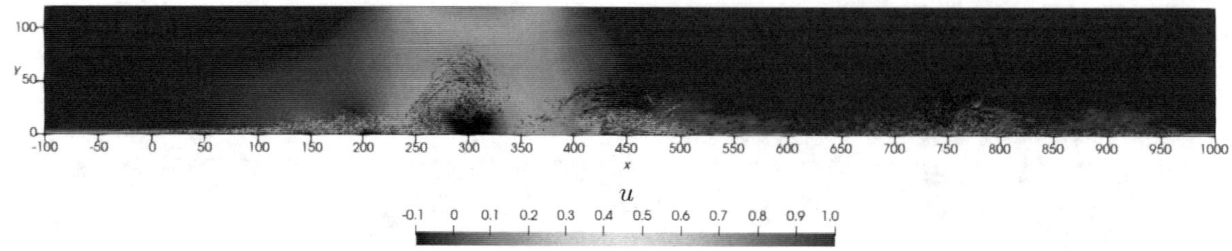

Figure 8: Instantaneous Q criterion isosurface (defined in figure 1) colored by the streamwise velocity component u on the periodic-forcing fine-grid case.

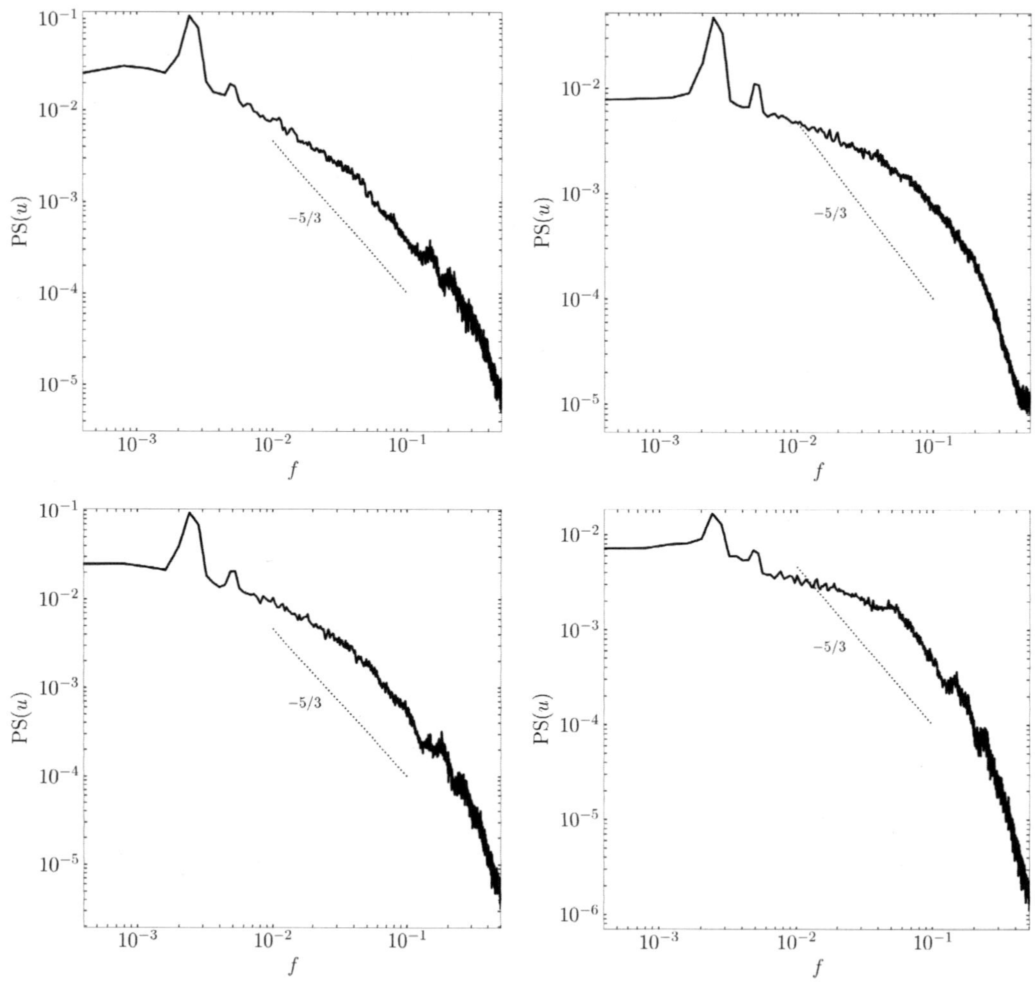

Figure 9: Power spectrum (PS) of the streamwise velocity component u of the periodic-forcing TSB sampled at 2 different locations. Left: $(x, y) = (324, 15)$. Right: $(x, y) = (372, 55)$. Top: fine-grid results. Bottom: coarse-grid results. See figure 2 caption for additional information on then calculation of the PS.

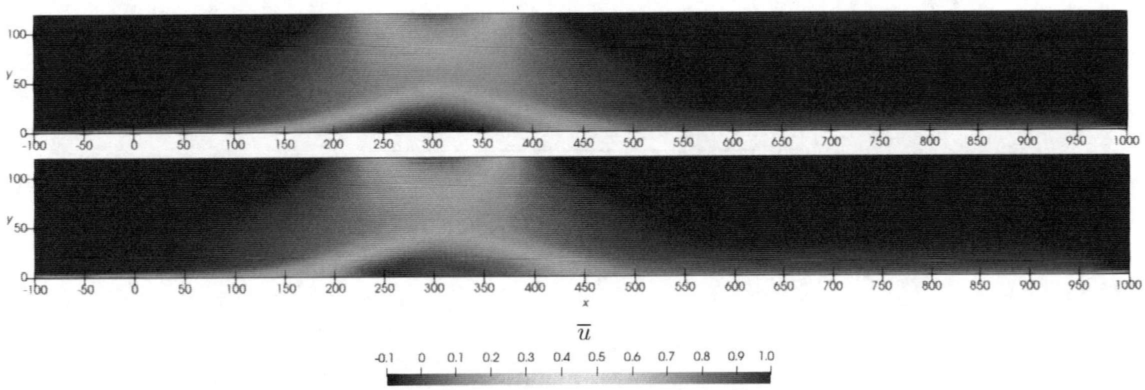

(a) Time-averaged streamwise velocity component $\overline{u}$ at $z = L_z/2$.

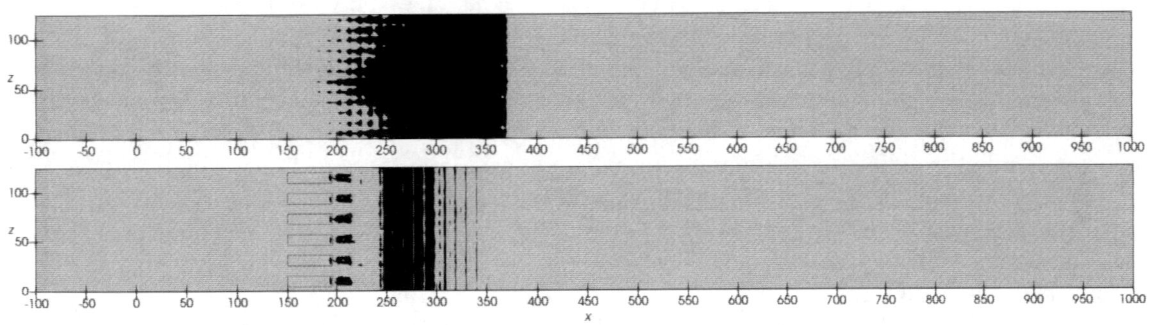

(b) Time-averaged recirculation region ($\overline{u} < 0$) at the xz-plane of the first off-wall node.

Figure 10: Coarse-grid time-averaged flow field. Both panels (a) and (b) show non-actuated (top) and periodic control (bottom) results.

3.2. DRL control

In the following, the results of the DRL control are discussed. First, the training of the DRL model is shown in figure 11. It can be observed that the model's loss correctly decreases with increasing number of episodes before it stagnates. The average reward metric is the average value of the time-averaged characteristic recirculation lengths provided by the pseudo-environments after an episode. It is observed that the average reward improves with training (bearing in mind that, during training, exploration noise is added to the controlling agent) and converges to an average reward lower than -1, hence reducing the TSB compared to the non-actuated baseline case. An improvement is also observed on the maximum reward, while the minimum reward remains stable. As previously explained, 24 trajectories are sampled in parallel since $N_{\mathrm{e}} = 8$ and $N_{\mathrm{pe}} = 3$, and an equal number of optimization steps (24) are performed thereafter. This process is repeated 96 times, hence accumulating a total of 2304 episodes and optimization steps. Noting that the training time for a single CFD environment takes 1.5 hours to simulate 1600 TU on a A100 NVIDIA GPU, the overall training walltime is 144 hours (6 days) when running 8 CFD environments in parallel (one per GPU). This results in a total of 1152 GPU-hours consumed.

Once the DRL agent has been trained, we test the learnt control strategy under deterministic behaviour, *i.e.* selecting the best possible action from the action probability distribution (its mean). The DRL model is run in deterministic mode for 20000 TU using a baseline (non-actuated) snapshot as initial condition. Table 3 shows the average characteristic recirculation length of the non-actuated, periodic control, and DRL control cases for the last 15000 TU,

	$\overline{l_x}$	$1 - \overline{l_x}/\overline{l_x^*}$
Periodic	129.5 ± 33.4	15.7%
DRL	114.7 ± 6.7	25.3%

Table 3: Time-averaged characteristic length $\overline{l_x}$ and standard deviation of periodic control and DRL control cases. The reduction of the recirculation length wrt. the non-actuated case is also shown.

The DRL control yields a larger reduction of the TSB compared to the classic periodic control, respectively -25.3% and -15.7%. The temporal signal l_x is plotted in figure 12 for the first 10000 TU. It is clearly appreciated that the DRL control quickly reduces the TSB length. Importantly, the control is performed smoothly, and no sudden oscillations are appreciated. While the periodic forcing also displays a reduction of the bubble length, spurious peaks can be observed in the signal (this is better appreciated in figure 7). This can arise from the fact the DRL control strategy is forced to conserve mass in space, *i.e.* instantaneously and therefore in time too, while the classical periodic forcing is only conserving mass in time. The instantaneous conservation of mass in incompressible flow is important not only from the physical point of view, but also numerically in the pressure solver. Since we do not correct for the instantaneous mass imbalance of the classical periodic control, spurious oscillations can arise, which are then captured by the l_x signal.

Flow fields for both periodic control and DRL control are shown in figure 13a (results of the periodic control are shown again for comparison purposes). The DRL control yields a different flow-field structure. A strong recirculation region located at $x = 180$, on top of the actuators, is observed. Since DRL actuators are spanwise-paired with equally positive and negative mass flow rates, this can eventually generate streamwise structures that interact with the TBL resulting in these small separation bubbles. This phenomenon is confirmed in figure 13b, and it can also be noted that the bubble reduction of the DRL control is qualitatively more prominent than the periodic-forcing control.

Last, the time signals of the actions set by the DRL control agent are shown in figure 14. First, the actuators oscillate between the maximum and minimum allowed values, $A_{\text{ac}} = \pm 0.3$. Shorty after, the a_3 actuator saturates on $+A_{\text{ac}}$ and a_2 follows, while a_1 saturates on $-A_{\text{ac}}$. Then, actuators bounce between $+A_{\text{ac}}$ and $-A_{\text{ac}}$, *i.e.* bang–bang control, while short transitional phases arise as well. A physical interpretation of the DRL actions cannot be derived from the current results yet, and a thorough assessment of the control strategy learnt by the DRL agent will be considered in future work.

4. Conclusions

In the present work, we investigate the efficacy of classical periodic control and deep reinforcement learning (DRL) control in reducing a turbulent separation bubble (TSB) generated by a suction and blowing (SB) boundary condition. The framework that integrates the SOD2D CFD solver (a novel multi-GPU spectral element solver developed at Barcelona Supercomputing Center) with the DRL model via SmartSim is also presented.

The classical periodic control, employing harmonic time forcing, provides notable success in reducing the TSB length by approximately 6.8% and 15.7% on a fine and a coarse grid, respectively. Additionally, a spectral analysis highlights the impact of the actuators on the flow, with clear peaks at the actuation frequency and harmonics. The coarse-grid results demonstrate

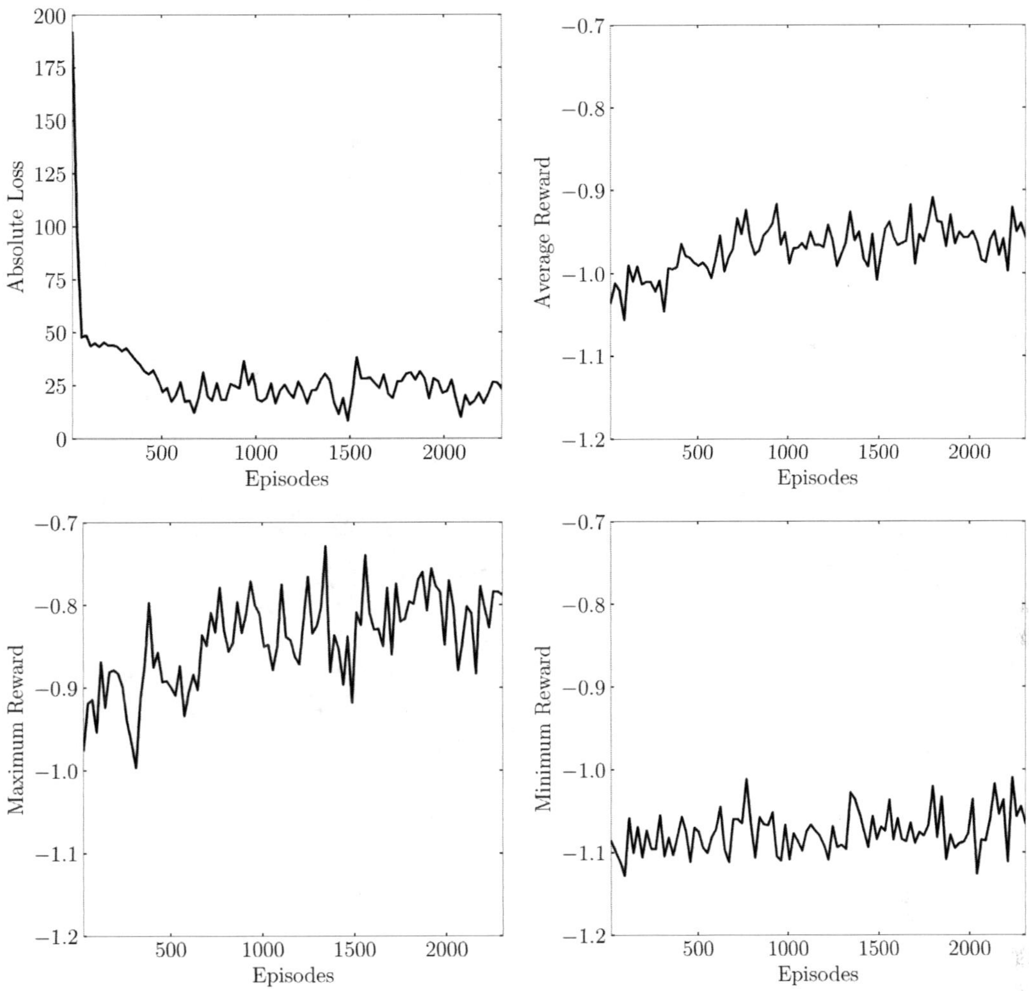

Figure 11: DRL training metrics. The average, maximum, and minimum reward signals are extracted from the 24 pseudo-environments running in parallel.

the preservation of the main flow features, validating the use of the coarse grid for the subsequent training of the DRL model. While the current periodic forcing considers a single configuration of frequency and amplitude, future work will expand this parametric space in order to maximize the periodic control efficacy. An investigation of the domain size shall also be explored to assess the confinement of the bubble for both non-actuated and actuated cases.

The DRL control demonstrated very promising results by successfully learning a control strategy that can effectively reduce the TSB length by 25.3% in the coarse grid. Compared with the classical periodic control, the DRL has the freedom to set the optimal mass flow rate of the actuators for a given environment state, hence constructing a complex control signal that can embed multiple frequencies. The training was performed using 24 parallel pseudo-environments running on 8 GPUs and took 144 hours (6 days) in total, so 1152 GPU-hours. The physical interpretation of the control strategy remains for future work, as well as the evaluation of the DRL strategy on the fine mesh. Nonetheless, to the best of our knowledge, the current results present a successful application of DRL-based control at one of the highest Reynolds numbers to this date.

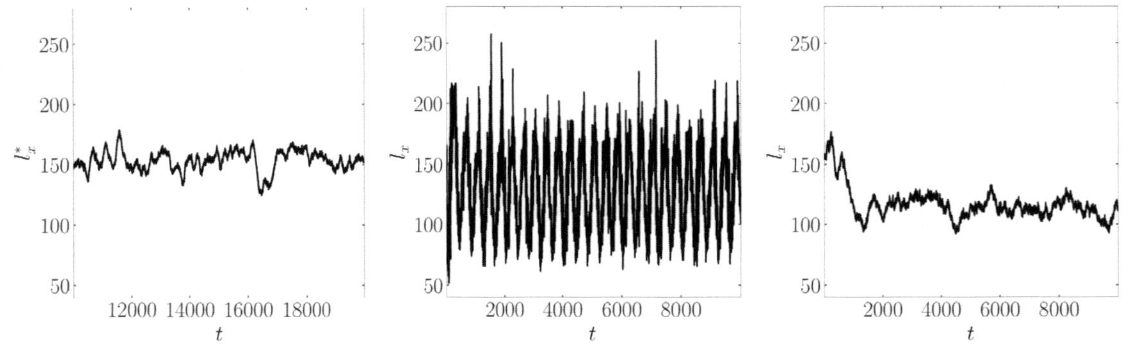

Figure 12: Temporal signal of the characteristic recirculation length for the non-actuated (left), periodic control (center), and the DRL control (right) coarse-grid cases. The periodic control and DRL control cases use the non-actuated snapshot at $t = 20000$ as initial condition.

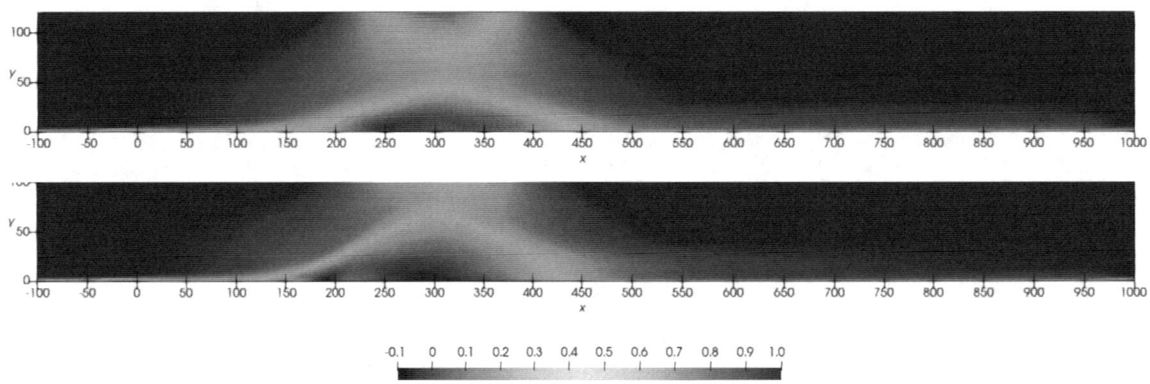

(a) Time-averaged streamwise velocity component $\overline{u}$ at $z = L_z/2$.

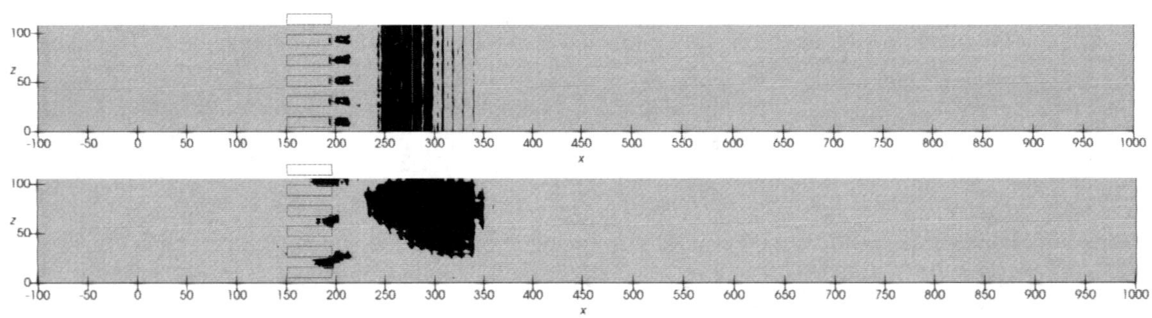

(b) Time-averaged recirculation region ($\overline{u} < 0$) at the xz-plane of the first off-wall node.

Figure 13: Coarse-grid time-averaged flow field. Both panels (a) and (b) show periodic control (top) and DRL control (bottom) results (periodic control results are shown again for better comparison).

Acknowledgments

This work was performer in part during the Fifth Madrid Summer Workshop, funded by the European Research Council (ERC) under the Caust grant ERC-AdG-101018287. OL has been partially funded by the European Commission's Horizon 2020 Framework

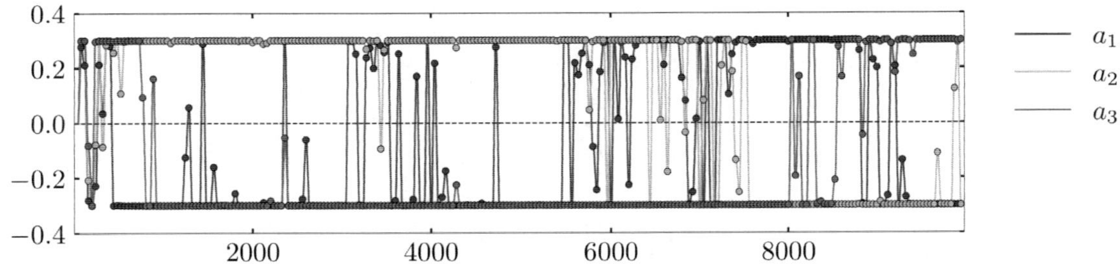

Figure 14: Temporal signal of the DRL-based actuators. The exponential smoothing between two discrete actions that sets the instantaneous actuators mass flow rate is also displayed.

program and the European High-Performance Computing Joint Undertaking (JU) under grant agreement no. 101093393, and by MCIN/AEI/10.13039/501100011033 and the European Union NextGenerationEU/PRTR (PCI2022-134996-2), project CEEC. OL has been partially supported by a Ramon y Cajal postdoctoral contract (Ref: RYC2018-025949-I). RV and FAA acknowledge financial support from ERC grant no. '2021-CoG-101043998, DEEPCONTROL'. The authors thank Laia Julió from the BSC support services and Dr. Andrew Shao from the SmartSim team for the help and fruitful discussions about the CFD–DRL software framework. The authors also thank the UPM Turbulence Summer School and our hosts Dr. Miguel Pérez Encinar and Dr. Adrián Lozano-Durán for the organization of the workshop. Finally, we acknowledge the National Academic Infrastructure for Supercomputing in Sweden (NAISS) for the computational time provided at the large-GPU systems Alvis and Berzelius.

References

[1] Gad-el-Hak M and Bushnell D M 1991 *J. Fluids Eng.* **113** 5–30
[2] Cattafesta L N and Sheplak M 2011 *Annu. Rev. Fluid Mech.* **43** 247–272
[3] Monty J, Harun Z and Marusic I 2011 *Int. J. Heat Fluid Flow* **32** 575–585
[4] Kitsios V, Atkinson C, Sillero J, Borrell G, Gungor A, Jiménez J and Soria J 2016 *Int. J. Heat Fluid Flow* **61** 129–136
[5] Abe H 2017 *J. Fluid Mech.* **833** 563–598
[6] Kitsios V, Sekimoto A, Atkinson C, Sillero J A, Borrell G, Gungor A G, Jiménez J and Soria J 2017 *J. Fluid Mech.* **829** 392–419
[7] Vinuesa R, Hosseini S M, Hanifi A, Henningson D S and Schlatter P 2017 *Flow Turbul. Combus.* **99** 613–641
[8] Wu W, Meneveau C and Mittal R 2020 *J. Fluid Mech.* **883** A45
[9] Bobke A, Vinuesa R, Örlü R and Schlatter P 2016 *J. Phys. Conf. Ser.* **708** 012012
[10] Bobke A, Vinuesa R, Örlü R and Schlatter P 2017 *J. Fluid Mech.* **820** 667–692
[11] Pozuelo R, Li Q, Schlatter P and Vinuesa R 2022 *J. Fluid Mech.* **939**
[12] Harun Z, Monty J P, Mathis R and Marusic I 2013 *J. Fluid Mech.* **715** 477–498
[13] Vila C S, Vinuesa R, Discetti S, Ianiro A, Schlatter P and Örlü R 2020 *Phys. Rev. Fluids* **5**
[14] You D and Moin P 2008 *J. Fluids Struct.* **24** 1349–1357
[15] Atzori M, Vinuesa R, Stroh A, Gatti D, Frohnapfel B and Schlatter P 2021 *Phys. Rev. Fluids* **6**
[16] Lehmkuhl O, Lozano-Durán A and Rodriguez I 2020 *J. Phys. Conf. Ser.* **1522** 012017
[17] Wu W, Meneveau C, Mittal R, Padovan A, Rowley C W and Cattafesta L 2022 *Phys. Rev. Fluids* **7** 084601
[18] Cho M, Choi S and Choi H 2016 *J. Fluids Eng.* **138**
[19] Vignon C, Rabault J and Vinuesa R 2023 *Phys. Fluids* **35**
[20] Rabault J, Kuchta M, Jensen A, Réglade U and Cerardi N 2019 *J. Fluid Mech.* **865** 281–302
[21] Tang H, Rabault J, Kuhnle A, Wang Y and Wang T 2020 *Phys. Fluids* **32**
[22] Xu H, Zhang W, Deng J and Rabault J 2020 *J. Hydrodyn.* **32** 254–258
[23] Li J and Zhang M 2021 *J. Fluid Mech.* **932**
[24] Varela P, Suárez P, Alcántara-Ávila F, Miró A, Rabault J, Font B, García-Cuevas L M, Lehmkuhl O and Vinuesa R 2022 *Actuators* **11** 359

[25] Suárez P, Alcántara-Ávila F, Miró A, Rabault J, Font B, Lehmkuhl O and Vinuesa R 2023 Active flow control for three-dimensional cylinders through deep reinforcement learning
[26] Wang Z, Fan D, Jiang X, Triantafyllou M S and Karniadakis G E 2023 *J. Fluid Mech.* **973**
[27] Yan L, Li Y, Hu G, li Chen W, Zhong W and Noack B R 2023 *Phys. Fluids* **35**
[28] Beintema G, Corbetta A, Biferale L and Toschi F 2020 *J. Turbul.* **21** 585–605
[29] Vignon C, Rabault J, Vasanth J, Alcántara-Ávila F, Mortensen M and Vinuesa R 2023 *Phys. Fluids* **35**
[30] Guastoni L, Rabault J, Schlatter P, Azizpour H and Vinuesa R 2023 *Eur. Phys. J. E* **46**
[31] Sonoda T, Liu Z, Itoh T and Hasegawa Y 2023 *J. Fluid Mech.* **960** ISSN 1469-7645
[32] Novati G, de Laroussilhe H L and Koumoutsakos P 2021 *Nat. Mach. Intell.* **3** 87–96
[33] Bae H J and Koumoutsakos P 2022 *Nat. Commun.* **13**
[34] Kurz M, Offenhäuser P and Beck A 2023 *Int. J. Heat Fluid Flow* **99** 109094
[35] Rabault J and Kuhnle A 2019 *Phys. Fluids* **31**
[36] Belus V, Rabault J, Viquerat J, Che Z, Hachem E and Reglade U 2019 *AIP Adv.* **9**
[37] Vreman A W 2004 *Phys. Fluids* **16** 3670–3681
[38] Gasparino L, Spiga F and Lehmkuhl O 2023 (accepted) *Comput. Phys. Commun.*
[39] Schlatter P and Örlü R 2012 *J. Fluid Mech.* **710** 5–34
[40] Hosseini S, Vinuesa R, Schlatter P, Hanifi A and Henningson D 2016 *Int. J. Heat Fluid Flow* **61** 117–128
[41] Hunt J, Wray A and Moin P 2099 *CTR, Proc. Summ. Progr.* 193–208
[42] Greenblatt D and Wygnanski I J 2000 *Prog. Aerosp. Sci.* **36** 487–545
[43] Aram S and Mittal R 2011 *Int. J. Flow Control* **3** 87
[44] Smith D R 2002 *AIAA J.* **40** 2277
[45] Buren T V, Leong C M, Whalen E and Amitay M 2016 *Phys. Fluids* **28** 037106
[46] Leschziner M A and Lardeau S 2011 *Philos. Trans. R. Soc.* **A369** 1495
[47] Dandois J, Garnier E and Sagaut P 2006 *AIAA J.* **44** 225
[48] Wu D K L and Leschziner M A 2009 *Int. J. Heat Fluid Flow* **30** 421
[49] Guadarrama S, Korattikara A, Ramirez O, Castro P, Holly E, Fishman S, Wang K, Gonina E, Wu N, Kokiopoulou E, Sbaiz L, Smith J, Bartók G, Berent J, Harris C, Vanhoucke V and Brevdo E 2018 TF-Agents: A library for Reinforcement Learning in TensorFlow
[50] Font B, Weymouth G D, Nguyen V T and Tutty O R 2021 *J. Comput. Phys.* **434** 110199
[51] Partee S, Ellis M, Rigazzi A, Shao A E, Bachman S, Marques G and Robbins B 2022 *J. Comput. Sci.* **62** 101707
[52] Kurz M, Offenhäuser P, Viola D, Resch M and Beck A 2022 *Softw. Impacts* **14** 100422
[53] Schulman J, Wolski F, Dhariwal P, Radford A and Klimov O 2017 Proximal policy optimization algorithms
[54] Mnih V, Kavukcuoglu K, Silver D, Rusu A A, Veness J, Bellemare M G, Graves A, Riedmiller M, Fidjeland A K, Ostrovski G, Petersen S, Beattie C, Sadik A, Antonoglou I, King H, Kumaran D, Wierstra D, Legg S and Hassabis D 2015 *Nature* **518** 529–533

Accelerating high order discontinuous Galerkin solvers using neural networks: Wall bounded flows

Oscar A. Mariño[1], David Mayoral[1], Adrián Juanicotena[1], Fernando Manrique de Lara[1], Esteban Ferrer[1,2]

[1] ETSIAE-UPM-School of Aeronautics, Universidad Politécnica de Madrid, Plaza Cardenal Cisneros 3, E-28040 Madrid, Spain

[2] Center for Computational Simulation, Universidad Politécnica de Madrid, Campus de Montegancedo, Boadilla del Monte, 28660 Madrid, Spain

E-mail: esteban.ferrer@upm.es

Abstract. High order solvers are accurate but computationally expensive as they require small time steps to advance the solution in time. In this work we include a corrective forcing to a low order solution to improve the accuracy while advancing in time with larger time steps, and achieve fast computations. The work uses a discontinuous Galerkin framework, where the polynomial order, inside each mesh element, can be varied to provide low or high accuracy. The corrective forcing is included for each high order Gauss nodal point in the mesh. This work is a continuation of [1, 2], where we extend the methodology to wall bounded flows. Namely, we adapt the methodology to a turbulent channel at $Re_\tau = 182$. In this case, we use three neural networks to correct different regions of the flow, which are distinguished by their $y+$ distance to the wall. The methodology is able to correct the low resolution simulation to attain flow statistics that are comparable to high order simulations. We include comparisons for the mean, Reynolds stresses and shear stress on the wall. We achieve good predictions using the corrected low order solution, in mean velocity and its corresponded fluctuations, as well as the shear stress on the wall.

1. Introduction

This paper is a continuation of [1, 2], where we presented neural network accelerations for the simulations of Burger's equations and the Navier-Stokes equations, using the 3D Taylor Green Vortex (TGV) problem. Our proposed methodology evolves a solution with a low-order polynomial while including a corrective forcing to obtain a more accurate solution. The corrective forcing is obtained using deep fully connected neural networks. The methodology proved to be useful, as it was able to correct the solution outside the time frame considered for training. Note that a high order solution was only necessary for training but not after, which allowed accelerations. In [2], we compared the accuracy provided by the methodology with simulations with varying polynomial orders (P) and showed an effective acceleration of 4-5, when correcting a P=3 solution using (only during training) a P=8 simulation. The resulting corrected order of accuracy was P=6.

Neural networks (NN) are gaining popularity in scientific computing as they can complement computational fluid dynamic methodologies; see reviews [3, 4, 5]. A fully connected neural network consists of multiple interconnected layers of artificial neurons, or units, organized into an input layer, one or more hidden layers, and an output layer. The information flows forward through the network, from the input layer to the output layer, without any loops or connections between neurons within the same layer.

Combining computational fluid dynamics (CFD) with neural networks (NN) has been explored predominantly in the context of low-order methods. Bar-Sinai *et al.* [6] utilized NNs to estimate spatial derivatives of the Burgers' equation on a low-resolution grid. In the field of turbulence modeling, Kochkove *et al.* [7] employed NNs to reconstruct low-fidelity simulations of 2D turbulence, highlighting the potential of CFD and NN integration. Guastoni *et al.* [8, 9] and Güemes *et al.* [10] have also applied NNs, specifically convolutional neural networks (CNNs) and generative adversarial networks (GANs), to spatially interpolate/extrapolate wall quantities in turbulent regimes and generate high-resolution models from low-resolution data. Moreover, researchers have employed NNs to develop turbulent closure models or as substitutes for conventional large eddy simulation (LES) models, as reviewed by Stachenfeld *et al.* [11], Beck and Kurz [12], and Duraisamy [13].

An alternative perspective on the methodology is to regard the low order simulation as a reduced order model, which is augmented through the utilization of a neural network (NN)-based corrective forcing, similar to the approach proposed in [14, 15]. In our research, the reduced order is low order simulation, and consequently retains the efficiency and physic-nature since we solve the Navier-Stokes equations on a coarse grid. This physics-based reduced order model offers flexibility in handling arbitrary boundary conditions, while achieving enhanced accuracy when combined with the corrective forcing provided by the neural network.

In this work, we first summarise the TGV problem, at $Re = 1600$, using more elements than in [2] and without using turbulent models. This problem has a simple set up but provides a complex flow behaviour including and initial laminar regime, followed by transition to turbulence, and finally a fully turbulent regime, as time progresses. In this case the neural network needs to extrapolate the solution in time, to obtain a useful corrective forcing.

Second, we extend the methodology to wall bounded flows. We test the methodology in a turbulent channel at $Re_\tau = 182$. This is a very well known case, that makes use of periodicity in the stream-wise and span-wise direction, but uses solid walls in the normal direction. In this second case, the neural network needs to interpolate (flow is homogeneous in time) to obtain the corrective forcing, but needs to face strong anisotropies in the wall normal direction.

The rest of the paper is organised as follows. First we summarise the methodology to obtain the corrective forcing, including details for wall bounded flows. In section 3, we present results for the TGV problem and the channel, to finalise with conclusions and outlooks.

2. Methodology

Consider a High Order (HO) Discontinuous Galerkin (DG) discretization of the 3D Navier-Stokes (NS) equations. The system is represented as:

$$\frac{d\boldsymbol{q}_{HO}}{dt} = \boldsymbol{p}^{HO}(\boldsymbol{q}_{HO}). \tag{1}$$

Here, $\boldsymbol{q}_{HO} \in \mathbb{R}^{dof^{HO}}$ contains the state variables of the compressible NS, including $(\rho, \rho u, \rho v, \rho w, \rho e)^T$ for all dof^{HO} in the DG mesh. The discrete operator $\boldsymbol{p}^{HO} : \mathbb{R}^{dof^{HO}} \to \mathbb{R}^{dof^{HO}}$ approximates the NS operator. The degrees of freedom are given by $dof^{HO} = N_{el} \times (P^{HO} + 1)^3$, where N_{el} is the number of elements in the mesh and P is the polynomial order.

We introduce a filter operator $\boldsymbol{G} : \mathbb{R}^{dof^{HO}} \to \mathbb{R}^{dof^{LO}}$ in (1), to obtain:

$$\frac{d\overline{\boldsymbol{q}}_{HO}}{dt} = \boldsymbol{G}\boldsymbol{p}^{HO}(\boldsymbol{q}_{HO}), \tag{2}$$

where $\overline{\boldsymbol{q}}_{HO} \in \mathbb{R}^{dof^{LO}}$ is a low order solution obtained once filtered, and is expected to be more accurate than a low order solution; and $dof^{LO} = N_{el} \times (P^{LO}+1)^3$. We also define a low order evolution equation:

$$\frac{d\boldsymbol{q}_{LO}}{dt} = \boldsymbol{p}^{LO}(\boldsymbol{q}_{LO}). \tag{3}$$

In general, $\boldsymbol{q}_{LO}$ and $\boldsymbol{q}_{HO}$ do not provide the same solution or accuracy, as $\boldsymbol{G}\boldsymbol{p}^{HO}(\boldsymbol{q}_{HO}) \neq \boldsymbol{p}^{LO}(\boldsymbol{q}_{LO})$, due to the non-linearity of the NS operator. To approximate the high order filtered solution $\boldsymbol{q}_{NN} \approx \boldsymbol{q}_{HO}$ using a low order evolution operator, we introduce a corrective forcing term. We rewrite Eq. (3) as:

$$\frac{d\boldsymbol{q}_{NN}}{dt} = \boldsymbol{p}^{LO}(\boldsymbol{q}_{NN}) + \boldsymbol{s}(\boldsymbol{q}_{NN}, \boldsymbol{q}_{HO}). \tag{4}$$

Here, $\boldsymbol{q}_{NN} \in \mathbb{R}^{dof^{LO}}$ represents the new variable obtained by adding the corrective forcing $\boldsymbol{s}(\boldsymbol{q}_{NN}, \boldsymbol{q}_{HO})$. The corrective forcing $\boldsymbol{s}(\boldsymbol{q}_{NN}, \boldsymbol{q}_{HO}) : \mathbb{R}^{dof^{LO}} \times \mathbb{R}^{dof^{HO}} \rightarrow \mathbb{R}^{dof^{LO}}$ depends on both the low and high order solutions and corrects the low order operator to retrieve a high order filtered evolution in time.

The explicit expression for the corrective forcing is given by:

$$\boldsymbol{s}(\boldsymbol{q}_{NN}, \boldsymbol{q}_{HO}) = \boldsymbol{G}\boldsymbol{p}^{HO}(\boldsymbol{q}_{HO}) - \boldsymbol{p}^{LO}(\boldsymbol{q}_{NN}). \tag{5}$$

An alternative expression to Eq. (5) can be derived by expanding the time derivatives in Eq. (3) and Eq. (4): $\boldsymbol{s}(\boldsymbol{q}_{NN}, \boldsymbol{q}_{HO}) = (\overline{\boldsymbol{q}}_{HO}^{n+1} - \boldsymbol{q}_{NN}^{n+1})/\Delta t_n$, and can used to train the neural network. The corrective forcing captures the high order modes and the nonlinear interactions between the low and high order solutions. It is modeled using a deep neural network in this work.

To advance the high and low order corrected systems in time, we consider a temporal discretization. For example, the corrected system Eq. (4) becomes:

$$\boldsymbol{q}_{NN}^{n+1} = \boldsymbol{q}_{NN}^{n} + \Delta t \left[\boldsymbol{p}^{LO}(\boldsymbol{q}_{NN}^{n}) + \boldsymbol{s}^{n}(\boldsymbol{q}_{NN}^{n}, \boldsymbol{q}_{HO}^{n})\right]. \tag{6}$$

Here, $\Delta t = t_{n+1} - t_n$ represents the time step size. The methodology can be applied to other explicit time marching schemes, although a third-order Runge-Kutta scheme is used in this work. This allows us to improve the accuracy of a low order solution by training a neural network with the high order solution and using it to correct the low order solution at future times when the high order solution is unknown.

2.1. High order discontinuous Galerkin solver: HORSES3D

The methodology has been implemented in our high-order spectral element solver HORSES3D [16]. HORSES3D is a 3D parallel framework developed at ETSIAE-UPM (the School of Aeronautics of the Polytechnic University of Madrid). This framework uses the high-order discontinuous Galerkin spectral element method (DGSEM) and is written in modern Fortran 2003. It targets simulations of fluid-flow phenomena such as those governed by compressible Navier-Stokes equations (used in this case) and high order polynomials of arbitrary order in each mesh element.

2.2. Corrective Forcing from Neural Networks

A corrective forcing $\boldsymbol{s}$ is constructed for each degree of freedom in the low order mesh, corresponding to each Gauss point resulting in $(P_{LO}+1)^3$ degrees of freedom per element.

We choose a fully connected Neural Network (NN) for its efficiency and simplicity, although other alternatives such as convolutional or recursive networks could have been considered. The

Journal of Physics: Conference Series **2753** (2024) 012023 doi:10.1088/1742-6596/2753/1/012023

NN consists of N_{la} layers ($N_{la} = 5$ in the figure), with the last layers using a linear activation function, the first and all intermediate hidden layers using the rectified linear unit (ReLU) activation function, as shown in Figure 1.

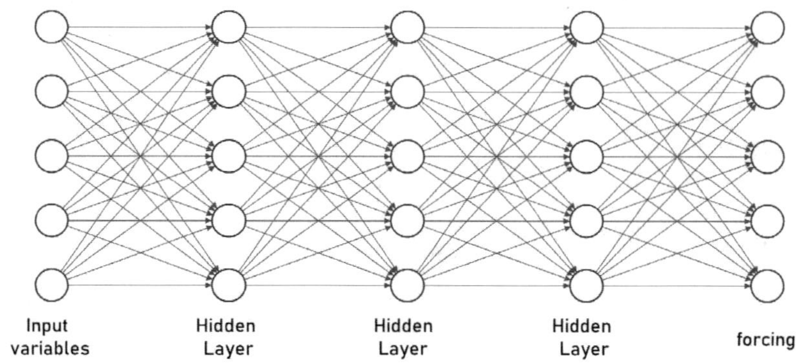

Figure 1: Fully connected deep neural network scheme. The input to the NN is a vector containing the nodal values for ρu, ρv, and ρw inside the element, and the output is a vector containing the nodal values for the forcing (s_2, s_3, and s_4) inside the element.

Regarding the state vectors, we conducted a preliminary test where all variables (density, momentum, and energy) were included in the training (we used the compressible solver HORSES3D). However, we found that a valid corrective forcing could not be obtained. In our test case, which considers incompressible flows with Mach number $M_0 = 0.08$ and 0.2 for the TGV and channel respectively, there exists a large disparity of scales between density/energy and momentum. To address this issue, we only use the three components of momentum for training and re-scale the values at each node. Consequently, the corrective forcing only consists of three components: $\boldsymbol{s} = (0, s_2, s_3, s_4, 0)^T$.

2.3. Isotropic algorithm
We choose to have the same NN architecture for every element and use all elements to train the network. This approach is advantageous in transitional/turbulent cases as in the Taylor Green Vortex case, that does not exhibit large variations between elements (e.g., no walls). Therefore, all elements can contribute to the training, as an element can represent the flow field of another element shifted in time or space.

However, if different flow structures are present in each mesh element, as in the channel case, then using all elements for training may not provide an accurate corrective forcing. In these cases, we divide the channel in layers (separate by $y+$ values) to enhance the training. Thus, while the methodology presented is independent of the flow physics, the NN training needs to consider the underlying flow physics.

The number of time steps used for each phase of the algorithm are summarized in the following table 1.

Regarding the remaining training hyperparameters, our primary emphasis was not on enhancing the overall efficiency of the algorithm, but rather on refining its precision. To achieve this, we employed an extensive number of training epochs, implemented a gradual reduction in learning rates over time, all with the overarching objective of minimizing the loss function to its utmost extent.

Table 1: TGV cases studied

Case	t= 4.5s	t=9s	t= 15s
HO phase			
Start(s)	4.5	9	15
HO iterations	600	600	600
LO & filtered HO iterations	200	200	200
Finish(s)	4.62	9.12	15.12
Training Phase			
Batches size	20	20	20
Epochs	30	30	30
Training time-steps	200	200	200
Training percentage (%)	90	90	90
LO correction phase			
Start(s)	4.5	9	15
LO iterations	1500	1000	1000
Finish(s)	5.4	9.6	15.6

2.4. Wall bounded flow training algorithm

While the Taylor–Green Vortex present time evolving flow structures (laminar to turbulent flow), the channel flow presents flow structures are consistent in time but significant variations can now be observed in space and between mesh elements. Indeed, anisotropy is introduced by the walls in the normal direction.

Previous knowledge about channel flows tells us that turbulence in boundary-layer tends to follow the same pattern and exhibits similar velocity profiles in the normal direction. Experimental evidence and DNS data show that three different regions can typically be identified within the boundary layer, namely: viscous sublayer, buffer layer and a "logarithmic-law region". Taking this into consideration, we divide the channel into several layers in the normal direction to the wall to train each section of the channel with a separate NN, so that we can capture specific physics within the corrective forcing related to the region. This resulted in effective corrective forcings. Note that the strong time-evolution of flow structures characteristic of the Taylor–Green Vortex is not present in the channel, and therefore there is no need for constant re-training of the NNs. Once trained, the corrective forcings are valid for all times.

In the following sections, we use the vertical coordinate to limit the three different layers for the whole height of the channel. We set the limits at $y^+_{limtit_1} \sim 8.6$ and, $y^+_{limtit_2} \sim 36$ in the first half of the channel. We replicate the same limits at $y^{top}_{limtit_i} = 2\delta - y^{bottom}_{limtit_i}$, such that the same wall distance in maintained. Thus, the channel is divided in 6 layers, 3 for each half. As the NN have no information on the spatial position, we combine the elements of the layers symmetrically to train each network, such that we have 3 NN, the first one for the elements near both walls, the second for the elements at mid distance, and the last one for the centre of the channel.

Due to the strong anisotropy observed in the channel case, the MLP architecture being used for the Taylor–Green Vortex problem has not proven to be able to capture the complex flow phenomena near the walls, even after assigning one dedicated NN to each boundary-layer region. The solution proposed here is to increase the number layers in each NN. The NNs with 20 layers provide reliable results while, in terms of computational cost, they do not involve much more training time than the previous 4-layer NNs.

3. Results

3.1. The Taylor Green Vortex Problem

Numerical experiments were conducted on the Taylor–Green Vortex (TGV) [17] for Reynolds numbers: $Re = 1000$. The TGV problem has been extensively used to study numerical methods and analyze their ability to reproduce various flow regimes[18]. The TGV configuration consists of a three-dimensional periodic box $[-\pi, \pi]^3$ with the following initial condition:

$$\begin{aligned}
\rho &= \rho_0, \\
v_1 &= V_0 \sin x \cos y \cos z, \\
v_2 &= -V_0 \cos x \sin y \cos z, \\
v_3 &= 0, \\
p &= \frac{\rho_0 V_0^2}{\gamma M_0^2} + \frac{\rho_0 V_0^2}{16}(\cos 2x + \cos 2y)(\cos 2z + 2),
\end{aligned} \tag{7}$$

where ρ_0, V_0, γ, and M_0 are constants representing density, velocity, specific heat ratio, and Mach number, respectively.

All simulations were performed using a coarse h-mesh with 16^3 elements and polynomial orders ranging from $P = 2$ to 8, as illustrated in Figure 2. Further details about our specific DG formulations can be found in [16]. We employed the Roe scheme for advective fluxes, the Bassi-Rebay 1 scheme for viscous fluxes, and a third-order three-stage Runge-Kutta scheme for time integration. The flow is considered incompressible with a Mach number of $M_0 = 0.08$ in all simulations. The reported times are non-dimensional and have been scaled with the characteristic velocity V_0 and the reference length $L = 1$.

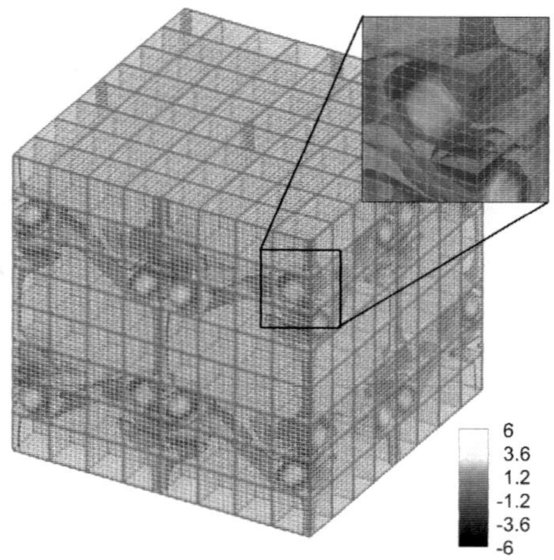

Figure 2: TGV problem: Mesh and Q-criterion iso-surfaces for Re=1600 at $t = 7$.

To validate the methodology, we show the time evolution of the L_2 norm for the difference between the high order solution $P_{HO} = 8$ and low order solutions (with and without corrective forcing) $P_{LO} = 2$. To compute this norm, we filter the high order solution and compare it with the low order solution as time progresses.

We check our methodology for three non-dimensional times. The first one corresponds to laminar flow structure ($t^* = 4.5$), the second one to a transitional regime $t^* = 9$ and the third one to a fully turbulent regime $t^* = 15$. For all times we test two NN architectures by increasing the number of layer from 4 to 20. In all cases, we observe that the NN-corrected simulation provides lower errors after training, but there is a clear dependency on the length of the correction depending on flow regime. In addition, using 20 layer clearly improves the predictions. We have also tested 30 layers but did not see a clear advantage (results not shown). It is noteworthy that the case where $t = 4.5s$ stands out as the most optimal scenario for our architecture. Conversely, when we scrutinized the outcomes across other time intervals, particularly in the instance where $t = 15s$, we discerned a marked decline in accuracy. We have observed improvements when using more complex neural network architectures (CNN or LSTM), and are currently preparing a detailed publication on the topic.

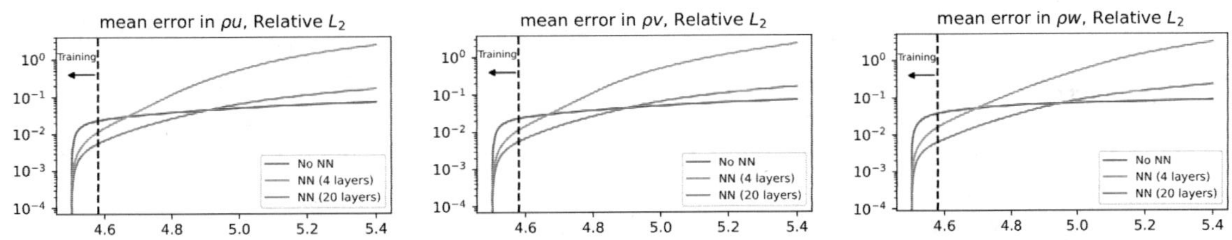

Figure 3: $\mathcal{L}_2$ error curves for different MLP architectures for $t^* = 4.5$.

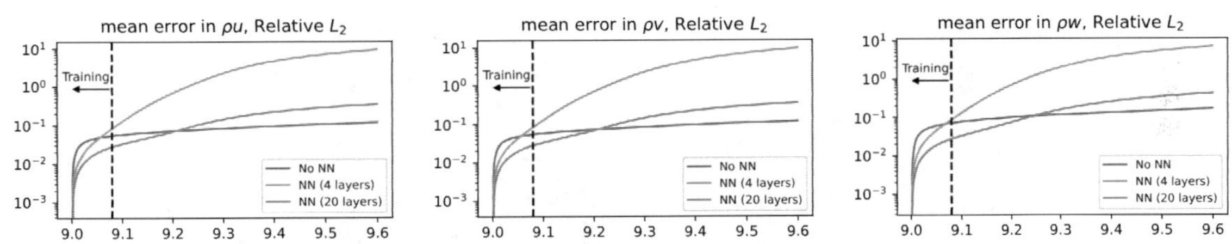

Figure 4: $\mathcal{L}_2$ error curves for different MLP architectures for $t^* = 9$.

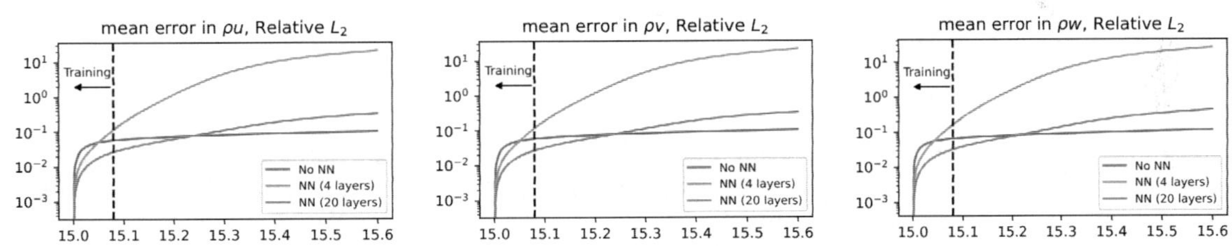

Figure 5: $\mathcal{L}_2$ error curves for different MLP architectures for $t^* = 15$.

3.2. Channel

The second case is a turbulent channel flow at friction Reynolds number $Re_\tau = 182$ and Mach number $M_0 = 0.2$. The friction Reynolds number is defined as $Re_\tau = u_\tau \delta/\nu$, where u_τ is the characteristic friction velocity, δ is channel half height and ν is the kinematic viscosity. The channel geometry is set as $6\delta \times 2\delta \times 3\delta$ in the x, y, z axis respectively. We employ adiabatic no-slip wall boundary conditions at the top and bottom of the domain ($y = 0$ and $y = 2\delta$), and periodic boundary conditions in the stream-wise and span-wise direction. To force the motion of the fluid in the stream-wise direction, we apply a constant force in this direction, equal to the theoretical shear stress at the wall. The settings for the numerical discretization are the same as the TGV case, namely, Roe and BR1 for the convective and viscous fluxes respectively and RK3 for the time integration.

We use a wall-resolving grid (first node from the wall is located at $y^+ = 1$) to avoid turbulence models or wall functions. The h-mesh consists of 1078 elements distributed as $11 \times 14 \times 7$ in the domain. The elements are evenly distributed in the x and z direction, while having stretching in the y direction, such that there are more elements in the wall regions. The fine simulation uses a polynomial order $P = 4$, while the low solution $P = 2$. We set a constant time step throughout the entire simulation, at a value of $\Delta t = 2 \times 10^{-3}\delta/U$ for the low order solution, while for the high order we use $\Delta t = 5 \times 10^{-4}\delta/U$, such that there are exactly 4 high order time steps per one low order step.

An initial stage of 30 flow-through units ($T = L_x/U$) is simulated to ensure a fully developed turbulent flow. Then, we train for 1000 low order time steps each NN corresponding to one of the three layers of the channel, as explained in Section 2.4. Finally, we run it during a time span of $33T$, where the turbulence statistics are recorded. A snapshot of the stream-wise velocity is presented in Figure 6, along with the h-mesh. Turbulence fluctuations can be observed in the figure.

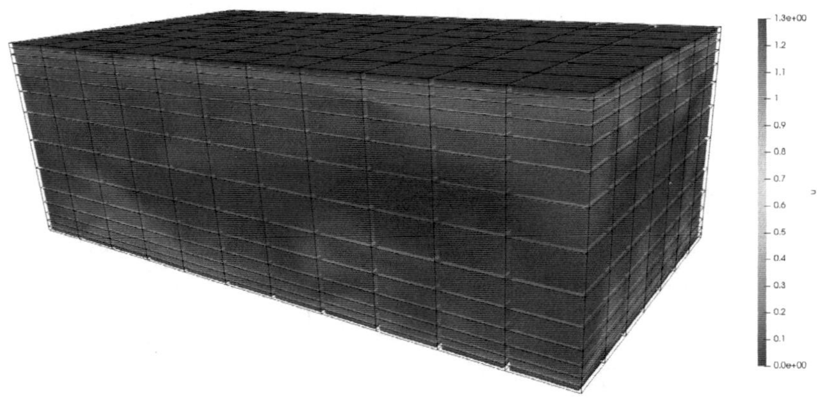

Figure 6: Channel flow: Instantaneous snapshot of stream-wise velocity for the high order $P = 4$ solution.

The mean stream-wise velocity profile in wall units is shown in Figure 7, for the high order and low order solution, as well as the corrected low order using 4 and 20 neuron layers. The high order solution is in very good agreement with the very well resolved DNS results of [19], while the low order solution deviates for $y^+ > 10$. In addition, the solution corrected with 4 neuron layers shows little difference with respect to the uncorrected low order solution. The solution corrected with 20 neuron layers is able to predict with accuracy the mean velocity, obtaining small disagreement after $y^+ = 30$, obtaining a close match to the high order $P = 4$ solution.

Similar behavior is found for the Reynolds stresses, depicted in Figure 8. Good agreement between the reference DNS simulation and the high order solution is apparent. When considering the variance of stream-wise velocity fluctuations (left figure), the low order solution is very different from the DNS solution while the NN corrected solution provides reasonable agreement. When considering the covariance of stream and wall-normal fluctuations of velocity (right figure), the low order corrected solution approaches the high order solution, improving significantly the low order results. Further improvements can be achieved by using more elaborated architectures or by increasing the number of layers/neurons, but the presented results show clearly the high potential of the proposed methodology also for wall bounded flows.

Another variable of interest is the wall shear stress (τ_w), related with the friction velocity (u_τ) and thus with the actual numerical friction Reynolds number (Re_τ). In table 2 the values obtained by each simulation are summarised and compared with the theoretical one. Very good

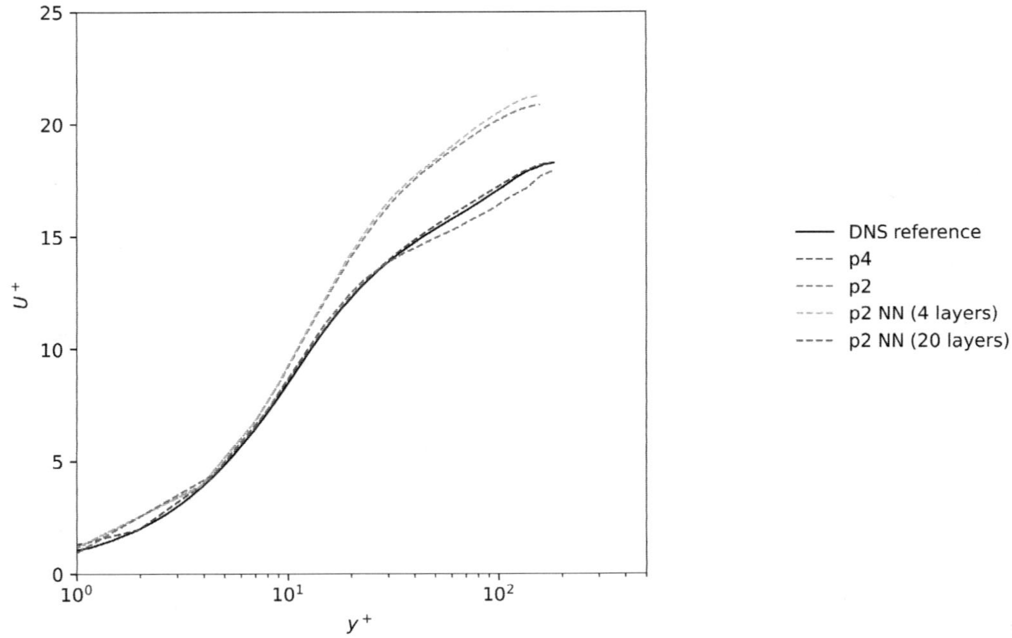

Figure 7: Mean stream-wise velocity profile for high order, low order and corrected low order compared with reference DNS results [19]

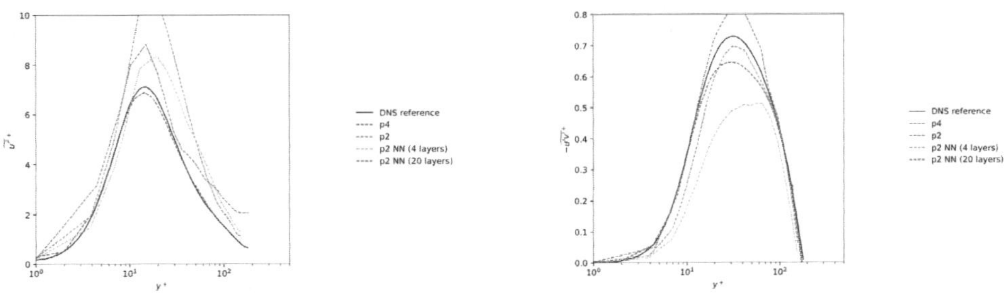

Figure 8: Variance of stream-wise velocity (left) and covariance of stream-wise and wall-normal velocity (right) for high order, low order and corrected low order compared with reference DNS results [19]

agreement is found for the high order solution and the one corrected with 20 neuron layers, with less than 3% error in both cases, while for the low order solution and the corrected using 4 layers, the friction Reynolds number is under-predicted, showing an error of 11%.

Table 2: Theoretical and calculated Reynolds numbers

Theoretical	HO (p4)	LO (p2), uncorrected	LO (p2), 4-layer NN (p2)	LO (p2), 20-layer NN
182	178	163	160	186

To assess the performance of the three neural networks, the s_2 component of the corrective forcing, associated to stream-wise momentum has been plotted along the wall-normal direction. The high order filtered solution has been plotted as a reference, since this is the aim to be achieved by the corrective forcing. Dashed lines divide the regions allocated to each neural network.

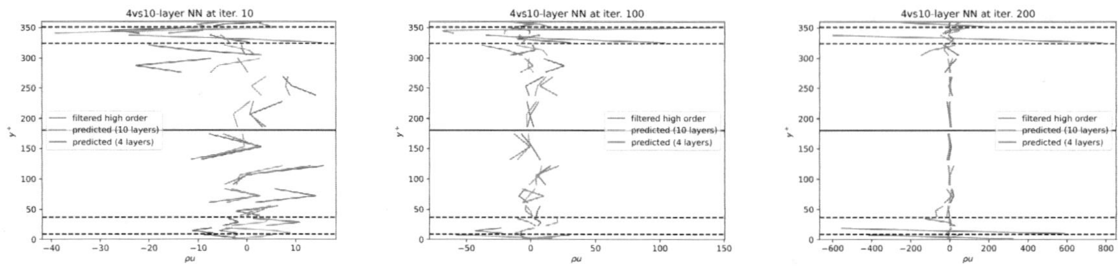

Figure 9: Resolved, filtered ρu and predicted ρu by the 4- and 10-layer NNs along the wall-normal direction, at iterations 10, 100 and 200. Dashed lines divide the regions allocated to each neural network.

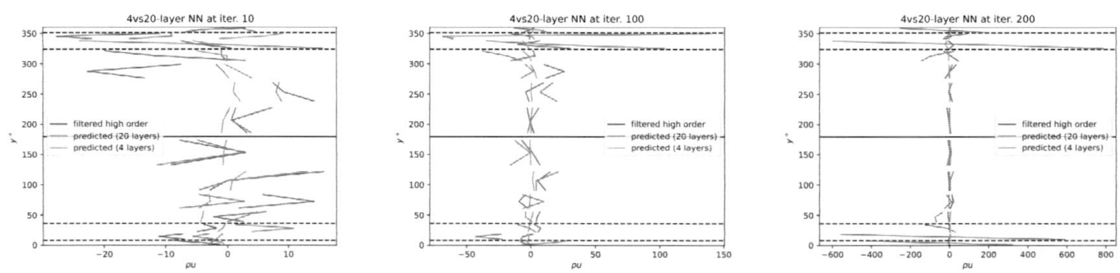

Figure 10: Resolved, filtered ρu and predicted ρu by the 4- and 20-layer NNs along the wall-normal direction, at iterations 10, 100 and 200. Dashed lines divide the regions allocated to each neural network.

While 4-layer NNs were already delivering satisfactory predictions in the center of the channel, results near walls were far from accurate and were being responsible for the divergence of the simulation after training. When using larger layers in the neural network, a substantial improvement near walls is achieved, although the performance at the channel center is not necessarily enhanced. Even though 10-layer neural networks seem to provide promising results, the simulations were diverging after long times. Non-divergence of the simulations is only ensured when using 20 neuron layers. Performance with larger neural networks is to be further investigated.

For completeness, Figure 11 shows the training and validation losses of the different neural networks used. A global loss decrease both during the training and validation proves a correct training process.

4. Conclusions

The application of neural networks (NNs), has shown great potential in accelerating high-order solvers for turbulent flows. NNs provide a flexible framework for learning complex relationships between inputs and outputs, making them suitable for constructing corrective forcings in high-order simulations. By training NNs on high-fidelity data, they can effectively capture the underlying flow physics and provide accurate corrections to low-order simulations.

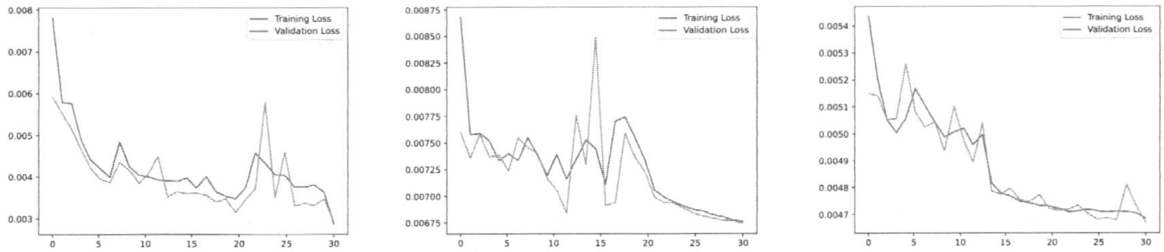

Figure 11: Training and validation losses of each neural network.

The use of NNs, offers several advantages in accelerating high-order solvers for turbulent flows. They can significantly reduce computational costs while enhancing accuracy. Furthermore, our proposed methodology can handle complex flow features, such as transitional flows, turbulence and now wall bounded flows.

Using the mathematical and physical knowledge of the flow provides a good basis to generate a good algorithm for the training of the NN. On the one hand, for temporal evolving flows it is necessary to retrain the network prove to be a good approach. On the other hand, for anisotropic features in space it proves useful to create a dedicated network for different regions.

It is important to note that the successful implementation of NN, in turbulent flow simulations relies on appropriate data generation, network architecture design, and training strategies. Proper validation and verification procedures are necessary to ensure the reliability and accuracy of the results. Continued research and development in this field are expected to further accelerate high order methods for more complex flows.

Acknowledgments
OM and EF would like to thank the support of Agencia Estatal de Investigación (for the grant "Europa Excelencia 2022" Proyecto EUR2022-134041/AEI/10.13039/501100011033) and the Mecanismo de Recuperación y Resiliencia de la Unión Europea. EF knowledges the help of the Comunidad de Madrid and Universidad Politécnica de Madrid for the Young Investigators award: APOYO-JOVENES-21-53NYUB-19-RRX1A0. This research has received funding from the European Commission through the H2020-MSCA-ITN-209 project zEPHYR (grant agreement No 860101) and the European Union (ERC, Off-coustics, project number 101086075). Views and opinions expressed here are those of the author(s) only and do not necessarily reflect those of the European Union or the European Research Council. Neither the European Union nor the granting authority can be held responsible for them. This work was performer in part during the Fifth Madrid Summer Workshop, funded by the European Research Council under the Caust grant ERCAdG-101018287. Finally, all authors gratefully acknowledge the Universidad Politécnica de Madrid (www.upm.es) for providing computing resources on Magerit Supercomputer.

References

[1] Manrique de Lara F and Ferrer E 2022 *Computers & Fluids* **235** 105274
[2] Manrique de Lara F and Ferrer E 2023 *J. Comput. Phys.* **489** 112253
[3] Brunton S L and Kutz J N 2019 *Data-Driven Science and Engineering: Machine Learning, Dynamical Systems, and Control* (Cambridge University Press)
[4] Vinuesa R and Brunton S L 2022 *Nature Computational Science* **2** 358–366
[5] Le Clainche S, Ferrer E, Gibson S, Cross E, Parente A and Vinuesa R 2023 *Aerosp. Sci. Technol.* **138** 108354
[6] Bar-Sinai Y, Hoyer S, Hickey J and Brenner M P 2019 *P. Nat. Acad. Sci. USA* **116** 15344–15349

[7] Kochkov D, Smith J A, Alieva A, Wang Q, Brenner M P and Hoyer S 2021 *P. Nat. Acad. Sci. USA* **118** e2101784118

[8] Guastoni L, Güemes A, Ianiro A, Discetti S, Schlatter P, Azizpour H and Vinuesa R 2021 *J. Fluid Mech.* **928** A27

[9] Balasubramanian A G, Guastoni L, Schlatter P, Azizpour H and Vinuesa R 2023 (*Preprint* ArXiv: 2303.00706)

[10] Güemes A, Discetti S, Ianiro A, Sirmacek B, Azizpour H and Vinuesa R 2021 *Phys. Fluids* **33** 075121

[11] Stachenfeld K, Fielding D B, Kochkov D, Cranmer M, Pfaff T, Godwin J, Cui C, Ho S, Battaglia P and Sanchez-Gonzalez A 2021 Learned coarse models for efficient turbulence simulation (*Preprint* ArXiv:2112.15275)

[12] Beck A and Kurz M 2021 *GAMM-Mitteilungen* **44** e202100002

[13] Duraisamy K 2021 *Phys. Rev. Fluids* **6**(5) 050504

[14] Fabra A, Baiges J and Codina R 2022 *Computer Methods in Applied Mechanics and Engineering* **399** 115280

[15] Dar Z, Baiges J and Codina R 2023 *Computer Methods in Applied Mechanics and Engineering* **415** 116232

[16] Ferrer E, Rubio G, Ntoukas G, Laskowski W, Mariño O, Colombo S, Mateo-Gabín A, Marbona H, Manrique de Lara F, Huergo D, Manzanero J, Rueda-Ramírez A, Kopriva D and Valero E 2023 *Computer Physics Communications* **287** 108700

[17] Taylor G I and Green A E 1937 *Proceedings of the Royal Society of London Series A* **158** 499–521

[18] Brachet M E, Meiron D I, Orszag S A, Nickel B G, Morf R H and Frisch U 1983 *J. Fluid Mech.* **130** 411–452

[19] Lee M and Moser R D 2015 *J. Fluid Mech.* **774** 395–415

Drag reduction in a minimal channel flow with scientific multi-agent reinforcement learning

D Wälchli[1], L Guastoni[2], R Vinuesa[2] and P Koumoutsakos[3]

[1] Computational Science and Engineering Laboratory, ETH Zürich, CH-8092, Switzerland
[2] FLOW, Engineering Mechanics, KTH Royal Institute of Technology, Stockholm, SE-100 44, Sweden
[3] Computational Science and Engineering Laboratory, Harvard University, Cambridge, MA 02138, USA

E-mail: petros@seas.harvard.edu

Abstract. We study drag reduction in a minimal turbulent channel flow using scientific multi-agent reinforcement learning (SMARL). The flow is controlled by blowing and suction at the wall of an open channel, with observable states derived from flow velocities sensed at adjustable heights. We explore the actions, state, and reward space of SMARL using the off-policy algorithm V-RACER. We compare single- and multi-agent setups, and compare the identified control policies against the well-known mechanism of opposition-control. Our findings demonstrate that off-policy SMARL reduces drag in various experimental setups, surpassing classical opposition-control by up to 20 percentage points.

1. Introduction

Turbulent flows are ubiquitous in natural phenomena and engineering applications that hold significant relevance for environmental and economic considerations. The control of such flows is of particular importance. For example, promoting turbulence can enhance mixing and increase efficiency in combustion engines while inhibiting it can reduce drag of airplanes and other vehicles, decreasing the required fuel consumption.

Passive, active and feedback control strategies for turbulent flows [1, 2, 3, 4] have been proposed over the years. In several cases, control strategies are first tested in numerical simulations before they are translated to experimental and real-world setups. Two challenges hinder this endeavor. First, turbulent flow simulations can be very costly prohibiting extensive iterations. Most importantly, control strategies developed in simulations in idealized configurations may not be translatable to actual experimental set ups and real world settings. We argue for the discovery of flow-control mechanisms through reinforcement learning that may exploit underlying patterns in the flow which

Journal of Physics: Conference Series **2753** (2024) 012024 doi:10.1088/1742-6596/2753/1/012024

can be translated across settings. Moreover, flow-control mechanism discovery is a process that can be translated from simulations to experimental setups [5, 6, 7, 8].

Reinforcement learning (RL) has increasingly found applications in designing models and learning control policies in fluid mechanics [9]. It was introduced a decade ago for flow-control problems [10] and more recently it has been deployed in several fluid-dynamics problems [11] and drag-reduction applications [12, 13, 14, 15]. Other applications include fish bio-locomotion [16], optimization of aerial/aquatic vehicles' path and motion [17, 18, 19], active flow-control for bluff bodies [20, 21, 22], shape optimization [23], and learning closures [11] and wall models [24] for turbulent flows.

This paper examines multi-agent RL for flow-control in a minimal open channel [13]. We study scientific single agent and multi-agent reinforcement learning (SMARL) and explore different state configurations. The minimal channel offers computational efficiency over the full channel simulations allowing to explore the effects of various control strategies and RL settings. The flow environment is based on the numerical solver `SIMSON` [25], which employs an efficient pseudo-spectral method for solving the Navier–Stokes equations in incompressible wall-bounded flows. The solver carries out direct numerical simulations (DNSs), resolving all time and length scales. Deploying control strategies in DNS can be expensive but has the benefit that the control strategy that does not depend on the particular limitations introduced by closures and reduced-order models. The proposed methodologies are compared with classical formulations on feedback control of turbulent flows, such as the opposition-control [26] and it is shown that it performs favorably. The paper is organized as follows: In section 2, we introduce the concepts of SMARL. In section 3 we explain the flow simulation's physics and highlight some details about the solver. In section 4 we describe the experimental setup, and in section 5, we present our results. In section 6 we summarize the present work.

2. Scientific Multi-Agent Reinforcement Learning (SMARL)

RL provides a computational approach for tackling control problems, where an agent actively interacts with its environment in pursuit of a specific goal [27]. Within RL, the agent executes actions that impact its environment, and its performance is assessed through scalar reward functions. Through the accumulation of experiences, the agent learns a policy $\pi(a \mid s)$, mapping states (s) to probability distributions and enabling it to sequentially sample actions (a) that maximize long-term, cumulative rewards selectively. Notably, RL has recently extended its application domains into the field of flow-control, where trained agents interact with the flow, acquiring pertinent information and optimizing decision-making processes to collectively swim [16], minimize drag [13], or reach a designated target location [19].

A crucial element for the success of RL is the representation of the policy function. Deep RL algorithms train neural networks to represent the policy $\pi_{\boldsymbol{\vartheta}}(a \mid s)$ with parameters $(\boldsymbol{\vartheta})$, eliminating the requirement for tabular or expertly designed state representations. In this work, a policy network is employed to sample the boundary conditions of the minimal channel to mimic blowing and suction. The choice of states and actions is critical. For example, the agent's state could be defined as the complete flow field at a specific time step and the action as the boundary condition for all grid points

Journal of Physics: Conference Series **2753** (2024) 012024 doi:10.1088/1742-6596/2753/1/012024

at the wall. With such an architecture, both the state and action spaces' dimensionality would scale with the simulation's degrees of freedom, and learning a policy can be difficult. Consequently, the control policy would be contingent on mesh size, requiring extensive neural network training and prohibitively large memory for storing the agent's experiences. To address these challenges, we employ multi-agent reinforcement learning and compare it to single agent RL.

In SMARL, a set of N_{agents} agents is distributed across a plane in the simulation domain. Each agent executes a localized action based on the local flow field. Enhancing the number of agents contributes to the adaptability of SMARL to localized flow patterns. However, the actions taken by other agents introduce complexities that can amplify uncertainties. For instance, if one agent's action increases drag, all agents receive negative feedback, irrespective of their own decisions. To mitigate these challenges, we employ policy optimization through the Remember and Forget Experience Replay for SMARL (Ref-ER SMARL) algorithm [28].

ReF-ER SMARL is an off-policy algorithm, *i.e.*, it leverages experience replay (ER) to utilize experiences across multiple policy iterations, enhancing data usage. Additionally, ReF-ER SMARL demonstrates inherent stability and has been shown to match or even exceed the performance of state-of-the-art RL algorithms and optimal-control methods on various single agent and multi-agent benchmark problems. Crucially for SMARL, ReF-ER explicitly governs the pace of policy changes.

In this study, SMARL seeks to find the parameters (denoted as $\boldsymbol{\vartheta}$) of the stochastic policy $\pi_{\boldsymbol{\vartheta}}$ that maximize the expected sum of rewards $r_t^{(i)}$ across each simulation $J(\boldsymbol{\vartheta}) = 1/N_{\text{agents}}\mathbb{E}_{\pi_{\boldsymbol{\vartheta}}}\left[\sum_{t=0}^{T}\sum_{i=1}^{N_{\text{agents}}} r_t^{(i)}\right]$ of a specified time horizon T. Notably, these parameters are shared among all agents, adhering to the centralized training and decentralized execution (CTDE) framework [29]. The collective experiences of agents are pooled into a shared experience replay, and updates are computed based on the ReF-ER MARL algorithm. For a detailed explanation of ReF-ER MARL, please refer to Appendix A.

3. Channel Flow

The velocity $\boldsymbol{u} : \Omega \times [0,T] \to \mathbb{R}^3$ and pressure $p : \Omega \times [0,T] \to \mathbb{R}$ of the fluid with kinematic viscosity $\nu \in \mathbb{R}$ and constant density $\rho \in \mathbb{R}$ is described by the Navier–Stokes equations, here written in convective form:

$$\begin{aligned} \frac{\partial \boldsymbol{u}}{\partial t} + (\boldsymbol{u} \cdot \nabla)\boldsymbol{u} &= -\frac{\nabla p}{\rho} + \nu \Delta \boldsymbol{u} + \boldsymbol{F}\,, \\ \nabla \cdot \boldsymbol{u} &= 0\,. \end{aligned} \tag{1}$$

The forcing term $\boldsymbol{F}$ is employed to keep the flow from decelerating due to the upper and lower plate that impose a zero boundary condition in both wall-parallel directions. A constant mass-flow rate determines the value of $\boldsymbol{F}$. The friction Reynolds number of a turbulent channel is

$$Re_\tau = u_\tau h/\nu\,, \tag{2}$$

where $u_\tau = \sqrt{\tau_w/\rho}$ denotes the friction velocity with wall-shear stress τ_w. The size of the solution domain in physical space is $\Omega = x_L \times y_L \times z_L$ in the streamwise, wall-normal, and span-wise directions respectively. For channel flow we set $y_L = 2h$ with h being the channel half-width. The velocity components are denoted by $\boldsymbol{u} = (u, v, w)$. The boundary conditions in the horizontal directions are periodic. The wall-normal boundary conditions however are inhomogenous and vary over time due to actuation (wall blowinging and suction) from the RL.

To advance the Navier–Stokes equation in time, a second-order implicit Crank–Nicholson scheme (CN) is being used for the linear terms, and an explicitly third order three stage Runge–Kutta (RK3) scheme. The discretization in horizontal direction (x- and z- directions) uses a Fourier-series expansion which assumes that the solution is periodic. The wall-normal discretization (y-direction) uses a Chebyshev expansion. A more in-depth description on the numerics can be found in Ref. [25].

4. Reinforcement Learning in Minimal Channel

The simulation rely on the interface presented by Guastoni *et al.* [13]. We consider $n_x = n_z = \{1, 2, 4, 8\}$ agents per dimension arranged equidistantly on the horizontal grid points. The agents interact with the channel flow at discrete times $t = 0, \ldots, T$ by imposing blowing and suction at the wall ($y^+ = 0$). The superscript denotes the inner-scaled wall-normal coordinates $y^+ = yu_\tau/\nu$. We study two configurations, the agents sense the velocity field u and v either at the first grid points above the wall ($y_s^+ = 0.1$) or at height $y_s^+ = 15$. When applying a control value at grid point (x_k, z_l) and time t_i, we subtract the mean of all actuations to guarantee zero-net-mass-flux, such that the actuation does not add or subtract fluid mass from the flow. Then the wall-normal boundary condition at said location changes linearly from $a_{t_{i-1}}^{(k,l)}$ to $a_{t_i}^{(k,l)}$ over the next simulation period $t \in [t_i, t_{i+1})$.

The reward of an agent is the negative friction at the wall averaged over the nearest grid points of the agent. More specifically, for the single agent case it is the average friction over all grid points, and for the 64-agent case ($n_x = n_z = 8$) it is the average friction over the nearest 4 adjacent (2×2) grid points. The friction is assumed to be proportional to the downstream flow velocity u at the first grid point above the wall ($y^+ = 0.1$). During training, the actions are sampled from a clipped Gaussian distribution $\pi_{\boldsymbol{\vartheta}}(a|s)$ that is fully determined by a mean $\mu_{\boldsymbol{\vartheta}}(s)$ and a standard deviation $\sigma_{\boldsymbol{\vartheta}}(s)$ [30]. The clipped Gaussian distribution allows to define an upper and lower bound for the actions. In Ref-ER SMARL, the mean and standard deviation of the optimal policy are learned together with the value function $V_{\boldsymbol{\vartheta}}(s)$ using a single neural network $\mathrm{NN}(s; \boldsymbol{\vartheta})$, with parameters $\boldsymbol{\vartheta}$ [31, 28]. Finding the optimal policy $\pi^\star$ means maximizing the expected cumulative reward:

$$V^\pi(\boldsymbol{\vartheta}) = \mathbb{E}_{a^{(t)} \sim \pi(\cdot|s^{(t)}; \boldsymbol{\vartheta})} \left[\sum_{t=0}^{T} \gamma^t r^{(t)} \right] , \tag{3}$$

and thus reducing the accumulated drag over the period $t \in [0, T)$, whereas the initial state is sampled from some distribution $s^{(0)} \sim p(\cdot)$, the actions are sampled from the learned policy $a^{(t)} \sim \pi(\cdot|s^{(t)}; \boldsymbol{\vartheta}))$, and the system follows the dynamics of the Navier–Stokes equations $s^{(t+1)}, r^{(t)} = D(s^{(t)}, a^{(t)})$. The discount factor $\gamma \in [0, 1)$ allows weighting short-term rewards more than long-term rewards.

Journal of Physics: Conference Series **2753** (2024) 012024 doi:10.1088/1742-6596/2753/1/012024

The size of the minimal channel is $\Omega = 2.67h \times h \times 0.8h$. The minimal channel is large enough to adequately simulate wall turbulence while reducing computational costs [32]. We discretize the domain with $N_x = N_z = 16$ and $N_y = 65$ grid points. In order to reduce the state and action space for RL, we compress the field (u and v) by a factor of 4 by averaging the velocity field over 4 adjacent (2×2) grid points, and we apply the same control to the four adjacent grid points respectively. This helped learning the policy for the single agent experiment. Using a compressed state yields a state and action space of 128 dimensions for the single agent case and a state/action space of 2 dimensions for the 64-agent case. The friction Reynolds number is set to $Re_\tau = 180$. The time discretization of the computational fluid dynamics (CFD) solver is $\delta t^+ \approx 0.03$ and the interactions with the environment happen every $\Delta t^+ = 0.6$ time steps. The simulation length is $T = 600$ equivalent to 1000 interactions per episode. The inner-scaled wall-normal coordinates are given by $y^+ = yu_\tau/\nu$ and the inner-scaled time units are $t^+ = tu_\tau^2/\nu$.

We compare our results with the classical opposition-control strategy [26]. Opposition-control serves as a straightforward and widely acknowledged control benchmark, offering a well-established theoretical understanding. It can be utilized as a standard for evaluating RL control laws. The wall-normal actuation, denoted as v_{wall} is expressed as follows:

$$v_{wall}(x, z, t) = -v(x, y_s, z, t) \tag{4}$$

where y_s denotes the sensing height. As in Ref. [13], we set $y_s^+ = 15$ for opposition-control and we clip the actuations of the RL agents to the maximal absolute velocities observed in opposition-control $v^+_{\text{wall}} = \pm 0.0429$ for a fair comparison.

We coupled `korali` with the `SIMSON` solver through the *Python* interface provided by Ref. [13]. `korali` is a framework for optimization, uncertainty quantification and (inverse) reinforcement learning, and it implements the Ref-ER SMARL algorithm. The framework targets execution on super-computers and offers built-in work distribution via the *message-passing interface (MPI)* for a parallel execution of the channel environments in order speed up experience collection and training. In fig. 1 we show a schematic of the setup.

5. Results

In our study, a policy is derived for a single agent acting on all grid points at the wall and for localized policies using 4, 16, and 64 agents that cover only a fraction of the grid points. Additionally, we compare two configurations in which the agents sense the flow field (u and v) close to the wall $y_s^+ = 0.1$ and at height $y_s^+ = 15$, similar to the opposition-control strategy. The comparison is depicted in fig. 2. We train the policies using Ref-ER SMARL, whereby each simulation of the environment generates 1000 new experiences that are stored in the experience replay memory. We stop the training when accumulating 10^6 experiences and then we evaluate the learned policies on five simulations and compare the drag reduction. The drag reduction R is calculated from an uncontrolled case $R = 1 - u_\tau/u_{\tau,\text{uncontrolled}}$. We document the configuration of the Ref-ER SMARL algorithm in Appendix B.

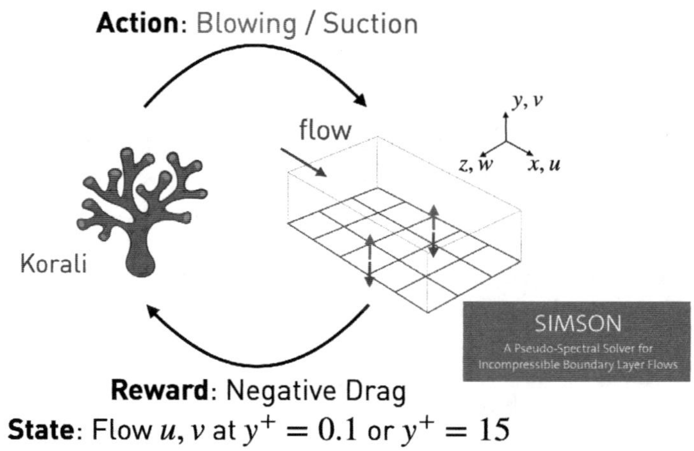

Figure 1: Schematic of the experimental setup. We couple the RL framework `korali` with the CFD solver `SIMSON`.

We evaluate the (stochastic) policies on five simulations of length $T = 600$ starting from 5 different initial flow fields, and we average the results. For reference, we include the drag reduction obtained with the opposition-control strategy. We observe that in both setups, the localized policy using 64 agents obtains the most significant drag reduction of almost 40%. The obtained drag reduction in the open channel is in agreement with the multi-agent results from Guastoni *et al.* [13] using 256 agents, where the sensing plane is also set at $y_s^+ = 15$. The result is a two-fold improvement compared to the opposition-control strategy with a drag reduction of almost 20%. The single agent policy performs subpar when sensing the flow near the wall. When sensing the flow at $y_s^+ = 15$, the single and multi-agent policies obtain comparable drag reduction. Interestingly, the control policies initially increase the drag up to $t^+ = 300$ and only in the second half of the episode obtain notable drag reduction.

During evaluation, we found that the stochastic policies obtain a more significant drag reduction than the deterministic policies. The deterministic policy is obtained when setting the control equal to the mean of the clipped Gaussian distribution $a = \mu_\vartheta(s)$ and ignoring the stochasticity of the policy. In fig. 3, we plot the drag reduction for the stochastic and deterministic policy for 64 agents with state sensed near the wall. The deterministic approach obtains almost no improvement in drag reduction, and performs subpar than the opposition-control strategy and exhibits increased variance.

In fig. 4, we plot the actuation on the wall at four different locations for opposition-control, the stochastic policy, and the deterministic policy for 64 agents. The actuation points are chosen on the diagonal of the grid at equidistant locations. The stochastic policy resembles the two-step control found in Ref. [13], whereas the deterministic policy is similar to the opposition-control regarding actuation amplitudes and variance. Notably, the deterministic policy introduces almost no actuation during the final third of the simulation.

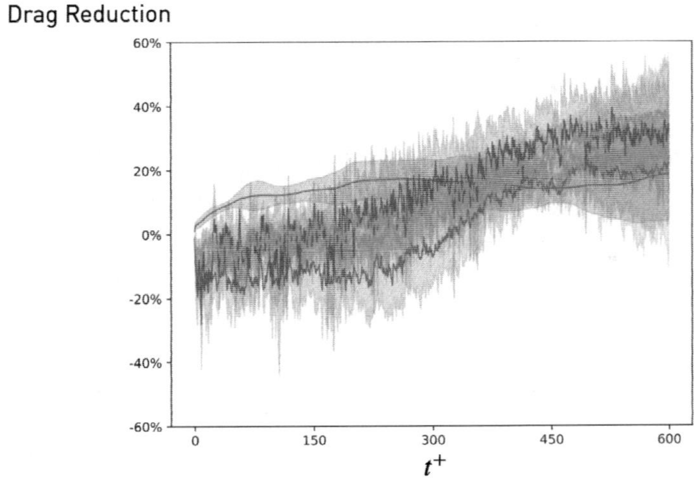

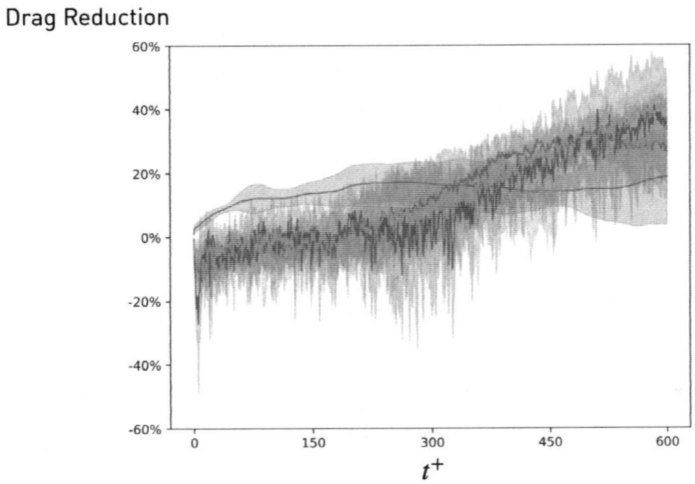

Figure 2: Comparison of drag reduction of RL policies with states sensed at different heights: $y_s^+ = 0.1$ (top) and $y_s^+ = 15$ (bottom). The colors indicate the instantaneous drag reduction for different numbers of agents: 1 agent (—), 4 agents (—), 16 agents (—), 64 agents (—). The red line (—) shows the drag reduction with opposition-control as a reference. The lines are the averages over five runs starting from different initial flow fields, and the shading indicates the confidence interval within one standard deviation of the mean.

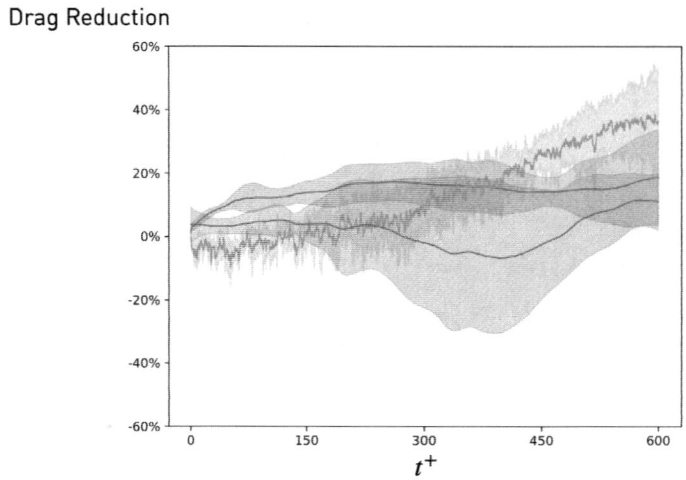

Figure 3: Drag reduction of stochastic (—) and deterministic (—) policy for 64 agents with state sensed at $y_s^+ = 0.1$. The red line (—) shows the drag reduction with opposition-control. The lines are the averages over five runs starting from different initial flow fields, and the shading indicates the confidence interval within one standard deviation of the mean.

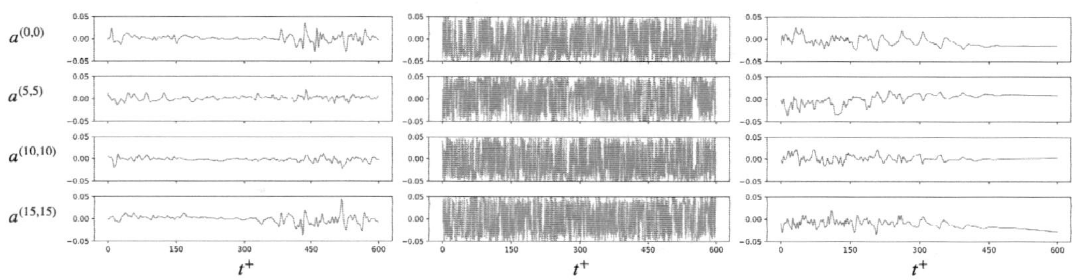

Figure 4: Control policies at four different locations on the wall for opposition-control (—), stochastic policy (—) and deterministic policy (—) with states sensed at $y_s^+ = 0.1$.

6. Discussion and Conclusions

This paper demonstrates the application of a SMARL framework to control the drag in a minimal channel flow [13]. Agents, located at the wall, use blowing and suction in order to minimize the flow's drag. We explore drag reduction with varying agent numbers and two different sensing heights for the state. The Ref-ER SMARL algorithm is utilized to optimize control policies, revealing that these policies result in improved drag reduction (up to 40%) compared to opposition-control (20%). This reduction is consistent with the findings of Guastoni *et al.* [13], where drag reduction in an open channel was studied using the RL algorithm DDPG [33] while sensing the flow at $y_s^+ = 15$.

Table A1: Hyperparameters for the Ref-ER MARL algorithm. For undocumented parameters we use the default ones presented in Refs [31, 28].

Parameter	Value
Neural Network Width	128
Neural Network Depth	2
Activation Function	SoftReLU
Batch Size	$512/N_{\text{agents}}$
Adam Learning Rate η	0.001
Updates Per Experience	1
Experience Replay Start Size	8192
Experience Replay Maximum Size	524288
Target Fraction Off-Policy Experiences $n^\star$	0.1
Discount Factor γ	0.995

When comparing policies trained with state sensing at the wall with those sensing the state at $y_s^+ = 15$, no significant improvements in drag reduction are observed. Similarly, varying the number of agents has minimal impact on evaluations. Future investigations are needed to explore the potential of localized policies with multiple agents for better generalization to larger channels. Finding a policy to reduce drag based on the flow field near the wall opens implementation possibilities for drag-reducing surfaces through active control using sensors attached to the object of interest.

Our findings indicate that stochastic policies outperform deterministic counterparts, suggesting that a jittering control mechanism can reduce drag in a channel. The stochastic policy resembles the two-step control found in Ref. [13], and further exploration is required to understand the impact of such control patterns and why they positively affect drag [34].

Acknowledgments
We are grateful to Professor Javier Jiménez and his team for hosting the Fifth Madrid Turbulence Workshop 2023. We would also like to thank C Lopez, O Arias, B Font, F Álcantara-Ávila, and M Lee for providing their support and invaluable discussions during the workshop. We thank A Lonzano-Durán for carefully reviewing the manuscript. This work was performer in part during the Fifth Madrid Summer Workshop, funded by the European Research Council under the Caust grant ERCAdG-101018287. RV and LG acknowledge financial support from the ERC Grant No. 2021-CoG-101043998 (DEEPCONTROL). DW was supported by the Swiss National Supercomputing Centre (CSCS) (project number s1160).

Appendix A. Methods: Remember and Forget Experience Replay for Multi-Agent Reinforcement Learning

Here we present technical description of the Ref-ER MARL algorithm. Multi-Agent Reinforcement Learning (MARL) aims to find the optimal policy in a Multi-Agent

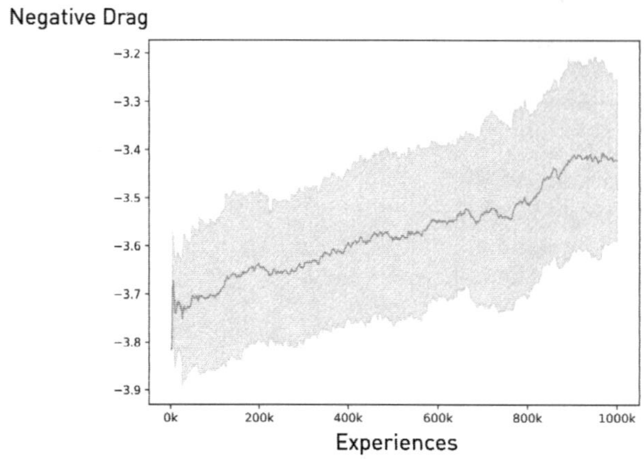

Figure A1: Exemplary training progress for the multi-agent setup with 64 agents and state sensed at $y_s^+ = 0.1$. The figure shows the cumulative reward of the episodes during training up to 10^6 experiences (equivalent to 1000 flow simulations). The line indicates the running mean and the shading includes the running standard deviation calculated from the past 100 cumulative rewards of the episodes.

Markov Decision Process (MAMDP). A MAMDP is a tuple $(\mathcal{S}, \mathcal{A}, \boldsymbol{r}, N, D)$ consisting of the states $\boldsymbol{s} \in \mathcal{S}^N$, actions $\boldsymbol{a} \in \mathcal{A}^N$, and rewards $\boldsymbol{r} \in \mathbb{R}^N$ observed by the $N \in \mathbb{N}_+$ agents in the environment. Here we consider a homogeneous setting, where all agents have the same state space, action space, and reward function. The agents' individual states, actions, and rewards are denoted by $s^{(i)} \in \mathcal{S}$, $a^{(i)} \in \mathcal{A}$, and $r^{(i)} \in \mathbb{R}$ for $i = 1, \dots, N$. The transition map is given by $\boldsymbol{r}, \boldsymbol{s}' = D(\boldsymbol{s}, \boldsymbol{a})$, *i.e.* it depends on the states and actions from all agents. This is contrary to environments in which agents take actions sequentially and the state is updated after each action. In MAMDP we specify the initial state distribution $p(\boldsymbol{s})$ and the stochastic policy $\pi(\boldsymbol{a}|\boldsymbol{s})$. The later is the probability distribution over the actions of the agents given the states. We assume that the policy factorizes as:

$$\pi(\boldsymbol{a}|\boldsymbol{s}) = \prod_{i=1}^{N} \pi(a^{(i)}|s^{(i)}) \,. \tag{A.1}$$

This allows for *decentralized execution*, where the agent samples an action solely based on its state. The vector-valued state-action-value $\boldsymbol{Q}$ function and state-value $\boldsymbol{V}$ are defined as:

$$\boldsymbol{Q}^{\pi}(\boldsymbol{s}, \boldsymbol{a}) = \mathbb{E}_{\pi}\left[\sum_{t=0}^{\infty} \gamma^t \boldsymbol{r}_t | \boldsymbol{s}_0 = \boldsymbol{s}, \boldsymbol{a}_0 = \boldsymbol{a}\right] , \quad \boldsymbol{V}^{\pi}(\boldsymbol{s}) = \mathbb{E}_{\pi}\left[\boldsymbol{Q}^{\pi}(\boldsymbol{s}, \boldsymbol{a})\right] . \tag{A.2}$$

where $\gamma \in [0, 1)$ is the discount factor and $\boldsymbol{r}_t = \boldsymbol{r}(\boldsymbol{s}_t, \boldsymbol{a}_t)$. Assuming equal importance for all agents yields the scalar state-value that is expressed in terms of the individual state value functions $V^\pi(s^{(i)})$:

$$P\left[\boldsymbol{V}^\pi(\boldsymbol{s})\right] = \frac{1}{N}\sum_{i=1}^{N} V^\pi(s^{(i)})\,. \tag{A.3}$$

The optimal policy $\pi^\star$ in the MAMDP maximizes the average of the state-values for all agents:

$$\pi^\star = \arg\max_\pi \frac{1}{N}\sum_{i=1}^{N} V^\pi(s^{(i)})\,,\ \forall \boldsymbol{s}\in\mathcal{S}^N\,. \tag{A.4}$$

In MARL the agents interact with the environment as depicted in fig. A2. At timestep t, the agents take actions $\boldsymbol{a}_t$ based on the states $\boldsymbol{s}_t$. Thereby the environment transitions into new states $\boldsymbol{s}_{t+1}$ and returns rewards $\boldsymbol{r}_t$. An *experience*, consists of states, actions, and rewards $(\boldsymbol{s}_t, \boldsymbol{a}_t, \boldsymbol{r}_t)$. Upon termination of the episode $\mathcal{E}_k = \{(\boldsymbol{s}_t, \boldsymbol{a}_t, \boldsymbol{r}_t)\}_{t=1}^{T_k}$ the experiences are stored in the replay memory $\mathcal{RM} = \{\mathcal{E}_k\}_{k=1}^{K}$. In the following d denotes the distribution of the experiences in the replay memory.

We extend V-RACER with ReF-ER [31] for MARL, training a neural-network that approximates the state-value and the policy parameters. Using the experiences in the replay memory, we train the neural-network with Adam [35]. In the following, $\pi_{\boldsymbol{\omega}}(a|s)$ and $V^\pi_{\boldsymbol{\vartheta}}(s)$ denote the policy and state-value function with the respective weights $\boldsymbol{\omega}$ and $\boldsymbol{\vartheta}$. Note, that the parameters $\boldsymbol{\omega}$ and $\boldsymbol{\vartheta}$ only differ in the output layer.

Appendix A.1. Value Learning

The relationships between the agents are included in the learning process by introducing a scalarization function $f:\mathbb{R}^N\to\mathbb{R}$ in the on-policy returns estimator Retrace [36]:

$$\hat{Q}^{\text{ret}}_{t,f} = f(\boldsymbol{r}_t) + \gamma V^\pi_f(\boldsymbol{s}_t) + \gamma\bar{\rho}_t(\hat{Q}^{\text{ret}}_{t+1,f} - Q^\pi_f(\boldsymbol{s}_t, \boldsymbol{a}_t))\,. \tag{A.5}$$

V-RACER approximates $Q^\pi_f(\boldsymbol{s}_t, \boldsymbol{a}_t) \approx V^\pi_f(\boldsymbol{s}_t)$, and hence the V-trace estimator [36] becomes:

$$\hat{V}^{\text{tbc}}_{t,f} = V^\pi_f(\boldsymbol{s}_t) + \bar{\rho}_t\left[f(\boldsymbol{r})_t + \gamma\hat{V}^{\text{tbc}}_{t+1,f} - V^\pi_f(\boldsymbol{s}_t)\right]\,, \tag{A.6}$$

and it can be related to the on-policy returns estimator Retrace via:

$$\hat{Q}^{\text{ret}}_{t,f} = f(\boldsymbol{r})_t + \gamma\hat{V}^{\text{tbc}}_{t+1,f}\,. \tag{A.7}$$

The truncated importance weight $\bar{\rho}_t$ will be defined in the following section Appendix A.2. It balances the impact from off-policy data on the current estimator of the state-value.

For the function f we distinguish two cases. The first case, which we refer to as *individual*, assumes that the value estimate depends solely on the reward observed by agent i:

$$f^{(i)}_{\text{individual}}(\boldsymbol{r}) = r^{(i)}\,,\quad V^\pi_{\text{individual},(i)}(\boldsymbol{s}) = V^\pi(s^{(i)})\,. \tag{A.8}$$

Journal of Physics: Conference Series **2753** (2024) 012024 doi:10.1088/1742-6596/2753/1/012024

In the second case, referred to as *cooperative*, the reward and the state-values are averaged:

$$f_{\text{cooporative}}(\boldsymbol{r}) = \frac{1}{N}\sum_{i=1}^{N} r^{(i)}, \quad V^{\pi}_{\text{cooporative}}(\boldsymbol{s}) = \frac{1}{N}\sum_{i=1}^{N} V^{\pi}(s^{(i)}). \tag{A.9}$$

The weights $\boldsymbol{\vartheta}$ of the neural-network $V^{\pi}_{\boldsymbol{\vartheta}}(s)$ are updated to minimize the loss:

$$\mathcal{L}(\boldsymbol{\vartheta}) = \mathbb{E}_d\left[\frac{1}{N}\sum_{i=1}^{N}\left(V^{\pi}_{\boldsymbol{\vartheta}}(s_t^{(i)}) - \hat{V}^{\text{tbc}}_{t,f}\right)^2\right]. \tag{A.10}$$

Note that both variants reduce to the original loss proposed in Ref. [31] when assuming a single agent.

Appendix A.2. Policy Gradient

Given the definition of $\hat{Q}^{\text{ret}}_{t,f}$, we learn the weights $\boldsymbol{\omega}$ of the policy $\pi_{\boldsymbol{\omega}}(s|a)$ by maximizing the expected advantage:

$$\mathcal{L}(\boldsymbol{\omega}) = \mathbb{E}_d\left[\frac{1}{N}\sum_{i=1}^{N}\rho_t(\boldsymbol{\omega})\left(\hat{Q}^{\text{ret}}_{t,f} - V^{\pi}(s_t^{(i)})\right)\right]. \tag{A.11}$$

Taking the gradient with respect to the weights $\boldsymbol{\omega}$ yields the off-policy gradient [37]. The importance weights $\rho_t(\boldsymbol{\omega})$ reflect the assumed dynamics (*full* or *local*) of the environment.

Adopting the *full dynamics* $\boldsymbol{r}^{t+1}, \boldsymbol{s}^{t+1} = D(\boldsymbol{s}^t, \boldsymbol{a}^t)$ and the factorization of the policy from eq. (A.1) yields the (full) importance weight:

$$\rho_t^N(\boldsymbol{\omega}) = \prod_{i=1}^{N}\frac{\pi_{\boldsymbol{\omega}}(a_t^{(i)}|s_t^{(i)})}{\pi(a_t^{(i)}|s_t^{(i)})}. \tag{A.12}$$

Here the denominator denotes the policy that was used when $s^{(i)}$ was sampled. When using this importance weight the value estimate and the policy update depend on the probability of the action of all agents.

In the *local dynamics model* we assume that interactions among the agents are negligible, the value estimate and the policy update should not be influenced by the probabilities of the other agents. This is achieved by using the (local) importance weight:

$$\rho_t(\boldsymbol{\omega}) = \frac{\pi_{\boldsymbol{\omega}}(a_t^{(i)}|s_t^{(i)})}{\pi(a_t^{(i)}|s_t^{(i)})}. \tag{A.13}$$

The truncated importance weight is given by $\bar{\rho}_t = \min(\rho_t, 1)$, respectively $\bar{\rho}_t = \min(\rho_t^N, 1)$.

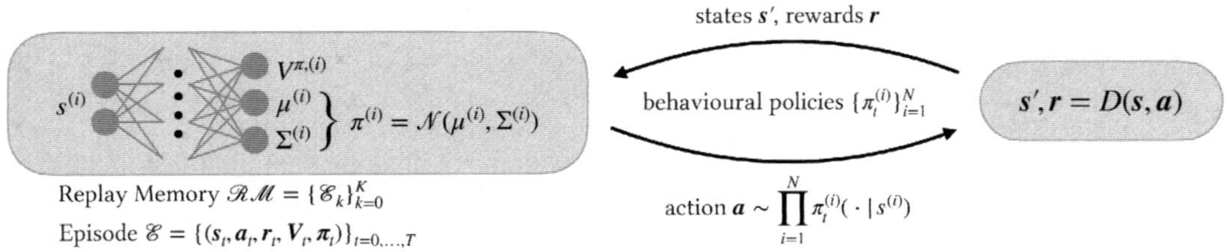

Figure A2: Schematic of the MARL loop. We employ one neural network to learn the value function V, the means μ, and the diagonal entries of the covariance matrix Σ.

Appendix A.3. Remember and Forget for Experience Replay

In order to update the policy, a mini-batch of experiences is sampled from the replay memory. For each of these experiences a gradient $\hat{\boldsymbol{g}}(\boldsymbol{\omega})$ is computed. ReF-ER avoids negative impact from off-policy data by classifying experiences based on the importance weight ρ. The criterion $1/c_{\max} < \rho < c_{\max}$ depends on the cut-off $c_{\max}$ that is annealed during training. The gradient is then regularized via:

$$\hat{\boldsymbol{g}}^{\mathrm{R}}(\boldsymbol{\omega}) = \begin{cases} \beta\hat{\boldsymbol{g}}(\boldsymbol{\omega}) - (1-\beta)\hat{\boldsymbol{g}}^{\mathrm{KL}}(\boldsymbol{\omega}), & \text{on-policy} \\ -(1-\beta)\hat{\boldsymbol{g}}^{\mathrm{KL}}(\boldsymbol{\omega}), & \text{else.} \end{cases} \tag{A.14}$$

The factor β is adjusted according to the fraction of off-policy samples $n_{\text{off-policy}}$ in the replay memory:

$$\beta \leftarrow \begin{cases} (1-\eta)\beta & \text{if } n_{\text{off-policy}} > n^\star \\ (1-\eta)\beta + \eta, & \text{otherwise.} \end{cases} \tag{A.15}$$

Here η denotes the learning rate used by Adam. This allows maintaining a target fraction $n^\star$ of off-policy experiences in the replay memory. The regularizer is the gradient $\nabla_{\boldsymbol{\omega}}$ of the Kullback–Leibler divergence between the current and the past policy [38]:

$$\hat{\boldsymbol{g}}^{\mathrm{KL}}(\boldsymbol{\omega}) = \nabla_{\boldsymbol{\omega}} D_{\mathrm{KL}}\left(\pi_k \| \pi_{\boldsymbol{\omega}}\right) . \tag{A.16}$$

For details we refer to the publication of the ReF-ER algorithm [31].

Appendix B. Training Details

In table A1 we present the hyperparameters used for the Ref-ER MARL algorithm that were used. Finally, in fig. A1 we show the training progress for the multi-agent setup with 64 agents and state sensed at $y_s^+ = 0.1$.

References

[1] Bechert D W and Bartenwerfer M 1989 *Journal of Fluid Mechanics* **206** 105–129
[2] Gad-el Hak M 2000 *Flow Control: Passive, Active, and Reactive Flow Management* (Cambridge University Press)
[3] Choi H, Jeon W P and Kim J 2008 *Annual Review of Fluid Mechanics* **40** 113–139
[4] Quadrio M 2011 *Philos. T. R. Soc. A* **369** 1428–1442

[5] Brunton S L and Noack B R 2015 *Applied Mechanics Reviews* **67** 050801
[6] Li R, Noack B R, Cordier L, Borée J and Harambat F 2017 *Exp. Fluids* **58** 103
[7] Minelli G, Dong T, Noack B R and Krajnović S 2020 *Journal of Fluid Mechanics* **893** A1
[8] Vinuesa R, Lehmkuhl O, Lozano-Durán A and Rabault J 2022 *Fluids* **7** 62
[9] Vignon C, Rabault J and Vinuesa R 2023 *Physics of Fluids* **35** 031301
[10] Gazzola M, Hejazialhosseini B and Koumoutsakos P 2014 *SIAM J. Sci. Compu.* **36** B622–B639
[11] Novati G, de Laroussilhe H L and Koumoutsakos P 2021 *Nature Machine Intelligence* **3** 87–96
[12] Sonoda T, Liu Z, Itoh T and Hasegawa Y 2023 *Journal of Fluid Mechanics* **960** A30
[13] Guastoni L, Rabault J, Schlatter P, Azizpour H and Vinuesa R 2023 *Eur. Phys. J. E* **46** 27
[14] Wang Z, Fan D, Jiang X, Triantafyllou M S and Karniadakis G E 2023 *J. Fluid Mech.* **973** A32
[15] Vignon C, Rabault J, Vasanth J, Alcántara-Ávila F, Mortensen M and Vinuesa R 2023 *Physics of Fluids* **35** 065146
[16] Verma S, Novati G and Koumoutsakos P 2018 *P. Natl. Acad. Sci. USA* **115** 5849–5854
[17] Reddy G, Celani A, Sejnowski T J and Vergassola M 2016 *P. Natl. Acad. Sci. USA* **113** E4877–E4884
[18] Colabrese S, Gustavsson K, Celani A and Biferale L 2017 *Phys. Rev. Lett.* **118** 158004
[19] Novati G, Mahadevan L and Koumoutsakos P 2019 *Phys. Rev. Fluids* **4** 093902
[20] Vandiver J K, Ma L and Rao Z 2018 *Journal of Sound and Vibration* **433** 29–54
[21] Rabault J, Ren F, Zhang W, Tang H and Xu H 2020 *Journal of Hydrodynamics* **32** 234–246
[22] Ren F, Rabault J and Tang H 2021 *Phys. Fluids* **33** 037121
[23] Viquerat J, Rabault J, Kuhnle A, Ghraieb H, Larcher A and Hachem E 2021 *Journal of Computational Physics* **428** 110080
[24] Bae H J and Koumoutsakos P 2022 *Nature Communications* **13** 1443
[25] Chevalier M, Schlatter P, Lundbladh A and Henningson D S 2007 *KTH Technical Report* URL `https://api.semanticscholar.org/CorpusID:118144498`
[26] Rebbeck H and Choi K S 2001 *Physics of Fluids* **13** 2142–2145
[27] Mnih V, Kavukcuoglu K, Silver D, Rusu A A, Veness J, Bellemare M G, Graves A, Riedmiller M, Fidjeland A K, Ostrovski G, Petersen S, Beattie C, Sadik A, Antonoglou I, King H, Kumaran D, Wierstra D, Legg S and Hassabis D 2015 *Nature* **518** 529–533
[28] Weber P, Wälchli D, Zeqiri M and Koumoutsakos P 2022 *arXiv e-prints* arXiv:2203.13319
[29] Oliehoek F A, Spaan M T J and Vlassis N 2011 *CoRR* **abs/1111.0062**
[30] Fujita Y and Maeda S i 2018 *Proceedings of the 35th International Conference on Machine Learning* vol 80 pp 1597–1606 URL `http://proceedings.mlr.press/v80/fujita18a.html`
[31] Novati G and Koumoutsakos P 2019 *Proceedings of the 36th International Conference on Machine Learning* (*Proceedings of Machine Learning Research* vol 97) ed Chaudhuri K and Salakhutdinov R (PMLR) pp 4851–4860 URL `https://proceedings.mlr.press/v97/novati19a.html`
[32] Jiménez J and Moin P 1991 *Journal of Fluid Mechanics* **225** 213–240
[33] Lillicrap T P, Hunt J J, Pritzel A, Heess N, Erez T, Tassa Y, Silver D and Wierstra D 2019 Continuous control with deep reinforcement learning (*Preprint* `ArXiv:1509.02971`)
[34] Seyde T, Gilitschenski I, Schwarting W, Stellato B, Riedmiller M, Wulfmeier M and Rus D 2021 Is bang-bang control all you need? solving continuous control with Bernoulli policies (*Preprint* `Arxiv:2111.02552`)
[35] Kingma D P and Ba J 2015 *3rd International Conference on Learning Representations, ICLR 2015, San Diego, CA, USA, May 7-9, 2015, Conference Track Proceedings* ed Bengio Y and LeCun Y (*Preprint* `Arxiv:1412.6980`)
[36] Munos R, Stepleton T, Harutyunyan A and Bellemare M G 2016 *arXiv preprint arXiv:1606.02647*
[37] Degris T, White M and Sutton R S 2012 *Proceedings of the 29th International Conference on Machine Learning* ICML'12 (Madison, WI, USA: Omnipress) p 179–186 ISBN 9781450312851
[38] Schulman J, Levine S, Moritz P, Jordan M I and Abbeel P 2015 Trust region policy optimization (*Preprint* `ArXiv:1502.05477`)

AUTHOR INDEX

Institute of Physics
Dirac House, Temple Back
Bristol BS1 6BE UK

ISSN: 1742-6588
ISBN 978-1-7138-9738-5
9 781713 897385